Lecture Notes in Computer Science 6075

Commenced Publication in 1973
Founding and Former Series Editors:
Gerhard Goos, Juris Hartmanis, and Jan van Leeuwen

Paul De Bra   Alfred Kobsa   David Chin (Eds.)

User Modeling, Adaptation, and Personalization

18th International Conference, UMAP 2010
Big Island, HI, USA, June 20-24, 2010
Proceedings

 Springer

Volume Editors

Paul De Bra
Eindhoven University of Technology, Computer Science Department
5600 MB Eindhoven, The Netherlands
E-mail: debra@win.tue.nl

Alfred Kobsa
University of California, School of Information and Computer Sciences
Irvine, CA 92697-3440, USA
E-mail: kobsa@uci.edu

David Chin
University of Hawaii, Department of Information and Computer Sciences
Honolulu, HI 96822, USA
E-mail: chin@hawaii.edu

Library of Congress Control Number: 2010927592

CR Subject Classification (1998): H.2.8, H.4, H.5, I.2, C.2, I.5

LNCS Sublibrary: SL 3 – Information Systems and Application, incl. Internet/Web
and HCI

ISSN   0302-9743
ISBN-10   3-642-13469-6 Springer Berlin Heidelberg New York
ISBN-13   978-3-642-13469-2 Springer Berlin Heidelberg New York

springer.com

© Springer-Verlag Berlin Heidelberg 2010
Printed in Germany

Typesetting: Camera-ready by author, data conversion by Scientific Publishing Services, Chennai, India
Printed on acid-free paper   06/3180

Preface

The 18th International Conference on User Modeling, Adaptation and Personalization (UMAP 2010) took place on Big Island, Hawaii during June 20–24, 2010. It was the second conference after UMAP 2009 in Trento, Italy, which merged the successful biannual User Modeling (UM) and Adaptive Hypermedia (AH) conference series.

The Research Paper track of the conference was chaired by Paul De Bra from the Eindhoven University of Technology and Alfred Kobsa from the University of California, Irvine. They were assisted by an international Program Committee of 80 leading figures in the AH and UM communities as well as highly promising younger researchers. Papers in the Research Paper track were generally reviewed by three and sometimes even four reviewers, with one of them acting as a lead who initiates a discussion between reviewers and reconciles their opinions in a meta-review. The conference solicited Long Research Papers of up to 12 pages in length, which represent original reports of substantive new research. In addition, the conference accepted Short Research Papers of up to six pages in length, whose merit was assessed more in terms of originality and importance than maturity and technical validation. The Research Paper track received 161 submission, with 112 in the long and 49 in the short paper category. Of these, 26 long and 6 short papers were accepted, resulting in an acceptance rate of 23.2% for long papers and 19.9% overall. Many authors of rejected papers were encouraged to resubmit to the Poster and Demo track of the conference.

Following the example of UMAP 2009, the conference also had an Industry Paper track chaired by Bhaskar Mehta from Google, Zürich, Switzerland and Kurt Partridge from PARC, Palo Alto, USA. This track covered innovative commercial implementations or applications of UMAP technologies, and experience in applying recent research advances in practice. Submissions to this track were reviewed by a separate Industry Paper Committee with 32 leading industry researchers and practitioners. Five long and one short submissions were received, and four of the long papers were accepted

The conference also included a Doctoral Consortium, a forum for PhD students to get feedback and advice from a Doctoral Consortium Committee of 17 leading UMAP researchers. The Doctoral Consortium was chaired by Ingrid Zukerman from Monash University, Australia, and Liana Razmerita from the Copenhagen Business School, Denmark. This track received 19 submissions of which 7 were accepted.

The traditional Poster and Demo Session of the conference was this time chaired by Fabian Bohnert from Monash University, Australia, and Luz Quiroga from the University of Hawaii, Manoa. As of the time of writing, the late submission deadline is still 2 months away, and hence the number of acceptances is still unknown. We expect though that this session will again feature dozens of lively posters and system demonstrations. Summaries of these presentations will be published in online adjunct proceedings.

The UMAP 2010 program also included Workshops and Tutorials that were selected by Chairs Judith Masthoff, University of Aberdeen, UK, and Yang Wang, Carnegie Mellon University, Pittsburgh, USA. The following workshops and tutorials were announced:

- Intelligent Techniques for Web Personalization and Recommender Systems, chaired by Bamshad Mobasher, Dietmar Jannach and Sarabjot Singh Anand
- Pervasive User Modeling and Personalization, chaired by Shlomo Berkovsky, Fabian Bohnert, Francesca Carmagnola, Doreen Cheng, Dominikus Heckmann, Tsvika Kuflik, Petteri Nurmi, and Kurt Partridge
- User Models for Motivational Systems: The Affective and the Rational Routes to Persuasion, chaired by Floriana Grasso, Nadja De Carolis and Judith Masthoff
- Adaptive Collaboration Support, chaired by Alexandros Paramythis and Stavros Demetriadis
- Adaptation in Social and Semantic Web, chaired by Carlo Tasso, Federica Cena, Antonina Dattolo, Styliani Kleanthous, David Bueno Vallejo and Julita Vassileva
- Architectures and Building Blocks of Web-Based User-Adaptive Systems, chaired by Michael Yudelson, Myola Pechenizkiy, Eelco Herder, Geert-Jan Houben and Fabian Abel
- Adaptation and Personalization in e-b/Learning Using Pedagogic Conversational Agents, chaired by Diana Pérez-Marín, Ismael Pascual-Nieto and Susan Bull
- User Modeling and Adaptation for Daily Routines: Providing Assistance to People with Special and Specific Needs, chaired by Estefanía Martín, Pablo Haya and Rosa M. Carro
- Designing Adaptive Social Applications (Tutorial), presented by Julita Vassileva
- Evaluation of Adaptive Systems (Tutorial), chaired by Stephan Weibelzahl, Alexandros Paramythis, Judith Masthoff

Finally, the conference also featured two invited talks and an industry panel. The invited speakers were:

- Ed Chang, Director of Research, Google China: "AdHeat — An Influence-based Diffusion Model for Propagating Hints to Personalize Social Ads"
- Stacey Marcella, Institute for Creative Technologies, University of Southern California: "Modeling Emotion and Its Expression in Virtual Humans"

In addition to all the contributors mentioned, we would also like to thank the Local Arrangements Chair Keith Edwards from the University of Hawaii, Hilo, and the Publicity Chair Eelco Herder from the University of Hannover, Germany. We deeply acknowledge the conscientious work of the Program Committee members and the additional reviewers, who are listed on the next pages. The conference would also not have been possible without the work of many "invisible" helpers, including Riet van Buul from the Eindhoven University of Technology, who helped with language and formatting corrections in the final versions of several papers. We also gratefully acknowledge our sponsors who helped us with funding and organizational expertise and affiliation: User Modeling Inc., ACM SIGART, SIGCHI and SIGIR, the Chen Family Foundation, Microsoft Research, the U.S. National Science Foundation, Springer, and the University of Hawaii at Manoa. Finally, we want to acknowledge the use of EasyChair for the management of the review process and the preparation of the proceedings, and the help of its administrator Andrei Voronkov in implementing system enhancements that this conference had commissioned.

April 2010   Paul De Bra
                                                                   Alfred Kobsa
                                                                   David Chin

Organization

UMAP 2010 was organized by the University of Hawaii, Manoa, in cooperation with User Modeling Inc., ACM/SIGIR, ACM/SIGCHI and ACM/SIGART. The conference took place during June 20–24, 2010 on Big Island, Hawaii.

Organizing Committee

General Chair

David N. Chin   University of Hawaii, Manoa

Program Co-chairs

Alfred Kobsa   University of California, Irvine, USA
Paul De Bra   Eindhoven University of Technology,
                                    The Netherlands

Industry Track Co-chairs

Kurt Partridge   PARC, Palo Alto, USA
Bhaskar Mehta   Google, Zürich, Switzerland

Workshop and Tutorial Co-chairs

Judith Masthoff   University of Aberdeen, UK
Yang Wang   University of California, Irvine, USA

Doctoral Consortium Co-chairs

Ingrid Zukerman   Monash University, Melbourne, Australia
Liana Razmerita   Copenhagen Business School, Denmark

Demo and Poster Co-chairs

Luz Quiroga   University of Hawaii, Manoa
Fabian Bohnert   Monash University, Melbourne, Australia

Local Arrangements Chair

Keith Edwards   University of Hawaii, Hilo

Publicity Chair

Eelco Herder   L3S Research Center, Hannover, Germany

Research Track Program Committee

Sarabjot Anand	University of Warwick, UK
Kenro Aihara	National Institute of Informatics, Japan
Liliana Ardissono	University of Turin, Italy
Lora Aroyo	Free University of Amsterdam, The Netherlands
Helen Ashman	University of South Australia, Australia
Ryan Baker	Worcester Polytechnic Institute, USA
Mathias Bauer	Mineway GmbH, Germany
Joseph Beck	Worcester Polytechnic Institute, USA
Shlomo Berkovsky	CSIRO, Australia
Mária Bieliková	Slovak University of Technology, Slovakia
Peter Brusilovsky	University of Pittsburgh, USA
Susan Bull	University of Birmingham, UK
Robin Burke	DePaul University, USA
Sandra Carberry	University of Delaware, USA
Keith Cheverst	Lancaster University, UK
Li Chen	Hong Kong Baptist University, China
Sherry Chen	Brunel University, UK
Chien Chin Chen	NTU, Taiwan
David Chin	University of Hawaii, USA
Luca Chittaro	University of Udine, Italy
Cristina Conati	University of British Columbia, Canada
Owen Conlan	Trinity College Dublin, Ireland
Albert Corbett	Carnegie Mellon University, USA
Hugh Davis	University of Southampton, UK
Paul De Bra	Eindhoven University of Technology, The Netherlands
Michel Desmarais	Ecole Polytechnic de Montreal, Canada
Vania Dimitrova	University of Leeds, UK
Ben Du Boulay	University of Sussex, UK
Marco de Gemmis	University of Bari, Italy
Peter Dolog	Aalborg University, Denmark
Serge Garlatti	ENST Bretagne, GET, France
Elena Gaudioso	UNED, Spain
Christina Gena	University of Turin, Italy
Jim Greer	University of Saskatchewan, Canada
Dominikus Heckmann	Saarland University, Germany
Nicola Henze	Leibniz Universität Hannover, Germany
Eelco Herder	L3S Research Center Hannover, Germany
Eric Horvitz	Microsoft Research, USA
Geert-Jan Houben	Technische Universiteit Delft, The Netherlands
Bin Hu	Lanzhou University, China
Dietmar Jannach	TU Dortmund, Germany
Paul Kamsteeg	Katholieke Universiteit Nijmegen, The Netherlands

Judy Kay	University of Sydney, Australia
Alfred Kobsa	UC Irvine, USA
Antonio Krueger	UNI - SB, Germany
Tsvi Kuflik	University of Haifa, Israel
James Lester	North Carolina State University, USA
Henry Lieberman	MIT, USA
George Magoulas	Birckbeck College, UK
Brent Martin	University of Canterbury, New Zealand
Bhaskar Mehta	Google, USA
Alessandro Micarelli	University of Rome III, Italy
Eva Millan	Univerisity of Malaga, Spain
Antonija Mitrović	University of Canterbury, New Zealand
Dunja Mladenic	J. Stefan Institute, Slovenia
Bamshad Mobasher	DePaul University, USA
Wolfgang Nejdl	University of Hannover, Germany
Hien Nguyen	University of Wisconsin Whitewater, USA
Michael O'Mahony	University College Dublin, Ireland
Georgios Paliouras	NCSR Demokritos, Greece
Christos Papatheodorou	Ionian University, Greece
Alexandros Paramythis	FIM Institute Johannes Kepler University, Austria
Cécile Paris	CSIRO, Australia
Paolo Petta	Austrian Res. Inst. for Artificial Intelligence, Austria
Jose Luis Perez de la Cruz	University of Malaga, Spain
Dimitris Pierrakos	NCSR Demokritos, Greece
Francesco Ricci	Free University of Bozen-Bolzano, Italy
Katharina Reinecke	University of Zurich, Switzerland
Charles Rich	Worcester Polytechnic Institute, USA
Cristóbal Romero Morales	Universidad de Cordoba, Spain
Gustavo Gonzalez-Sanchez	Mediapro R&D, Spain
Daniel Schwabe	University of Rio de Janeiro, Brazil
Barry Smyth	University College Dublin, Ireland
Marcus Specht	Open University Netherlands, The Netherlands
Myra Spiliopoulou	University of Magdeburg, Germany
Julita Vassileva	Univeristy of Saskatchewan, Canada
Sebastián Ventura	University of Cordoba, Spain
Vincent Wade	Trinity College Dublin, Ireland
Gerhard Weber	PH Freiburg, Germany
Stephan Weibelzahl	National College of Ireland, Ireland
Michael Yudelson	Univeristy of Pittsburgh, USA
Massimo Zancanaro	FBK - cit, Italy
Diego Zapata-Rivera	Educational Testing Service, USA
Ingrid Zukerman	Monash University, Australia

Industry Track Program Committee

Mauro Barbieri	Philips Research, The Netherlands
Stefano Bertolo	European Commission
Elizabeth Churchill	Yahoo! Research, USA
William Clancey	NASA, USA
Thomas Colthurst	Google, USA
Doree Duncan Seligmann	Avaya Labs, USA
Enrique Frias-Martinez	Telefonica Research, Spain
Mehmet Goker	Strands Labs, USA
Rich Gossweiler	Google, USA
Ido Guy	IBM Research, Israel
Gustavo Gonzalez-Sanchez	University of Girona, Spain
Thomas Hofmann	Google, Switzerland
Ashish Kapoor	Microsoft Research
Paul Lamere	Sun Microsystems, USA
Shoshana Loeb	Telcordia Applied Research, USA
Jiebo Luo	Kodak Research Lab, USA
Francisco Martin	MyStrands, USA
Andreas Nauerz	IBM, Germany
Nuria Oliver	Telefonica, Madrid
Daniel Olmedilla de la Calle	Telefonica Research, Spain
Igor Perisic	LinkedIn, USA
Jeremy Pickens	FXPAL, USA
Kunal Punera	Yahoo! Research, USA
Prakash Reddy	HP Labs, USA
Christoph Rensing	HTTC, Germany
John Riedl	University of Minnesota, USA
Xuehua Shen	Google, USA
Malcolm Slaney	Yahoo! Research, USA
Neel Sundaresan	E-bay Laboratories, USA
Ryen White	Microsoft Research, USA

Doctoral Consortium Committee

Liliana Ardissono	University of Turin, Italy
Anne Boyer	LORIA - Université Nancy 2, France
Armelle Brun	LORIA - Université Nancy 2, France
Peter Brusilovsky	University of Pittsburgh, USA
David Chin	University of Hawaii, USA
Alexandra Cristea	University of Warwick, UK
Paul De Bra	Eindhoven University of Technology, The Netherlands
Vania Dimitrova	University of Leeds, UK
Eric Horvitz	Microsoft Research, USA

Geert-Jan Houben   Technische Universiteit Delft,
                                    The Netherlands
Anthony Jameson   FBK-irst, Italy
Judy Kay   University of Sydney, Australia
Alfred Kobsa   UC Irvine, USA
Diane Litman   University of Pittsburgh, USA
Joseph Konstan   University of Minnesota, USA
Alfred Kobsa   UC Irvine, USA
Liana Razmerita   Copenhagen Business School, Denmark
Julita Vassileva   Univeristy of Saskatchewan, Canada
Ingrid Zukerman   Monash University, Australia

Additional Reviewers

Fabian Abel   L3S Research Center Hannover, Germany
Mohd Anwar   University of Saskatchewan, Canada
Michal Barla   Slovak University of Technology, Slovakia
Claudio Biancalana   University of Rome III, Italy
Kristy Elizabeth Boyer   North Carolina State University, USA
Janez Brank   J. Stefan Institute, Slovenia
Christopher Brooks   University of Saskatchewan, Canada
Stefano Burigat   University of Udine, Italy
Fabio Buttussi   University of Udine, Italy
Alberto Cabas Vidani   University of Udine, Italy
Annalina Caputo   University of Bari, Italy
Giuseppe Carenini   University of British Columbia, Canada
Francesca Carmagnola   University of Leeds, UK
Carlos Castro-Herrera   DePaul University, USA
Federica Cena   University of Turin, Italy
Fei Chen   Avaya Labs, USA
Karen Church   University College Dublin, Ireland
Ilaria Corda   University of Leeds, UK
Declan Dagger   Trinity College Dublin, Ireland
Theodore Dalamagas   Ionian University, Greece
Lorand Dali   J. Stefan Institute, Slovenia
Carla Delgado-Battenfeld   TU Dortmund, Germany
Eyal Dim   University of Haifa, Israel
Frederico Durao   Aalborg University, Denmark
Jill Freyne   University College Dublin, Ireland
Fabio Gasparetti   University of Rome III, Italy
Mouzhi Ge   TU Dortmund, Germany
Fatih Gedikli   TU Dortmund, Germany
Jonathan Gemmell   DePaul University, USA
Yaohua Ho   NTU, Taiwan
Daniel Krause   L3S Research Center Hannover, Germany
Joel Lanir   University of Haifa, Israel

Table of Contents

Kcynotc Speakers

Modeling Emotion and Its Expression in Virtual Humans
(Extended Abstract) 1
  Stacy Marsella

AdHeat — An Influence-Based Diffusion Model for Propagating Hints
to Personalize Social Ads (Abstract)............................... 3
  Edward Y. Chang

Full Research Papers

Can Concept-Based User Modeling Improve Adaptive Visualization? ... 4
  Jae-wook Ahn and Peter Brusilovsky

Interweaving Public User Profiles on the Web....................... 16
  Fabian Abel, Nicola Henze, Eelco Herder, and Daniel Krause

Modeling Long-Term Search Engine Usage 28
  Ryen W. White, Ashish Kapoor, and Susan T. Dumais

Analysis of Strategies for Building Group Profiles 40
  Christophe Senot, Dimitre Kostadinov, Makram Bouzid,
  Jérôme Picault, Armen Aghasaryan, and Cédric Bernier

Contextual Slip and Prediction of Student Performance After Use of an
Intelligent Tutor 52
  Ryan S.J.d. Baker, Albert T. Corbett, Sujith M. Gowda,
  Angela Z. Wagner, Benjamin A. MacLaren, Linda R. Kauffman,
  Aaron P. Mitchell, and Stephen Giguere

Working Memory Span and E-Learning: The Effect of Personalization
Techniques on Learners' Performance 64
  Nikos Tsianos, Panagiotis Germanakos, Zacharias Lekkas,
  Costas Mourlas, and George Samaras

Scaffolding Self-directed Learning with Personalized Learning Goal
Recommendations.. 75
  Tobias Ley, Barbara Kump, and Cornelia Gerdenitsch

Instructional Video Content Employing User Behavior Analysis: Time
Dependent Annotation with Levels of Detail........................ 87
  Junzo Kamahara, Takashi Nagamatsu, Masashi Tada,
  Yohei Kaieda, and Yutaka Ishii

A User-and Item-Aware Weighting Scheme for Combining Predictive
User Models 99
   Fabian Bohnert and Ingrid Zukerman

PersonisJ: Mobile, Client-Side User Modelling 111
   Simon Gerber, Michael Fry, Judy Kay, Bob Kummerfeld,
   Glen Pink, and Rainer Wasinger

Twitter, Sensors and UI: Robust Context Modeling for Interruption
Management 123
   Justin Tang and Donald J. Patterson

Ranking Feature Sets for Emotion Models Used in Classroom Based
Intelligent Tutoring Systems..................................... 135
   David G. Cooper, Kasia Muldner, Ivon Arroyo,
   Beverly Park Woolf, and Winslow Burleson

Inducing Effective Pedagogical Strategies Using Learning Context
Features 147
   Min Chi, Kurt VanLehn, Diane Litman, and Pamela Jordan

"Yes!": Using Tutor and Sensor Data to Predict Moments of Delight
during Instructional Activities 159
   Kasia Muldner, Winslow Burleson, and Kurt VanLehn

A Personalized Graph-Based Document Ranking Model Using a
Semantic User Profile... 171
   Mariam Daoud, Lynda Tamine, and Mohand Boughanem

Interaction and Personalization of Criteria in Recommender Systems ... 183
   Shawn R. Wolfe and Yi Zhang

Collaborative Inference of Sentiments from Texts.................... 195
   Yanir Seroussi, Ingrid Zukerman, and Fabian Bohnert

User Modelling for Exclusion and Anomaly Detection: A Behavioural
Intrusion Detection System...................................... 207
   Grant Pannell and Helen Ashman

IntrospectiveViews: An Interface for Scrutinizing Semantic User
Models 219
   Fedor Bakalov, Birgitta König-Ries, Andreas Nauerz, and
   Martin Welsch

Analyzing Community Knowledge Sharing Behavior 231
   Styliani Kleanthous and Vania Dimitrova

A Data-Driven Technique for Misconception Elicitation 243
   Eduardo Guzmán, Ricardo Conejo, and Jaime Gálvez

Modeling Individualization in a Bayesian Networks Implementation of
Knowledge Tracing   255
    Zachary A. Pardos and Neil T. Heffernan

Detecting Gaming the System in Constraint-Based Tutors   267
    Ryan S.J.d. Baker, Antonija Mitrović, and Moffat Mathews

Bayesian Credibility Modeling for Personalized Recommendation in
Participatory Media   279
    Aaditeshwar Seth, Jie Zhang, and Robin Cohen

A Study on User Perception of Personality-Based Recommender
Systems   291
    Rong Hu and Pearl Pu

Compass to Locate the User Model I Need: Building the Bridge
between Researchers and Practitioners in User Modeling   303
    Armelle Brun, Anne Boyer, and Liana Razmerita

Industry Papers

myCOMAND Automotive User Interface: Personalized Interaction with
Multimedia Content Based on Fuzzy Preference Modeling   315
    Philipp Fischer and Andreas Nürnberger

User Modeling for Telecommunication Applications: Experiences and
Practical Implications   327
    Heath Hohwald, Enrique Frías-Martínez, and Nuria Oliver

Mobile Web Profiling: A Study of Off-Portal Surfing Habits of Mobile
Users   339
    Daniel Olmedilla, Enrique Frías-Martínez, and Rubén Lara

Personalized Implicit Learning in a Music Recommender System   351
    Suzana Kordumova, Ivana Kostadinovska, Mauro Barbieri,
    Verus Pronk, and Jan Korst

Short Research Papers

Personalised Pathway Prediction   363
    Fabian Bohnert and Ingrid Zukerman

Towards a Customization of Rating Scales in Adaptive Systems   369
    Federica Cena, Fabiana Vernero, and Cristina Gena

Eye-Tracking Study of User Behavior in Recommender Interfaces   375
    Li Chen and Pearl Pu

Recommending Food: Reasoning on Recipes and Ingredients   381
  Jill Freyne and Shlomo Berkovsky

Disambiguating Search by Leveraging a Social Context Based on the
Stream of User's Activity   387
  Tomáš Kramár, Michal Barla, and Mária Bieliková

Features of an Independent Open Learner Model Influencing Uptake
by University Students   393
  Susan Bull

Doctoral Consortium Papers

Recognizing and Predicting the Impact on Human Emotion (Affect)
Using Computing Systems   399
  David G. Cooper

Utilising User Texts to Improve Recommendations   403
  Yanir Seroussi

Semantically-Enhanced Ubiquitous User Modeling   407
  Till Plumbaum

User Modeling Based on Emergent Domain Semantics   411
  Marián Šimko and Mária Bieliková

"Biographic Spaces": A Personalized Smoking Cessation Intervention
in Second Life   415
  Ana Boa-Ventura and Luís Saboga-Nunes

Task-Based User Modelling for Knowledge Work Support   419
  Charlie Abela, Chris Staff, and Siegfried Handschuh

Enhancing User Interaction in Virtual Environments through Adaptive
Personalized 3D Interaction Techniques   423
  Johanna Renny Octavia, Karin Coninx, and Chris Raymaekers

Author Index   427

Modeling Emotion and Its Expression in Virtual Humans

Stacy Marsella

Institute for Creative Technologies
University of Southern California

Extended Abstract of Keynote Talk

A growing body of work in psychology and the neurosciences has documented the functional, often adaptive role of emotions in human behavior. This has led to a significant growth in research on computational models of human emotional processes, fueled both by their basic research potential as well as the promise that the function of emotion in human behavior can be exploited in a range of applications. Computational models transform theory construction by providing a framework for studying emotion processes that augments what is feasible in more traditional laboratory settings. Modern research in the psychological processes and neural underpinnings of emotion is also transforming the science of computation. In particular, findings on the role that emotions play in human behavior have motivated artificial intelligence and robotics research to explore whether modeling emotion processes can lead to more intelligent, flexible and capable systems. Further, as research has revealed the deep role that emotion and its expression play in human social interaction, researchers have proposed that more effective human computer interaction can be realized if the interaction is mediated both by a model of the user's emotional state as well as by the expression of emotions.

Our lab considers the computational modeling of emotions from the perspective of a particular application area, virtual humans. Virtual humans are autonomous virtual characters that are designed to act like humans and interact with them in shared virtual environments, much as humans interact with humans. As facsimiles of humans, virtual humans can reason about the environment, simulate the understanding and expression of emotion, and communicate using speech and gesture. A range of application areas now use this technology, including education, health intervention and entertainment. They are also being used as virtual confederates for experiments in social psychology.

The computational modeling of emotions has emerged as a central challenge of virtual human architectures. In particular, researchers in virtual characters for gaming and teaching environments have sought to endow virtual characters with emotion-related capabilities so that they may interact more naturally with human users.

Computational models of emotion used in virtual humans have largely been based on appraisal theory, the predominant psychological theory of emotion. Appraisal theory argues that emotion arise from patterns of individual assessments concerning the relationship between events and an individual's beliefs, desires and intentions, sometimes referred to as the person-environment relationship. These assessments, often called appraisal variables, characterize aspects of the personal significance of events (e.g., was

P. De Bra, A. Kobsa, and D. Chin (Eds.): UMAP 2010, LNCS 6075, pp. 1–2, 2010.

this event expected in terms of my prior beliefs? is this event congruent with my goals; do I have the power to alter the consequences of this event?). Patterns of appraisal are associated with specific emotional responses, including physiological and behavioral reactions. In several versions of appraisal theory, appraisals also trigger cognitive responses, often referred to as coping strategies—e.g., planning, procrastination or resignation—feeding back into a continual cycle of appraisal and re-appraisal.

In this talk, I will give an overview of virtual humans. I will then go into greater detail on how emotions is modeled computationally in virtual humans, including the theoretical basis of the models in appraisal theory, how we validate models against human data and how human data is also used to inform the animation of the virtual human's body.

AdHeat — An Influence-Based Diffusion Model for Propagating Hints to Personalize Social Ads

Edward Y. Chang

Director of Research
Google

Abstract of Keynote Talk

AdHeat is our newly developed social ad model considering user influence in addition to relevance for matching ads. Traditionally, ad placement employs the relevance model. Such a model matches ads with Web page content, user interests, or both. We have observed, however, on social networks that the relevance model suffers from two shortcomings. First, influential users (users who contribute opinions) seldom click ads that are highly relevant to their expertise. Second, because influential users' contents and activities are attractive to other users, hint words summarizing their expertise and activities may be widely preferred. Therefore, we propose AdHeat, which diffuses hint words of influential users to others and then matches ads for each user with aggregated hints. Our experimental results on a large-scale social network show that AdHeat outperforms the relevance model on CTR (click through rate) by significant margins. In this talk, the algorithms employed by AdHeat and solutions to address scalability issues are presented.

P. De Bra, A. Kobsa, and D. Chin (Eds.): UMAP 2010, LNCS 6075, p. 3, 2010.
© Springer-Verlag Berlin Heidelberg 2010

Can Concept-Based User Modeling
Improve Adaptive Visualization?

Jae-wook Ahn and Peter Brusilovsky

School of Information Sciences
University of Pittsburgh
Pittsburgh, PA, 15260
{jahn,peterb}@mail.sis.pitt.edu

Abstract. Adaptive visualization can present user-adaptive information in such a way as to help users to analyze complicated information
spaces easily and intuitively. We presented an approach called Adaptive VIBE, which extended the traditional reference point-based visualization algorithm, so that it could adaptively visualize documents of
interest. The adaptive visualization was implemented by separating the
effects of user models and queries within the document space and we
were able to show the potential of the proposed idea. However, adaptive visualization still remained in the simple *bag-of-words* realm. The
keywords used to construct the user models were not effective enough
to express the concepts that need to be included in the user models.
In this study, we tried to improve the old-fashioned keyword-only user
models by adopting more concept-rich named-entities. The evaluation
results show the strengths and shortcomings of using named-entities as
conceptual elements for visual user models and the potential to improve
the effectiveness of personalized information access systems.

1   Introduction

Personalized information access is one of the most important keys to user satisfaction in today's information environment. Numerous information services
and applications are producing new information every second and it is getting
more and more complicated to access relevant items in time. Personalization
plays a role in that challenge. It tries to solve the problem by understanding
a user's needs and providing tailored information efficiently. There are several
approaches for this personalized information access: personalized information retrieval [16], information filtering [10], and adaptive visualization [13, 21]. Among
them, adaptive visualization is an attempt to improve information visualization
by adding an adaptation component. Through adaptation, users can modify the
way in which the system visualizes a collection of documents [21]. It combines
algorithm-based personalization with user interfaces in order to better learn
about users and to provide personalized information more efficiently. It also
shares the spirit of exploratory search [15]. Both attempt to enhance users' own
intelligence by providing more interactive and expressive user interfaces so that

P. De Bra, A. Kobsa, and D. Chin (Eds.): UMAP 2010, LNCS 6075, pp. 4–15, 2010.
© Springer-Verlag Berlin Heidelberg 2010

they can achieve better search results. However, adaptive visualization is even more evolved than simple exploratory searching, because it actively endeavors to estimate users' search context and help them to discover optimal solutions.

In order to implement the adaptive visualization, we extended a well-known visualization framework called VIBE (Visual Information Browsing Environment) [18] and created Adaptive VIBE. VIBE is a reference point (called POI, meaning Point Of Interest)-based visualization method and we extended it to visualize the user models and the personalized search results. We have begun to evaluate this idea [1] and are currently studying user behaviors with the system. However, the user models adopted in previous study were constructed using the old-fashioned, keyword-based *bag-of-words* approach. We have always suspected the limitation of the keyword-based user modeling for dealing with large amount of data; therefore, we decided to address this problem in the current study by extending the user models and enriching them with more semantic-rich elements. We chose to use named-entities (NEs, henceforth) as alternatives to the simple keywords. They were expected to be semantically richer than keywords and could better represent concepts.

This paper investigates whether the use of NEs in the user models – especially in the Adaptive VIBE visualization – can lead us to build better personalized information access services. In the next section, the ideas of concept-based user modeling and NE-based information systems are introduced (Sect. 2). In Sect. 3, the proposed adaptive visualization and the concept-based user modeling are described. The following sections explain the methodology and the results of our experiments with the NE-based adaptive visualization. The concluding section discusses the implications of this study and future plans.

2 Concept-Based User Modeling and Named-Entities

Keyword-based user modeling is a traditional approach widely used for content-based personalization and other related areas. Even though this simple *bag-of-words* approach has been working relatively well, its limitations such as the polysemy problem or the independence assumption among keywords were consistently noted too. Therefore, there have been many attempts to build user models to overcome the limitations and they are classified into two categories: network-based and ontology-based user models [8]. Networked user models adopted in projects like [9, 14] were constructed in a way that connected the concept nodes included in the user model networks and tried to represent the relationships among the concepts. Ontology-based approaches [5, 20] incorporated more sophisticated methods. Unlike the network user models where the relationships were flat, they tried to build user models hierarchically by making use of already-existing ontologies.

Despite all of these efforts, we still could see the chance to enrich the meanings of the user model elements themselves. Therefore, we tried to use NEs as conceptual elements in our user models and to extend the semantics and expressive power of the user models. As a semantic category, NEs act as pointers to real world entities such as locations, organizations, people, or events [19]. NEs can provide much richer semantic content than most vocabulary keywords and many researchers argued that semantic features were able to better model essential document content. Therefore, the application of NEs was considered to improve a user's ability to find and access the right information [19]. They have been studied extensively in various language processing and information access tasks such as document indexing [17] and topic detection and tracking [12]. At the same time, NEs have been successfully adopted by analytic systems such as [4], where user interaction and feedback plays a key role similar to that in the personalized information access systems.

To our knowledge, however, there has been no attempt to directly incorporate NEs into user model construction. We have utilized NEs as conceptual elements for news articles (where NEs can be particularly useful for catching concepts) in one of our previous studies and found that NEs organized into the editor's 4Ws (Who, What, Where, and When) could assist users in finding relevant information in a non-personalized information retrieval setting [2]. With this experience, we could expect NEs to be high-quality semantic elements which would enhance the user model representation.

3   Adaptive Concept-Based Visualization: The Technology

3.1   Adaptive VIBE Visualization

VIBE was first developed at the University of Pittsburgh and Molde College in Norway [18]. It is a reference point (called POI, Point Of Interest)-based visualization, which displays documents according to their similarity ratios to the POIs, so that more similar documents are located closer to the POIs (for more details about the visualization algorithm, see [11] and [18]). Figure 1 shows examples of the VIBE visualization. On top of this general idea, we attempted to add adaptivity by separating the originally-equivalent POIs into multiple groups. The traditional VIBE usually arranged the POIs in a circle, where POIs with different layers of meaning were treated equally (like a round table) and which required further user intervention to organize the different groups of POIs. For Adaptive VIBE, we grouped the different POIs into different locations from the beginning. That is, we separated the two groups of POIs – query and user model POIs. By separating them, we were able to spatially distinguish the documents which were more related to the query or the user model, respectively.

This method is similar to the usual personalized searching method, where documents are re-ranked according to their similarities to user models. The documents more related to user models are brought higher to the top of the ranked list, while less related ones are at the bottom. In Adaptive VIBE, the one-dimensional ranked list is now replaced with a two-dimensional spatial visualization. The documents that used to be scattered all over the screen (located according to their similarities to POIs or query terms) are now organized by their similarities closer to the query or user model. In order to implement this separation, we added two new adaptive layouts of POIs (Hemisphere and Parallel) to the old circular layout (Radial) as shown in Fig. 1. There, it can be seen that the document space is separated into two parts: the one that is closer to the query side and the other closer to the user model side. This separation is the result of the effect of the user model POIs (using the adaptive layouts).

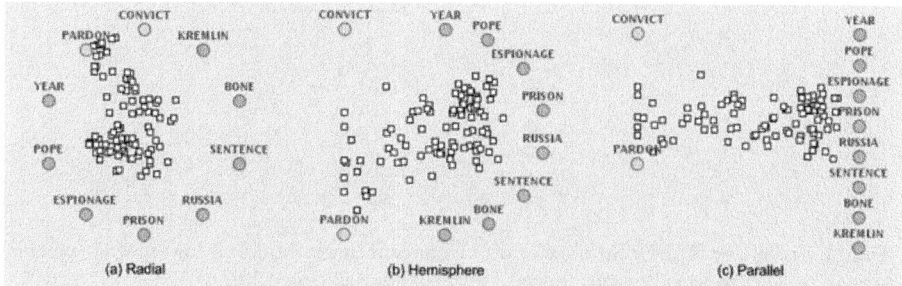

Fig. 1. Adaptive VIBE layouts (a) Radial, (b) Hemisphere, and (c) Parallel. Yellow (CONVICT and PARDON) and blue (YEAR, POPE, and so on) POIs are query terms and user model keywords, respectively. White squares are retrieved documents.

We could extend the visual user models even further by incorporating conceptual NEs into them. Figure 2 shows an example of this extension. Originally, there were only keyword-based POIs (POPE, YEAR, ESPIONAGE, and CHARGE) but we added five more NE-based POIs to the model (lowercased in the figure). With the addition of these NEs, the user model could express more information. It was not just increasing the number of POIs, but adding more meanings to the user model. For example, *united_states_of_america* is usually split into 4 words and expressed as *unite, state,* and *america* (after stemmed and stopwords are removed) in keyword-based approaches. *Russia* and *russian* are reduced into one stemmed word, *russia.* However, these lost meanings were recovered in NE-based user models and we expected that it would help users to access relevant information. The following sections describe the NE-based user model construction process in more detail.

3.2 Named Entity Extraction

We first needed to extract NEs from texts in order to build NE-based user models. For this task, we used software developed by our partner at IBM [7]. With the help of the NE annotator, we could extract the NEs to construct the user models and calculate the similarity between documents and the entities as inputs to the Adaptive VIBE system. The NE annotation process was based on a statistical maximum-entropy model that recognized 32 types of named, nominal and pronominal entities (such as person, organization, facility, location, occupation, etc), and 13 types of events (such as event_violence, event_communication, etc). Among them, we selected the nine most frequent entity types.

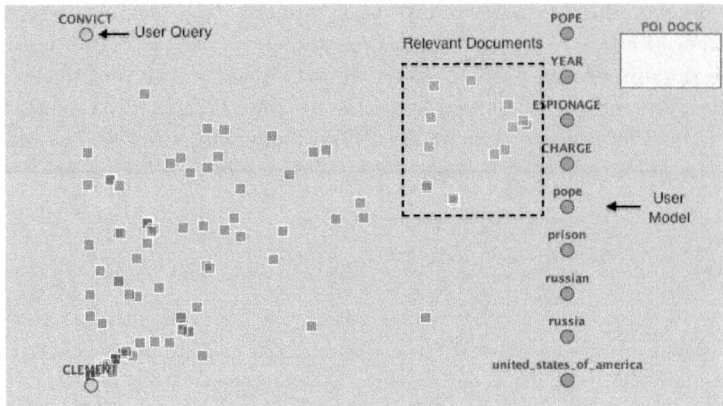

Fig. 2. Adaptive VIBE enriched with a concept user model – lowercased elements (pope, prison, russian, russia, united_states_of_america) are NEs

One very important characteristic of the NE annotator we used was that it could distinguish between different forms of the same entities within- and between-documents. For example, it was able to detect "ski lovers" and "who", which were pointing to the same group of people and could give them the same ID within a single document. They were marked as ZBN20001113.0400.0019-E75 which represented the 75th entity in document ZBN20001113.0400.0019. Therefore, those two entities with different forms ("ski lovers" and "who") could be assessed as having the same meaning (E75) by the system. At the same time, the annotator could do the same thing across the documents. It could endow a single ID "XDC:Per:wolfgang_schussel" to the words/phrases in the text like "Schussel", "director", "Chancellor", and "him", so that users could grasp the fact that they represented a single person. This capability was considered very promising, since it could deliver the semantics of the entities in the text regardless of the varying textual representations. For more details about the NE annotation algorithm and the selection process, please see [2].

3.3   Construction of Concept-Based User Models Using NEs

As discussed in the previous sections, we assumed that NEs were semantically richer than vocabulary keywords and would contribute greatly to accessing relevant information. This expectation was grounded on our previous study [2] which used NEs as pseudo-facets for browsing information. However, we had no idea about the best method for constructing NE-based user models. Is it a better approach to use NEs only in user models? What fraction of NEs should be used with keywords, if we choose to mix them? Therefore, we prepared seven combinations of the keyword and NE mixtures, in the spectrum between the extreme "keywords-only" mixture and the "NE-only" mixture. They are as follows: **k20n0, k10n0, k5n5, n8n8, k10n10, k0n10, k0n20**. Here, $kxny$ represents

x keywords and y NEs. Therefore, k5n5 means 5 keywords and 5 NEs, while k10n0 means 10 keywords only. We chose these combinations considering the optimal number of user model POIs displayed in the visualization. Because we didn't want to place too many user model POIs on the screen and make users frustrated, we configured $x + y$ (total number of user model POIs displayed at the same time on the screen) to be no more than 20. Using those combinations of keyword/NE mixtures, we could test various conditions such as equivalent importance (e.g. k5n5), keyword only (e.g. k10n0), and NE only (e.g. k0n10).

The NEs were extracted from user feedback information (notes saved by users in our prototype system) just like the case of keywords [1, 3]. Among the candidates, NEs with higher TF-IDF values were selected for constructing the NE-based user models. When calculating the TF-IDF values, the NE normalization process introduced in the previous section was utilized. That is, "ski lover" and "who" were recognized as the same terms and counted as TF=2.

4 Study Design

4.1 Hypotheses and Measures

We defined two hypotheses in this study in order to test the validity of the NE-based adaptive visualization.

H1) The proposed NE-based adaptive visualization will better separate relevant and non-relevant documents in the visualization.
H2) In the NE-based adaptive visualization, the relevant documents will be more attracted by the user models.

They were defined considering the nature of an ideal information access system. An ideal information access system has to have the ability to sort out valuable information from noise and to provide such information to users efficiently. The hypotheses exactly reflect those characteristics. Adaptive VIBE aims to distinguish relevant documents and then locate them spatially close to the user models. In order to measure the separation of relevant documents from non-relevant ones, we adopted the Davies-Bouldin Validity Index (DB-index). It determines the quality of clustering by measuring the compactness and separation of the clusters of those two types of documents [6]. It is a ratio of the spread of elements in clusters and the distances between those clusters (Equation 1). Therefore, it produces smaller scores as the clusters become compact and as the clusters are far from each other, which means better clusterings.

$$DB = \frac{1}{n} \sum_{i=1}^{n} max_{i=j} \left\{ \frac{S_n(Q_i) + S_n(Q_j)}{S(Q_i, Q_j)} \right\} \tag{1}$$

$S(Q)$ = average distance within a cluster Q
$S(Q_1, Q_2)$ = distance between two cluster centroids

4.2   Dataset

As mentioned briefly earlier, we constructed a dataset from the log data of our text-based personalized information retrieval study [3]. It aimed to help users to search the TDT4[1] news corpus for information by mediating the user query and the user model with a text-based user interface. The TDT4 corpus was built for constructing a news understanding systems and is comprised of 96,260 news articles. We chose TDT4 because NEs could represent important concepts appearing in news texts. From the log file of the study, we could extract the information as below.

1. `userid` and `query`
2. `retrieved documents` and the `relevance` of each document
3. `user notes` – explicit user feedbacks from which user model keywords and NEs would be extracted

That is, we had stored a snapshot of every users' search activity, the output from the system, and the user model constructed by the system (or the source of user model). Using this data, we were able to rebuild the user models using keywords and NEs (as shown in the previous section), and then re-situate the Adaptive VIBE visualizations according to each user model. Moreover, we had the relevance information of each document and could observe how they were represented in the visualizations. This relevance information was not available to the users during the user study, but we could take advantage of its availability to evaluate the quality of the adaptive visualizations (as in Fig. 2). The next section shows the analysis result of those adaptive visualizations and discusses their properties.

5   Experimental Results

5.1   Separation of Relevant and Non-relevant Documents by Concept-Based Adaptive Visualization

In our previous study, we found that the adaptive visualization was able to produce clusters of relevant and non-relevant documents and that the relevant document cluster was more attracted to the user model side [1]. However, the user model of the study made use of keywords only and the power of the user model in the visualization was assumed to be limited. Therefore, we prepared various combinations of user model elements (keywords plus NEs as introduced in Sect. 3) and tested them with our adaptive visualization system. The first step of the analysis was to examine how well the relevant and non-relevant document clusters were formed. Using the DB-index, we could calculate the quality of the clusterings. Table 1 and Fig. 3 show the DB-indices of three different POI layouts of Adaptive VIBE using eight different mixtures of keywords and NEs. From this data, it can be easily seen that using only keywords or NEs

---

[1] Topic Detection and Tracking Project, <http://www.ldc.upenn.edu/TDT>

Table 1. Comparison of cluster validity of adaptive visualization ($k x n y$ means x keywords and y NEs combination in the user models)

Layout	k20n0	k10n0	k5n5	k8n8	k10n10	k0n10	k0n20
Radial	3.22	2.37	2.08	2.20	2.25	2.37	2.62
Parallel	1.89	1.57	**1.37**	1.40	1.55	2.62	2.03
Hemisphere	3.37	2.91	2.12	1.99	2.00	3.61	3.03

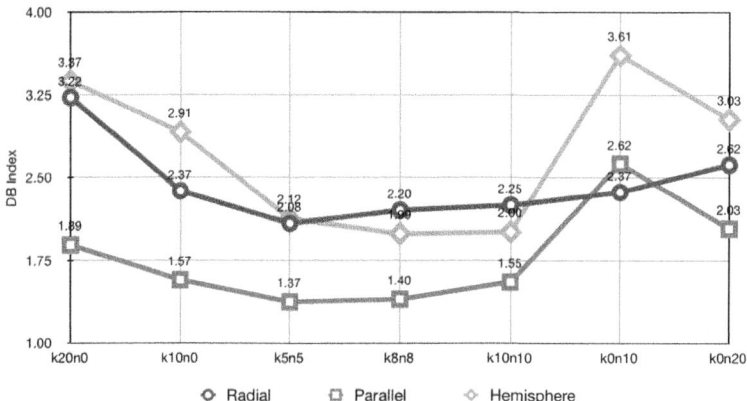

Fig. 3. Comparison of cluster validity of adaptive visualization

for user models (k20n0, k10n0, k0n10, k0n20) generally resulted in low clustering quality. However, when the keywords and the NEs were mixed within the user models, the clustering quality improved (k5n5, k8n8, and k10n10). Among the three POI layouts, the Parallel layout exhibited the best clustering quality. This result supports our first hypothesis, because the personalized adaptive visualization method (Parallel layout) and the use of NEs in the user models outperformed other combinations. It can be understood as more powerful user models (equipped with NEs) were able to stretch the space out and separated the relevant documents from the others. In order to examine the significance of the differences among keyword/NE mixtures, we conducted the Kruskal-Wallis rank sum tests on the three most representative mixtures (k10n0, k5n5, and k0n10) per each layout. These mixtures were chosen in order to compare the best keyword+NE mixture (k5n5) with keyword/NE only mixtures (k10n0 and k0n10) that have the same number of POIs (=10). The result shows that the clustering quality was significantly different among the mixtures when the Parallel layout was used (Table 2). Regarding the Parallel layout, the DB-index scores of three mixtures were all significantly different (Table 3).

Cluster Compactness vs. Between-Cluster Distance. DB-Index is the ratio between the within-cluster compactness and between-cluster distance. We found that the Adaptive VIBE layout and equally-mixed keyword/NE user

Table 2. Comparison of mean DB-index among three mixtures (k10n0, k5n5, k0n10)

Layout	Radial	Parallel	Hemisphere
Kruskal-Wallis χ^2	0.0462	8.93	3.3834
p	0.9772	**0.0115**	0.1842

Table 3. Pairwise Wilcox signed rank tests by keyword/NE mixture

Layout=Parallel	k5n5	k10n0
k5n5	-	p=0.002
k0n10	$p < 0.001$	p=0.007

Table 4. Comparing within-cluster spread and between-cluster distance

Layout	Within-cluster spread			Between-cluster distance		
	k5n5	k10n0	k0n10	k5n5	k10n0	k0n10
Radial	87.67	81.25	87.67	55.23	49.94	47.68
Parellel	154.47	152.13	161.44	**151.63**	138.05	125.31
Hemisphere	110.54	109.50	110.54	82.97	70.76	74.78

models could produce good results but we needed deeper analysis. By separating the nominator and denominator of the DB-index equation, we could compare the within-cluster spreads and between-cluster distances in terms of two other variables: keyword/NE mixture and Adaptive VIBE POI layout (Table 4). It shows that the differences among the keyword/NE mixtures were not evident when we observed the cluster spreads, but that there were bigger differences in terms of the between-cluster distances across the three mixtures. In all cases, the k5n5 mixture showed the largest distance and the differences between other mixtures were always statistically significant (Wilcox signed rank test, $p < 0.01$). This result suggests that the significant differences of overall DB-indices among the mixtures (where k5n5 was the best) observed in the previous section were caused by the cluster distance, rather than the different inner-compactness of clusters.

5.2   User Model Effects on Adaptive Visualization

So far, we have seen that the adaptive visualization could separate the relevant and non-relevant document clusters. It could also work more effectively when the user models were constructed using the mixture of keywords and NEs. However, this just tells us that there were separations and cannot let us know what they really looked like. Therefore, the following analysis focused on the distribution of relevant and non-relevant document clusters in the visualization. Table 5 compares the horizontal positions of the cluster centroids in various conditions. The relevant document clusters were always located closer to the user models (larger in their horizontal positions). Particularly, the distances between the cluster centroids were largest when the Parallel layout (which separates the user model and

Table 5. Comparing horizontal positions of cluster centroids (in pixels)

Keyword/NE Mixture	Clusters	Radial	Parallel	Hemisphere
k10n0	Relevant	313.58	318.56	350.35
	Non-relevant	302.99	188.19	304.74
	Distance	10.59*	130.37*	45.61*
k5n5	Relevant	315.73	332.02	361.77
	Non-relevant	301.46	192.71	308.44
	Distance	14.27*	**139.31***	53.33*
k0n10	Relevant	300.68	269.44	328.31
	Non-relevant	294.71	161.63	291.77
	Distance	5.96**	107.80*	36.54*

$(^*p < 0.01, \ ^{**}p = 0.038)$

the query space the most) was used whereas the Radial layout (non-personalized) produced very small between-cluster distances. The differences between relevant and non-relevant clusters' horizontal positions (or distances) were all statistically significant (Wilcox signed rank test). This result confirms our second hypothesis that the user model attracts more the relevant documents than the query side. We should note that the mixture of five keywords and five NEs shows the biggest distance in the Parallel layout (139.31) and thus supports the validity of concept-based user modeling for adaptive visualization. The mean differences of cluster distances across three mixtures were all statistically significant (Wilcox signed rank test, $p < 0.001$).

6   Conclusions

In this paper, we introduced our innovative approach for adaptive visualization and concept-based user modeling. Adaptive visualization is a promising personalized information access method that can efficiently guide users to relevant information. Concept-based user modeling is an alternative to old keyword-based approaches, which can enrich user models by adding more semantics. We integrated named-entities into user models and examined the quality of the adaptive visualization method equipped with the concept-based user models.

An experiment was conducted using the proposed approach and the result showed that the mixture of keywords and NEs provided the best results in terms of separating relevant documents from non-relevant ones. We also discovered that the cluster separation was due more to the between-cluster distances rather than cluster compactness. Moreover, the effect that user models could attract relevant documents around them was seen, which supports the utility of our idea that adaptive visualization can help users to access relevant information more easily.

In our future work, we plan to conduct a large-scale user study and examine user behaviors about our adaptive visualization and the concept-based user model. We are going to determine if the systems will work as expected and will observe under what situations users can benefit from the potential of the systems. We are also planning to migrate the adaptive visualization concept into other types of user interfaces and visualizations, including force-directed visualization and NE-based personalized browsing/searching. At the same time, more sophisticated concept-based user modeling ideas are being investigated.

References

1. Ahn, J., Brusilovsky, P.: Adaptive visualization of search results: Bringing user models to visual analytics. Information Visualization 8(3), 167–179 (2009)
2. Ahn, J., Brusilovsky, P., Grady, J., He, D., Florian, R.: Semantic annotation based exploratory search for information analysts. In: Information Processing and Management (in press, 2010)
3. Ahn, J., Brusilovsky, P., He, D., Grady, J., Li, Q.: Personalized web exploration with task models. In: Huai, J., Chen, R., Hon, H.W., Liu, Y., Ma, W.Y., Tomkins, A., Zhang, X. (eds.) Proceedings of the 17th International Conference on World Wide Web, WWW 2008, Beijing, China, April 21-25, pp. 1–10. ACM, New York (2008)
4. Bier, E.A., Ishak, E.W., Chi, E.: Entity workspace: An evidence file that aids memory, inference, and reading. In: Mehrotra, S., Zeng, D.D., Chen, H., Thuraisingham, B.M., Wang, F.Y. (eds.) ISI 2006. LNCS, vol. 3975, pp. 466–472. Springer, Heidelberg (2006)
5. Chen, C.C., Chen, M.C., Sun, Y.: Pva: a self-adaptive personal view agent system. In: KDD '01: Proceedings of the seventh ACM SIGKDD international conference on Knowledge discovery and data mining, pp. 257–262. ACM Press, New York (2001)
6. Davies, D.L., Bouldin, D.W.: A cluster separation measure. IEEE Trans. Pattern Anal. Mach. Intell. PAMI 1(2), 224–227 (2009)
7. Florian, R., Hassan, H., Ittycheriah, A., Jing, H., Kambhatla, N., Luo, X., Nicolov, H., Roukos, S., Zhang, T.: A statistical model for multilingual entity detection and tracking. In: Proceedings of the Human Language Technologies Conference (HLT-NAACL'04), Boston, MA, USA, May 2004, pp. 1–8 (2004)
8. Gauch, S., Speretta, M., Chandramouli, A., Micarelli, A.: User profiles for personalized information access. In: Brusilovsky, P., Kobsa, A., Nejdl, W. (eds.) Adaptive Web 2007. LNCS, vol. 4321, pp. 54–89. Springer, Heidelberg (2007)
9. Gentili, G., Micarelli, A., Sciarrone, F.: Infoweb: An adaptive information filtering system for the cultural heritage domain. Applied Artificial Intelligence 17(8-9), 715–744 (2003)
10. Hanani, U., Shapira, B., Shoval, P.: Information filtering: Overview of issues, research and systems. User Modeling and User-Adapted Interaction 11(3), 203–259 (2001)
11. Korfhage, R.R.: To see, or not to see – is that the query? In: SIGIR '91: Proceedings of the 14th annual international ACM SIGIR conference on Research and development in information retrieval, pp. 134–141. ACM, New York (1991)
12. Kumaran, G., Allan, J.: Text classification and named entities for new event detection. In: SIGIR '04: Proceedings of the 27th annual international ACM SIGIR conference on Research and development in information retrieval, pp. 297–304. ACM, New York (2004)
13. Leuski, A., Allan, J.: Interactive information retrieval using clustering and spatial proximity. User Modeling and User-Adapted Interaction 14(2-3), 259–288 (2004)

14. Magnini, B., Strapparava, C.: User modelling for news web sites with word sense based techniques. User Modeling and User-Adapted Interaction 14(2), 239–257 (2004)
15. Marchionini, G.: Exploratory search: from finding to understanding. Commun. ACM 49(4), 41–46 (2006)
16. Micarelli, A., Gasparetti, F., Sciarrone, F., Gauch, S.: Personalized search on the world wide web. In: Brusilovsky, P., Kobsa, A., Nejdl, W. (eds.) Adaptive Web 2007. LNCS, vol. 4321, pp. 195–230. Springer, Heidelberg (2007)
17. Mihalcea, R., Moldovan, D.L.: Document indexing using named entities. Studies in Informatics and Control 10(1), 21–28 (2001)
18. Olsen, K.A., Korfhage, R., Sochats, K.M., Spring, M.B., Williams, J.G.: Visualization of a document collection: The vibe system. Information Processing and Management 29(1), 69–81 (1993)
19. Petkova, D., Croft, B.W.: Proximity-based document representation for named entity retrieval. In: CIKM '07: Proceedings of the sixteenth ACM conference on Conference on information and knowledge management, pp. 731–740. ACM, New York (2007)
20. Pretschner, A., Gauch, S.: Ontology based personalized search. In: 11th IEEE Intl. Conf. on Tools with Artificial Intelligence (ICTAI'99), Chicago, IL, pp. 391–398 (1999)
21. Roussinov, D., Ramsey, M.: Information forage through adaptive visualization. In: DL '98: Proceedings of the third ACM conference on Digital libraries, pp. 303–304. ACM, New York (1998)

Interweaving Public User Profiles on the Web

Fabian Abel, Nicola Henze, Eelco Herder, and Daniel Krause

IVS – Semantic Web Group & L3S Research Center,
Leibniz University Hannover, Germany
{abel,henze,herder,krause}@l3s.de

Abstract. While browsing the Web, providing profile information in social networking services, or tagging pictures, users leave a plethora of traces. In this paper, we analyze the nature of these traces. We investigate how user data is distributed across different Web systems, and examine ways to aggregate user profile information. Our analyses focus on both explicitly provided profile information (name, homepage, etc.) and activity data (tags assigned to bookmarks or images). The experiments reveal significant benefits of interweaving profile information: more complete profiles, advanced FOAF/vCard profile generation, disclosure of new facets about users, higher level of self-information induced by the profiles, and higher precision for predicting tag-based profiles to solve the cold start problem.

1   Introduction

In order to adapt functionality to the individual users, systems need information about their users [1]. The Web provides opportunities to gather such information: users leave a plethora of traces on the Web, varying from profile data to tags. In this paper we analyze the nature of these distributed user data traces and investigate the advantages of interweaving publicly available profile data originating from different sources: social networking services (Facebook, LinkedIn), social media services (Flickr, Delicious, StumbleUpon, Twitter) and others (Google). The main research question that we will answer in this paper is the following: what are the benefits of aggregating these public user profile traces?

In our experiments we analyze the characteristics of both traditional profiles – which are explicitly filled by the end-users with information about their names, skills or homepages (see Sect. 3) – as well as rather implicitly generated tag-based profiles (see Sect. 4). We show that the aggregation of profile data reveals new facets about the users and present approaches to leverage such additional information gained by profile aggregation. We made all approaches and findings presented in this paper available for the public via the *Mypes*[1] service: it enables users to inspect their distributed profiles and provides access to the aggregated and semantically enriched profiles via a RESTful API.

---

[1] http://mypes.groupme.org/

P. De Bra, A. Kobsa, and D. Chin (Eds.): UMAP 2010, LNCS 6075, pp. 16–27, 2010.
© Springer-Verlag Berlin Heidelberg 2010

2   Related Work

Connecting data from different sources and services is in line with today's Web 2.0 trend of creating *mashups* of various applications [2]. Support for the development of interoperable services is provided by initiatives such as the data-portability project[2], standardization of APIs (e.g. OpenSocial) and authentication and authorization protocols (e.g. OpenID, OAuth), as well as by (Semantic) Web standards such as RDF, RSS and specific Microformats. Further, it becomes easier to connect distributed user profiles—including social connections—due to the increasing take-up of standards like FOAF [3], SIOC[3], or GUMO [4]. Conversion approaches allow for flexible user modeling [5]. Solutions for user identification form the basis for personalization across application boundaries [6]. Google's Social Graph API[4] enables application developers to obtain the social connections of an individual user across different services. Generic user modeling servers such as CUMULATE [7] or PersonIs [8] as well as frameworks for mashing up profile information [9] appear that facilitate handling of aggregated user data. Given these developments, it becomes more and more important to investigate the benefits of user profile aggregation in context of today's Web scenery.

In [10], Szomszor et al. present an approach to combine profiles generated in two different tagging platforms to obtain richer interest profiles; Stewart et al. demonstrate the benefits of combining blogging data and tag assignments from Last.fm to improve the quality of music recommendations [11]. In this paper we do not only analyze the benefits of aggregating tag-based user profiles [12, 13], which we enrich with Wordnet[5] facets, but also consider explicitly provided profiles coming from five different social networking and social media services.

3   Traditional Profile Data on the Web

Currently, users need to manually enter their profile attributes in each separate Web system. These attributes—such as the user's *full name*, current *affiliations*, or the *location* they are living at—are particularly important for social networking services such as LinkedIn or Facebook, but may be considered as less important in services such as Twitter. In our analysis, we measure to which degree users fill in their profile attributes in different services. To investigate the benefits of profile aggregation we address the following questions.

1. How detailed do users fill in their public profiles at social networking and social media services?
2. Does the aggregated user profile reveal more information about a particular user than the profile created in some specific service?

---

[2] http://www.dataportability.org/
[3] http://rdfs.org/sioc/spec/
[4] http://socialgraph.apis.google.com
[5] http://wordnet.princeton.edu/

3. Can the aggregated profile data be used to enrich an incomplete profile in an individual service?
4. To which extent can the service-specific profiles and the aggregated profile be applied to fill up standardized profiles such as FOAF [3] and vCard [14]?

3.1   Dataset

To answer the questions above, we crawled the public profiles of 116032 distinct users via the Social Graph API. People who have a Google account can explicitly link their different accounts and Web sites; the Social Graph API allows developers to look up the different accounts of a particular user. On average, the 116032 users linked 1.26 accounts while 70963 did not link any account.

For our analysis on traditional profiles we were interested in popular services where users can have public profiles. We therefore focused on the social networking services Facebook and LinkedIn, as well as on Twitter, Flickr, and Google. Figure 1(a) lists the number of public profiles and the concrete profile attributes we obtained from each service. We did not consider private information, but only crawled attributes that were publicly available. Among the users for whom we crawled the Facebook, LinkedIn, Twitter, Flickr, and Google profiles were 338 users who had an account at all five different services.

3.2   Individual Profiles and Profile Aggregation

The completeness of the profiles varies from service to service. The public profiles available in the social networking sites Facebook and LinkedIn are filled more accurately than the Twitter, Flickr, or Google profiles—see Fig. 1(b). Although Twitter does not ask many attributes for its user profile, users completed their profile up to just 48.9% on average. In particular the *location* and *homepage*— which can also be a URL to another profile page, such as MySpace—are omitted most often. By contrast, the average Facebook and LinkedIn profile is filled up to 85.4% and 82.6% respectively. Obviously, some user data is replicated at multiple services: name and profile picture are specified at nearly all services, location was provided at 2,9 out of five services. However, inconsistencies can be found in the data: for example, 37.3% of the users' *full names* in Facebook are not exactly the same as the ones specified at Twitter.

For each user we aggregated the public profile information from Facebook, LinkedIn, Twitter, Flickr, and Google, i.e. for each user we gathered attribute-value pairs and mapped them to a uniform user model. Aggregated profiles reveal more facets (17 distinct attributes) about the users than the public profiles available in each separate service. On average, the completeness of the aggregated profile is 83.3%: more than 14 attributes are filled with meaningful values. As a comparison, this is 7.6 for Facebook, 8.2 for LinkedIn and 3.3 for Flickr. Aggregated profiles therewith reveal significantly more information about the users than the public profiles of the single services.

Further, profile aggregation enables completion of the profiles available at the specific services. For example, by enriching the incomplete Twitter profiles with

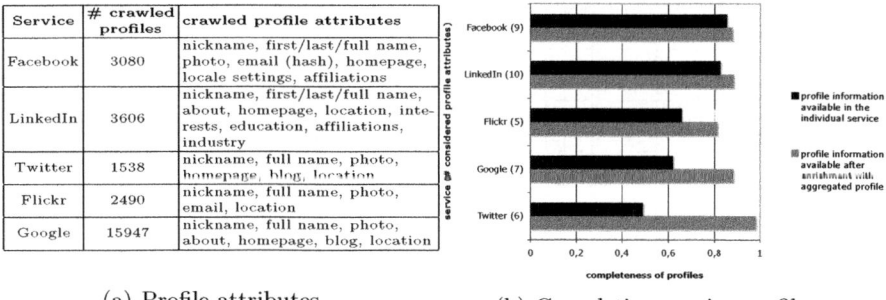

Service	# crawled profiles	crawled profile attributes
Facebook	3080	nickname, first/last/full name, photo, email (hash), homepage, locale settings, affiliations
LinkedIn	3606	nickname, first/last/full name, about, homepage, location, interests, education, affiliations, industry
Twitter	1538	nickname, full name, photo, homepage, blog, location
Flickr	2490	nickname, full name, photo, email, location
Google	15947	nickname, full name, photo, about, homepage, blog, location

(a) Profile attributes   (b) Completing service profiles

Fig. 1. Service profiles: (a) number of public profiles as well as the profile attributes that were crawled from the different services and (b) completing service profiles with aggregated profile data. Only the 338 users who have an account at each of the listed services are considered.

information gathered from the other services, the completeness increases to more than 98% (see Fig. 1(b)): profile fields that are often left blank, such as location and homepage, can be obtained from the social networking sites. Moreover, even the rather complete Facebook and LinkedIn profiles can benefit from profile aggregation: LinkedIn profiles can, on average, be improved by 7%, even though LinkedIn provides three attributes—*interests*, *education* and *industry*—that are not in the public profiles of the other services (cf. Fig. 2(a)).

In summary, profile aggregation results in an extensive user profile that reveals more information than the profiles at the individual services. Moreover, aggregation can be used to fill in missing attributes at the individual services.

3.3   FOAF and vCard Generation

In most Web 2.0 services, user profiles are primarily intended to be presented to other end-users. However, it is also possible to use the profile data to generate FOAF [3] profiles or vCard [14] entries that can be fed into applications such as Outlook, Thunderbird or FOAF Explorer.

Figure 2(a) lists the attributes each service can contribute to fill in a FOAF or vCard profile, if the corresponding fields are filled out by the user. Figure 2(b) shows to which degree the real service profiles of the 338 considered users can actually be applied to fill in the corresponding attributes with adequate values.

Using the aggregated profile data of the users, it is possible to generate FOAF profiles and vCard entries to an average degree of more than 84% and 88% respectively—the corresponding attributes are listed in Fig. 2(a). Google, Flickr and Twitter profiles provide much less information applicable to fill the FOAF and vCard details. Although Facebook and LinkedIn both provide seven attributes that can potentially be applied to generate the vCard profile, it is interesting to see that the actual LinkedIn user profiles are more valuable and produce vCard entries with average completeness of 45%; using Facebook as

Attribute	vCard	FOAF	Fa	L	T	Fl	G
nickname	x	x	x	x	x	x	x
first name		x	x	x			
last name		x	x	x			
full name	x	x	x	x	x	x	x
profile photo	x	x	x		x	x	x
about	x			x			x
email	x	x	x			x	
homepage	x	x	x	x	x		x
blog	x	x			x		x
location	x	x		x	x	x	x
locale settings	x			x			
interests		x			x		
education		x	x				
affiliations	x	x	x	x			
industry	x				x		

(a) Services and available attributes

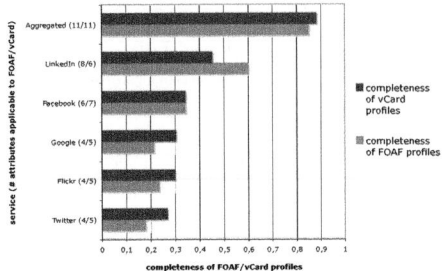

(b) Completing FOAF/vCard profiles

Fig. 2. FOAF/vCard profile generation: (a) services and attributes available in the the public profiles of Facebook (Fa), LinkedIn (L), Twitter (T), Flickr (Fl), and Google (G) that can be applied to fill in a FOAF profile or a vCard entry and (b) completing FOAF and vCard profiles with the actual user profiles

a data source this is only 34%. In summary, the aggregated profiles are thus a far better source of information to generate FOAF/vCard entries than the service-specific profiles.

3.4   Synopsis

Our analysis of the user profiles distributed across the different services point out several advantages of profile aggregation and motivate the intertwining of profiles on the Web. With respect to the key questions raised at the beginning of the section, the main outcomes can be summarized as follows.

1. Users fill in their public profiles at social networking services (Facebook, LinkedIn) more extensively than profiles at social media services (Flickr, Twitter) which can possibly be explained by differences in purpose of the different systems.
2. Profile aggregation provides multi-faceted profiles that reveal significantly more information about the users than individual service profiles can provide.
3. The aggregated user profile can be used to enrich incomplete profiles in individual services, to make them more complete.
4. Service-specific profiles as well as the aggregated profiles can be applied to generate FOAF profiles and vCard entries. The aggregated profile represents the most useful profile, as it completes the FOAF profiles and vCard entries to 84% and 88% respectively.

As user profiles distributed on the Web describe different facets of the user, profile aggregation brings some advantages: users do not have to fill their profiles over and over again; applications can make use of more and richer facets/attributes of the user (e.g. for personalization purposes). However, our analysis shows also the risk of intertwining user profiles. For example, users who deliberately leave out some fields when filling their Twitter profile might not be aware that the corresponding information can be gathered from other sources.

Table 1. Tagging statistics of the 139 users who have an account at Flickr, Stumble-Upon, and Delicious

	Flickr	StumbleUpon	Delicious	Overall
tag assignments	3781	12747	61884	78412
distinct tags	691	2345	11760	13212
tag assignments per user	27.2	91.71	445.21	564.12
distinct tags per user	5.22	44.42	165.83	71.82

4   User Activity Data on the Web

Most social media systems enable users to organize content with tags (freely chosen keywords). The tagging activities of a user form a valuable source of information for determining the interests of a user [12, 13]. In our analysis we examine the nature of the tag-based profiles in different systems. Again, we investigate the the benefits of aggregating profile data and answer the following questions.

1. What kind of tag-based profiles do individual users have in the different systems?
2. Does the aggregation of tag-based user profiles reveal more information about the users than the profiles available in some specific service?
3. Is it possible to predict tag-based profiles in a system, based on profile data gathered from another system?

4.1   Individual Tagging Behavior across Different Systems

From the 116032 users , 139 users were randomly selected who linked their Flickr, StumbleUpon, and Delicious accounts. Table 1 lists the corresponding tagging statistics. For these users, we crawled 78412 tag assignments that were performed on the 200 latest images (Flickr) or bookmarks (Delicious and StumbleUpon). Overall, users tagged more actively in Delicious than in the other systems: more than 75% of the tagging activities originate from Delicious, 16.3% from Stum-bleUpon and 5% from Flickr. The usage frequency of the distinct tags shows a typical power-law distribution in all three systems, as well as in the aggregated set of tag assignments: while some tags are used very often, the majority of tags is used rarely or even just once.

On average, each user provided 564.12 tag assignments across the different systems. The user activity distribution corresponds to a gaussian distribution: 26.6% of the users have less than 200 tag assignments, 10.1% have more than 1000 and 63.3% have between 200 and 1000 tag assignments. Interestingly, people who actively tagged in one system do not necessarily perform many tag assign-ments in another system. For example, none of the top 5% taggers in Flickr or StumbleUpon is also among the top 10% taggers in Delicious. This observation of unbalanced tagging behavior across different systems again reveals possible

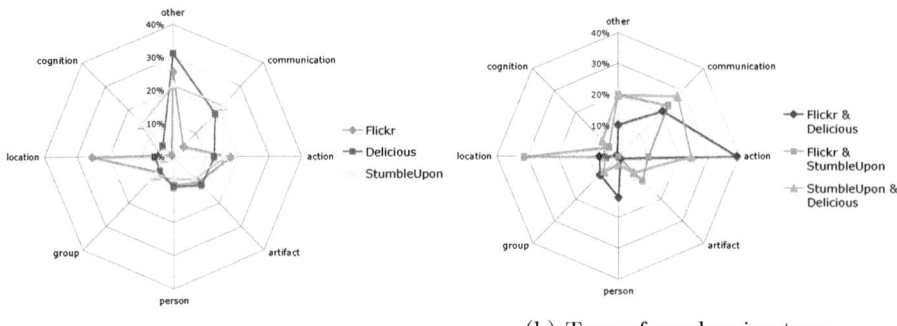

(a) Type of tags in the systems   (b) Type of overlapping tags

Fig. 3. Tag usage characterized with Wordnet categories: (a) Type of tags users apply in the different systems and (b) type of tags individual users apply in two different systems

advantages of profile aggregation for current tagging systems: given a sparse tag-based user profile, the consideration of profiles produced in other systems might be used to tackle sparsity problems.

4.2   Commonalities and Differences in Tagging Activities

In order to analyze commonalities and differences of the users' tag-based profiles in the different systems, we mapped tags to Wordnet categories and considered only those 65% of the tags for which such a mapping exists. Figure 3(a) shows that the type of tags in StumbleUpon and Delicious are quite similar, except for *cognition* tags (e.g., research, thinking), which are used more often in StumbleUpon than in Delicious. For both systems, most of the tags—21.9% in StumbleUpon and 18.3% in Delicious—belong to the category *communication* (e.g., hypertext, web). By contrast, only 4.4% of the Flickr tags refer to the field of communication; the majority of tags (25.2%) denote locations (e.g., Hamburg, tuscany). *Action* (e.g., walking), *people* (e.g., me), and *group* tags (e.g., community) as well as words referring to some *artifact* (e.g., bike) occur in all three systems with similar frequency. However, the concrete tags seem to be different. For example, while artifacts in Delicious refer to things like "tool" or "mobile device", the artifact tags in Flickr describe things like "church" or "painting". This observation is supported by Fig. 3(b), which shows the average overlap of the individual category-specific tag profiles. On average, each user applied only 0.9% of the Flickr artifact tags tags also in Delicious. For Flickr and Delicious, action tags allocate the biggest fraction of overlapping tags. It is interesting to see that the overlap of location tags between Flickr and StumbleUpon is 31.1%, even though location tags are used very seldomly in StumbleUpon (3.3%, as depicted in Fig. 3(a)). This means that if someone utilizes a location tag in StumbleUpon, it is likely that she will also use the same tag in Flickr.

Having knowledge on the different (aggregated) tagging facets of a user opens the door for interesting applications. For example, a system could exploit

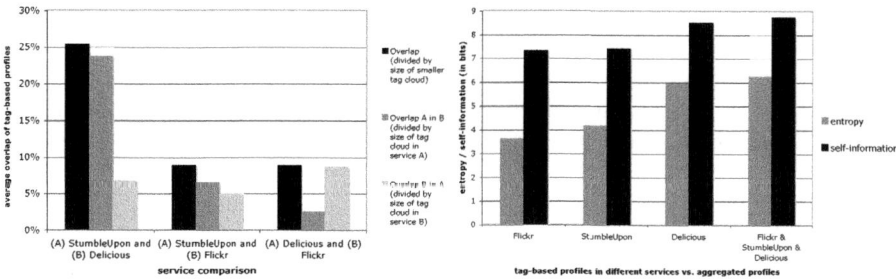

(a) Overlap of tag-based profiles   (b) Entropy and self-information

Fig. 4. Aggregation of tag-based profiles: (a) average overlap and (b) entropy and self-information of service-specific profiles in comparison to the aggregated profiles

StumbleUpon tags referring to locations to recommend Flickr pictures even if the user's Flickr profile is empty. In Sect. 4.4 we will present an approach that takes advantage of the faceted tag-based profiles for predicting tagging behavior.

4.3   Aggregation of Tagging Activities

To analyze the benefits of aggregating tag-based profiles in more detail we measure the information gain, entropy and overlap of the individual profiles. Figure 4(a) describes the average overlap with respect to three different metrics: given two tag-based profiles A and B, the overlap is (1) $overlap = \frac{A \cap B}{min(|A|,|B|)}$, (2) $overlap_{AinB} = \frac{A \cap B}{|A|}$, or (3) $overlap_{BinA} = \frac{A \cap B}{|B|}$. For example, $overlap_{AinB}$ denotes the percentage of tags in A that also occur in B.

The overlap of the tag-based profiles produced in Delicious and Stumble-Upon is significantly higher than the overlap of service combinations that include Flickr. However, on average, a user still just applies 6.8% of her Delicious tags also in StumbleUpon, which is approximately as high as the percentage of tags a StumbleUpon user also applies in Flickr. Overall, the tag-based user profiles do not overlap strongly. Hence, users reveal different facets of their profiles in the different services.

Figure 4(b) compares the averaged entropy and self-information of the tag-based profiles obtained from the different services with the aggregated profile. The entropy of a tag-based profile T, which contains of a set of tags t, is computed as follows.

$$entropy(T) = \sum_{t \in T} p(t) \cdot \textit{self-information}(t) \qquad (1)$$

In Equation 1, $p(t)$ denotes the probability that the tag t was utilized by the corresponding user and $\textit{self-information}(t) = -log(p(t))$. In Fig. 4(b), we summarize self-information by building the average of the mean self-information of the users' tag-based profiles. Among the service-specific profiles, the tag-based profiles in Delicious, which also have the largest size, bear the highest entropy and average self-information. By aggregating the tag-based profiles, self-information increases clearly by 19.5% and 17.7% with respect to the Flickr and Stumble-Upon profiles respectively. Further, the tag-based profiles in Delicious can benefit from the profile aggregation as the self-information would increase by 2.7% (from 8.53 bit to 8.76 bit) which is also considerably higher, considering that self-information is measured in bits (e.g., with 8.53 bits one could describe 370 states while 8.76 bits allow for decoding of 434 states).

Aggregation of tag-based profiles thus reveals more valuable new information about individual users than focusing just on information from single services. However, some fraction of the profiles also overlap between different systems, as depicted in Fig. 4(a). In the next section we analyze whether it is possible to predict those overlapping tags.

4.4 Prediction of Tagging Behavior

Systems that rely on user data usually have to struggle with the *cold start problem*; especially those systems that are infrequently used or do not have a large base of users require solutions to that problem. In this section we investigate the applicability of profile aggregation. Therefore, we evaluate different approaches with respect to the following task.

Tag prediction task. *Given a set of tags that occur in the tag-based profile of user u in system A, the task of the tag prediction strategy is to predict those tags that will also occur in u's profile in system B.*

We measure the performance by means of *precision* (= correctly classified as overlapping tags / classified as overlapping tags), *recall* (= correctly classified as overlapping tags / overlapping tags), and *f-measure* (= harmonic mean of precision and recall). Our intention is not to find the best prediction algorithm, but to examine the impact of features extracted from profile aggregation. Hence, we apply a *Naive Bayes classifier*, which we feed with different features. The benchmark tag prediction strategy (*without profile aggregation*) bases its decision on a single feature: (F1) overall usage frequency of t in system B. In contrast, the strategy that makes use of *profile aggregation* also applies (F2) u's usage frequency of t in system A and (F3) size of u's profile in system A.

Figure 5(a) compares the average performance of both tag prediction strategies. For each of the 139 users and each service combination (Flickr → Delicious, Delicious → Flickr, StumbleUpon → Delicious, etc.) the strategies had to tackle the prediction task specified above. The benefits of the profile aggregation features are significant. The profile aggregation strategy performs—with respect to the f-measure—96.1% better than the strategy that does not benefit from profile aggregation (correspondingly, the improvement of precision and recall is explicit). Further, it is important to notice that the average percentage of overlapping tags is less than 4%. Thus, a random strategy, which simply guesses whether tag t will overlap or not (probability of 0.5), would fail with a precision lower than 2%.

On average, the profile aggregation strategy can thus detect 57.4% of the tags in system A that will also be part of the tag-based profile in system B. The

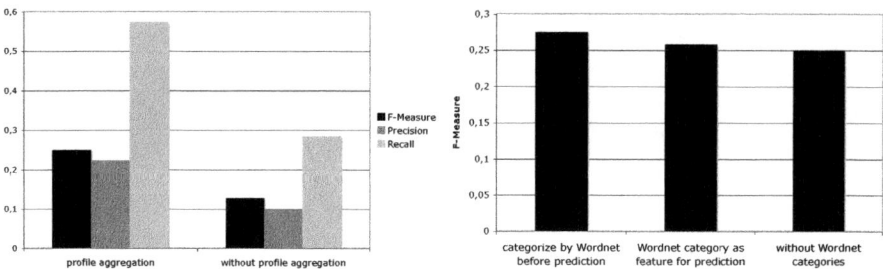

(a) Average performance of tag prediction   (b) Impact of Wordnet categorization

Fig. 5. Performance of tag prediction: (a) with and without aggregation of tag-based profiles and (b) improving prediction performance (with profile aggregation) by means of Wordnet categorization

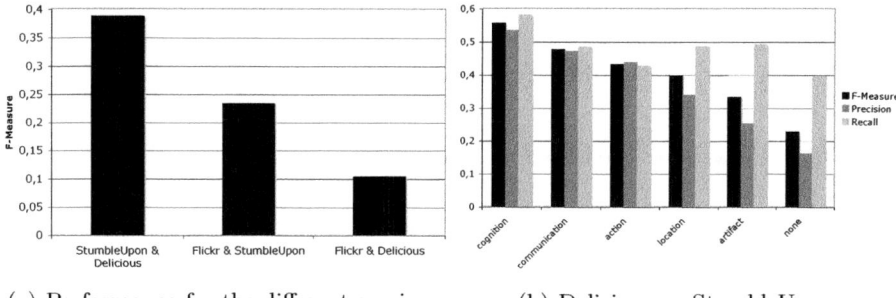

(a) Performance for the different services   (b) Delicious → StumbleUpon

Fig. 6. Tag prediction performance for specific services

performance can further be improved by clustering the tag-based profiles according to Wordnet categories. Figure 5(b) shows that the consideration of Wordnet features—(F4) Wordnet category of t and (F5) relative size of corresponding Wordnet category cluster in u's profile—leads to a small improvement from 0.25 to 0.26 regarding the f-measure. However, if tag predictions are done for each Wordnet cluster of the profiles separately, the improvement is considerably high as the f-measure increases from 0.25 to 0.28.

Figure 6 shows the tag prediction performance (using features F1-5) focusing on specific service combinations. While tag predictions for Flickr/Delicious based on tag-based profiles from Delicious/Flickr perform quite weak, the predictions between Flickr and StumbleUpon show a much better performance (f-measure: 0.23). For the two bookmarking services, StumbleUpon and Delicious, which also have the highest average overlap (cf. Fig. 4(a)), tag prediction works best with f-measure of 0.39 and precision of 0.36. Figure 6(b) illustrates for what kind of tags prediction works best between Delicious and StumbleUpon. For tags that cannot be assigned to a Wordnet category (*none*), the precision is just 16% while recall of 40% might still be acceptable. However, given tags that can be mapped to Wordnet categories, the performance is up to 0.57 regarding f-measures. Given *cognition* tags (e.g., search, ranking) of a particular user u, the profile aggregation strategy, which applies the features F1-5, can predict the cognition tags u will use in StumbleUpon with a precision of nearly 60%: even if a user has not performed any tagging activity in StumbleUpon, one could recommend 10 cognition tags out of which 6 are relevant for u.

4.5 Synopsis

The results of our analyses and experiments indicate several benefits of aggregating and interweaving tag-based user profiles. We showed that users reveal different types of facets (illustrated by means of Wordnet categories) in the different systems. By combining tag-based profiles from Flickr, StumbleUpon, and Delicious, the average self-information of the profiles increases significantly. Although the tag-based service-specific profiles overlap just to a small degree, we proved that the consideration of profile data from other sources can be applied to solve cold start problems. In particular, we showed that the profile aggregation strategy for predicting tag-based profiles significantly outperforms the benchmark that does not incorporate profile features from other sources.

5   Conclusions and Future Work

In this paper we analyzed the benefits of interweaving public profile data on the Web. For both explicitly provided profile information (e.g. name, hometown, etc.) and rather implicitly provided tag-based profiles (e.g. tags assigned to bookmarks), the aggregation of profile data from different services (e.g, LinkedIn, Facebook, Flickr, etc.) reveals significantly more facets about the individual users than one can deduce from the separated profiles. Our experiments show the advantages of interweaving distributed user data for various applications, such as completing service-specific profiles, generating FOAF or vCard profiles, producing multi-faceted tag-based profiles, and predicting tag-based profiles to solve cold start problems. End-users and application developers can immediately benefit from our research by using the Mypes service (http://mypes.groupme.org/).

In our future work we will focus on possible correlations between traditional and tag-based profiles. For example, in initial experiments we analyzed whether tag-based profiles conform to the skills users specified at LinkedIn. Given the dataset described in Sect. 3, 76.2% of the users applied at least one of the, on average, 8.56 LinkedIn skills also as a tag in Delicious. Further, we found first evidence that for users, who belong to the same group based on their social networking profile (in particular location and industry), the similarities between the tag-based profiles is higher than for users belonging to different groups. In the future, we will continue these experiments and investigate how explicitly provided profile data can be exploited in social media systems.

Acknowledgments. This work is partially sponsored by the EU FP7 project GRAPPLE (http://www.grapple-project.org/).

References

1. Jameson, A.: Adaptive interfaces and agents. In: The HCI handbook: fundamentals, evolving technologies and emerging applications, pp. 305–330 (2003)
2. Zang, N., Rosson, M.B., Nasser, V.: Mashups: Who? What? Why? In: Czerwinski, M., Lund, A., Tan, D. (eds.) Proceedings of Conference on Human factors in computing systems (CHI '08), pp. 3171–3176. ACM, New York (2008)
3. Brickley, D., Miller, L.: FOAF Vocabulary Specification 0.91. Namespace document, FOAF Project (November 2007), http://xmlns.com/foaf/0.1/
4. Heckmann, D., Schwartz, T., Brandherm, B., Schmitz, M., von Wilamowitz-Moellendorff, M.: GUMO – the general user model ontology. In: Ardissono, L., Brna, P., Mitrović, A. (eds.) UM 2005. LNCS (LNAI), vol. 3538, pp. 428–432. Springer, Heidelberg (2005)
5. Aroyo, L., Dolog, P., Houben, G., Kravcik, M., Naeve, A., Nilsson, M., Wild, F.: Interoperability in pesonalized adaptive learning. Journal of Educational Technology & Society 9 (2), 4–18 (2006)
6. Carmagnola, F., Cena, F.: User identification for cross-system personalisation. Information Sciences: an International Journal 179(1-2), 16–32 (2009)
7. Yudelson, M., Brusilovsky, P., Zadorozhny, V.: A user modeling server for contemporary adaptive hypermedia: An evaluation of the push approach to evidence propagation. In: Conati, C., McCoy, K.F., Paliouras, G. (eds.) UM 2007. LNCS (LNAI), vol. 4511, pp. 27–36. Springer, Heidelberg (2007)
8. Assad, M., Carmichael, D., Kay, J., Kummerfeld, B.: PersonisAD: Distributed, active, scrutable model framework for context-aware services. In: LaMarca, A., Langheinrich, M., Truong, K.N. (eds.) Pervasive 2007. LNCS, vol. 4480, pp. 55–72. Springer, Heidelberg (2007)
9. Abel, F., Heckmann, D., Herder, E., Hidders, J., Houben, G.J., Krause, D., Leonardi, E., van der Slujis, K.: A framework for flexible user profile mashups. In: Dattolo, A., Tasso, C., Farzan, R., Kleanthous, S., Vallejo, D.B., Vassileva, J. (eds.) UMAP 2009. LNCS, vol. 5535, pp. 1–10. Springer, Heidelberg (2009)
10. Szomszor, M., Alani, H., Cantador, I., O'Hara, K., Shadbolt, N.: Semantic modelling of user interests based on cross-folksonomy analysis. In: Sheth, A.P., Staab, S., Dean, M., Paolucci, M., Maynard, D., Finin, T.W., Thirunarayan, K. (eds.) ISWC 2008. LNCS, vol. 5318, pp. 632–648. Springer, Heidelberg (2008)
11. Stewart, A., Diaz-Aviles, E., Nejdl, W., Marinho, L.B., Nanopoulos, A., Schmidt-Thieme, L.: Cross-tagging for personalized open social networking. In: Cattuto, C., Ruffo, G., Menczer, F. (eds.) Hypertext, pp. 271–278. ACM, New York (2009)
12. Firan, C.S., Nejdl, W., Paiu, R.: The benefit of using tag-based profiles. In: Almeida, V.A.F., Baeza-Yates, R.A. (eds.) LA-WEB, pp. 32–41. IEEE Computer Society, Los Alamitos (2007)
13. Michlmayr, E., Cayzer, S.: Learning User Profiles from Tagging Data and Leveraging them for Personal(ized) Information Access. In: Golder, S., Smadja, F. (eds.) Proceedings of the Workshop on Tagging and Metadata for Social Information Organization at WWW '07 (May 2007)
14. Dawson, F., Howes, T.: vCard MIME Directory Profile. Request for comments, IETF, Network Working Group (September 1998)

Modeling Long-Term Search Engine Usage

Ryen W. White, Ashish Kapoor, and Susan T. Dumais

Microsoft Research,
One Microsoft Way, Redmond WA 98052, USA
{ryenw,akapoor,sdumais}@microsoft.com

Abstract. Search engines are key components in the online world and the choice of search engine is an important determinant of the user experience. In this work we seek to model user behaviors and determine key variables that affect search engine usage. In particular, we study the engine usage behavior of more than ten thousand users over a period of six months and use machine learning techniques to identify key trends in the usage of search engines and their relationship with user satisfaction. We also explore methods to determine indicators that are predictive of user trends and show that accurate predictive user models of search engine usage can be developed. Our findings have implications for users as well as search engine designers and marketers seeking to better understand and retain their users.

Keywords: Search Engine, Predictive Model.

1 Introduction

Search engines such as Google, Yahoo!, and Bing facilitate rapid access to the vast amount of information on the World Wide Web. A user's decision regarding which engine they should use most frequently (their *primary* engine) can be based on factors including reputation, familiarity, effectiveness, interface usability, and satisfaction [14], and can significantly impact their overall level of search success [18, 20]. Similar factors can influence a user's decision to switch from one search engine to another, either for a particular query if they are dissatisfied with search results or seek broader topic coverage, or for specific types of tasks if another engine specializes in such tasks, or more permanently as a result of unsatisfactory experiences or relevance changes, for example [18]. The barrier to switching engines is low and multiple engine usage is common. Previous research suggests that 70% of searchers use more than one engine [20].

Research on engine switching has focused on characterizing short-term switching behavior such as predicting when users are going to switch engines within a search session [7, 10, 18], and promoting the use of multiple engines for the current query if another engine has better results [20]. Given the economic significance of multiple engine usage to search providers, and its prevalence among search engine users, it is important to understand and model engine usage behavior more generally. Narrowly focusing on the use of multiple search engines for a single query or within a single search session provides both limited insight into user preferences and limited data

P. De Bra, A. Kobsa, and D. Chin (Eds.): UMAP 2010, LNCS 6075, pp. 28–39, 2010.

from which to model multiple engine usage. There has been some research on modeling engine usage patterns over time, such as studies of switching to develop metrics for competitive analysis of engines in terms of estimated user preference and user engagement [8] or building conceptual and economic models of search engine choice [13, 17]. Rather than characterizing and predicting long-term engine usage, this previous work has focused on metric development or has specifically modeled search engine loyalty.

In this paper we model user behaviors and determine key variables that affect search engine usage. In particular, we study engine usage of over ten thousand consenting users over a period of six months using log data gathered from a widely-distributed browser toolbar. We perform analysis of the data using machine learning techniques and show that there are identifiable trends in the usage of search engines. We also explore methods to determine indicators of user trends and show that accurate predictive user models of search engine usage can be developed. Knowledge of the key trends in engine usage and performant predictive models of this usage are invaluable to the designers and marketers of search engines as they attempt to understand and support their users and increase market share.

The rest of this paper is structured as follows. Section 2 presents related work on search engine switching, engine usage, and consumer loyalty/satisfaction. Section 3 presents our research and experiments on modeling and predicting engine usage. Section 4 discusses our approach and its implications. We conclude in Sect. 5.

2 Related Work

The most significant related work lies in the areas of search engine switching, consumer choice regarding search engine usage, and studies of consumer loyalty with a product or brand and their associated levels of satisfaction. In this section we describe work in each of these areas and relate it to the research described in this paper.

Some research has examined engine switching behavior within a search session. Heath and White [7] and Laxman et al. [10] developed models for predicting switching behavior within search sessions using sequences of user actions. White and Dumais [18] used log analysis and a survey to characterize search engine switching behavior, and used features of the active query, the current search session, and the user to predict engine switching. White et al. [20] developed methods for predicting which search engine would produce the best results for a query. One way in which such a method could be used is to promote the use of multiple search engines on a query-by-query basis, using the predictions of the quality of results from multiple engines. Studying switching within a session is useful for understanding and predicting isolated switching events. However, to develop a better understanding of users' engine preferences and model multi-engine usage effectively we must look beyond isolated sessions. The research presented in this paper leverages engine usage statistics aggregated weekly over a six-month period to build models of search engine usage across many thousands of search engine users.

Mukhopadhyay et al. [13] and Telang et al. [17] used economic models of choice to understand whether people developed brand loyalty to a particular search engine, and how search engine performance (as measured by within-session switching) affected user choice. They found that dissatisfaction with search engine results had both short-term and long-term effects on search engine choice. The data set is small by modern log analysis standards (6,321 search engine switches from 102 users), somewhat dated (data from June 1998 – July 1999 including six search engines but not Google), and only summary level regression results were reported. Juan and Chang [8] described some more recent research in which they summarize user share, user engagement and user preferences using click data from an Internet service provider. They identify three user classes (loyalists to each of the two search engines studied and switchers), and examine the consistency of engine usage patterns over time. We build on this work to identify key trends in search engine usage, reason about why certain behaviors were observed, and develop predictive models of search engine usage based on observed usage patterns and user satisfaction estimates.

There is a large body of work in the marketing community regarding product or brand switching and the relationship between satisfaction and loyalty. Research in these areas is typically concerned with identifying factors that influence customer defections [5] or developing models of satisfaction or loyalty that make it easier for businesses to understand customer rationale and take corrective action if needed [9, 11]. In this paper we model the usage of multiple search engines over time and base part of our model on estimates of searcher satisfaction gleaned from log data. Although satisfaction is the predominant metric used by companies to detect and manage defections to competitors [1,3], more recent research has found that knowledge of competitors and attitudinal and demographic factors, among other influences, can also play an important role [14].

To summarize, the research presented in this paper differs from earlier work in that we study patterns of usage for multiple search engines over a long period of time, identify key trends, and develop predictive models for these trends.

3   Modeling Search Engine Usage

Understanding and predicting retention and switching behavior is important for search engine designers and marketers interested in satisfying users and growing market share. To model long-term patterns of search engine use, we used data from a widely-distributed browser toolbar (described in more detail below). We examine the patterns of search engine usage for tens of thousands of users over a six-month period of time. We first identify key trends in people's usage patterns (e.g., sticking with the same engine over time, switching between engines, etc.). We then summarize features that distinguish among the different usage trends. Finally, we develop models to predict which trend a particular individual will follow over time.

Data Collection. We used six months of interaction logs from September 2008 through February 2009 inclusive, obtained from hundreds of thousands of consenting users of a widely-distributed browser toolbar. These log entries include a unique identifier for the user, a timestamp for each Web page visited, and the URL of the Web page visited. Intranet and secure (https) URL visits were excluded at the source. To remove variability caused by geographic and linguistic variation in search behavior, we only include log entries generated in the English speaking United States locale.

From these logs, we extracted search queries issued to three popular Web search engines (which we call A, B and C) for each of the 26 weeks. We selected users who issued at least 10 queries per week in all weeks of the study. Applying this threshold gave us a sufficient number of queries each week to reliably study engine usage. These users formed the pool from which study subjects were randomly selected and for whom engine usage models were constructed.

Features. We extracted several different features to describe user interaction with the search engines. These features are shown in Table 1. We summarize the proportion of queries issued to each search engine. We also use the number of queries and the average length of the queries sent to each engine. Finally we use measures of a user's satisfaction with the search engine. Satisfaction estimates for each of the engines are determined using the fraction of queries issued to that engine that have post-query dwell times equaling or exceeding 30 seconds. This estimate of user satisfaction is based on results from Fox et al. [6], in which they learned models of which implicit user interactions (such as page dwell time) are predictive of explicit judgments of user satisfaction. We further broke down satisfaction according to whether queries were navigational or not because poor performance on this usually straightforward class of queries may be especially likely to result in engine switches. Navigational queries were defined as queries for which the same search result was clicked at least 95% of the time. For each user we have three values for each of these features (one corresponding to each search engine) for each of the 26 weeks of our study.

Table 1. Features of user interaction with each search engine (per week)

Feature	Description
fractionEngine	Fraction of queries issued to search engine
queryCountEngine	Number of queries issued to search engine
avgEngineQueryLength	Average length (in words) of queries to search engine
fractionEngineSAT	Fraction of search engine queries that are satisfied
fractionNavEngine	Fraction search engine queries defined as navigational
fractionNavEngineSAT	Fraction of queries in fractionNavEngine satisfied

3.1 Key Trends in Long-Term Search Engine Usage

We first analyze the data using dimensionality reduction techniques to develop insights about key trends. In particular, we use non-negative matrix factorization (NNMF) [10] of the data to summarize key behavioral patterns using a small number of basis vectors. NNMF is a method to obtain a representation of data using non-negativity constraints. This leads to part-based representation such that the original data can be represented as an additive combination of a small number of basis vectors.

Formally, we construct **A**, an $M{\times}N$ matrix, where the columns represent users ($N=$ 10,000 users) and the rows are the vector representation of 26 weeks of search engine usage statistics. The observations we used for our initial analysis are the proportion of queries that each user issued to each of the three search engines, thus $M=26\times3$.

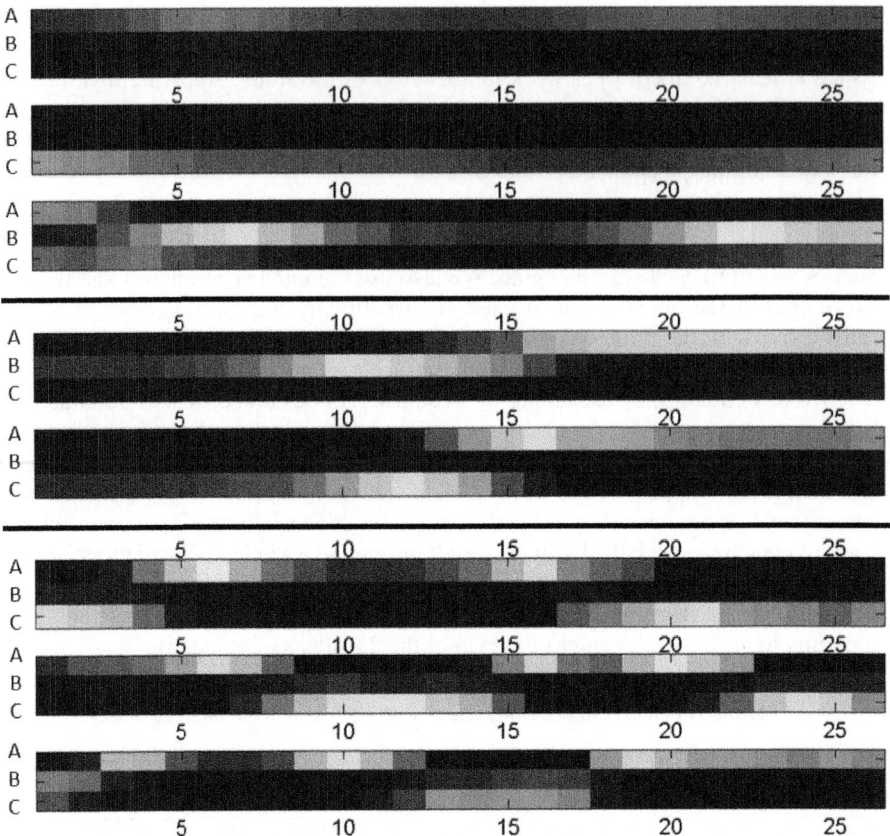

Fig. 1. Basis vectors learned from non-negative matrix factorization. We see three main patterns: sticking to a particular engine (first, second and third rows), switches that persist (fourth and fifth rows) and oscillations between different choices (sixth, seventh and eighth rows).

This matrix is decomposed into non-negative vectors matrix $\mathbf{W}$ and $\mathbf{H}$ of dimensions $M \times R$ and $R \times N$ respectively. Each column of $\mathbf{W}$ represents a basis vector while each row of the matrix $\mathbf{H}$ encodes the weight of the corresponding basis vector for each of the N users. The non-negative factorization means that we can interpret each individual observation vector as an additive combination of basis vectors (or parts). Consequently, each of the basis vectors represents the key trends that make up the behaviors that we observe in all the users.

Figure 1 shows the results of this analysis where we depict the top eight recovered basis vectors. Each image in the figure depicts one vector comprising the 26 weeks (x-axis) usage of three search engines (y-axis, labeled A, B, C). The red end of the spectrum represents high usage of an engine and the violet end represents low usage. A light blue coloring represents roughly equal usage of all three engines. In the first row, for example, we see a group of users who consistently use search engine A throughout the six-month period we studied. The fifth row shows users who initially use search engine C almost all the time, gradually decrease their usage of C (color transition from red to light blue), and slowly increase their usage of engine A. The seventh row show a pattern of behavior in which search engine A is used initially, followed by a gradual move to engine C (although it is not used all of the time, as indicated by the green rather than red color), then back to A, and finally back to C.

In general, this NNMF analysis identifies three key behavioral patterns:

1) *No Switch*: Sticking to one search engine over time (first, second, third rows),
2) *Switch*: Making a switch to another search engine and then persisting with the new engine (fourth and fifth rows), and
3) *Oscillation*: Oscillating between different engines (sixth, seventh, eighth rows).

While a majority of users indeed stick to one particular search engine, from the user modeling perspective it is interesting to analyze the other two general trends in user interactions with search engines (switches and oscillations). In particular, persisting switches are interesting as they indicate a significant evolution in user preference. From a search engine's perspective it is invaluable to understand the reasons for such switches and a predictive user model that would warn of such a switch could have significant implications for the search providers involved. Similarly, an oscillating preference in search engines might represent complementary properties of engines and understanding the causal factors can help improve engine performance.

Given these key patterns of long-term search engine usage, we now seek to understand causal variables and build predictive models that can detect such behaviors.

3.2 Indicators of User Behaviors

The analyses in the previous section were based on search engine usage. We also have several other features that represent user satisfaction and query behavior: e.g., number of queries, average query length, proportions of navigational queries, etc. In this section we are interested in finding features / indicators that can distinguish between usage trends that constitute the three key behavioral trends.

To explore this we build on top of the NNMF analysis. We use the decomposition to filter and partition users that clearly exhibit the three key trends in their long-term usage. Specifically, we use the weights from the encoding matrix H to rank-order users. The weight in each row of H corresponds to the weight of the particular basis vector in each of the users. For example, we can sort by the values of the weight corresponding to the fourth basis vector (i.e., the fourth row in H) to rank order users that exhibit a persisting switch from engine B to A. Once the users are sorted we considered the first 500 users as representatives who distinctively exhibit the key trend.

We use the aforementioned method to construct three sets of 500 users who correspond to the three key trends of 1) sticking with one engine, 2) making a switch that persists, and 3) oscillating between engines. Given these sets of users, we can now analyze indicators / features that differ across groups. Statistical testing is performed using analyses of variance (ANOVA) with Tukey post-hoc testing (Q) as appropriate.

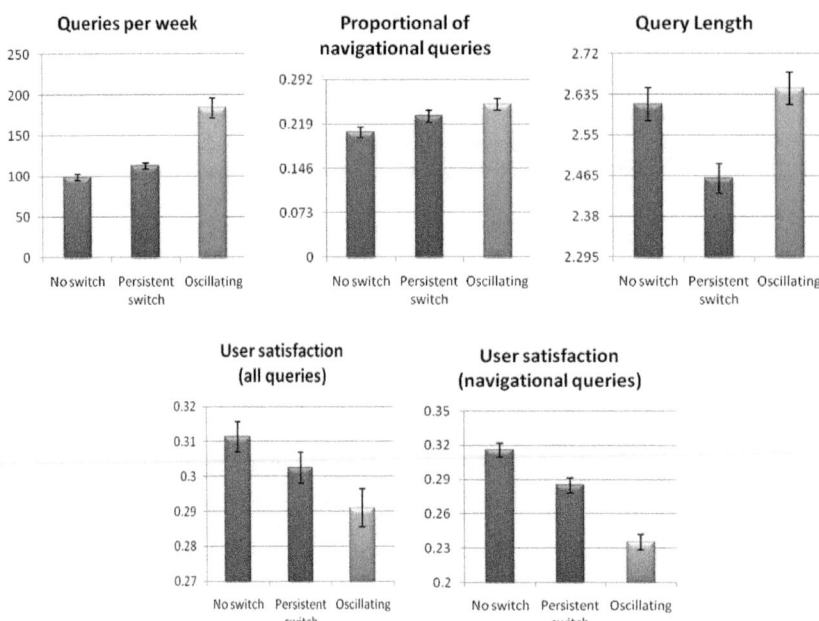

Fig. 2. Comparison of different features for the three different groups. We show means across 500 users for each group. The error bars denote the standard error.

Figure 2 shows results for different variables across the three different groups. We first observe that the users in the oscillating group issue a significantly larger number of queries than the other two groups (F(2,1497)= 36.5, p < .001; Q ≥ 6.65, p ≤ .001). Query frequency has previously been shown to be related to searcher expertise, and this suggests that oscillating users may be more skilled [19]. This heightened sophistication is also consistent with their awareness and usage of multiple search engines. Second, we see that these oscillating users issue a higher proportion of navigational queries than the others (F(2,1497)= 6.2, p = .002; Q ≥ 3.73, p ≤ .02). This may be because oscillating users are more familiar with Web resources and issue more queries to search engines requesting access to those resources. Third, we observe that oscillating users are less satisfied, both in general and for navigational queries (both F(2,1497) ≥ 4.6, both p ≤ .01; Q ≥ 3.44, p ≤ .03). These sophisticated users may pose harder queries which search engines do not perform well on or they may be more demanding in terms of what information is required to satisfy them. Finally, we notice that the query length is significantly smaller for persistent switchers (F(2,1497)= 9.6, p < .001; Q ≥ 3.56, p ≤ .03). If we consider the query length as a proxy for query complexity then we can hypothesize that those who are switching and sticking with the new search engine issue simple queries and might represent a population that is less familiar with search engines, a conjecture supported by [19].

Since user satisfaction has been shown to be important in brand loyalty, we examined the correlation between the frequency of usage and satisfaction. Pearson's correlation coefficients (r) between engine usage and user satisfaction for all queries were statistically significant although generally low (-.14 ≤ r ≤ .36, for the three engines), and the correlations between usage and satisfaction for navigational queries were slightly lower (-.11 ≤ r ≤ .24). Interestingly, the correlation between usage and user satisfaction varied dramatically between search engines. Engine A's usage patterns even exhibited a negative correlation with satisfaction, suggesting that people who use Engine A more often are less satisfied than those who use it less often. This engine was the most popular engine in our sample and this finding suggests that factors beyond satisfaction (e.g., brand loyalty, familiarity, or default search provider settings) may be more important in determining its usage than on the other engines.

3.3 Predicting Search Engine Usage

We aim to predict the search engine usage trend for every user from their past behavior. Specifically motivated by the discussion in the previous sections, we are interested in determining if the user will: (i) stick to their choice of search engine, (ii) switch to another engine and persist with the switch or (iii) oscillate between different search engines. Further we want to make these predictions as early as possible, since early prediction can be used by search engines to prevent unwanted switches or oscillations.

In order to explore this question, we construct a dataset consisting of 500 users from each of the three key user trends (1,500 in total). Since a majority of users belonged to the trivial *No Switch* condition, to obtain a reasonably balanced classification set we selected these 1,500 users (which was also the maximal set where all three classes were balanced). We consider examples as belonging to the *No Switch* class if the user had a single dominating search engine for more than 22 weeks in the 26 week period (i.e., same search engine for more than 85% of the weeks). Similarly, we consider users as belonging to the *Persistent Switch* class if following at least three weeks of use of a particular engine they switch to another and persist for at least eight weeks. Note that while there might be some minor oscillating characteristics in these cases, we still consider them as users belonging to persistent switch category as there was at least one switch that did persist for a significant period. The remainder of the examples are considered to belong to the *Oscillating* class. Our aim, thus, is to see if we can build a good predictive model to discriminate amongst these three classes.

The analysis from the previous section provides insights about how such predictions can be made. Specifically, Table 1 summarizes features that capture general query characteristics as well as satisfaction levels of users. Given these observations about users' past behavior we compute statistics (mean, max, min) that summarizes the usage over past weeks for all the features described in Table 1. In addition, we also compute binary features that indicate: (i) if the user has a single search engine that is dominant for more than 90% of the time observed so far (isOneEngineDominant), and (ii) if user had already made any persisting switch (observedPersistSwitch).

We performed experiments where we first split the 1,500 examples into random 50-50 train-test splits. The training data is used to learn linear classifiers with a one-vs.-all design using a Gaussian Process Regression (GPR) approach [15]. For all our experiments we used a linear kernel and set the noise parameter as $\sigma^2 = 0.1$. Once the model is trained, we can then predict class labels on the test points. Further, we are

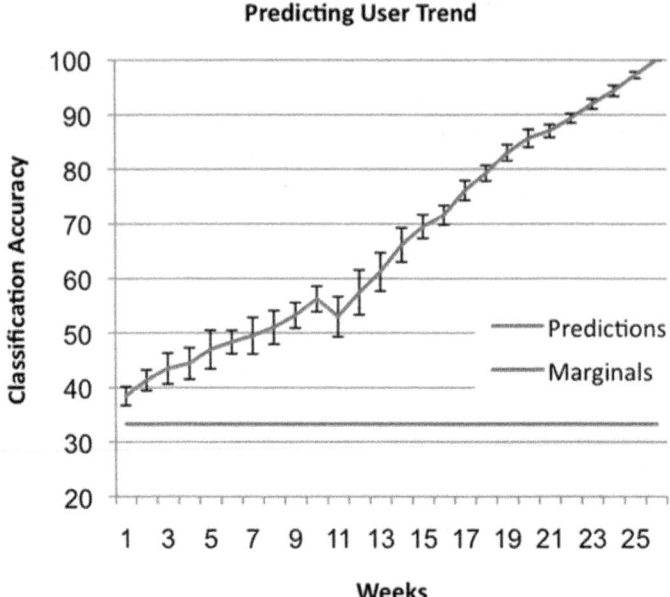

Fig. 3. Classification accuracy as we vary the number of past weeks observed. Accuracy improves as more and more past data is observed. The error bars denote standard deviation over 10 random train-test split.

also interested in determining how soon the model can predict the user trend; hence, we run these experiments for varying the number of past weeks as input. This whole methodology is run for 10 different random train-test splits. Figure 3 shows the mean recognition accuracy on the test set as we vary the number of weeks for which the data has been observed. We observe that even with one week of data, the model can predict better than random (marginal model with accuracy 33.3%). As the model observes an increasing amount of data the prediction accuracy improves constantly, eventually reaching an accuracy of 100% when all the information is observed at week 26. Note that achieving 100% accuracy at week 26 is not surprising as the test labels for these data points were originally generated by looking at 26 weeks of data. However, the experiment described above shows the promise of user modeling in predicting user trend much earlier before 26 weeks (for example even predicting based on one week's observations leads to better than chance accuracy).

Next, we analyzed the features that are helping most in classification by looking at the absolute weights in the learned linear model. A high absolute value of a weight corresponding to a feature indicates high importance in the classification. In particular, for each of the one-vs.-all classifiers the most likely prediction takes the following form [15]: $y = \mathbf{w}^T \cdot \mathbf{x}$

Here, $\mathbf{x}$ denotes the observation vector and $\mathbf{w}$ is the learned classifier. The magnitude of every ith component w_i in $\mathbf{w}$ represents the contribution on the feature x_i in the classification. Consequently, we can sort the features used in prediction by the absolute

Table 2. Most discriminatory features for classifying user trends via observations to week 10

No Switch vs. All	Switch vs. All	Oscillate vs. All
isOneEngineDominant	min fractionEngine A	min fractionEngine C
min fractionEngine A	min fractionEngine C	isOneEngineDominant
observedPersistSwitch	min fractionEngine B	observedPersistSwitch
max fractionEngine A	max fractionEngine A	min fractionEngineSAT C
min fractionEngine B	max fractionEngine C	mean fractionEngineSAT A
mean fractionEngineSAT A	isOneEngineDominant	min fractionEngine B
mean fractionEngine A	max queryCountEngine C < 50	mean fractionEngineSAT B
min fractionNavEngine A	min fractionEngineSAT C	mean fractionEngineSAT C
mean fractionNavEngine A	mean fractionNavEngine A	max queryCountEngine B < 50
max fractionEngine C	observedPersistSwitch	min fractionEngineSAT B

value of the corresponding weight in the linear classifier. Table 2 shows the top 10 features selected for classification at week 10 using the average absolute value of weights across the 10 random splits. The two computed binary features (isOneEngine-Dominant and observedPersistSwitch) are important across all three classifiers. Further, it is interesting to note that these important features constitute not only the fraction of search engine usage but also the statistics about the satisfaction as well as characteristics about the queries being issued.

4 Discussion

In this work, we present evidence that there are characteristic trends in search engine usage. We also show that it is possible to develop predictive user models of these trends using a small number of features about previous engine usage and satisfaction estimates based on user interactions with search results. These findings can be used in several ways to impact the domain of search engines.

The methods general methods that we present in this paper to characterize usage patterns over time are applicable to other data sets, but the specific parameter values (e.g., the number of weeks used to help define classes of switching in our case) may need to be adapted for other data sets. We believe that our findings of a small number of consistent and predictable patterns have implications for improving the design, marketing, and user experience with search engines.

The ability to identify key trends provides insights into the different usage patterns of individual search engines as well as differences across engines. By understanding the strengths and weaknesses of different engines, appropriate resources could be directed towards general search engine improvements. In addition, the ability to predict what a searcher might do could help design adaptive search interfaces that improve relevance for particular query types, provide additional user support for query specification, or develop alternative result presentation methods.

From a marketing perspective, the ability to accurately identify different classes of users could allow search providers to target resources (e.g., marketing campaigns, incentives not to switch, etc.) to users who are the most likely to switch from (or to) their search engine. The development of richer user models could support finer-grained long-term usage analysis and more specific modeling of users' behaviors and interests. All these measures can help the search engine grow its customer base and be more sensitive to user interaction patterns and satisfaction levels.

The key method that enables the above scenarios is predictive user modeling that classifies users based on their past trends of search interactions. There is ample potential in exploring methodologies that would exploit the temporal structure of the data with richer features to provide more accurate classification as early as possible. Another possibility is to improve the classification accuracy by using a more powerful classification method, perhaps involving complex weighted committees [3].

5 Conclusions and Future Work

In this paper we have modeled aspects of multiple search engine usage over a period of six months. Our analysis identified three main classes of search engine users: those who stick with the same engine, those who switch then stick, and those who oscillate between search engines. We observed differences in how users in each of these classes interact with search engines and offered some explanations for their behavior. We also showed a small but significant correlation between search engine switching and our measure of user satisfaction. Finally, we developed a classifier to predict usage trends given historic engine usage data and satisfaction estimates, and showed that it could rapidly improve its accuracy with more data, surpassing the marginal model after only one week.

Future work involves incorporating more features into our models, such as user demographics and more detailed information about the types of queries they are issuing. This will allow us to identify additional classes of search engine usage and predict usage trends with greater levels of accuracy. Further, we are also interested in incorporating the predictive user models in order to improve search engine design and user experience.

References

1. Anderson, E.W., Sullivan, M.W.: The Antecedents and Consequences of Customer Satisfaction for Firms. Marketing Science 12, 125–143 (1993)
2. Anderson, R.E., Srinivasan, S.S.: E-Satisfaction and E-Loyalty: A Contingency Framework. Psychology and Marketing 20(3), 123–138 (2003)
3. Blum, A., Burch, C.: On-line Learning and the Metrical Task System Problem. Machine Learning 39(1), 35–58 (2000)
4. Bolton, R.N.: A Dynamic Model of the Duration of the Customer's Relationship with a Continuous Service Provider: The Role of Satisfaction. Marketing Science 17(1), 45–65 (1998)
5. Capraro, A.J., Broniarczyk, S., Srivastava, R.K.: Factors Influencing the Likelihood of Customer Defection: The Role of Consumer Knowledge. Journal of the Academy of Marketing Science 31(2), 164–175 (2003)

6. Fox, S., Karnawat, K., Mydland, M., Dumais, S.T., White, T.: Evaluating Implicit Measures to Improve the Search Experience. ACM Transactions on Information Systems 23(2), 147–168 (2005)
7. Heath, A.P., White, R.W.: Defection Detection: Predicting Search Engine Switching. In: Ma, W.Y., Tomkins, A., Zhang, X. (eds.) World Wide Web Conference, pp. 1173–1174 (2008)
8. Juan, Y.F., Chang, C.C.: An Analysis of Search Engine Switching Behavior Using Click Streams. In: Deng, X., Ye, Y. (eds.) WINE 2005. LNCS, vol. 3828, pp. 806–815. Springer, Heidelberg (2005)
9. Keaveney, S.M., Parthasarathy, M.: Customer Switching Behavior in Online Services: An Exploratory Study of the Role of Selected Attitudinal, Behavioral, and Demographic Factors. Journal of the Academy of Marketing Science 29(4), 374–390 (2001)
10. Laxman, S., Tankasali, V., White, R.W.: Stream Prediction Using a Generative Model Based on Frequent Episodes in Event Sequences. In: Li, Y., Liu, B., Sarawagi, S. (eds.) ACM SIGKDD Conference on Knowledge Discovery and Data Mining, pp. 453–461 (2008)
11. Lee, D.D., Seung, S.: Learning the Parts of Objects by Non-negative Matrix Factorization. Nature 401, 788–791 (1999)
12. Mittal, B., Lassar, W.M.: Why do Customers Switch? The Dynamics of Satisfaction versus Loyalty. Journal of Services Marketing 12(3), 177–194 (1998)
13. Mukhopadhyay, T., Rajan, U., Telang, R.: Competition Between Internet Search Engines. In: Hawaii International Conference on System Sciences (2004)
14. Pew Internet and American Life Project: Search Engine Users (2005) (accessed December 2008)
15. Rasmussen, C.E., Williams, C.K.I.: Gaussian Processes for Machine Learning. MIT Press, Cambridge (2006)
16. Sasser, T.O., Jones, W.E.: Why Satisfied Customers Defect. Harvard Business Review (November-December 1995)
17. Telang, R., Mukhopadhyay, T., Wilcox, R.: An Empirical Analysis of the Antecedents of Internet Search Engine Choice. In: Workshop on Information Systems and Economics (1999)
18. White, R.W., Dumais, S.T.: Characterizing and Predicting Search Engine Switching Behavior. In: Cheung, D., Song, H., -Y., Chu, W., Hu, X., Lin, J., Li, J., Peng, Z. (eds.) ACM CIKM Conference on Information and Knowledge Management, pp. 87–96 (2009)
19. White, R.W., Morris, D.: Investigating the Querying and Browsing Behavior of Advanced Search Engine Users. In: Clarke, C.L.A., Fuhr, N., Kando, N., Kraaij, W., de Vries, A.P. (eds.) ACM SIGIR Conference on Research and Development in Information Retrieval, pp. 255–261 (2007)
20. White, R.W., Richardson, M., Bilenko, M., Heath, A.P.: Enhancing Web Search by Promoting Multiple Search Engine Use. In: Myaeng, S.-H., Oard, D.W., Sebastiani, F., Chua, T.S., Leong, M.K. (eds.) ACM SIGIR Conference on Research and Development in Information Retrieval, pp. 43–50 (2008)

Analysis of Strategies for Building Group Profiles

Christophe Senot, Dimitre Kostadinov, Makram Bouzid, Jérôme Picault,
Armen Aghasaryan, and Cédric Bernier

Bell Labs, Alcatel-Lucent, Centre de Villarceaux, Route de Villejust,
91620 Nozay, France
{Christophe.Senot,Dimitre_Davidov.Kostadinov,
Makram.Bouzid,Jerome.Picault,Armen.Aghasaryan,
Cedric.Bernier}@alcatel-lucent.com

Abstract. Today most of existing personalization systems (e.g. content recommenders, or targeted ad) focus on individual users and ignore the social situation in which the services are consumed. However, many human activities are social and involve several individuals whose tastes and expectations must be taken into account by the service providers. When a group profile is not available, different profile aggregation strategies can be applied to recommend adequate content and services to a group of users based on their individual profiles. In this paper, we consider an approach intended to determine the factors that influence the choice of an aggregation strategy. We present a preliminary evaluation made on a real large-scale dataset of TV viewings, showing how group interests can be predicted by combining individual user profiles through an appropriate strategy. The conducted experiments compare the group profiles obtained by aggregating individual user profiles according to various strategies to the "reference" group profile obtained by directly analyzing group consumptions.

Keywords: group recommendations, individual profile, group profiles, aggregation strategies, evaluation.

1 Introduction

In order to satisfy increasing needs for service personalization, mastering the knowledge of individual user profiles is no longer sufficient. Indeed, there exist numerous services which are consumed in a social or virtual environment. For example, to provide personalization services with a real added-value for interactive IPTV, its content needs to be adapted (through VoD/program mosaic, or targeted adverts) to different tastes and interests of the viewers' group (family members, friends, etc.). In a different context, to increase the ROI of advertisers, the digital billboards need to dynamically adapt their content to the surrounding group of individuals. Similar needs are also present in virtual spaces like web conferences, chat rooms or social networking applications. In all such environments, the personalization technology has to go beyond individual adaptive systems by bringing in group profiling and group recommendation systems, the intelligence that allows to conciliate

P. De Bra, A. Kobsa, and D. Chin (Eds.): UMAP 2010, LNCS 6075, pp. 40–51, 2010.

potentially conflicting user interests, needs and restrictions [2, 5-8, 13]. To cope with the complexity of social environments, different aggregation strategies for making group recommendation have been suggested in [10, 11]. The relevance of each strategy can vary from one group to another according to their characteristics, contexts and member preferences. Many questions related to strategy selection can then be asked: how to select the right strategy? Which group recommendation strategies provide the best results? Can we determine some factors that influence the choice of a group recommendation strategy? We herein present an approach that tries to answer these questions through a preliminary evaluation made on a real large-scale dataset of TV viewings. The conducted experiments compare the group profiles obtained by aggregating individual user profiles according to various strategies to the "reference" group profile obtained by directly analyzing the group consumptions.

The paper is structured as follows. Section 2 describes the related work on group recommendations approaches. Section 3 describes the prerequisites and motivations for the current work. Section 4 presents a methodology for comparing group recommendation strategies. Section 5 discusses the results of conducted evaluations, and finally, Sect. 6 provides conclusions and perspectives for future research.

2 Related Work

There are basically two main approaches for providing recommendations for a group of users when a "real" group profile is not available. The first combines individual recommendations to generate a list of group recommendations [2], while the second computes group recommendations using a group profile obtained by aggregating individual profiles (e.g. [6, 13]). In this paper, we focus on the second method.

In the last decade, several strategies allowing the aggregation of individual preferences for building a group profile have been proposed [10, 14]. We classified them into three categories [4]: majority-based, consensus-based, and borderline strategies.

The majority-based strategies use the most popular items (or item categories) among group members. For example, with the Plurality Voting strategy, each member votes for his preferred item (or item category) and the one with the highest votes is selected. Then, this method is reiterated on the remaining items (item categories) in order to obtain a ranked list. For example, GroupCast [7] displays content that suits the intersection of user profiles when the persons are close to a public screen.

The consensus-based strategies consider the preferences of all group members. Examples include the Utilitarian strategy which averages the preferences of all the group members, the Average without Misery, the Fairness, or the Alternated Satisfaction. As an example, MusicFX [6] recommends the most relevant music station in a fitness centre using a group profile computed by summing the squared individual preferences. By applying this strategy 71% of clients noticed a positive effect (as compared to the absence of the recommendation system). However, the authors did not conduct any evaluation with other strategies.

The borderline strategies consider only a subset of items (item categories), in individual profiles, based on user roles or any other relevant criteria. For example, the Dictatorship strategy uses the preferences of only one member, who imposes his tastes to the rest of the group. The Least Misery strategy and the Most Pleasure strategy keep for each preference respectively the minimum and maximum level of interest among group members. For example, PolyLens [13] uses the Least Misery strategy to recommend movies for small user groups based on the MovieLens [12] database. Their survey showed that 77% of PolyLens users found group recommendations more helpful than individual ones. Yet this system only works with a single strategy.

In summary, the existing approaches for group recommendations are based on a single profile aggregation strategy, which improves the users' satisfaction compared to individual recommendations, but there is a lack of comparison between possible strategies. Masthoff [9] compared strategies for constructing a group profile from individual ones. She proposed a sociological study of various strategies made on a small set of users. However, a large scale empirical comparison of strategies is still missing. Consequently, we propose an empirical study of profile aggregation strategies performed in the TV domain. This study will be used as basis for building a strategy selector which is able to find the most appropriate strategy according to several criteria such as application domains, group characteristics, context etc.

3   Prerequisites and Motivations

In this section, we first introduce the profiling approach used for the construction of the individual user profiles as well as the "reference" group profiles based on consumption traces. Then, we introduce a strategy selection mechanism based on group characteristics which motivates the evaluations presented in this paper.

3.1   The Profiling Approach

The user profile is basically represented by a set of <concept, value> pairs, where each value is taken from the interval [0,1] and reflects the level of interest in the given semantic concept (item category). More generally, the profiling engine manipulates three important types of information:

- *Quantity of Affiliation (QoA)* characterizes the degree of affiliation of a content item to a given semantic concept. Each content item is characterized by a set of QoA, e.g. the film "Shrek" is described by {Animation = 0.9, Comedy = 0.8}.
- *Quantity of Consumption (QoC)* characterizes the degree of intensity of a consumption act with respect to a given semantic concept. For example, the larger part of a movie (e.g. "Shrek") is viewed by the user, the higher is his interest in the semantic concepts Animation and Comedy. Thus, each consumption act can be characterized by a set of QoC.
- *Quantity of Interest (QoI)* characterizes the degree of interest of the user in a given semantic concept. The user profile is composed of a set of QoI.

The profiling algorithm consists first in estimating the QoC values for each user consumption trace, and then in updating iteratively the QoI values. An example of such update function is the sigmoid-based approach as described in [1]. In addition, a decay function is applied at fixed periods of time in order to account for non-consumption effect on the interest categories depending on the frequency or recentness of respective consumptions.

3.2 Strategy Selection Framework

In order to dynamically select an adequate strategy with a desirable behaviour among the large number of existing variants, we are trying to build an intelligent strategy selection procedure based on group characteristics, contextual data as well as group interaction traces [4] (see Fig. 1). In particular, to select the most appropriate strategy this procedure relies on different group characteristics like the nature of relations between the members, the group cohesiveness, its structure, its diversity, its size, etc. In this paper, we focus on the characteristics of TV viewer groups.

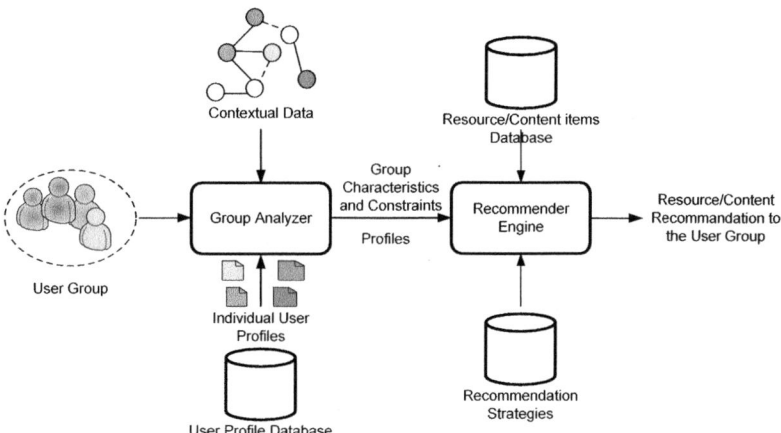

Fig. 1. Strategy selection based on group characteristics

4   Evaluation Methodology and Execution

This section presents the methodology we followed to compare different strategies of group profile aggregation as well as the dataset requirements. Then, it describes the used dataset and the main steps of the executed tests.

4.1 Test Methodology

In order to assess the relevance and the feasibility of automatic strategy selection based on group characteristics, we apply the following methodology (see Fig. 2):

1. group profiles are computed and used as reference for comparing the different profile aggregation strategies. These *reference profiles* are built on all the consumptions made by the group and are hence assumed to reflect its real preferences,
2. *user profiles* reflecting individual preferences of group members are computed,

3. profile aggregation strategies are used to estimate group profiles from individual ones, called *aggregated profiles*,
4. the obtained aggregated profiles are compared to the reference group profiles (i.e. the similarity between them is computed),
5. the obtained results are analyzed to find which strategy performed well in which cases and according to which group characteristics,
6. rules for selecting the most appropriate profile aggregation strategy could then be inferred according to group characteristics.

Applying this evaluation methodology requires a dataset containing or allowing computing user profiles and reference profiles. At the same time, this dataset should contain information characterizing the groups and their members (e.g. demographic data, user behaviour features, group composition, etc). User and group profiles can be either explicitly defined (already available as such in the dataset) or implicitly inferred from user and group consumptions, respectively. In the latter case, the dataset should provide information about the relevance of each consumed content item (for QoC computation) and a sufficient number of group and individual consumptions allowing a profiling algorithm to learn their preferences.

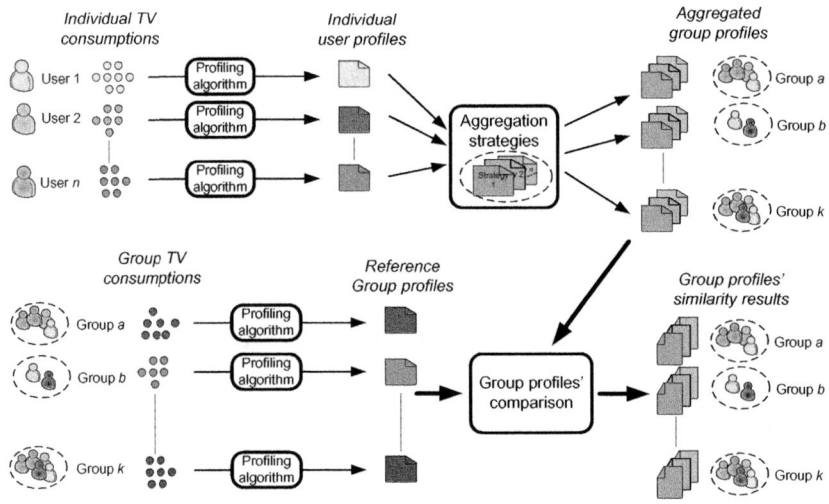

Fig. 2. Evaluation methodology for building a strategy selector

4.2 Dataset Description

We have processed 6 months of TV viewing data (from 1st September 2008 to 1st March 2009) from the BARB [3] dataset in order to build a new dataset that fulfils the abovementioned requirements. BARB provides estimates of the number and the characteristics of people watching TV programs in the UK. These estimates are built on a minute-by-minute viewing data produced by a panel of users and households which is representative of the UK audience. The BARB dataset contains 3 types of data: (i) information about users (e.g. demographic data, social category, etc.) and households (number of people in the home, number of TV sets etc.), (ii) program metadata restricted to the title and the genre and (iii) viewing data describing watching activity of users on programs.

On the 1st September 2008, the BARB panel was composed of 14,731 users forming 6,423 households. During this 6-months period, the users generated about 30 millions of viewing traces where each trace represents a viewing session of a given user on a given program. Information about groups of users in the same household watching the same program is provided by sharing the same session identifier among their traces. A new session begins when the group composition changes and/or the channel changes. Thus, several sessions may exist for the same program and/or the same user (group of users).

In order to make the BARB data conform to the requirements of our evaluations, several adaptations have been performed. First, groups in all households have been identified and their corresponding viewings have been constructed from the shared viewings of their members. Second, the viewings of the same program made by the same group have been aggregated in a single viewing trace containing information about:

- the duration of the program,
- the total number of minutes the group spent in watching the program,
- the number of sessions associated to the program (i.e. number of times the group changed the channel – zapped – during the program),
- the value of the first start offset, which is the moment the group started watching the program. It is equal to the period of time separating the beginning of the program and the beginning of the first watching session of the program,
- the value of the last end offset, which corresponds to the last moment the group watched the program. It is equal to the period of time separating the end of the last watching session and the end of the program,
- the percentage of the program viewed by the group.

The same process has been also applied to individual user viewings.

Third, in order to prevent noise in viewing traces, we filtered them by removing programs with a duration less than a certain threshold (e.g. 3 minutes) and programs belonging to very long sessions where the user probably forgot to switch off his TV (e.g. sessions whose duration is longer than 4 hours and which contain more than 3 programs successively without any zapping).

Finally, a relevance score (QoC) has been computed for each remaining viewed program in the dataset according to the group/user who watched it. As no information is available about the level at which genres are representative for programs, the values of QoA are binary (0 or 1). Thus, for a given user or group, the relevance of a program is function of the time spent in watching it, the moment of its discovery (first start offset), the moment the user stopped watching it (last end offset) and the channel changing activity between these two moments. Intuitively, the relevance of a program is assumed to be high when the user/group watched it until the end, didn't miss a minute since it discovered it, and watched a large part of it.

The final step of the dataset construction consists in selecting a subset of groups for the experiments. In order to have a sufficient number of viewings necessary to construct group and user profiles, the minimum number of viewings associated to each group and to each one of its members was fixed to 70; this corresponds to approximately 3 viewings per week. We obtain then 28 households offering at least one group of size 4 or higher satisfying the previous condition. As one of the goals of the experiments is to analyze the user behaviour in groups of different sizes and different compositions, all groups of size superior or equal to 2 in these households were selected. The features of the selected groups are summarized in Table 1.

Table 1. Statistics of groups selected for experiments

Group size	Group composition	NB of groups	NB total of groups
2	2 children	10	70
	1 teenager; 1 child	2	
	2 teenagers	2	
	1adult; 1 child	13	
	1 adult; 1 teenager	13	
	2 adults	30	
3	1 teenager; 2 children	1	38
	1 adult; 2 children	11	
	1 adult; 1 teenager; 1 child	2	
	1 adult; 2 teenagers	4	
	2 adults; 1 child	7	
	2 adults; 1 teenager	9	
	3 adults	4	
4	1 adult; 1 teenager; 2 children	1	27
	1 adult; 2 teenagers; 1 child	1	
	2 adults; 2 children	11	
	2 adults; 1 teenager; 1 child	3	
	2 adults; 2 teenagers	5	
	3 adults; 1 child	2	
	3 adults; 1 teenager	2	
	4 adults	2	
5	3 adults; 2 children	1	1

Total: 136 groups

4.3 Tests Description

This section presents the main steps of the evaluations performed on the dataset built from BARB data.

The first step of our analysis consisted in building a user profile for each family member among selected households. During this step only the consumptions where the user watched the TV alone were considered for computing the degree of interest (QoI) of each concept. The latter is inferred by using the profiling approach introduced in Sect. 3.1. Among the possible QoI update functions, we chose one having a sigmoid learning curve. This function avoids introducing casual interests in the user profile as it requires that a concept is consumed a certain number of times and with a certain intensity before considering it as relevant for the user. In addition, we used an exponential decay function with a 7-day periodicity to capture changes in user interests. This function decreases the QoI of concepts which are less frequently or no longer consumed. More details on this profiling approach can be found in [1].

Using the same profiling algorithm, we built a reference profile for each identified group, based on the real group consumption histories only.

In the second step, we built another set of group profiles for the identified groups by aggregating the individual profiles of the members composing the groups. This was done using different strategies taken from the three main categories of group recommendation strategies presented in Sect. 2:

- a consensus-based strategy: Utilitarian,
- a majority-based strategy: Plurality Voting, and
- three borderline strategies: Least Misery, Most Pleasure, and Dictatorship.

As most of the aggregation strategies described in the literature are based on user ratings (generally between 1 and 5), we slightly adapted them to our profile model based on a set of <concept, value> pairs. For the Utilitarian, Least Misery and Most Pleasure strategies, the aggregated QoI value of each concept corresponds respectively to the average, the minimum and the maximum of the user profiles QoIs. In the Plurality Voting strategy the aggregated QoI value of each concept is set to 1 if a majority of user profile QoIs are higher than a given threshold otherwise it is equal to 0. The result of the Dictatorship strategy is the closest user profile in comparison to the reference profile. Notice that our dataset allows detecting the dictator as we have for each individual his user profile and the reference profile of the group. The proximity between the profiles is computed according to a given similarity measure. We consider that highest is the similarity stronger is the dictator impact on the group.

At the end, for each group we compared the group profiles obtained by aggregation to the corresponding reference group profiles. This was done by using two similarity measures, the cosine and the Pearson correlation. The analysis of the obtained results is presented in the next section.

5 Results and Analysis

In this section, we describe the first set of results obtained according to the methodology described above and bring some initial responses to the two following questions: which aggregated profile has the highest proximity with the reference profile? And, can we determine some factors that influence the choice of an aggregation strategy (based on group characteristics)?

Figure 3 and Fig. 4 compare the reference profile to the aggregated profiles, using the cosine similarity and Pearson correlation respectively.

5.1 Which Strategy Provides the Highest Proximity?

The main lessons we can learn from this experiment testing five profile aggregation strategies are the following.

- *clear domination of the consensus-based strategy*: For a large majority of groups (Table 2), the *Utilitarian* strategy is the one that gives the best results, i.e. for which the aggregated profile obtained from user profiles is the closest to the reference profile obtained from learning group consumptions. These results confirm with a larger set of observations what has been studied and reported by Masthoff in the past (through some user studies) [9].
- *ide mocracy" does not seem to play an important role*: From the experiments, the concepts of misery (*Least Misery* strategy) or vote (*Plurality voting* strategy) are the ones which are the worst (highest distance between aggregated group profile and reference profile). This experimental result seems to contradict somehow what Masthoff reported in [9] about modelling a group of television users.

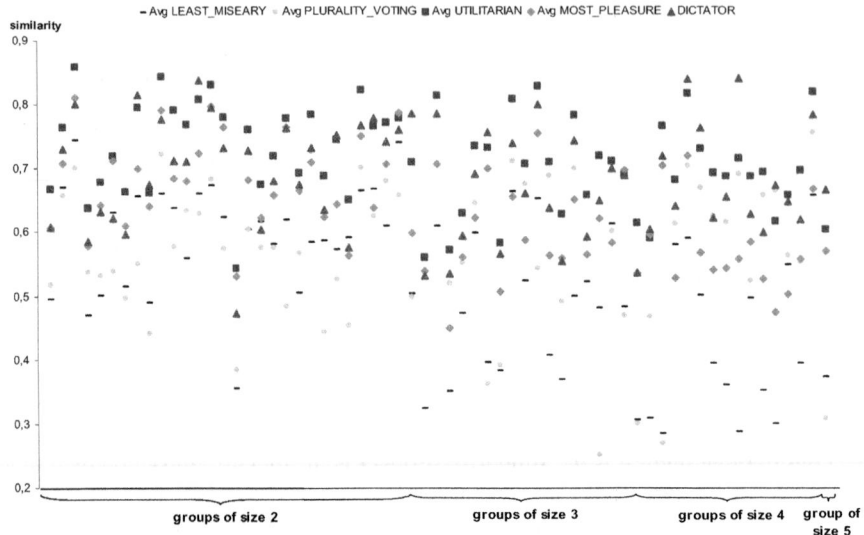

Fig. 3. Comparison of profile aggregation strategies (cosine similarity)

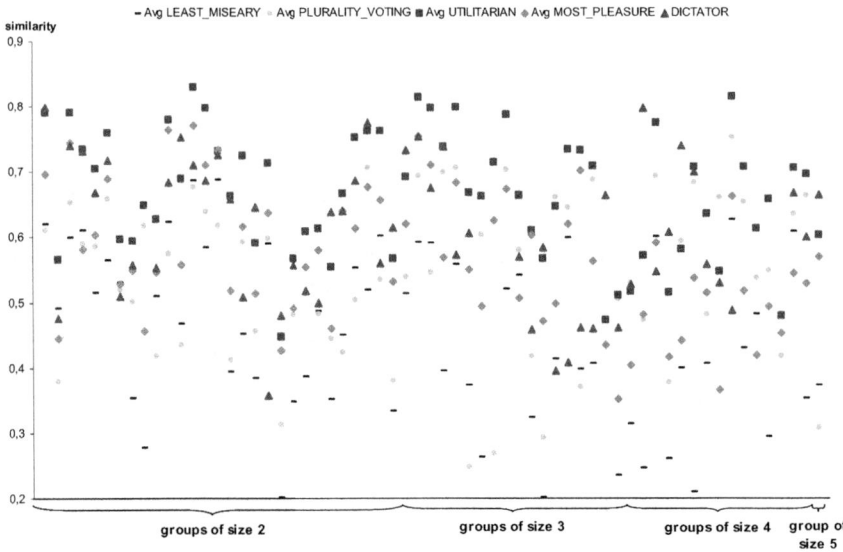

Fig. 4. Comparison of profile aggregation strategies (Pearson correlation)

- the *Dictatorship* strategy where one individual profile is imposed to the whole group provides good results, and even outperforms the *Utilitarian* strategy for 20% of the groups with the cosine similarity. In average, the *Dictatorship* strategy is the second best one. We have to put into perspective this result due to the type of data on which we based our experiment. It is certainly true for a TV service where groups are small (between 2 and 5) and members are used to watch TV together. But in the case of another service like the MusicFX [6] music recommender system, the same conclusion can not be inferred without tests.

Table 2. Summary of best aggregation strategies on 136 groups

	Percentage of groups for which the strategy is the best	
Strategy	Cosine similarity	Pearson correlation
least misery	0%	0%
plurality voting	0%	7.14%
utilitarian	76.20%	82.16%
most pleasure	3.17%	1.78%
dictatorship	20.63%	8.92%

5.2 Which Are the Factors That May Influence the Choice of a Strategy?

From Fig. 3 and Fig. 4 we can notice that there is:

– *a relative invariance according to the group size*: There does not seem to be any correlation between the choice of a strategy and the size of the group (at least for best and worst strategies). However, again this has to be interpreted carefully because we considered data from households where groups are small (2 to 5 members – we excluded higher size groups due to the lack of meaningful data),
– *an invariance according to the group composition:* We did not notice any significant difference in the results depending on the composition of the group (adults only, children only or mix of both).

As the *Dictatorship* strategy provides relevant results (second strategy after the Utilitarian Strategy) we performed additional evaluations on this strategy in order to find out how dictators could be characterized. The results of the evaluations are presented in Table 3 which shows for each heterogeneous group composition the corresponding type of dictator (adult, teenager and child). In most of group compositions the dictator is an adult except when the number of teenagers within the group is higher than the number of adults. The teenager exception should be handled with care since we have only a small number of groups having this composition.

We tried to go a step further in characterizing the dictator by checking if the gender is a factor of influence. For that we studied two group compositions containing only adults and noticed that there is no obvious correlation between the gender and the choice of the dictator (with groups of 2 adults 53% for men and 47% for women, with groups of 3 adults 50% for both).

To summarize, the evaluation results contribute to better understanding of how different categories of strategies behave in the case of TV viewer groups. In particular they suggest that while the *Utilitarian* strategy is the most appropriate for the majority of tested groups, the *Dictatorship* strategy provides very close and for some groups better results. Given that the latter requires much less knowledge on individual interests (only the profile of the leader/dictator needs to be known), it could be a good substitute to the Utilitarian strategy whenever those data are missing.

Table 3. Types of dictator according to group composition

Adult	Teenager	Child	Group Composition	Nb of Groups
92,31%	Ø	7,69%	1 adult, 1 child	13
63,64%	Ø	36,36%	1 adult, 2 children	11
61,54%	38,46%	Ø	1 adult, 1 teenager	13
50,00%	50,00%	0,00%	1 adult, 1 teenager, 1 child	2
0,00%	100,00%	Ø	1 adult, 2 teenagers,	4
0,00%	100,00%	0,00%	1 adult, 2 teenagers, 1 child	1
85,71%	Ø	14,29%	2 adults, 1 child	7
81,82%	Ø	18,18%	2 adults, 2 children	11
100,00%	0,00%	Ø	2 adults, 1 teenager	9
66,67%	33,33%	0,00%	2 adults, 1 teenager, 1 child	3
100,00%	0,00%	Ø	2 adults, 2 teenagers	5
100,00%	Ø	0,00%	3 adults, 1 child	2
100,00%	Ø	0,00%	3 adults, 2 children	1
100,00%	0,00%	Ø	3 adults, 1 teenager	2
Ø	100,00%	0,00%	1 teenager, 1 child	2
Ø	100,00%	0,00%	1 teenager, 2 children	1
0,00%	0,00%	100,00%	1 adult, 1 teenager, 2 children	1

6 Conclusions and Perspectives

In this paper, we presented an approach that makes use of group characteristics in order to select the most appropriate group recommendation strategy. Preliminary evaluation is made on a real large-scale dataset of TV viewings, showing how group interests can be predicted by combining individual user profiles through an appropriate strategy. These experiments compared the aggregated group profiles obtained by aggregating individual user profiles according to various strategies to the "reference" group profile obtained by directly analyzing group consumptions.

Although the initial results do not necessarily justify per se the creation of a strategy selector framework (as the resulting rules in case of TV would be quite simple), we believe this idea is still interesting, especially for other domains where the group dynamics is more complex as mentioned in the PolyLens study [13]. Thus, further work – either done through statistical analysis or through user studies - will also be dedicated to different strategy evaluations with other types of groups like visitors of a pub, users of a social network, or individuals arbitrarily gathering in public places with a digital screen. Other work perspectives could focus on studying the dynamics of TV groups: e.g. are there some recommendation strategies that can impact the group structure after a while (e.g. new members join or others leave)?

References

1. Aghasaryan, A., Betgé-Brezetz, S., Senot, C., Toms, Y.: A Profiling Engine for Converged Service Delivery Platforms. Bell Labs Tech. J. 13(2), 93–103 (2008)
2. Ardissono, L., Goy, A., Petrone, G., Segnan, M., Torasso, P.: INTRIGUE: Personalized Recommendation of Tourist Attractions for Desktop and Handset Devices. Applied Artificial Intelligence: Special Issue on Artificial Intelligence for Cultural Heritage and Digital, Libraries 17(8-9), 687–714 (2003)

3. BARB: Broadcaster Audience Research Board, http://www.barb.co.uk
4. Bernier, C., Brun, A., Aghasaryan, A., Bouzid, M., Picault, J., Senot, C., Boyer, A.: Topology of communities for the collaborative recommendations to groups. In: 3rd International Conference on Information Systems and Economic Intelligence (SIIE), Tunisia (2010)
5. Jameson, A.: More than the sum of its members: Challenges for group recommender systems. In: Proceedings of the International Working Conference on Advanced Visual Interfaces, Gallipoli, Italy, pp. 48–54. ACM Press, New York (2004)
6. McCarthy, J.F., Anagnost, T.D.: MusicFX: An Arbiter of Group Preferences for Computer Supported Collaborative Workouts. In: ACM 1998 Conference on CSCW (1998)
7. McCarthy, J.F., Costa, T.J., Liongosari, E.S.: UniCast, OutCast & GroupCast: Three Steps Toward Ubiquitous, Peripheral Displays. In: Abowd, G.D., Brumitt, B., Shafer, S. (eds.) UbiComp 2001. LNCS, vol. 2201, pp. 332–345. Springer, Heidelberg (2001)
8. McCarthy, K., Salamó, M., McGinty, L., Smyth, B.: CATS: A synchronous approach to collaborative group recommendation. In: Proceedings of the Nineteenth International Florida Artificial Intelligence Research Society Conference, Melbourne Beach, FL, pp. 86–91. AAAI Press, Menlo Park (2006)
9. Masthoff, J.: Modeling a group of television viewers. In: Proceedings of the Future TV: Adaptive instruction in your living room workshop, associated with ITS02 (2002)
10. Masthoff, J.: Group Modeling: Selecting a Sequence of Television Items to Suit a Group of Viewers. User Modeling and User-Adapted Interaction 14, 37–85 (2004)
11. Masthoff, J., Gatt, A.: In pursuit of satisfaction and the prevention of embarrassment: affective state in group recommender systems. User Modeling and User-Adapted Interaction 16(3-4), 281–319 (2006)
12. MovieLens, http://www.movielens.org
13. O'Connor, M., Cosley, D., Konstan, J.A., Riedl, J.: PolyLens: A Recommender System for Groups of Users. In: Proceedings of ECSCW, pp. 199–218 (2001)
14. Yu, Z., Zhou, X., Hao, Y., Gu, J.: A TV program recommendation for multiple viewers based on user profile merging. User Modeling and User-Adapted Interaction 16(1), 63–82 (2006)

Contextual Slip and Prediction of Student Performance after Use of an Intelligent Tutor

Ryan S.J.d. Baker[1], Albert T. Corbett[2], Sujith M. Gowda[1],
Angela Z. Wagner[2], Benjamin A. MacLaren[2], Linda R. Kauffman[3],
Aaron P. Mitchell[3], and Stephen Giguere[1]

[1] Department of Social Science and Policy Studies, Worcester Polytechnic Institute
100 Institute Road, Worcester MA 01609, USA
{rsbaker,sujithmg,sgiguere}@wpi.edu
[2] Human-Computer Interaction Institute, Carnegie Mellon University, 5000 Forbes Avenue,
Pittsburgh, PA 15213, USA
corbett@cmu.edu, awagner@cmu.edu, maclaren@andrew.cmu.edu
[3] Department of Biological Sciences, Carnegie Mellon University, 5000 Forbes Avenue,
Pittsburgh, PA 15213, USA
{lk01,apm1}@andrew.cmu.edu

Abstract. Intelligent tutoring systems that utilize Bayesian Knowledge Tracing have achieved the ability to accurately predict student performance not only within the intelligent tutoring system, but on paper post-tests outside of the system. Recent work has suggested that contextual estimation of student guessing and slipping leads to better prediction within the tutoring software (Baker, Corbett, & Aleven, 2008a, 2008b). However, it is not yet clear whether this new variant on knowledge tracing is effective at predicting the latent student knowledge that leads to successful post-test performance. In this paper, we compare the Contextual-Guess-and-Slip variant on Bayesian Knowledge Tracing to classical four-parameter Bayesian Knowledge Tracing and the Individual Difference Weights variant of Bayesian Knowledge Tracing (Corbett & Anderson, 1995), investigating how well each model variant predicts post-test performance. We also test other ways to utilize contextual estimation of slipping within the tutor in post-test prediction, and discuss hypotheses for why slipping during tutor use is a significant predictor of post-test performance, even after Bayesian Knowledge Tracing estimates are controlled for.

Keywords: Student Modeling, Bayesian Knowledge Tracing, Intelligent Tutoring Systems, Educational Data Mining, Contextual Slip.

1 Introduction

Since the mid-1990s, Intelligent Tutoring Systems have used Bayesian approaches to infer whether a student knows a skill, from the student's pattern of errors and correct responses within the software [6, 11, 18]. One popular approach, Bayesian Knowledge Tracing, has been used to model student knowledge in a variety of learning systems, including intelligent tutors for mathematics [10], genetics [7],

P. De Bra, A. Kobsa, and D. Chin (Eds.): UMAP 2010, LNCS 6075, pp. 52–63, 2010.
© Springer-Verlag Berlin Heidelberg 2010

computer programming [6], and reading skill [3]. Bayesian Knowledge Tracing has been shown to be statistically equivalent to the two-node dynamic Bayesian network used in many other learning environments [13]. Bayesian Knowledge Tracing keeps a running assessment of the probability that a student currently knows each skill. Each time a student attempts a problem step for the first time, the software updates its probability that the student knows the relevant skill, based on whether the student successfully applied that skill. In the standard four-parameter version of Bayesian Knowledge Tracing described in Corbett & Anderson [6], each skill has two learning parameters, one for Initial Knowledge, and one for the probability of Learning the skill at each opportunity, and two performance parameters, one for Guessing correctly, and one for Slipping (making an error despite knowing the skill). By assessing the student's latent knowledge, it is possible to tailor the amount of practice each student receives, significantly improving student learning outcomes [5, 6].

Recent work has suggested that a new variant of Bayesian Knowledge Tracing, called Contextual-Guess-and-Slip, may be able to predict student performance within the tutoring software more precisely than prior approaches to Bayesian Knowledge Tracing [1, 2]. The Contextual-Guess-and-Slip approach examines properties of each student response as it occurs, in order to assess the probability that the response is a guess or slip. However, while better prediction within the software is valuable, the real goal of Bayesian Knowledge Tracing is not to predict performance within the tutoring software, but to estimate the student's underlying knowledge – knowledge that should transfer to performance outside of the tutoring software, for example on post-tests.

Hence, in this paper, we investigate how well the Contextual-Guess-and-Slip model can predict student learning outside of the tutoring software, comparing it both to the canonical four-parameter version of Bayesian Knowledge Tracing, and to the Individual Difference Weights version of Bayesian Knowledge Tracing [6]. The Individual Difference Weights version finds student-level differences in the four parameters, and has been shown to improve the prediction of post-test performance for students who have reached mastery within the tutor. We also investigate other ways to utilize data on student slipping within the learning software, in order to study how to increase the accuracy of post-test prediction.

2 Data

The data used in the analyses presented here came from the Genetics Cognitive Tutor [7]. This tutor consists of 19 modules that support problem solving across a wide range of topics in genetics (Mendelian transmission, pedigree analysis, gene mapping, gene regulation and population genetics). Various subsets of the 19 modules have been piloted at 15 universities in North America.

This study focuses on a tutor module that employs a gene mapping technique called *three-factor cross*. The tutor interface for this reasoning task is displayed in Fig. 1. In this gene mapping technique a test cross is performed (in this case, of two fruit flies) that focuses on three genes. In Fig. 1 the three genes are labeled G, H and F. In the data table on the left of the figure, the first column displays the eight possible offspring phenotypes that can result from this test cross and the second column displays the number of offspring with each phenotype. The problem solution depends on the phenomenon of "crossovers" in meiosis, in which the chromosomes in homologous pairs exchange genetic material. In Fig. 1 the student has almost finished the problem. To the right of the table, the student has summed the offspring in each of the phenotype groups and identified the group which reflects the parental phenotype (no crossovers), which groups result from a single crossover in meiosis, and which group results from two crossovers. The student has compared the phenotype groups to identify the middle of the three genes and entered a gene sequence below the table. Finally, in the lower right the student has calculated the crossover frequency between two of the genes, A and B, and the distance between the two genes. The student will perform the last two steps for the other two gene pairs.

In this study, 71 undergraduates enrolled in a genetics course at Carnegie Mellon University used the three-factor cross module as a homework assignment. Half the students completed a fixed curriculum of 8 problems and the other half completed between 6 and 12 problems under the control of Knowledge Tracing and Cognitive Mastery [6]. The 71 students completed a total of 19,150 problem solving attempts across 9259 problem steps in the tutor. Students completed a paper-and-pencil problem-solving pretest and posttest consisting of two problems. There were two test forms, and students were randomly selected to receive one version at pre-test and the other version at post-test, in order to counterbalance test difficulty. Each of the two problems on each test form consisted of 11 steps involving 7 of the 8 skills in the Three-Factor Cross tutor lesson, with two skills applied twice in each problem and one skill applied three times.

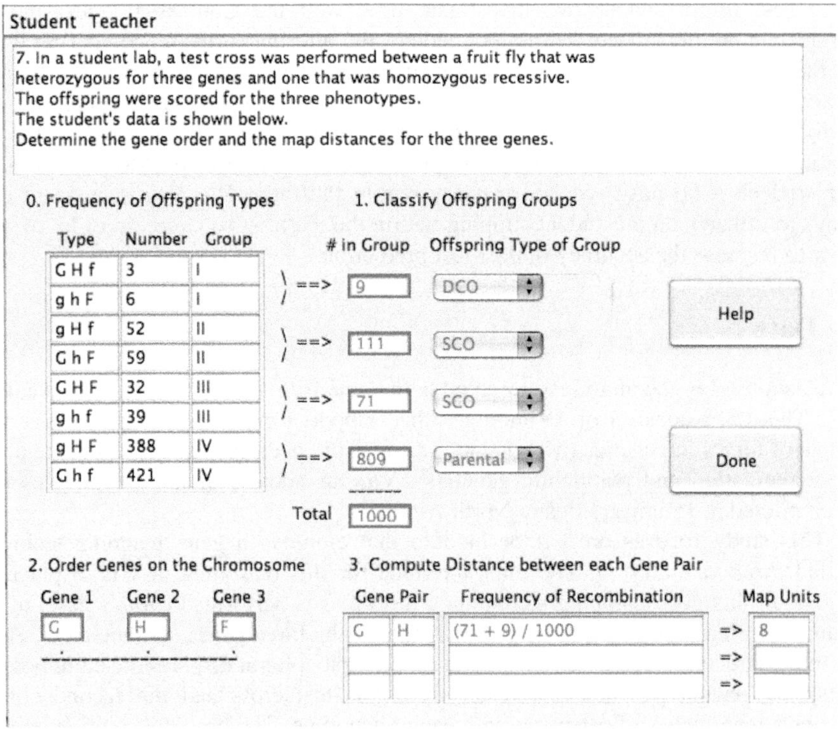

Fig. 1. The Three-Factor Cross lesson of the Genetics Cognitive Tutor

After the study, Bayesian Knowledge Tracing and Contextual Guess and Slip models were fit and applied to data from students' performance within the tutor.

3 Bayesian Knowledge Tracing Variants

All of the models discussed in this paper are variants of Bayesian Knowledge Tracing, and compute the probability that a student knows a given skill at a given time. The Bayesian Knowledge Tracing model assumes that at any given opportunity to demonstrate a skill, a student either knows the skill or does not know the skill, and may either give a correct or incorrect response (help requests are treated as incorrect by the model). A student who does not know a skill generally will give an incorrect response, but there is a certain probability (called **G**, the Guess parameter) that the student will give a correct response. Correspondingly, a student who does know a skill generally will give a correct response, but there is a certain probability (called **S**, the Slip parameter) that the student will give an incorrect response. At the beginning of using the tutor, each student has an initial probability (L_0) of knowing each skill, and at each opportunity to practice a skill the student does not know, the student has a certain probability (**T**) of learning the skill, regardless of whether their answer is correct.

The system's estimate that a student knows a skill is continually updated, every time the student gives an initial response (a correct response, error, or help request) to a problem step. First, the system applies Bayes' Theorem to re-calculate the probability that the student knew the skill before making the attempt, using the evidence from the current step. Then, the system accounts for the possibility that the student learned the skill during the problem step. The equations for these calculations are:

$$P(L_{n-1}|Correct_n) = \frac{P(L_{n-1}) * (1 - P(S))}{P(L_{n-1}) * (1 - P(S)) + (1 - P(L_{n-1})) * (P(G))}$$

$$P(L_{n-1}|Incorrect_n) = \frac{P(L_{n-1}) * P(S)}{P(L_{n-1}) * P(S) + (1 - P(L_{n-1})) * (1 - P(G))}$$

$$P(L_n|Action_n) = P(L_{n-1}|Action_n) + ((1 - P(L_{n-1}|Action_n)) * P(T))$$

Three variants on Bayesian Knowledge Tracing were applied to the data set. The first variant was the standard four-parameter version of Bayesian Knowledge Tracing described in [6], where each skill has a separate parameter for Initial Knowledge, Learning, Guessing, and Slipping. As in [6], the values of Guess and Slip were bounded, in order to avoid the "model degeneracy" problems [cf. 1] that arise when performance parameter estimates rise above 0.5 (When values of these parameters go above 0.5, it is possible to get paradoxical behavior where, for instance, a student who knows a skill is more likely to get it wrong than to get it right). In the analyses in this paper, both Guess and Slip were bounded to be below 0.3. However, unlike in [6], brute force search was used to find the best fitting parameter estimates – all potential parameter combinations of values at a grain-size of 0.01 were tried (e.g. 0.01 0.01 0.01 0.01, 0.01 0.01 0.01 0.02, 0.01 0.01 0.01 0.03... 0.01 0.01 0.02 0.01... 0.99 0.99 0.3 0.1). Recent investigations both in our group and among colleagues (e.g. [12, 14]) have suggested that the Bayesian Knowledge Tracing parameter space is non-convex [4] and that brute force approaches lead to better fit than previously-used algorithms such as Expectation Maximization (cf. [3]), Conjugate Gradient Search [cf. 6], and

Generalized Reduced Gradient Search (cf. [1]). These same investigations have suggested that brute force is computationally tractable for the data set and parameter set sizes seen in Bayesian Knowledge Tracing, since time increases linearly with the number of student actions but is constant for the number of skills (since only one skill applies to each student action, the number of mathematical operations is identical no matter how many skills are present). The four-parameter model's number of parameters is the number of cognitive rules * 4 – in this case, 32 parameters.

The second variant was Contextual-Guess-and-Slip [1, 2]. In this approach, as above, each skill has a separate parameter for Initial Knowledge and Learning. However, Guess and Slip probabilities are no longer estimated for each skill; instead, they are computed each time a student attempts to answer a new problem step, based on machine-learned models of guess and slip response properties in context (for instance, longer responses and help requests are less likely to be slips). The same approach as in [1, 2] was used, where 1) a four-parameter model is obtained, 2) the four-parameter model is used to generate labels of the probability of slipping and guessing for each action within the data set, 3) machine learning is used to fit models predicting these labels, 4) the machine-learned models of guess and slip are substituted into Bayesian Knowledge Tracing in lieu of skill-by-skill labels for guess and slip, and finally 5) parameters for Initial Knowledge and Learn are fit. Greater detail on this approach is given in [1, 2]. The sole difference in the analyses in this paper is that brute force was used to fit the four-parameter model in step 1, instead of other methods for obtaining four-parameter models. In [1, 2], Contextual-Guess-and-Slip models were found to predict student correctness within three Cognitive Tutors for mathematics (Algebra, Geometry, and Middle School Mathematics) significantly better than four-parameter models obtained using curve-fitting (cf. [6]) or Expectation Maximization [3]. As Contextual-Guess-and-Slip replaces two parameters *per skill* (G and S) with a smaller number of parameters across all skills (contextual G and contextual S), Contextual-Guess-and-Slip has a smaller total number of parameters, though only slightly so in this case, given the small number of skills; the parameter reduction is much greater when a larger number of skills are fit at once – in the mathematics tutors, the Contextual-Guess-and-Slip models only had 55% as many parameters as the four-parameter models.

The third variant was Individual Difference Weights on Bayesian Knowledge Tracing. With Individual Difference Weights, a four-parameter model is fit, and then a best-fitting weight for each student is computed for each parameter (e.g. student74 has one weight for L0 for all skills, one weight for T for all skills, one weight for G for all skills, and one weight for S for all skills, and parameter values are a function of the skill parameters and student weights). The student's individualized parameter values for a given skill are computed as a function of their individual difference weights and the skill-level parameters, using a formula given in [6]. That paper found that as students approached cognitive mastery, the Individual Difference Weights model was more accurate at predicting post-test scores than the four-parameter model. As the Individual Difference Weights approach has four parameters for each skill and four parameters for each student, it has substantially more parameters than the other approaches – in this case, 316 parameters.

4 Modeling Student Performance

4.1 Predicting Performance in the Tutor

While knowledge tracing models the student's knowledge, the underlying assumptions also yield an accuracy prediction in applying a rule:

$$P(correct_n)=P(L_{n-1})*(1-P(S))+(1-P(L_{n-1}))*P(G)$$

However, applying this rule to compare the three models' fit to the tutor data biases in favor of the Contextual-Guess-and-Slip model, since that model examines properties of each student response in order to generate a contextualized estimate of $p(S)$ and $p(G)$ for that response. To compare the three models' fit to student tutor performance in an unbiased fashion, we predict student correctness at time N just from the model's knowledge estimate at time N-1. This approach underestimates accuracy for all models (since it does not include the probability of guessing and slipping when answering), but does not bias in favor of any type of model.

We evaluate model goodness using A' [8], the probability that the model can distinguish correct responses from errors, because A' is an appropriate metric when the predicted value is binary (correct or not correct), and the predictors are numerical (probability of knowing the skill, or probability of getting the skill correct). We determine whether a model is statistically significantly better than chance by computing the A' value for each student, comparing differences between A' values and chance (cf. [8]) (giving a Z value for each student), and then using Stouffer's method [15] to aggregate across students. We determine whether the difference between two models is statistically significant by computing A' values for each student, comparing differences between A' values with a Z test [8], and then aggregating across students using Stouffer's method [15]. Both of these methods account for the non-independence of actions within each student.

Each model's effectiveness at predicting student performance within the tutor is shown in Fig. 2. The four-parameter model achieved A' of 0.758 in predicting student performance at time N from the students' knowledge estimate at time N-1. The Contextual-Guess-and-Slip model achieved A' of 0.755. The Individual Difference Weights model performed more poorly, with an A' of 0.734. All three models were significantly better than chance, Z=48.69, Z=44.85, Z=45.33, p<0.0001. The difference between the four-parameter model and the Individual Difference Weights model was significant, Z=-2.12, p=0.03, but the other two differences were not significant, Z= -1.36, p= 0.17, Z= -0.78, p= 0.43.

If we instead predict student performance at time N using the guess and slip parameters, the four-parameter model and the individual difference weights models achieve much closer performance (As previously mentioned, it is not valid to use this approach with the Contextual-Guess-and-Slip model). In this case, the four-parameter model achieves an A' of 0.769, and the Individual Difference Weights model achieves an A' of 0.768. Both models are significantly better than chance, Z=50.44, Z= 51.12, p<0.001. These two models are not statistically significantly different from each other, Z=0.127, p=0.89.

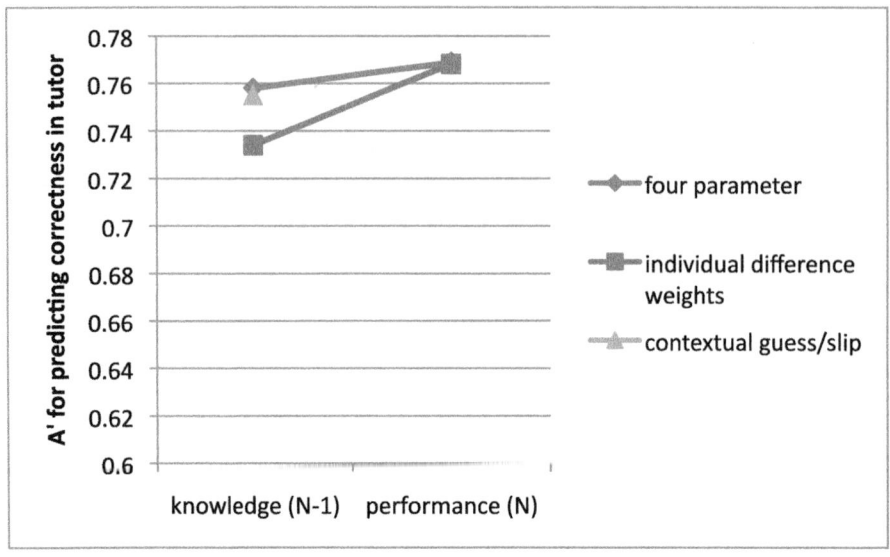

Fig. 2. The ability of each model to predict performance within the tutor

4.2   Predicting Post-Test Solely from Final Knowledge Estimates

Beyond predicting performance within the tutor, it is important to see how well the different methods predict student performance outside of the tutor. If any method sees significant degradation outside of the tutor, it may be over-fit to student behavior within the tutor, rather than capturing indicators of learning that will persist even outside of the tutor.

Again, the simplest way to use tutor estimates of student knowledge to predict the post-test, is simply to look at the correlation between just the models' estimate of student knowledge and the student's post-test performance. This approach is unlikely to be the most precise approach, as it ignores the possibility that the student will guess or slip on the test. However, it is equally feasible for all three approaches.

In predicting the post-test, we account for the number of times each skill will be utilized on the test (assuming perfect performance). Of the eight skills in the tutor lesson, one is not exercised on the test, and is eliminated from the model predicting the post-test. Of the remaining seven skills, four are exercised once, two are exercised twice and one is exercised three times, in each of the two posttest problems. These first two skills are each counted twice and the latter skill three times in our attempts to predict the post-test. We utilize this approach in all attempts to predict the post-test in this paper (including in later sections). As post-test scores represent the average correct on the test, we average the estimates of student skill together rather than multiplying them.

The full pattern of results for each model's ability to predict the post-test, based on the different assumptions in Sect. 4.1, 4.2, and 4.3, is shown in Fig. 3. We predict the post-test using each model's estimate of student knowledge of each skill, assessing the goodness of prediction with correlation, since the model estimates and the actual test scores are both numerical.

The four-parameter approach achieves a correlation of 0.430 to the post-test. The contextual-guess-and-slip approach achieves a correlation of 0.289 to the post-test. The individual difference weights approach achieves a correlation of 0.412 to the post-test. All three of the correlations are statistically significant ($p = .02$ for the contextual-guess-and-slip approach, and $p<0.01$ for the other two), while none of the differences among the correlations were statistically significant, although the difference between the four-parameter model and the contextual-guess-and-slip approach approached significance, $t(68)=1.63$, $p=0.12$, for a two-tailed test of the significance of the difference between two correlation coefficients for correlated samples.

4.3   Predicting Post-Test from Final Knowledge Estimates and Non-contextual Guess/Slip Estimates

One limitation to the approach above is that performance, whether in the tutor or on a paper post-test, is not simply a function of the student's knowledge. It is also a function of the probability that the students guesses (giving a correct answer despite not knowing the skill), or slips (making an error despite knowing the skill), as described in Sect. 4.1. Determining appropriate guess and slip rates for the paper post-test is not a trivial problem, since the students are working in a different environment. For instance, they may be more or less prone to physical slips (such as mis-typing or mis-writing) on paper than in the tutor, and they may be more or less cautious when the tutor is not providing immediate accuracy feedback. However, the performance parameter estimates derived from tutor behavior with both the standard four-parameter model and the Individual Difference Weights version have been shown to predict test data quite accurately [6, 7].

Within the four-parameter and Individual Difference Weights approaches, we can compute the probability that the student will get each answer right on the post-test based on both the final knowledge estimates, and the model's parameters for guess and slip for each skill (or for Individual Difference Weights, the parameters for each skill and student):

$$P(\text{correct}_n)=P(L_{n-1})*(1-P(S))+(1-P(L_{n-1}))*P(G)$$

When we do this, the correlation between the estimates of the probability of getting the skill correct in the four parameter model, and the post-test score rises very slightly, from 0.430 to 0.434. The difference between this model, and the earlier fit where four-parameter model estimates of final knowledge are used, is not statistically significant, $t(68)=0.63$, $p=0.53$, for a two-tailed test of the significance of the difference between two correlation coefficients for correlated samples.

By contrast, the correlation between the estimates of the probability of getting the skill correct in the Individual Difference Weights model, and the post-test score appears to drop within this approach, from 0.412 to 0.352. But as above, this model does not differ significantly in correlation from the model using the estimates of the probability of knowledge in the individual difference weights model, $t(68)=0.66$, $p=0.51$, for a two-tailed test of the significance of the difference between two correlation coefficients for correlated samples. In addition, there is not a statistically significant difference between the two models, t(68)= 1.19, p=0.23, for a two-tailed test of the significance of the difference between two correlation coefficients for correlated samples.

4.4   Predicting Post-Test from Final Knowledge Estimates and Contextual Guess/Slip Estimates

Contextual models of guess and slip assess the probability that a student slipped at a specific time, within the tutoring software. These models cannot be used as-is to predict guess and slip on a paper post-test, as behavioral indicators such as timing are not available. Instead, the contextual estimates of guess and slip from within the tutor can be used as an indicator of how much each student guessed and slipped while using the tutor. The most straightforward way to do this is to average the contextual slip and guess values at each problem step.

Hence, one option for using these estimates is to use the average guess and slip for each student and skill in lieu of the non-contextual parameter estimates, within equation 1 above. A model which does this achieves a poor correlation to post-test score, 0.181. It is not statistically significantly worse than the earlier fit using the contextual guess-and-slip model's estimates of final knowledge, t(68)=-0.70, p=0.48, for a two-tailed test of the significance of the difference between two correlation coefficients for correlated samples. It is, however, marginally statistically significantly worse than the four-parameter model's prediction of performance, t(68)=-1.74, p=0.09.

Although the contextual guess-and-slip model's estimates of performance, used in this fashion, led to poor prediction of the post-test score, there may still be useful information in the contextual estimates of guess and slip themselves. The correlation between average contextual slip and post-test score, r=0.272, is statistically significantly higher than chance, F(1,69)=5.521, p=0.02, and the correlation between average contextual guess and post-test score, r=-0.325, is also statistically significantly higher than chance, F(1,69)=8.17, p<0.01. In addition, if we restrict analysis to values of contextual slip over 0.5 (e.g., where there is a probability over 50% that the action is a slip), the correlation to post-test score is particularly strong, r=0.453, and is statistically significantly higher than chance, F(1,69)=17.801, p<0.001.

One way to determine whether this information is potentially useful is to generate a post hoc prediction of post-test performance with a combination of the average contextual slip, and the four-parameter prediction of post-test performance (e.g. the model including skill-level estimates of guess and slip, from Sect. 4.3), using linear regression. While the average contextual slip and average contextual guess are not statistically significant predictors in a model already containing the four-parameter model prediction, average contextual slip over 0.5 **is** statistically significant in a model already containing the four-parameter model prediction, F(2,68)=10.81, p<0.001. In other words, a linear regression model including both average contextual slip over 0.5 and the four-parameter prediction, has statistically significantly better fit than the model just containing the brute-force prediction, achieving a correlation to post-test of 0.491.

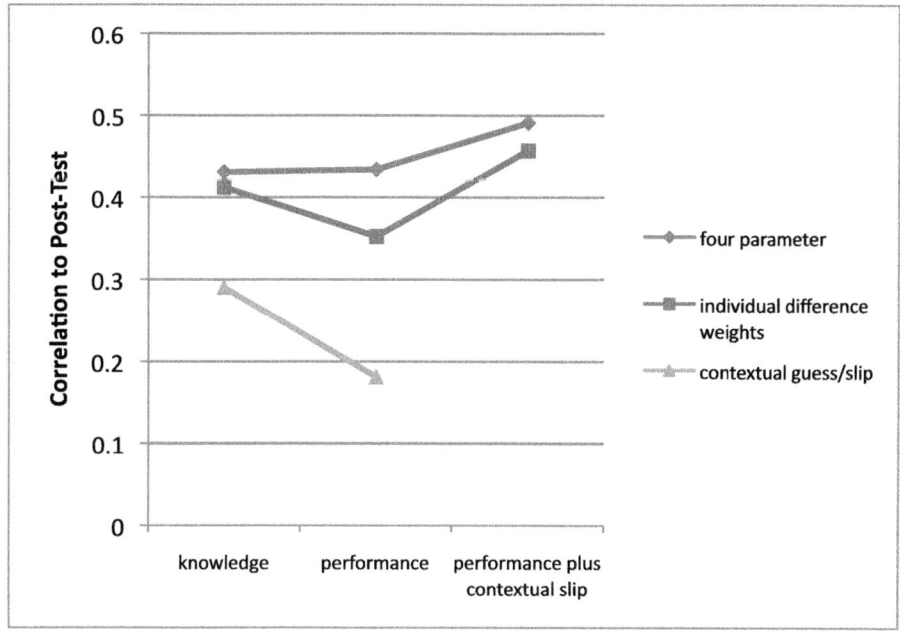

Fig. 3. The correlation of each model to the post-test

This finding indicates that slipping within the tutor is an indicator of some aspect of student learning that is associated with the failure to transfer knowledge to a cognitively identical problem in a different setting (outside the tutor).

A similar pattern is seen with the Individual Difference Weights model. Again, average contextual slip over 0.5 is statistically significant in a model already containing the Individual Difference Weights model prediction, $F(2,68)=9.00, p<0.01$.

A clear implication can be seen from this pattern of results. Though the current formulation of Contextual-Guess-and-Slip does not port to the post-test, there is clear evidence that future models integrating evidence on contextual slip have the potential to do better at predicting the post-test than the current generation of Bayesian Knowledge Tracing prediction. Determining how to integrate contextual slip information in a replicable fashion will be an important area of future work.

5   Discussion and Conclusions

Overall, the findings here suggest that the Contextual-Guess-and-Slip approach, in its current form, does a fine job of predicting performance within the tutoring system, performing comparably to or slightly better than the four-parameter approach and Individual Difference Weights approach. However, the Contextual-Guess-and-Slip approach predicted performance much more poorly outside of the tutor than within the tutor. One possible explanation is that Contextual-Guess-and-Slip is over-fit to aspects of student performance within the tutor; that said, given that Contextual-Guess-and-Slip has fewer parameters, it is unlikely that it is over-fit in general, compared to the other models (cf. [9]). Another explanation is that by allowing a student who slips a great deal to still be assessed as having mastery, Contextual-Guess-and-Slip discards evidence of incomplete or non-robust knowledge.

Note that the Individual Difference Weights approach also failed to predict the post-test better than the standard four-parameter approach. In Corbett and Anderson's earlier work [6], the advantage of the Individual Difference Weights model emerged when students reached very high levels of $P(L_n)$ estimates, higher than students reached in this study. The results of this study provides converging evidence that the benefit of individual different weights only emerges for high $P(L_n)$ levels.

Despite the failure of Contextual-Guess-and-Slip to predict performance on the post-test, estimates of Contextual Slip appear to be a valuable addition to the knowledge and performance prediction obtained in the four-parameter approach. A post-hoc model combining average contextual slip among actions where $P(S)$ was over 0.5, and the performance predictions from Bayesian Knowledge Tracing, performs significantly better than the performance predictions alone. This finding indicates that slipping during the tutor is an indicator of some aspect of student learning that is not captured by Bayesian Knowledge Tracing. However, $P(S)$ can have several meanings, including indicating shallow knowledge or general carelessness during tutor usage. It may be possible to disentangle these possibilities with measures of robust learning (cf. [16, 17]), where shallow learning is likely to compromise performance to a greater degree, and with questionnaire assessments of carelessness.

Hence, it appears that potential remains for utilizing contextual estimates of slipping in predicting student performance outside of intelligent tutoring systems. This is important, because better prediction of post-test scores is likely to lead to more effective adaptation within intelligent tutoring systems – in particular, understanding *why* slip predicts post-test will determine which type of adaptation is most appropriate for a student who appears to know the skill within the software, but who has frequently slipped during the process of knowledge acquisition.

Acknowledgements. This research was supported by the National Science Foundation via grant "Empirical Research: Emerging Research: Robust and Efficient Learning: Modeling and Remediating Students' Domain Knowledge", award number DRL0910188.

References

1. Baker, R.S.J.d., Corbett, A.T., Aleven, V.: More Accurate Student Modeling Through Contextual Estimation of Slip and Guess Probabilities in Bayesian Knowledge Tracing. In: Proceedings of the 9th International Conference on Intelligent Tutoring Systems, pp. 406–415 (2008)
2. Baker, R.S.J.d., Corbett, A.T., Aleven, V.: Improving Contextual Models of Guessing and Slipping with a Truncated Training Set. In: Proceedings of the 1st International Conference on Educational Data Mining, pp. 67–76 (2008)
3. Beck, J.E., Chang, K.-m.: Identifiability: A fundamental problem of student modeling. In: Conati, C., McCoy, K., Paliouras, G. (eds.) UM 2007. LNCS (LNAI), vol. 4511, pp. 137–146. Springer, Heidelberg (2007)

4. Boyd, S., Vandenberghe, L.: Convex Optimization. Cambridge University Press, Cambridge (2004)
5. Corbett, A.: Cognitive computer tutors: Solving the two- sigma problem. In: Bauer, M., Gmytrasiewicz, P.J., Vassileva, J. (eds.) UM 2001. LNCS (LNAI), vol. 2109, pp. 137–147. Springer, Heidelberg (2001)
6. Corbett, A.T., Anderson, J.R.: Knowledge Tracing: Modeling the Acquisition of Procedural Knowledge. User Modeling and User-Adapted Interaction 4, 253–278 (1995)
7. Corbett, A., Kauffman, L., Maclaren, B., Wagner, A., Jones, E.: A Cognitive Tutor for Genetics Problem Solving: Learning Gains and Student Modeling. Journal of Educational Computing Research 42, 219–239 (2010)
8. Fogarty, J., Baker, R., Hudson, S.: Case Studies in the use of ROC Curve Analysis for Sensor-Based Estimates in Human Computer Interaction. In: Proceedings of Graphics Interface, pp. 129–136 (2005)
9. Hawkins, D.M.: The Problem of Overfitting. Journal of Chemical Information and Computer Sciences 44(1), 1–12 (2004)
10. Koedinger, K.R., Corbett, A.T.: Cognitive tutors: Technology bringing learning sciences to the classroom. In: Sawyer, R.K. (ed.) The Cambridge handbook of the learning sciences, pp. 61–77. Cambridge University Press, New York (2006)
11. Martin, J., VanLehn, K.: Student Assessment Using Bayesian Nets. International Journal of Human-Computer Studies 42, 575–591 (1995)
12. Pavlik, P.I., Cen, H., Koedinger, J.R.: Performance Factors Analysis – A New Alternative to Knowledge Tracing. In: Proceedings of the 14th International Conference on Artificial Intelligence in Education, pp. 531–540 (2009)
13. Reye, J.: Student Modeling based on Belief Networks. International Journal of Artificial Intelligence in Education 14, 1–33 (2004)
14. Ritter, S., Harris, T., Nixon, T., Dickinson, D., Murray, R.C., Towle, B.: Reducing the Knowledge Tracing Space. In: Proceedings of the 2nd International Conference on Educational Data Mining, pp. 151–160 (2009)
15. Rosenthal, R., Rosnow, R.L.: Essentials of Behavioral Research: Methods and Data Analysis, 2nd edn. McGraw-Hill, Boston (1991)
16. Schmidt, R.A., Bjork, R.A.: New conceptualizations of practice: common principles in three paradigms suggest new concepts for training. Psychological Science 3(4), 207–217 (1992)
17. Schwartz, D.L., Martin, T.: Inventing to prepare for future learning: The hidden efficiency of encouraging original student production in statistics instruction. Cognition and Instruction 22, 129–184 (2004)
18. Shute, V.J.: SMART: Student modeling approach for responsive tutoring. User Modeling and User-Adapted Interaction 5(1), 1–44 (1995)

Working Memory Span and E-Learning: The Effect of Personalization Techniques on Learners' Performance

Nikos Tsianos[1], Panagiotis Germanakos[2,3], Zacharias Lekkas[1],
Costas Mourlas[1], and George Samaras[2]

[1] Faculty of Communication and Media Studies, National and Kapodistrian
University of Athens, Stadiou Str., GR 105-62, Athens, Hellas
[2] Department of Computer Science, University of Cyprus, CY-1678 Nicosia, Cyprus
[3] Department of Management and MIS, University of Nicosia, 46 Makedonitissas Ave.,
P.O. Box 24005, 1700 Nicosia, Cyprus
{ntsianos,mourlas}@media.uoa.gr,
{pgerman,cssamara}@cs.ucy.ac.cy, zlekkas@gmail.com

Abstract. This research paper presents the positive effect of incorporating individuals' working memory (WM) span as a personalization factor in terms of improving users' academic performance in the context of adaptive educational hypermedia. The psychological construct of WM is robustly related to information processing and learning, while there is a wide differentiation of WM span among individuals. Hence, in an effort to examine the role of cognitive and affective factors in adaptive hypermedia along with psychometric user profiling considerations, WM has a central role in the authors' effort to develop a user information processing model. Encouraged by previous findings, a larger scale study has been conducted with the participation of 230 university students in order to elucidate if it is possible through personalization to increase the performance of learners with lower levels of WM span. According to the results, users with low WM performed better in the personalized condition, which involved segmentation of the web content and aesthetical annotation, while users with medium/high WM span were slightly negatively affected by the same techniques. Therefore, it can by supported it is possible to specifically address the problem of low WM span with significant results.

Keywords: Adaptive Hypermedia, Working Memory, User Profiling, Cognitive Psychology, Individual Differences.

1 Introduction

Learning is related to a number of individual cognitive and affective trait and state-like characteristics, which account for the corresponding variability in learning performance. Constructs at different levels, such as IQ, fluid intelligence, personality and approaches to learning, have been reported as predictors of academic performance [1]; motivation along with numerical, verbal and spatial cognitive abilities have been related to specific patterns of academic performance [2], while state anxiety has been found to mediate trait-like individual differences [3]. The construct of working

P. De Bra, A. Kobsa, and D. Chin (Eds.): UMAP 2010, LNCS 6075, pp. 64–74, 2010.

memory (WM) has also been identified as a predictor of learning performance [4, 5], while there are numerous studies that relate WM with learning and cognitive processes.

The research that is presented in this paper is focused on measuring learners' WM capacity, on examining the differences in performance in relation to WM resources, and finally on improving the performance of learners with lower levels of WM span. It should be mentioned that the authors have previously conducted relevant research, in an effort to build an adaptive educational system that incorporates psychological constructs that reflect individual differences. These differences, both trait and state-like, are represented by a three-dimensional user model, which consists of: a) cognitive style, b) speed of processing, visual attention, WM, and c) emotional processing [6]. This model aims to coherently combine preferences, abilities, trait and state-like characteristics, and to optimize the learning performance of users through mapping these characteristics on the instructional method.

The constructs of intelligence and fluid intelligence have deliberately been excluded, since it would be very complex, if not impossible, to establish personalization rules; the user profiling procedure would also be very burdensome, and perhaps assigning learners in groups according to their intelligence would raise ethical issues. On the other hand, WM is indicative of the cognitive abilities that are related to learning and correlated at some extent to general intelligence [7, 8].

As it concerns the empirical evaluation of the aforementioned user model, personalization on the basis of cognitive style, visuospatial WM and anxiety was proven to increase the performance of learners [9]. Still, the construct of WM was initially only partially approached and measured, while the methodology of the following experimental approach needed to be improved [10].

Within this context of ongoing experimental evaluation, this paper presents an extensive empirical study that was conducted in order to evaluate the role of WM span in educational hypermedia and, mainly, to assess the effectiveness of corresponding personalization techniques in terms of actually assisting learners with low levels of WM span in improving their performance.

2 Theoretical Background

One of the predominant theories of WM is Baddeley and Hitch's multicomponent model [11]. According to Baddeley, "the term working memory refers to a brain system that provides temporary storage and manipulation of the information necessary for such complex cognitive tasks as language comprehension, learning, and reasoning" [12].

Baddeley also refers to individual differences in the WM (digit) span of the population, thus providing a very good argument for using this construct as a personalization factor. Since WM is considered to be a predictor of academic performance, it would be of high importance to alleviate learning difficulties of learners with low levels of WM.

A brief description of the WM system is that is consisted of the central executive (CE) that controls two slave systems: a) the visuospatial sketchpad and b) the phonological loop. A later addition to the model is the episodic buffer that provides a temporary interface between the slave systems and the long term memory [13]. Baddeley's diagrammatical representation of the system is illustrated in Fig. 1.

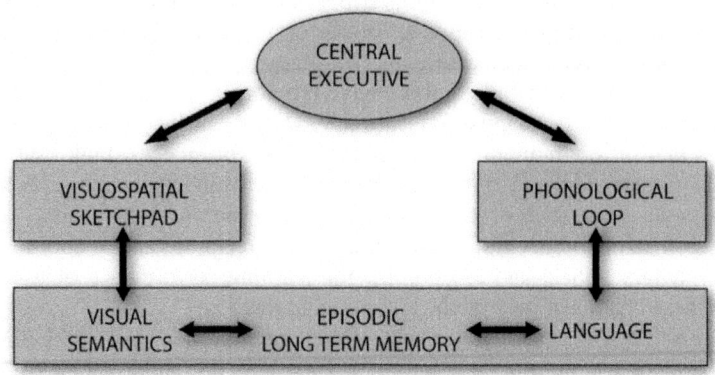

Fig. 1. Multicomponent model of WM

The CE is assumed to be an attentional-controlling system of limited resources; the visuospatial sketch pad manipulates visual images and spatial information, while the phonological loop stores and rehearses speech-based information and is necessary for the acquisition of both native and second-language vocabulary. The role of the episodic buffer is out of the scope of our research, which essentially is based on the original version of the model.

Both subsystems and the CE, which are generally independent from each other [14], have limited capacity. Though the number of items (or chunks) that can be stored in WM storage is dependent on various factors (such as length of words for example), individuals vary in their storage capacity (as mentioned above), the same way they vary in intelligence. In line with the notion of user profiling and satisfying users' needs, it could be argued that learners with low WM and CE capacity should be identified and instructed in a way that does not require manipulation of large chunks of information at the same time.

As it concerns the field of educational hypermedia, recent studies seem to establish a relation between WM resources and hypertextual learning. DeStefano and LeFevre [15] reviewed 38 studies that mainly address the issue of cognitive load in hypertext reading, showing that WM is often considered as a significant factor even at the level of explaining differences in performance. Lee and Tedder [16] examine the role of WM in different computer texts, and their results show that low WM span learners do not perform equally well in hypertext environments. Accordingly, McDonald and Stevenson [17] argue that non-linear hypertextual learning spaces are more demanding in WM resources in comparison to hierarchically structured environments. Also, Dutke and Rinck [18] have reported that certain tasks in multimedia learning require more WM resources, while individuals with lower levels of verbal and visuospatial WM capacity face increased difficulties.

Naumann et al [19] found that cognitive and metacognitive strategy training benefits learners with large WM capacity, "whereas the learning outcomes of participants with a small working memory capacity were deteriorated by both types of training." Also, in relation to WM capacity, the term Cognitive Load Theory is often used especially when providing guidelines for designing hypermedia applications [20]; for example, in a very recent study that involved EEG measurements [21], it was found that leads in hypertext nodes may assist in decreasing cognitive load and on acquisition of domain and structural knowledge.

Based on the above, it could be argued that WM capacity (or span) may predict learning performance in hypertext environments, and that certain structures or methods of presentation are more demanding in WM resources. Consequently, in the context of adaptive educational hypermedia [22–25], WM span could constitute a significant user profiling and personalization factor, since: a) there are distinct differences with measurable effects among the learner population, and b) different hypertext structures and methods of presentation may benefit (or hamper) the performance of learners.

3   Research Questions and Design Implications

Learning in a hypermedia environment requires cognitive processing of visual and verbal content, involving both WM slave systems and CE resources. Hence, the first step would be the measurement of each learner's visual and verbal working memory capacity with corresponding psychometric tests. Subsequently, an empirical evaluation of the performance of learners grouped according to their WM span levels and the use of personalization techniques would reveal if there are any significant differences.

It should be noted that our research interest is focused on learners with low levels of WM span; the main aim is to assist in the development of personalized instructional techniques that would ensure the effectiveness of adaptive educational hypermedia regardless of individual differences and abilities.

3.1   Research Questions

In the broader context of our research on WM and adaptive educational hypermedia, our research questions were the following:

  i)   Are WM capacity psychometric tools appropriate for the context of hypermedia learning?
  ii) Do low WM learners perform worse than those with higher levels of memory capacity and CE function?
  iii) Is it possible with the use of personalized instructional techniques to increase the performance of low WM learners?

3.2   Classification and Personalization

The classification of users according to WM span tests (visual memory and CE/verbal storage) was a main issue of concern. First of all, since these two measurements are independent, it would be possible for a user to perform significantly better in only one of the tests. However, considering that an e-learning course may as well contain both visual and verbal material, a more holistic approach in WM capacity would be more suitable for the needs of our approach. Consequently, the system profiled users on the basis of the aggregated performance in both tests, albeit with some additional considerations.

First of all, it should be reminded once more that our main concern is to identify users with low WM. The threshold that distinguishes medium from high WM individuals was known for the case of the visual test, but the modified CE/verbal storage test was not tested across a standard population. As a result, we adopted a threshold that relatively identifies low WM individuals, after conducting a pilot study.

In terms of scoring, there was a complete analogy between the two tests by transforming the scores. Those that did not exceed the 1/3 of the aggregated score were classified as low WM learners. Regardless however of the total score, users that scored very low in one of the two tests were also classified as low WM learners, assuming that they lack the corresponding WM resources.

As it concerns the low WM personalized condition, which is a challenge of its own, the learning content was altered in two ways. Firstly, the content presented simultaneously on one webpage was segmented. A decreased number of learning objects (images and paragraphs of text) was assumed to require less cognitive resources from users with limited storage capacity and attentional control, allowing them to keep a more gradual pace on information processing. Initially, light-weighted versions of the pages are given to the users with low WM span and then, by clicking on the screen, the page unfolded at its full extent, with the remaining learning objects being presented to the user. This rather simple approach was proven effective in our previous experiments, possibly due to the fact that this gradual assimilation of information reduces the risk of cognitive overload.

The second method of personalization was the annotation of textual objects. This approach is partially derived from studies exploring the relationship of hypertext and WM [15]. It seems that diagrammatical representations and highly structured texts assist low WM users; thus, at the level of better structuring the text, different colors were used for annotating paragraph titles, distinguishing in parallel different sections of the page. Bold text and colors were used for important concepts, links and titles, in an effort to help learners organize information. In a sense, the system imposes on low WM learners a strategy of reading and organizing information; the assumption that this would be proven beneficial is related, though not very closely, to the fact that strategies such as rehearsal have a positive effect on low WM learners [26].

It should be clarified that both these methods are innovative and quite explorative, in the absence of well defined guidelines for improving the performance of low WM individuals. The literature over the implications of WM in every aspect of information processing is truly exhaustive, but the idea of leveling the performance of individuals despite their differences in cognitive abilities seems out of the scope, to our knowledge, of most prior research.

We definitely acknowledge that our approach is assumptive, but considering the lack of previous endeavors in exploring adaptive educational hypermedia and WM and in optimizing the performance of low WM learners, we rely on our experimental results in order to validate our methods.

4 Experimental Method

4.1 Design and Procedure

The design of the study was a single-factor, between-participants design, involving four groups of users: a) a group of low WM users that received a personalized course, presumably suitable for them, b) a group of low WM users who received a standard, non-personalized course, c) a control group of users with normal/high levels of WM who received the standard on-line course, and d) a control group of normal/high WM learners who received an on-line course that was personalized on the needs of low WM learners (same environment with group a). The dependent variable was learners' scores in an exam that followed the on-line course.

All versions of the learning environment were personalized on learners' cognitive style, in order to control for the impact of this factor on performance; our previous experimental results demonstrated that matching the instructional style to users' cognitive style positively affects performance, while mismatching has an adverse effect [9]. Hence, in order to control for any possible effects of matching or mismatching the instructional style to learners' preferences, the system provided personalized on style environments to all participants, based on Riding's Cognitive Style Analysis [27].

The participants were Greek speaking students from the Universities of Athens and Cyprus, 65% female and 35% male, with their age varying from 18 to 21 years. The number of valid participants was 230 out of a total of 260 users; 30 participants were excluded due to very poor performance (near zero scores) in the WM tests and the exam, which could imply either failure to follow the tests' rules or complete lack of interest. Participation in the experiment was voluntary.

The mean duration of the procedure was approximately one-hour, though there were not any time constraints imposed on learners. The data were gathered from three consecutive identical experiments: two were conducted in a computer science laboratory in Cyprus and one in Athens, with approximately 15 participants in each session.

Each user logged in the system, took the cognitive style and WM assessment tests, and was quasi-randomly assigned into one of the aforementioned groups; thereafter the learner was navigated to the e-learning course. The subject of the e-learning procedure was an introductory course on algorithms. This course has also been used in our previous experiments, mainly because participants lack any previous knowledge of computer science. Immediately after the completion of the course, participants were asked to take a comprehension on-line test about what they had been taught. Their scores on this test was the dependent variable indicating academic performance (maximum possible score=100).

4.2 WM Span Measurement Considerations and Tools

The first step in setting up our experiments was to measure users' WM with the appropriate psychometric tools. Integrating such measurements in an adaptive hypermedia system through a user profiling procedure essentially requires the development of electronic versions of pencil and paper tests. In the case of visuospatial

WM span, a tool was already available [28]; it only had to be implemented in the .NET platform of our environment.

The authors however were not aware of an electronic version of a phonological loop span and CE test. For that reason, we were provided with an extended Greek version of the listening sentence recall test of the WMTB-C [29]. This test measures both the CE function and the verbal storage ability, providing an indication of individuals' WM ability. In its original form, it is a pencil and paper listening test; in the case of e-learning though information is usually conveyed through written text. For that reason, we were mainly interested in learners' ability to manipulate written and not acoustic verbal information; this is why in the electronic version of the test we opted for on-screen presentation of written sentences rather than auditory articulation.

This probably leads to a differentiated form of the original test, addressing perhaps different aspects of WM than those originally intended; still, by experimentally assessing the validity of the measurements, we expected that the relative classification of learners would be more appropriate for a web-environment, focusing on storage of written verbal material and CE function in front of a computer screen. A brief description of test follows, for the purposes of clarifying how the test was adapted in our system.

Users are required to store the last word of a series of consecutively presented (written) sentences, while deciding at the same time whether the meaning of each sentence makes sense or not. The test gradually becomes more difficult, since the number of sentences increases from two (first level) to nine (last level). There are six series of sentences in each level, and users have to remember correctly the last words of four at least series in order to proceed to the next level.

At the third level, for example, four sentences are presented one after the other, each remaining on screen for two seconds. Users have to decide if the meaning of each sentence is true of false, by pressing the corresponding key, triggering the presentation of the next sentence. When all sentences are presented, users are asked to fill a corresponding number of text fields with the last word of each sentence. Scoring is the same as in the original test.

5   Results

The mean scores of the four groups of learners demonstrate that the personalization techniques that were employed (segmentation of the content and annotation) benefited learners with low WM span; in contrast, these techniques had a slightly negative effect on learners of the control group (see table 1). A one-way analysis of variance was performed on the data (since the assumptions of normality and homogeneity of variances were met), revealing that this difference is statistically significant: $F_{(3,226)}=3.930$, $p=0.009$.

A post hoc analysis (Tukey HSD) revealed that the difference in scores is statistically significant between the three first groups (see table 2); the personalized control group did not differ significantly from any other group, which was expected since learners' scores in this condition were close to the total mean.

Table 1. Mean Scores of Learner Groups

Condition	N	Mean Score	Std. Deviation
Personalized Low WMS	46	59.17	15.71
Non-personalized Low WMS	47	50.27	14.06
Non-Personalized Control Group	87	59.46	16.15
Personalized Control Group	50	55.94	16.40
Total	230	56.76	16.01

Table 2. Post Hoc Analysis of Learner Groups' Scores

Tukey HSD

(I) Condition	(J) Condition	Mean Difference (I-J)	Significance
Personalized Low WMS	Non-personalized Low WMS	8.90*	0.034
	Non-Personalized Control Group	-0.29	1.000
	Personalized Control Group	3.23	0.745
Non-personalized Low WMS	Non-Personalized Control Group	-9.18*	0.008
	Personalized Control Group	-5.66	0.289
Non-Personalized Control Group	Personalized Control Group	3.52	0.588

According to these findings, it is shown that:

- Learners with medium/ high levels of WM performed better than those with low levels of WM, in the same non personalized environment (+9.2 points). Thus, WM has an effect on users' performance in educational hypermedia.
- Learners with low WM improved their performance in the personalized condition by 8.9 points, reaching the performance of medium/high WM learners (only -0.29 points difference).
- The personalization method that was employed had no positive effect on learners with medium/high levels of WM; on the contrary, though statistically non significant, these learners had lower scores than those of the non-personalized control group (-5.7 points), though still better than non-personalized low WM learners (+3.5 points). Hence, it may be argued that the segmentation and annotation techniques address directly the low WM span issue and do not generally improve the method of presentation.

Additionally, the scores of the two WM span tests were not correlated. This is in line with the fact that the components of Baddeley and Hitch's model are relatively independent; otherwise, the validity of the measurements would be questioned.

6 Discussion

According to the findings of this study, our research questions were answered as follows: i) the measurement of WM with electronic versions of psychometric tools reflects users' cognitive ability in hypermedia environments, ii) low WM learners perform worse than those with higher levels of memory capacity, and iii) certain personalization techniques may assist low WM learners in optimizing their performance, reaching the levels of those with higher WM.

Therefore, it seems that less simultaneously presented learning content and structuring the text with annotations seemed to address the issue of limited storage and attentional control efficiently. It should be noted that these techniques do not positively affect all learners, but specifically address the limitations of low WM span.

There are however some limitations in our study. First of all, the personalization rules were based on our assumptions; even if the results justify this approach, there should be a large scale evaluation of the proposed adaptation techniques. Simple ideas often work, but considering the depth and numerous implications of WM, further research is needed to establish a robust set of adaptive educational hypermedia design guidelines.

Also, it remains ambiguous whether low WM learners were assisted more by the segmentation of the content or the annotation of the text. Both techniques were employed in the personalized condition, and it is impossible to distinguish separate effects. Segmentation of the web page was proven significant in our previous work with visual WM span, albeit with smaller effect. Annotation of the text may also have been useful, but since in this experiment we also measured verbal storage and CE capacity, perhaps identifying a larger number of low WM learners increased the positive effect of segmentation; the effect of annotating the text should be separately examined.

The way we incorporated WM measurement tools in our system was mainly affected by the needs of our research in adaptive hypermedia. First of all, we focused on written text verbal storage and CE function, than auditory; additionally, instead of using a battery of WM tests that examine this construct in depth, we measured what we believed was adequate for our exploratory approach, without posing difficult and time consuming challenges to users. Still, we consider that there is room for improvement in capturing electronically the WM capacity of users. For example, a backward word span task (demanding users to recall words in the reverse order) would increase the validity of the measurements and provide a better insight on learners' abilities.

Nevertheless, all our research questions were answered in a way that supports our approach, and the notion that WM is a key factor in e-learning was validated. Moreover, instead of simply acknowledging this effect, it was shown that it is possible to assist learners effectively, putting into meaningful practice the theoretical background of this construct. This encourages us to continue research on our model, incorporating individual differences theories in the field of adaptive e-learning. Future work on this line of research includes the measurement of state-like user characteristics, especially those related to emotional processing. Real time biometric techniques have already been included in our experiments, and in parallel with the aforementioned WM findings, further optimization of learners' performance is anticipated.

Acknowledgments. The project is co-funded by the EU project CONET (INFSO-ICT-224053) and by the Cyprus Research Foundation under the project MELCO (ΤΠΕ/OP120/0308 (BIE)/14).

References

1. Tomas Chamorro-Premuzic, T., Furnham, A.: Personality, intelligence and approaches to learning as predictors of academic performance. Personality and Individual Differences 44, 1596–1603 (2008)
2. Lau, S., Roeser, R.W.: Cognitive abilities and motivational processes in science achievement and engagement: A person-centered analysis. Learning and Individual Differences 18(4), 497–504 (2008)
3. Chen, G., Gully, S.M., Whiteman, J., Kilcullen, R.N.: Examination of Relationships Among Trait-Like Individual Differences, State-Like Individual Differences, and Learning Performance. Journal of Applied Psychology 85(6), 835–847 (2000)
4. Colom, R., Escorial, S., Shih, P.C., Privado, J.: Fluid intelligence, memory span, and temperament difficulties predict academic performance of young adolescents. Personality and Individual Differences 42(8), 1503–1514 (2007)
5. Alloway, T.P.: Working memory, but not IQ, predicts subsequent learning in children with learning difficulties. European Journal of Psychological Assessment 25(2), 92–98 (2009)
6. Germanakos, P., Tsianos, N., Lekkas, Z., Mourlas, C., Samaras, G.: Capturing Essential Intrinsic User Behaviour Values for the Design of Comprehensive Web-based Personalized Environments. Computers in Human Behavior 24(4), 1434–1451 (2008)
7. Colom, R., Abad, F.J., Quiroga, A., Shih, P.C., Flores-Mendoza, C.: Working memory and intelligence are highly related constructs, but why? Intelligence 36(6), 584–606 (2008)
8. Lynn, R., Irwing, P.: Sex differences in mental arithmetic, digit span, and g defined as working memory capacity. Intelligence 36(3), 226–235 (2008)
9. Tsianos, N., Lekkas, Z., Germanakos, P., Mourlas, C., Samaras, G.: An Experimental Assessment of the Use of Cognitive and Affective Factors in Adaptive Educational Hypermedia. IEEE Transactions on Learning Technologies (TLT) 2(3), 249–258 (2009)
10. Tsianos, N., Germanakos, P., Lekkas, Z., Mourlas, C., Belk, M., Samaras, G.: Working Memory Differences in e-Learning Environments: Optimization of Learners' Performance through Personalization. In: Houben, G.-J., McCalla, G., Pianesi, F., Zancanaro, M. (eds.) UMAP 2009. LNCS, vol. 5535, pp. 385–390. Springer, Heidelberg (2009)
11. Baddeley, A.: The concept of working memory: A view of its current state and probable future development. Cognition 10(1-3), 17–23 (1981)
12. Baddeley, A.: Working Memory. Science 255, 556–559 (1992)
13. Baddeley, A.: The episodic buffer: a new component of working memory? Trends in Cognitive Sciences 11(4), 417–423 (2000)
14. Loggie, R.H., Zucco, G.N., Baddeley, A.D.: Interference with visual short-term memory. Acta Psychologica 75(1), 55–74 (1990)

15. DeStefano, D., Lefevre, J.: Cognitive load in hypertext reading: A review. Computers in Human Behavior 23(3), 1616–1641 (2007)
16. Lee, M.J., Tedder, M.C.: The effects of three different computer texts on readers' recall: based on working memory capacity. Computers in Human Behavior 19(6), 767–783 (2003)
17. McDonald, S., Stevenson, R.J.: Disorientation in hypertext: the effects of three text structures on navigation performance. Applied Ergonomics 27(1), 61–68 (1996)
18. Dutke, S., Rinck, M.: Multimedia learning: Working memory and the learning of word and picture diagrams. Learning and Instruction 16, 526–537 (2006)
19. Naumann, J., Richter, T., Christmann, U., Groeben, N.: Working memory capacity and reading skill moderate the effectiveness of strategy training in learning from hypertext. Learning and Individual Differences 18, 197–213 (2008)
20. Kirschner, P.A.: Cognitive load theory: implications of cognitive load theory on the design of learning. Learning and Instruction 12(1), 1–10 (2002)
21. Antonenko, P.D., Niederhauser, D.S.: The influence of leads on cognitive load and learning in a hypertext environment. Computers in Human Behavior 26, 140–150 (2010)
22. Cristea, A., Stewart, C., Brailsford, T., Cristea, P.: Adaptive Hypermedia System Interoperability: a 'real world' evaluation. Journal of Digital Information 8(3) (2007), http://journals.tdl.org/jodi/article/view/235/192
23. Papanikolaou, K.A., Grigoriadou, M., Kornilakis, H., Magoulas, G.D.: Personalizing the Interaction in a Web-based Educational Hypermedia System: the case of INSPIRE. User-Modelling and User-Adapted Interaction 13(3), 213–267 (2003)
24. Carver Jr., C.A., Howard, R.A., Lane, W.D.: Enhancing student learning through hypermedia courseware and incorporation of student learning styles. IEEE Transactions on Education 42(1), 33–38 (1999)
25. Gilbert, J.E., Han, C.Y.: Arthur: A Personalized Instructional System. Journal of Computing in Higher Education 14(1), 113–129 (2002)
26. Turley-Ames, K.J., Whitfield, M.M.: Strategy training and working memory task performance. Journal of Memory and Language 49, 446–468 (2003)
27. Riding, R.J., Cheema, I.: Cognitive Styles – an overview and integration. Educational Psychology 11(3&4), 193–215 (1991)
28. Demetriou, A., Christou, C., Spanoudis, G., Platsidou, M.: The development of mental processing: Efficiency, working memory, and thinking. Monographs of the Society for Research in Child Development 67(1), 1–155 (2002)
29. Pickering, S., Gathercole, S.: The Working Memory Test Battery for Children. The Psychological Corporation (2001)

Scaffolding Self-directed Learning with Personalized Learning Goal Recommendations

Tobias Ley[1,2], Barbara Kump[3], and Cornelia Gerdenitsch[2]

[1] Know-Center, Inffeldgasse 21a, 8010 Graz, Austria
tley@know-center.at
[2] Cognitive Science Section, University of Graz, Universitätsplatz 2, 8010 Graz, Austria
{tobias.ley,cornelia.gerdenitsch}@uni-graz.at
[3] Knowledge Management Institute, Graz University of Technology, Inffeldgasse 21a,
8010 Graz, Austria
bkump@tugraz.at

Abstract. Adaptive scaffolding has been proposed as an efficient means for supporting self-directed learning both in educational as well as in adaptive learning systems research. However, the effects of adaptation on self-directed learning and the differential contributions of different adaptation models have not been systematically examined. In this paper, we examine whether personalized scaffolding in the learning process improves learning. We conducted a controlled lab study in which 29 students had to solve several tasks and learn with the help of an adaptive learning system in a within-subjects control condition design. In the learning process, participants obtained recommendations for learning goals from the system in three conditions: fixed scaffolding where learning goals were generated from the domain model, personalized scaffolding where these recommendations were ranked according to the user model, and random suggestions of learning goals (control condition). Students in the two experimental conditions clearly outperformed students in the control condition and felt better supported by the system. Additionally, students who received personalized scaffolding selected fewer learning goals than participants from the other groups.

Keywords: Adaptive scaffolding, Personalization, Adaptive Learning Systems, Self directed learning, Layered Evaluation, APOSDLE.

1 Adaptive Scaffolding in Self-directed Learning

Self-directed learning (SDL) has gained importance both in higher education as well as in the workplace [1], where it is seen as an essential part of discretionary use of knowledge [2]. SDL is a self-initiated action that involves goal setting and regulating one's efforts to reach the goal, and can be seen as a continuous engagement in acquiring, applying and creating knowledge and skills in the context of an individual learner's unique problems [1].

Simons [3] differentiates three types of psychological learning functions that need to be carried out in SDL: *preparatory* (e.g. choosing learning goals and sub goals),

P. De Bra, A. Kobsa, and D. Chin (Eds.): UMAP 2010, LNCS 6075, pp. 75–86, 2010.
© Springer-Verlag Berlin Heidelberg 2010

executive (e.g. selecting information) and *closing* functions (e.g. thinking about future use and transfer conditions). Any of these can be carried out by a learner alone or with the help of others, like teachers, fellow students, supervisors or computers. In SDL, the starting point is the perception of a knowledge need of the learners arising in their actions. Based on this, learners determine the goals of learning, initiate purposive information seeking behaviour by identifying and choosing possible sources, and interact with the sources to obtain the desired information [4].

SDL has often been studied in open-ended learning contexts, like problem-based or experiential learning where the curriculum is not predefined, but driven by individual learner interest and goals. Typical systems that have been studied are hypertext or hypermedia systems. In these contexts, several researchers have alluded to the potential difficulties posed by SDL. Above all, the absence of appropriate instructional support has been found detrimental for learning [5]. SDL puts additional meta-cognitive demands on learners with which they have to cope in addition to learning about the topic [6, 7]. Learners have been found to get lost or distracted from their primary goal, and experience cognitive overload and disorientation [8].

In educational research, *scaffolding* has been suggested as a way to support learners in SDL. Scaffolding involves providing assistance to students on an as-needed basis, fading the assistance as their competence increases [9]. A computer-based learning environment can provide such scaffolds, thereby taking over some of the learning functions mentioned by Simons. In the context of hypermedia systems, a distinction can be made between conceptual, metacognitive, procedural and strategic scaffolds [10]. In our research, we are mainly considering conceptual scaffolds which provide guidance on what knowledge to consider during problem solving [11]. In [7], the term *semantic scaffolding* was introduced to refer to guidance that supports the creation of conceptual knowledge from unstructured texts. Semantic scaffolding (e.g. presenting advance organizers or learning objectives before learners engage in a text) have been found to positively enhance learning, e.g. through activating prior knowledge, by directing cognitive activities or forming semantic macrostructures.

Besides this fixed scaffolding, Azevedo and colleagues [12] experimentally tested the effect of *adaptive scaffolds* in a self-directed learning task in a hypermedia environment. Besides a traditional fixed scaffolding condition where a fixed list of learning goals for the task (provided by a domain expert) was given to the learners, they introduced an adaptive scaffolding condition where a human tutor provided learners with advice on several self-regulatory strategies (like planning and monitoring learning progress). The tutor adapted these hints dynamically to the current state of the learner. Both conditions were compared to a control condition in which no scaffolding was provided. The findings suggested that students in the adaptive scaffolding condition learned better than students in the other two conditions, both in terms of gains in conceptual understanding and declarative knowledge. The authors attribute the results to more effective use of self regulatory behaviours in the adaptive condition which allowed students to learn more effectively.

Whereas prior educational research has mainly concentrated on researching fixed conceptual scaffolding or adaptive scaffolding provided by human tutors, research into adaptive hypermedia systems (AHS) and intelligent tutoring systems (ITS) have been addressing adaptive learner support for a long time. ITS usually structure the learning task either in terms of a predefined curriculum or by a fine grained analysis of solution behaviours to offer curriculum sequencing, problem solving support, or intelligent solution analysis [13]. AHS guide learners by employing adaptive presentation or adaptive navigation support, e.g. possibilities to hide, annotate, generate or sequence links depending on the user model [14].

In order to realize adaptation in a learning system, at least two different types of models are needed, namely a domain model and a user model. The domain model structures the learning domain, e.g. by specifying the concepts and their relationships. It is a potential scaffold as it provides an expert view on the domain and can potentially guide novices in the process of goal setting and information acquisition. This is comparable to the fixed scaffolding condition used by Azevedo et al. [12]. In addition, the user model represents the knowledge and other characteristics of a user within a learning system. It provides the potential to adapt scaffolds to the current knowledge state of the user. This can be termed personalized scaffold as it takes into account the current knowledge state of the user.

Taken together, educational research has employed only fixed scaffolding or adaptive scaffolding provided by human tutors. Azevedo et al. [12] conclude from a review that more research on adaptive scaffolds is needed to establish which ones are effective and why. In adaptive learning systems research (AHS and ITS), the effect of automatic adaptations on SDL effectiveness has not been studied in a systematic way. In particular, the contribution of each of the models, domain model and user model, to successful scaffolding is not clear. The aim of our work is to gain a better understanding of the effectiveness of personalized scaffolds on self-directed learning when scaffolds are dynamically generated in an adaptive learning environment. Rather than being generated by a human tutor, we are looking at scaffolds generated by models of an adaptive learning environment.

In the next section, we will present APOSDLE, a self-directed learning system that can serve as an example of scaffolding self-directed learning with adaptive technologies. In the study presented here, we have used the learning goal recommendation mechanisms.

2 APOSDLE: Scaffolding in Work-Integrated Learning

APOSDLE is an adaptive system that follows the work-integrated learning approach [15]. It has been developed over four years by a European consortium in an attempt to support self-directed learning at knowledge intensive workplaces. Because APOSDLE should support learning in work domains not structured by a curriculum or by algorithmic tasks, we are pursuing an open ended learning approach. This means that learners receive learning hints in their normal work context allowing learning as part of the usual working activities. In contrast to ITS, these hints are not prefabricated learning materials to teach a certain skill, but are automatically derived from the available and extending knowledge-base in a particular organisation. It is for this reason that diagnosis and adaptive scaffolding are particularly challenging.

For the present study, the second APOSDLE prototype was used which has been instantiated in five different domains (like aircraft simulation and innovation consulting). This prototype was realized as a set of widgets that run on the desktop and display learning hints according to the task the user is currently performing

(Fig. 1). The task is either automatically detected through analysis of keystrokes and opened desktop applications, or selected manually from a list (I). For a task at hand, learning goals are then automatically suggested to a user in an adaptive manner (II), taking into account the user's knowledge state as stored in the user model (see below).

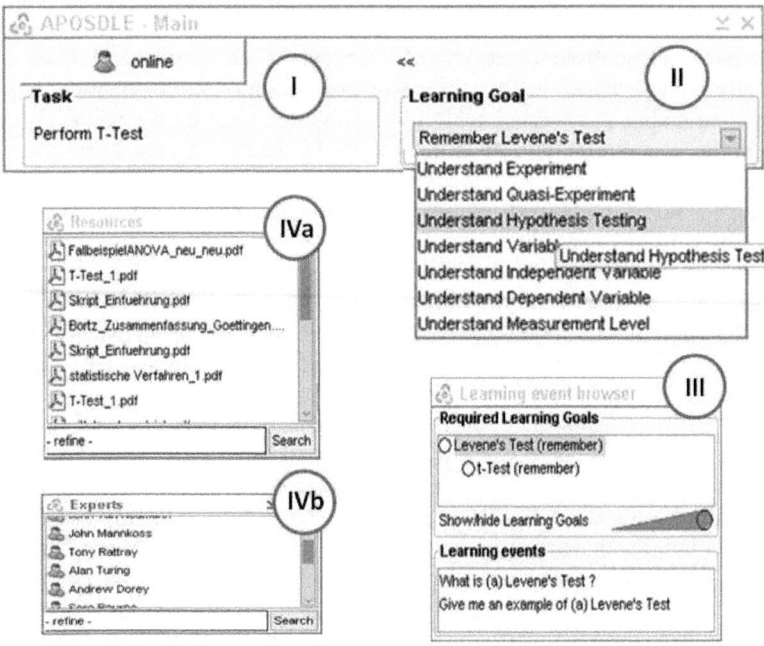

Fig. 1. Graphical user interface of some of the APOSDLE widgets used in the study

Figure 1 presents an example from the learning domain of statistical data analysis: APOSDLE has determined that the current user most likely needs to *Remember Levene's Test* for the current task (*Perform a t-test*). This learning goal is displayed together with the prerequisite learning goals and available learning events (III), as well as several resources for learning purposes (IVa) and experts that may be contacted (IVb). Learning events (III) are concise learning materials which address a specific learning goal. They are generated using a predefined pedagogical structure that is automatically filled with existing materials available in the organization. By choosing one of the learning events, APOSDLE presents the learning event and marks the relevant sections in the materials.

To realize the kind of adaptation described above, APOSDLE makes use of a domain model and a user model. A detailed description of the *APOSDLE domain model* is given in [16]. Most important for the present study are *tasks* that users work on during their daily work, like for example "Perform a statistical test". For each of those tasks, APOSDLE assumes a set of *skills* (e.g. "understand t-test") required to solve the task. This information is provided through the *task-skill assignment*. The skills are recommended to users when performing a task as *learning goals*. These allow the user to access learning material which is related to the respective learning goal. The APOSDLE domain model is structured according to the principles of Competence-based Knowledge Space Theory (CbKST) [17]. A prerequisite relation between the skills can be directly derived from task-skill assignment [18]. The sets of tasks and skills, the mapping between task and skills and the prerequisite relation on the set of skills together represent the APOSDLE domain model.

Also corresponding to the basic ideas of CbKST, *APOSDLE user models* are represented in terms of sets of skills that represent the current knowledge state of the user. In order to make inferences on a user's skills, APOSDLE observes the tasks a user has worked on in the past, their frequency and success, and adds all assigned skills for the particular tasks to the instance of the user model (*task-based skill assessment*, see [18]). To allow the user model to reflect also more open ended interactions, we have recently extended this conception by observing all user interactions (such as opening a document or starting a collaboration) and inferring from this information the level of competence in a certain skill [19].

Scaffolding in APOSDLE can be based on the domain model by simply presenting a user a list of all learning goals assigned to the current task he or she is performing (the *task demand*). Alternatively, a mechanism exists to adapt these recommendations to the current knowledge state by means of the user model. Therefore, the task demand is compared to the set of skills possessed by the user. If there is a discrepancy (*learning need*), APOSDLE recommends the missing skills as learning goals in a ranked list (the dropdown box II in Fig. 1). The algorithm for ranking the learning goals has the following characteristics: Learning goals that never have been applied, or that have been applied less frequently are ranked higher than learning goals that have been applied more frequently. Learning goals that are "more important" in the learning domain, i.e. learning goals that are assigned to more tasks than others, are ranked higher. The latter implicitly takes into account the previously mentioned prerequisite relation that exists for learning goals.

3 An Experimental Study of Learning Goal Recommendation

Our research question was whether personalized scaffolding, i.e. recommendations for learning goals ranked according to the user model, would increase performance in a self-directed learning task, decrease the time spent on these tasks and increase perceived support. We compared this condition to a condition where only the domain model was used to generate learning goals (fixed scaffolding), as well as to a condition where learning goals were generated randomly (control condition).

To address these questions, we follow a layered evaluation approach suggested by several authors [20, 21]. This approach breaks up system complexity of adaptive systems into assessable, self-contained functional units. Typically, systems are broken up into (a) the *inference mechanisms* and (b) the *adaptation decision*. While endeavors related to (a) seek to answer the question if user characteristics are successfully detected by the adaptive system, evaluations of (b) ask if the adaptation decisions are valid and meaningful. In the present study, the layered approach is realized by a two phase procedure (see Sect. 3.3). First, the inference mechanism was controlled for by means of an estimate of each participant's prior knowledge using a paper and pencil pre-test. Second, the adaptation decision was then tested in an experimental phase where participants interacted with the APOSDLE system.

3.1 Building the Domain Model: Statistical Data Analysis Learning Domain

In order to customize APOSDLE for a learning domain, the domain model needs to be created. For the present study the domain model was modelled in terms of the tasks in the statistical data analysis domain (e.g. *Defining the Research Sample, Selecting a Statistical Measure or Test*), as well as the skills needed to perform these tasks (e.g *Understanding of the Cohen's Kappa, Understanding of Statistical Significance*). The domain model had been constructed, validated and refined according to a modelling methodology described in [22] based on the procedure suggested by [18]. It consisted of 22 tasks, 64 skills, and the mapping between them. The number of skills assigned to tasks ranged from 1 to 48 (mean = 14, SD = 12.17).

3.2 Experimental Design and Participants

The study has been conducted as a laboratory experiment using a balanced one-factorial, multivariate repeated measure design. Type of scaffolding was used as within-subjects factor, meaning that each of the conditions was given to every participant. There were three scaffolding conditions which differed in the list of learning goal recommendations given to the learners after they had selected a task:

- *Personalized scaffolding (experimental condition 1): Learning goals are ranked according to domain model and user model.* Participants obtained a list of learning goals in accordance with the domain model (learning goals assigned to the current task). These learning goals were ranked according to the user model by the algorithm described in Sect. 2.
- *Fixed scaffolding (experimental condition 1): Learning goals are chosen according to the domain model, but randomly ranked.* Participants received a list of learning goals in accordance with the domain model (learning goals assigned to the current task). These learning goals were randomly ranked in the list.
- *Control Condition: learning goals are randomly chosen.* Participants obtained a random sample from the set of all learning goals in the domain. To control for the length of the list, the number of learning goals in the list was equal to the number of learning goals assigned to the specific task in the domain model.

Dependent variables measured the *perceived support* (summing over five items each using a 4-point Likert scale), the required *time to solve the task* (measured in seconds from onset to termination of task), and overall *task performance* (measuring whether the exercise was solved correctly or not). To be able to collect qualitative data, participants were asked to think aloud during the whole experimental session.

We tested 29 subjects, all of which were students of psychology of various semesters at Karl-Franzens University of Graz. The three conditions were given to each participant in three trials. To avoid training effects, conditions were randomized across trials. A double-blind experimental design was used to control for experimenter effects. We hypothesized that there would be differences regarding perceived support, time to solve a task and task performance depending on whether the APOSDLE system recommends learning goals ranked according to the user model, only selected through the domain model, or presented at random.

3.3 Experimental Procedure and Materials

The present study consists of two phases. In the pre-test phase, the aim was to capture participants' knowledge about the learning domain (the knowledge state of the users). In the experimental phase, participants interacted with the APOSDLE system and tried to solve exercises they had not been able to solve in the previous session.

Pre-test Phase. In this phase, a combination of single and group sessions were held. A paper-based task test was used to identify the actual knowledge state of the participants and thereby obtain reliable estimates of knowledge state of each user to be represented in the user model. For each of the 22 defined tasks in the domain model, an exercise had been formulated. Of these, sixteen were multiple choice items and six were in a free answer format. They were designed so as to be achievable in about one minute. An example of an exercise for the task Designing a Study is given in Fig. 2. The order of exercises in the pre-test was randomized across participants.

Designing a Study
Imagine you want to design a study, which should investigate if young people with bulimia nervosa differ from young people with anorexia nervosa in personality. How would you design this study? Independent Variable:_____   Dependent Variable:_____ What kind of sample would you choose to test the research question? ☐ Independent Samples   ☐ Dependent Samples Can this study be classified as an experiment or a quasi-experiment? ☐ Experiment   ☐ Quasi-experiment   ☐ I do not know

Fig. 2. Exercise for the task *Designing a Study* (translated from the original German version)

All exercises in the *pre-test* were coded as either correctly or incorrectly solved for each participant. For the six questions with open answer format, two raters independently rated each response as either correct or incorrect with an interrater agreement (Cohen's Kappa) exceeding 0.80. Participants were able to solve $M = 9.62$ $(SD = 3.07)$ exercises on average. The large variance indicates a broad coverage of the domain by the participants. The average item difficulty (relative frequency of correct solutions) was $p = .44$ $(SD = .28)$. For each participant, only those exercises were then chosen for the experimental phase which they had not been able to solve in the pre-test. Out of this set, three exercises were chosen according to the following criteria: the outer fringe concept (i.e. tasks from a knowledge state immediately following the subjects given knowledge state [23]), exercises with a minimum learning transfer (i.e. with a minimal overlap in learning goals assigned to them), and exercises with similar item difficulties.

Experimental Phase. In the experimental phase, participants interacted with APOSDLE and tried to solve the three exercises that had been selected for them. In the session, the experimenter logged on the participant (initializing his or her user profile) and selected the task in APOSDLE. Then, participants could freely browse the list of learning goals (see Fig. 1) and access the learning events and contents that were displayed. The list of the learning goals was varied by the three experimental conditions. Participants were requested to think-aloud while working on the tasks. Neither the participants nor the experimenter were aware of the experimental condition. To measure perceived support, participants had to provide ratings to five 4-point scale items after each exercise (Cronbach's α = .892).

4   Results

Looking at *task performance* in the three groups, there is a clear effect of learning goal recommendations. Figure 3 presents the frequency of task that were solved and tasks that were not solved. In the control group, only 6 problems were solved correctly, while in each of the two experimental groups (*personalized scaffolding and fixed scaffolding*) participants solved 18 tasks correctly.

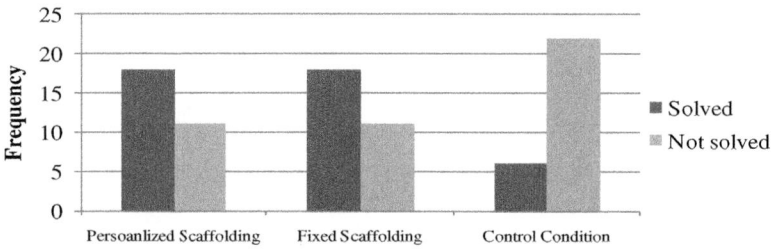

Fig. 3. Frequencies of solved task for the three experimental conditions (maximum = 29)

For testing the differences between the groups, a multidimensional Chi-square test was computed. The significance test (corrected for alpha error accumulation with the Bonferroni method) confirmed the statistically significant difference between the three experimental conditions ($\chi^2_{(2, n=87)}$ = 13.257, p = .001). The standardised residuals show that this difference occurs in the random condition where more learners could not solve the exercise than in the two experimental conditions.

As for the *solution time,* participants in the *personalized scaffolding* condition took on average 348.2 seconds (SD = 143.8s) to solve the exercises. In the *fixed scaffolding* condition, a solution time of 355.4 seconds (SD = 158.4s) was obtained. In the control condition, the time was 301.3 seconds (SD = 192.4s). We computed a one-factorial, univariate repeated measure ANOVA which showed no statistically significant difference between the conditions ($F_{(2,48)}$ = .843, p = .437).

A measure of *perceived support* was obtained by summing over five items in the questionnaire which had been presented after each exercise. Hence, the value ranged from zero (no perceived support) to 15 (maximum support). Average values in the three conditions are given in Table 1. Participants felt less supported in the control condition than in the other two which was also confirmed by a univariate repeated measure ANOVA ($F_{(2,54)} = 35.959$, $p = .000$).

Table 1. Perceived support in the three conditions

	Mean	SD
Personalized Scaffolding	8.89	3.89
Fixed Scaffolding	8.89	2.83
Control Condition	3.04	3.36

We then used the log data and think aloud protocols to analyze in more depth the process of *selecting between the recommendations*. We counted the number of learning goals selected from the recommended list (Table 2). On average, participants chose more learning goals to solve an exercise in the control condition than in the experimental conditions. Furthermore, people tended to choose more learning goals in the *personalized scaffolding* condition (M=1.87) than in the *fixed scaffolding* condition (M=1.53).

Table 2. Number of selected learning goals in the three conditions

	Mean	SD	Min.	Max.
Personalized Scaffolding	1.53	0.73	1	3
Fixed Scaffolding	1.87	1.11	1	5
Control Condition	2.21	1.17	1	4

5 Discussion

In answering our research questions about the effects of personalized scaffolding, we found some evidence for the fact that effective learning goal recommendations can scaffold self-directed learning and increase task performance. Recommending learning goals, either adapted to the users' task through the domain model (*fixed scaffolding*), or in addition adapted to the users' skills by taking into account the user model (*personalized scaffolding*), resulted in statistically significant effects in relation to the control condition for two of the dependent variables (task performance and perceived support). For both dependent variables, not only were the results of statistic significance, but also highly significant from a practical perspective. As to the performance, participants in the experimental conditions solved three times as many exercises correctly as in the control condition. In terms of perceived support, the difference between the means on the measurement scale was larger than 5.8 points which equates to a mean difference of over one point on a four point Likert scale.

An important question is why the two experimental conditions did not differ with regard to these dependent variables. Clearly, we were not able to produce as strong effects with the recommendations based on the user model as human tutors were able to (e.g. in [12]). Of course, a human tutor might be more sensitive to the exact feedback to give, tutors also serve additional functions (like motivational support), and in the Azevedo case, the tutor was also asked to only provide procedural and strategic scaffolding. When looking at number of learning goals selected in our study, however, there was a small but significant effect in that personalized scaffolding led to a smaller number of selections than the other two conditions. We take this as promising evidence that personalization was successful (as it reduced search), but that it did not produce effects on task performance or perceived support.

Stronger effects for personalized scaffolding might be expected in conditions where higher degrees of cognitive load are present [7]. For example, all of the following conditions could contribute to the fact that personalizing scaffolds might be beneficial over relying on recommendations from the domain model alone: in the case where lists of recommendations are longer, where time pressure is a stronger factor (which it might be outside the laboratory), or where tasks are more difficult or learners less experienced in the domain. In fact, we may have observed larger effects if we had not chosen tasks from within the learners' outer fringe, which has also been related to the Zone of Proximal Development [23].

Additionally, validity of the domain model may have impaired the effects. While tests concerning validity of the domain model have been performed in this as well as in other APOSDLE domains, and all modelling was done according to a specified modelling methodology [18], we did observe some violations of the prerequisite relation in the performance data. This shows once again that validity of the domain model cannot be taken for granted, and needs constant checking and revision [24].

Contrary to our expectations, conditions did not differ to a significant degree with regard to the solution time. We attribute this to the fact that this variable as we measured it had too many intervening factors not attributable to actual solution time of our participants. Things like response time of the software (which sometimes took a long time to return large documents) may have had an effect, or the fact that participants took different amounts of time until they decided to terminate an exercise.

A methodological issue may have been the setup of the control condition which is a major challenge in evaluating adaptive systems. Learners in our control condition received the same amount of hints, but not contingent on neither their actual task nor their state of knowledge. An alternative would have been to present hints according to their knowledge state, but not contingent on their task. Yet a further alternative would have been a yoked control condition (e.g. [25]) in which participants are paired with learners in the experimental condition to receive the same recommendations.

Finally, while a detailed analysis of self regulation in self-directed learning was out of the scope of this study (see e.g. [12]), we could observe some informative differences in the learning strategies employed by the participants in our study by analysing the think aloud protocols. It seems that depending on the exact task, learners were either more inclined to locate factual knowledge (in the case of trying to perform a statistical analysis), or gaining a deeper understanding (in the case of research methods). This shows the potential that lies in differentiating between different types of learning goals (e.g. [16]) when supporting self-directed learning.

6   Conclusions and Future Work

With this research, we have found evidence for the fact that automatically generated scaffolds can enhance performance and reduce search in a self-directed learning environment. With our study, we have differentiated the contributions of different models in our adaptive learning environment, namely the domain and the user model.

We are currently applying some of the techniques mentioned here in a large scale field evaluation with the APOSDLE system. Additionally, we are planning to research further conditions that may impact effectiveness of scaffolding in adaptive workplace learning, such as time pressure or cognitive load.

Acknowledgements. APOSDLE (http://www.aposdle.org) has been partially funded under grant 027023 in the IST work programme of the European Community.The Know-Center is funded within the Austrian COMET Program - Competence Centers for Excellent Technologies - under the auspices of the Austrian Federal Ministry of Transport, Innovation and Technology, the Austrian Federal Ministry of Economy, Family and Youth and by the State of Styria. COMET is managed by the Austrian Research Promotion Agency FFG.

References

1. Fischer, G., Scharff, E.: Learning technologies in support of self-directed learning. Journal of Interactive Media in Education 98(4), 1–32 (1998)
2. Lindstaedt, S., de Hoog, R., Ähnelt, M.: Supporting the Learning Dimension of Knowledge Work. In: Cress, U., Dimitrova, V., Specht, M. (eds.) EC-TEL 2009. LNCS, vol. 5794, pp. 639–644. Springer, Heidelberg (2009)
3. Simons, P.R.: Towards a constructivistic theory of self-directed learning. In: Straka, G.A. (ed.) Conceptions of self-directed learning, Waxmann, pp. 155–169 (2000)
4. Choo, C.W.: The knowing organization. How organizations use information to construct meaning, create knowledge, and make decision. Oxford University Press, New York (1998)
5. Mayer, R.E.: Should there be a three-strikes rule against pure discovery learning? The case for guided methods of instruction. American Psychologist 59(1), 14–19 (2004)
6. Narciss, S., Proske, A., Koerndle, H.: Promoting self-regulated learning in web-based environments. Computers in Human Behavior 23(3), 1126–1144 (2007)
7. Schnotz, W., Heiß, A.: Semantic scaffolds in hypermedia learning environments. Computers in Human Behavior 25(2), 371–380 (2009)
8. Müller-Kalthoff, T., Möller, J.: The Effects of Graphical Overviews, Prior Knowledge, and Self-Concept on Hypertext Disorientation and Learning Achievement. J. of Educational Multimedia and Hypermedia 12(2), 117–134 (2003)
9. Hogan, K., Pressley, M.: Scaffolding student learning: Instructional approaches and issues. Brookline Books, Cambridge (1997)
10. Hannafin, M., Land, S., Oliver, K.: Open learning environments: Foundations, methods, and models. In: Reigeluth, C.M. (ed.) Instructional design theories and models, pp. 115–140. Erlbaum, Mahwah/N.J (1999)

11. Vye, N., Schwartz, D., Bransford, J., Barron, B., Zech, L.: SMART environments that support monitoring, reflection, and revision. In: Hacker, D., Dunlosky, J., Graesser, A. (eds.) Metacognition in educational theory and practice, pp. 305–346. Erlbaum, Mahwah/N.J (1998)
12. Azevedo, R., Cromley, J., Seibert, D.: Does adaptive scaffolding facilitate students' ability to regulate their learning with hypermedia? Contemp. Educ. Psych. 29, 344–370 (2004)
13. Brusilovsky, P., Peylo, C.: Adaptive and Intelligent Web-based Educational Systems. Int. J. of Artificial Intelligence in Education 13, 159–172 (2003)
14. Brusilovsky, P.: Adaptive Hypermedia. User Modeling and User-Adapted Interaction 11(1-2), 87–110 (2001)
15. Lindstaedt, S.N., Ley, T., Scheir, P., Ulbrich, A.: Applying Scruffy Methods to Enable Work-integrated Learning. Europ. J. of the Informatics Professional 9(3), 44–50 (2008)
16. Ley, T., Kump, B., Ulbrich, A., Scheir, P., Lindstaedt, S.N.: A Competence-based Approach for Formalizing Learning Goals in Work-integrated Learning. In: EdMedia 2008, pp. 2099–2108. AACE, Chesapeake/VA (2008)
17. Korossy, K.: Extending the theory of knowledge spaces: A competence-performance approach. Zeitschrift für Psychologie 205, 53–82 (1997)
18. Ley, T., Kump, B., Albert, D.: A methodology for eliciting, modelling, and evaluating expert knowledge for an adaptive work-integrated learning system. Int. J. of Human-Computer Studies 68(4), 185–208 (2010)
19. Lindstaedt, S., Beham, G., Kump, B., Ley, T.: Getting to Know Your User – Unobtrusive User Model Maintenance within Work-Integrated Learning Environments. In: Cress, U., Dimitrova, V., Specht, M. (eds.) EC-TEL 2009. LNCS, vol. 5794, pp. 73–87. Springer, Heidelberg (2009)
20. Brusilovsky, P., Karagiannidis, C., Sampson, D.: The Benefits of Layered Evaluation of Adaptive Applications and Services. In: Weibelzahl, S., Chin, D., Weber, G. (eds.) Empirical evaluation of adaptive systems, Workshop at the UM 2001, pp. 1–8 (2001)
21. Weibelzahl, S., Lauer, C.U.: Framework for the evaluation of adaptive CBR-systems. In: Vollrath, I., Schmitt, S., Reimer, U. (eds.) Experience Management as Reuse of Knowledge, GWCBR 2001, Baden-Baden, Germany, pp. 254–263 (2001)
22. Ghidini, C., Rospocher, M., Serafini, L., Faatz, A., Kump, B., Ley, T., Pammer, V., Lindstaedt, S.: Collaborative enterprise integrated modelling. In: EKAW 2008, pp. 40–42, INRIA, Grenoble (2008)
23. Falmagne, J., Cosyn, E., Doble, C., Thiery, N., Uzun, H.: Assessing mathematical knowledge in a learning space: Validity and/or reliability. Paper Presented at the Annual Meeting of the Am. Educational Research Association (2007)
24. Kump, B.: A Validation Framework for Formal Models in Adaptive Work-Integrated Learning. In: Nejdl, W., Kay, J., Pu, P., Herder, E. (eds.) AH 2008. LNCS, vol. 5149, pp. 416–420. Springer, Heidelberg (2008)
25. Kalyuga, S., Sweller, J.: Rapid dynamic assessment of expertise to improve the efficiency of adaptive e-learning. Educ. Technol. Research & Development 53(3), 83–93 (2005)

Instructional Video Content Employing User Behavior Analysis:
Time Dependent Annotation with Levels of Detail

Junzo Kamahara[1], Takashi Nagamatsu[1], Masashi Tada[1],
Yohei Kaieda[1], and Yutaka Ishii[2]

[1] Graduate School of Maritime Sciences, Kobe University,
5-1-1 Fukae-minami, Higashi-Nada, Kobe 658-0022, Japan
kamahara@maritime.kobe-u.ac.jp, nagamatu@kobe-u.ac.jp
[2] Information Science and Technology Center, Kobe University,
1-1 Rokkoudai, Nada, Kobe 657-8501, Japan
ishii@kobe-u.ac.jp

Abstract. We develop a multimedia instruction system for the inheritance of skills. This system identifies the difficult segments of video by analyzing user behavior. Difficulties may be inferred by the learner's requiring more time to fully process a portion of video; they may replay or pause the video during the course of a segment, or play it at a slow speed. These difficult video segments are subsequently assumed to require the addition of expert, instructor annotations, in order to enable learning. We propose a time-dependent annotation mechanism, employing a level of detail (LoD) approach. This annotation is superimposed upon the video, based on the user's selected speed of playback. The LoD, which reflects the difficulty of the training material, is used to adapt whether to display the annotation to the user. We present the results of an experiment that describes the relationship between the difficulty of material and the LoDs.

Keywords: User Behavior, Level of Detail, Timed Annotation.

1 Introduction

To enable effective self-learning, the subject should be provided with video-based, annotated multimedia content, where the annotations provide a description of the instructions to perform some task. Acquiring a professional skill typically requires that the learner practice the same task repeatedly, under the guidance of an expert. Video-based multimedia content is capable of supporting this repeated practice. Based on this, we have developed a skill acquisition support system [1]; a multimedia system that displays instructions as text superimposed upon a video recording of an expert performing some task. These written instructions, intended to aid self-learning, are annotations generated based upon an interview with an expert on the subject. We employ a skill inheritance model comprised of three steps: capturing a recording of a skill's execution, adding annotations obtained by interviewing an expert and the pursuit of self-learning using the generated instructional video content. In keeping with this

P. De Bra, A. Kobsa, and D. Chin (Eds.): UMAP 2010, LNCS 6075, pp. 87–98, 2010.

three stage model, our system consists of three subsystems that perform the following operations: recording the expert during performance of the task, authoring annotations based upon the expert's advice, and replaying the instructional content using a player. We describe our model and system, in detail, in the following section.

We propose an identification method [2], which is capable of identifying the difficult portions of a task depicted in an instructional video by analyzing the operating logs of the video player being used by the learner. The portions of the task that are identified as difficult are those where the expert's movements are perceived by the learner as being difficult to emulate. According to [3], the user's perceptions can be inferred based upon his or her browsing behavior.

In a previous study [2], we analyzed the operating logs from a video player, assessing user browsing behavior to identify those video segments that were replayed more than once or where the pause button was pressed at least once, inferring the difficult portions of the task. The results of the experiment, conducted with 10 subjects, were good; we successfully identified 93.1% of the total number of video segments that the learners verbally indicated they had found difficult.

Extending upon this method, we consider that a learner can also modify the speed of playback of an instructional video. The segments of video that a learner replays at a slow speed can reasonably be considered to be difficult from the user's perspective, as well. Thus, we propose a behavioral analysis incorporating the cumulative playback time (CPT), which takes into account variation in playback speed, as well as the user's pausing of the video. CPT is a simple indicator of the duration of time over which a learner watches a video in each media period. Furthermore, we propose a normalized CPT, which, when considered, allows for a comparison of the results of different learners through normalization of the time spent by the learners observing each media period.

The CPT and normalized CPT are generic calculations representing the time consumed by the user in watching a video. CPT can be used to represent complex operational conditions, including play, stop, pause, etc. Later, we carry out an analysis of CPT values for various portions of a video, using experimental data. Each period is subsequently labeled as difficult, or not. The average result of the CPT analysis from our experiment, involving 9 subjects, suggests that it is capable of identifying almost all aspects of difficulty from an instruction video. The F-measure of our result was 0.8 when the threshold was set to the average value of CPT; we were able to successfully classify the majority of difficult periods within the video. As such, we suggest that we can use the value of normalized CPT to accurately determine the level of difficulty for each video segment.

In this paper, we propose the concept of time dependent annotation. Our proposed annotations employ the concept of level of detail (LoD). This concept, which pertains to identification of the complexity of objects, originates from the field of 3D computer graphic animation [4]. The complexity of an object depicted in a given 3D space is adjusted based on the virtual distance between the object and the observer.

We incorporate the concept of LoD into time dependent media. A time dependent annotation will be superimposed over specified video segments, as required; an annotation with a high level of detail will not be employed when the video is viewed by the user at normal speed. A higher LoD annotation is displayed in the video segment when it is determined to be more difficult, because it is assumed that the learner requires more detailed annotations to facilitate their comprehension of this material.

In other words, our use of LoDs in the selection and display of annotations allows our system to adapt to the learner's learning ability. The system achieves this by operating under the assumption that a learner who has an easy time understanding a given video segment will watch the video at normal speeds.

To support the legitimacy of our assumption, we compare the result of questionnaires completed by experiment participants, which indicate their perception of the difficulty associated with different segments of video instruction, as well as the calculation of normalized CPT values.

In the next section, we describe our multimedia instruction system, which presents multimedia instructional content for imparting knowledge and skills pertaining to the performance of some task. In Sect. 3, we explain the concepts of CPT and LoD in detail. In Sect. 4, we present and discuss a comparison between the questionnaire responses provided by subjects regarding video segment difficulty and the calculated CPT values. Finally, in Sect. 5, we present our conclusions.

1.1 Related Work

Analyzing user behavior is not a new concept. For video summarization, there are a number of prior studies [5, 6] that seek to identify different segments of video based upon logs of the user's browsing. These studies aim to identify *important* or *interesting* segments. Yu et al. [5] treat segments as shots, dividing them in advance and analyzing the linkages between these shots without considering the user's browsing time. He *et al.* [6] propose a method of summarizing informational multimedia content by referring to slide pitch and pause behavior. The ultimate goal of video summarization is to determine whether segments can be skipped or not.

In the aspect of video surrogates, there are some researches [7, 8] which concentrate on the fast forward surrogates. These studies depict the users' response with their claimed fast forward interface. However, they do not mind the slow forward for watching a video in detail.

We were unable to locate a prior study that incorporates LoDs in the presentation of annotations during playback. With regard to video editing, Casares et al. [9] introduce the concept of hierarchical levels of detail with respect to metadata; they do not employ the concept in regard to playback content directly. Deherty et al. [10] proposed detail-on-demand on which the user can watch a multi-level video summary that includes summaries of different levels of detail. However, their interface requires clicking to follow the hyperlink for watching more detailed video.

2 Multimedia Instruction System

Multimedia instruction system [1] is a multimedia authoring and presentation system. The system consists of three subsystems, including the recording of video of an expert performing the task at hand, the authoring of annotations based upon expert advice, and the playing of the instructional content using a multimedia player. We present an overview of our system in Fig. 1.

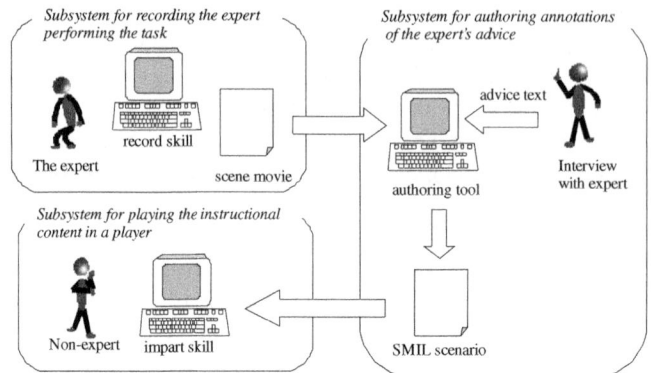

Fig. 1. Overview of proposed Multimedia Instruction System

A. Subsystem for recording of the expert performing the task

The recording subsystem ("record skill") captures a video ("scene video") of the expert performing the target task. The scenes of the video are shown from the perspective of the expert by using a head-mounted camera. This is done because the expert's view provides the viewer with a close-up of the task objective; by providing them with this view of the actions being performed, the learner will be able to imitate the expert more easily.

B. Subsystem for authoring annotations of the expert's advice

Before authoring the instructional content, we must first collect a volume of component material that supports multimedia self-learning; the process upon which our research focuses. This component material consist of video segments, taken scene-by-scene from the video, as mentioned above, the instructional text based upon expert interviews, and the anchor points that tie the instructional text to a given scene of video. This material constitutes a form of multimedia metadata that must be created manually.

The authoring subsystem ("authoring tool") is used to create content from these materials. The output of the authoring subsystem is the multimedia presentation written in SMIL [11]. We present a screenshot of the tool in Fig. 2.

The timeline of the video segments requiring instructional text is displayed in the bottom left of Fig. 2. The video segment that is currently being displayed is the represented by the colored rectangle on the video segment timeline. The beginning and ending times of the segment are selected using the "|<" or ">|" buttons, located above the video segment timeline. It should be noted that, in terms of authoring, this specification is tedious work.

Identifying the video segment through analysis of user behavior produces these segment timelines automatically. Once this has been done, the author of the content can add the instructional text and anchor points manually.

C. Subsystem for playing the instructional content in a multimedia player

The player subsystem ("impart skill") is an SMIL multimedia player. The implemented player is a streamlined version of the SMIL player, which can present video, images,

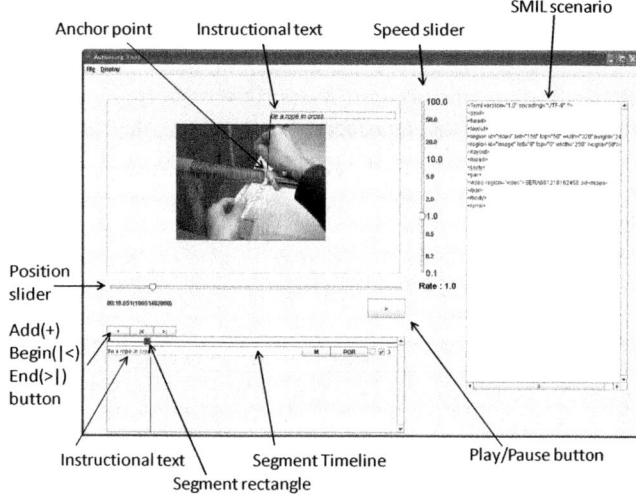

Fig. 2. Screenshot of authoring tool

and superimposed text. This player is used to collect user behavior and to facilitate the learning of instructional content. Fig. 3 depicts the instructional video player interface employed in our study.

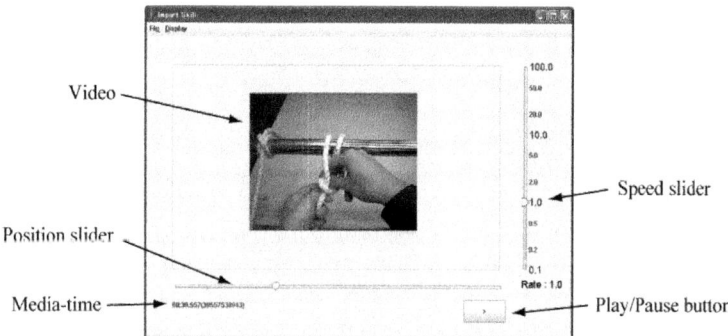

Fig. 3. Instructional video player

The instructional video is displayed at the center of the window. The user is able to start and stop the video using the play/pause button, as well as fast-forward or rewind the video using the position slider. Further, the user is able to change the playback speed using the speed slider, which can be manipulated using the mouse cursor, or by mouse scroll (trackball or wheel).

The state of the play/pause button and the movement of the sliders reflect the user's perception of the material's difficulty, as they watch the video. We assume that, if the user finds a given segment of video difficult to understand, he or she will utilize one of the above player functions to view that portion in greater detail. Therefore, analyzing and collecting user behavior data in this manner helps us to determine the portions of the task that are found difficult by the user. In order to achieve this, we have incorporated functions into the player to support the recording of operating logs, which include events handled by the interface, the speed of playback relative to normal speeds, the time position of the video ("media time"), and the actual time elapsed from the beginning of the video ("actual time"). We provide an example of operating log data in Table 1.

Table 1. An example operating log segment

No.	Event	Speed	Media Time	Actual Time
1	Start	1.0	0	154360000000
2	Stop	1.0	726500000	161656000000
3	Rate change	0.7943282	726500000	163563000000
4	Rate change	0.19952624	726500000	163875000000
.	...	...	...	...
9	Slider move	0.19952624	12092719133	183860000000
.	...	...	...	...

3 CPT and LoD

In this section, we introduce the concepts of Cumulative Playback Time, Normalized CPT and Level of Detail.

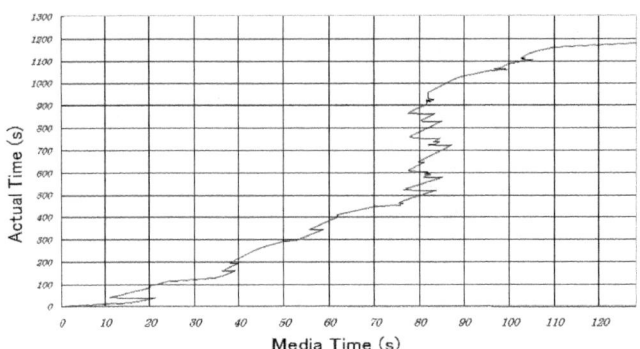

Fig. 4. Example playback curve observed for one subject

Before explaining CPT, we first present a playback curve observed for one viewer, in Fig. 4. The horizontal and vertical axes indicate the media time and the actual time, respectively. The positive and negative slopes of the curve represent the viewer's forward and backward progress through the video, respectively.

3.1 Cumulative Playback Time

Earlier, we made reference to Cumulative Playback Time (CPT), a new metric we propose that represents the volume of time consumed by a viewer in watching the video over a given media playback period. Each media playback period has a common interval (e.g., 0.1 s).

We formally define the CPT as follows:

$$M = \{m_1, m_2, \ldots, m_n\}$$
$$A = \{a_1, a_2, \ldots, a_n\}$$
$$R = \{r_1, r_2, \ldots, r_n\}$$
$$Periods = \{p_1, p_2 \ldots, p_{M_{total}}\}$$

$$f(k) = \begin{cases} a_{k+1} - a_k & (if \quad m_k = m_{k+1}) \\ \dfrac{1}{r_k} |m_{k+1} - m_k| & (if \quad m_k \neq m_{k+1}) \end{cases}$$

$$CPT_{[p_t, p_{t+1})} = \sum f(k)_{[p_t, p_{t+1})}$$

(1)

M, A, and R refer to the media time, the actual time, and the speed of playback in each event, respectively. n represents the number of events. M_{total} represents the length of the video in seconds. The term "*Periods*" represents the set of periods, which comprise single second divisions of the media. The function $f(k)$ determines the elapsed time for event k. Finally, $CPT[p_t, p_{t+1})$ is defined as the total elapsed time of each event k in period p_t.

If one wished to employ an interval period of 0.1 s, you would employ $M_{total}/0.1$ instead of M_{total}. Note that M_{total} as also indicates the number of time intervals, computed by dividing the total time of the video by the interval size.

3.2 Normalized CPT

The total elapsed time for each subject will vary. As we cannot reasonably compare each CPT value for the different subjects directly, we propose a method of normalizing the value. The normalized CPT is calculated by dividing the base CPT value for a given period by the total actual viewing time elapsed. The formal calculation of the normalized CPT is as follows:

$$NormalizedCPT_p = \frac{CPT_p}{A_{total}} M_{total}$$

(2)

A_{total} is the total elapsed viewing time for the subject, and is calculated as $A_{total} = a_n - a_1$. CPT_p is the CPT value in period p.

When the value of the normalized CPT is equal to 1, this indicates that the value is equal to the average elapsed time of each period, as the period interval is 1s. Of course, when the period interval is 0.1s, the average CPT becomes approximately 0.1.

We present only a brief introduction to the metrics of CPT and normalized CPT because these calculations are not the primary topic of this paper; these concepts will be discussed in greater detail in future work.

3.3 Level of Detail

The LoD of an annotation is a specified threshold that determines whether the annotation is presented when the video is played back at a given speed. The number of levels specified is dependent upon the content material.

An example image incorporating annotations with LoDs is depicted in Fig. 5.

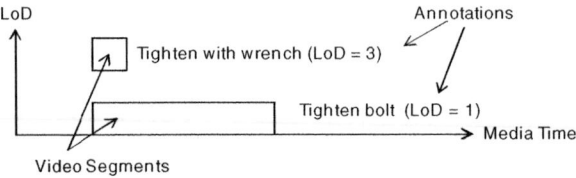

Fig. 5. Example image of annotations with LoD

Typically, annotations with higher levels of detail are shown for only short durations. This is because the learner will usually only require detailed elaboration for very difficult portions of the video, which tend to be infrequent and short. If all such short duration annotations are presented, they flicker upon the video and are not able to be read by the learner. If, however, these sections are viewed in slow motion, this is not an issue.

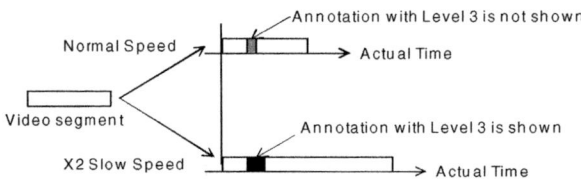

Fig. 6. Extension of duration for presenting annotation

When the learner watches the video segment at slow playback speeds, as they seek to understand the performance of a difficult task component, the duration of the video segment is extended in actual time. This results in the associated extension of the duration of time during which the annotation is presented to the user. We demonstrate this extension of duration in the presentation of annotations in Fig. 6.

The playback speed depends on the user's operation of the multimedia player during their watching of an instructional video. In other words, the user can shift the level of detail upon the video naturally when he or she changes the playback speed. The use of annotation with LoDs enables the system's adaptation based upon the user's ability to understand the presented task.

We provide an example determination of the LoD setting in table 2, below:

Table 2. An example determination of the LoD setting (Normal speed ratio=1.0)

	Playback Speed Ratio	Level of Detail	Level of Presented Annotations
Fast	>2.0	0	0
Normal	0.5–2.0	1	0, 1
Slow	<0.5	2	0, 1, 2

4 Comparison between User Perceptions and CPT Identification

In a prior study, we conducted experiments, the results of which revealed that the analysis of the operating logs is an effective method for identifying segments of video depicting complex tasks. In this experiment, we obtained two sets of data pertaining to video segments, which were recorded based upon user behavior and questionnaire responses provided by users about their perception of task complexity. The instructional task we selected for this experiment pertained to rope work. This task is relatively difficult to learn independently because it involves a series of subtasks.

We carried out CPT analysis of the experimental data obtained in our previous study. The setting of the experiment was as follows [2]: the expert selected for this task was an associate professor that taught rope work at our university. We recorded a video of the expert performing the rope work task. After recording the video, the expert was asked to provide verbal instructions for the task; these instructions were then used to author the annotation text. During the authoring process, the expert was allowed to watch his video recording and asked to indicate those instances of video where instructional text was likely required. We then extracted the relevant portions of video based upon the interview.

After completion of our experiment with the rope work expert, we conducted an experiment with ten learners. The objective of this experiment was to record the operating logs of the instructional video player as learners viewed the instructional video regarding rope work. During the course of this experiment, no text was displayed. The subjects selected were 10 male university students, who had no prior knowledge of or experience in rope work. The subjects were asked to complete the task that was explained on the video; this video demonstrated how to tie five different knots. After completing the task, the subjects were asked to answer a questionnaire in which they indicated those segments of the task that they felt were particularly difficult to perform.

Figure 7 depicts the CPT result for one subject. The horizontal and vertical axes indicate the media time and the CPT value, respectively. The duration of the media segments was set to 1 sec because a value of 0.1s was determined to be too brief for a subject to distinguish. In Fig. 7, the dotted line represents the average CPT for each period. The red short horizontal lines near the 20 sec CPT mark are portions of video deemed difficult by the subject on their questionnaire. The peak CPT values correspond approximately to the difficult portions of video indicated in the figure. The peak CPT values that were not indicated to be difficult portions by this subject were identified as such by other subjects.

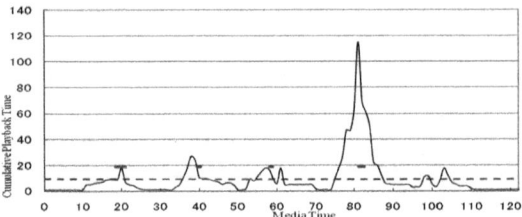

Fig. 7. CPT result obtained from one subject

We depict the average results of the 9 subjects in Fig. 8. One subject was not factored into the calculated CPT because the subject completed watching the video very early. The temporal resolution of Fig. 8 is 0.1s. The left vertical axis employs a log scale. The data pertaining to those video segments where the expert offered advice to the learners and to those segments where the learners indicated they had experienced difficulty is also included in Fig. 8. The upper blue horizontal line is the portion of video that was supported by expert advice, and the lower red horizontal line is union of the difficult portions that were indicated by the responses of the subjects' on their questionnaires. The dotted line depicts the number of subjects that claimed a given part was difficult on the questionnaires.

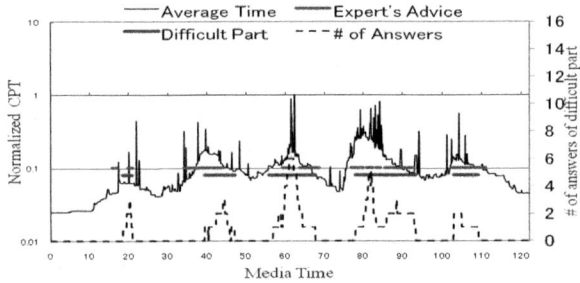

Fig. 8. Average CPT obtained from 9 subjects

In addition, we present a comparison between the number of answers indicating perceived difficulty, and the CPT values, in Fig. 9. The time resolution of Fig. 9 is 1s. The number of subject answers indicates the average for the 1s duration.

The correlation coefficient between the number of questionnaire answers and the CPT values is 0.658. There are five peak segments based on the questionnaire answers, which equate with five constituent subtasks of the overall rope work task. In particular, the 3rd and 4th peaks, reading from the left, would indicate that these two subtasks are more difficult than the other subtasks. The CPT value peaks appear to be almost the same. Based upon these peaks, we assign a higher LoD to the video segments that exceed the threshold of 2.

In table 3, we present the number of periods identified as difficult video segments. The interval of each period is 1s. The total number of periods for the video is 122. Table 3 also presents the results of the recall/precision ratio, when the answer average is assumed to be an exact representation of the true value.

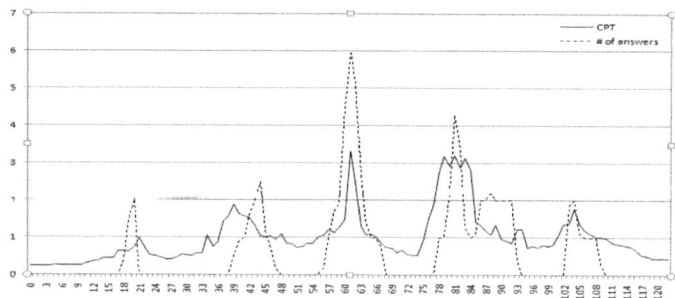

Fig. 9. Comparison between the number of answers and the CPT

Table 3. Total counted periods

Threshold		# of retrieved	# of matched	Recall/ Precision	
>2	CPT	9	5	Precision	55.6%
	Average of answers	9		Recall	55.6%
>1	CPT	48	23	Precision	47.9%
	Average of answers	28		Recall	82.1%

5 Conclusion

In this paper, we detail a multimedia instruction system, introduce the concept of CPT and normalized CPT and propose time dependent annotation using LoDs. The average CPT values obtained for the experiment's 9 subjects were in line with the majority of those aspects of the task that were found to be difficult by the participants. When we employed the normalized CPT equation, we were able to classify the difficult segments of the task video, using a line value of 0.1 as the CPT average.

Time dependent annotation using LoDs is capable of presenting an expert's advice at appropriate, difficult points in the instructional video, based upon the learner's behavior, which is reflective of their ability to comprehend the presented material. This annotation is superimposed on the video based on the user's selected playback speed. The LoD concept, which reflects the difficulty of the instruction at any given point in a video, is used to determine whether to show the annotation to the viewer.

Through a comparison between the user's verbal results and the calculated CPT values, we can classify the different subtasks based upon their level of difficulty. We demonstrate that the recall/precision ratio of the CPT has two thresholds. We suggest that this result demonstrates that the CPT value can be used to determine the LoD of a segment. We do not discuss about the audio annotation because the audio annotation is not robust to change the pitch of audio and not suitable for presenting details.

In future work, we plan to consider the method by which the LoD of the displayed annotation is determined, based upon the user's chosen video playback speed.

References

1. Nagamatsu, T., Kaieda, Y., Kamahara, J., Shimada, H.: Development of a Skill Acquisition Support System Using Expert's Eye Movement. In: Smith, M.J., Salvendy, G. (eds.) HCII 2007. LNCS, vol. 4558, pp. 430–439. Springer, Heidelberg (2007)
2. Kamahara, J., Nagamatsu, T., Fukuhara, Y., Kaieda, Y., Ishii, Y.: Method for Identifying Task Hardships by Analyzing Operational Logs of Instruction Videos. In: SAMT 2009. LNCS, vol. 5887, pp. 161–164. Springer, Heidelberg (2009)
3. Syeda-Mahmood, T., Ponceleon, D.: Learning Video Browsing Behavior and its Application in the Generation of Video Previews. In: Proc. Multimedia '01, vol. 9, pp. 119–128. ACM, New York (2001)
4. Clark, J.H.: Hierarchical Geometric Models for Visible Surface Algorithms. Communications of the ACM 19(10), 547–554 (1976)
5. Yu, B., Ma, W.-Y., Nahrstedt, K., Zhang, H.-J.: Video Summarization Based on User Log Enhanced Link Analysis. In: Proc. Multimedia '03, pp. 382–391. ACM, New York (2003)
6. He, L., Sanocki, E., Gupta, A., Grudin, J.: Auto-summarization of Audio-Video Presentations. In: Proc. of Multimedia '99. ACM, New York (1999)
7. Drucker, S.M., Glatzer, A., Mar, S.D., Wong, C.: SmartSkip: consumer level browsing and skipping of digital video content. In: Proc. of the SIGCHI conference on Human factors in computing systems, pp. 219–226. ACM, New York (2002)
8. Wildemuth, B.M., Marchionini, G., Yang, M., Geisler, G., Wilkens, T., Hughes, A., Gruss, R.: How fast is too fast? evaluating fast forward surrogates for digital video. In: Proc. of the 3rd ACM/IEEE-CS joint conference on Digital libraries, pp. 221–230. IEEE Computer Society, Los Alamitos (2003)
9. Casares, J., Long, C., Myers, B., Bhatnagar, R., Stevens, S., Dabbish, L., Yocum, D., Corbett, A.: Simplifying Video Editing using Metadata. In: Proc. of DIS'02, pp. 157–166. ACM, New York (2002)
10. Doherty, J., Girgensohn, A., Helfman, J., Shipman, F., Wilcox, L.: Detail-on-demand hypervideo. In: Proc. of the Multimedia '03, pp. 600–601. ACM, New York (2003)
11. SMIL–Synchronized Multimedia Integration Language,
    http://www.w3.org/AudioVideo/

A User-and Item-Aware Weighting Scheme for Combining Predictive User Models

Fabian Bohnert and Ingrid Zukerman

Faculty of Information Technology, Monash University
Clayton, VIC 3800, Australia
{fabian.bohnert,ingrid.zukerman}@infotech.monash.edu.au

Abstract. Hybridising user models can improve predictive accuracy. However, research on linearly combining predictive user models (e. g., used in recommender systems) has often made the implicit assumption that the individual models perform uniformly across the user and item space, using static model weights when computing a weighted average of the predictions of the individual models. This paper proposes a weighting scheme which combines user- and item-specific weight vectors to compute user- and item-aware model weights. The proposed hybridisation approach adaptively estimates online the model parameters that are specific to a target user as information about this user becomes available. Hence, it is particularly well-suited for domains where little or no information regarding the target user's preferences or interests is available at the time of offline model training. The proposed weighting scheme is evaluated by applying it to a real-world scenario from the museum domain. Our results show that in our domain, our hybridisation approach attains a higher predictive accuracy than the individual component models. Additionally, our approach outperforms a non-adaptive hybrid model that uses static model weights.

1 Introduction

Previous research has shown that user model hybridisation can improve the predictive accuracy of the individual models being hybridised, e. g., [1–5]. However, research on combining predictive user models has often made the implicit assumption that the performance of the individual models is uniform across the user and item space, using static model weights when computing a weighted average of the predictions of the models [2]. By contrast, this paper presents a weighting scheme for linearly combining the predictions of user models in a user- and item-aware fashion. Our approach is inspired by latent factor models for recommender systems, e. g., [6], which explain a user's *item ratings* as the inner product of user- and item-specific factor vectors inferred from rating patterns. We implement this approach by representing the *model weights* as the inner product of user- and item-specific weight vectors.

Our approach adaptively estimates *online* the model parameters that are specific to a target user as information about this user becomes available. Hence, our approach is particularly well-suited for domains where little or no information regarding the target user's preferences or interests is available at the time of *offline* model training. In this paper, we propose two online estimation approaches: (a) a cross validation approach

P. De Bra, A. Kobsa, and D. Chin (Eds.): UMAP 2010, LNCS 6075, pp. 99–110, 2010.

to learn a target user's weight vector from his/her available ratings; and (b) a nearest-neighbour collaborative approach, which estimates the weight vector from the weight vectors of other like-minded users.

We apply our hybridisation approach to a real-world scenario from the museum domain. This is done by combining two personalised collaborative Gaussian *Spatial Process Models* which predict a visitor's interest in museum exhibits from non-intrusive observations of the visitor's movements through a physical museum [7]. Our evaluation demonstrates that our hybridisation approach attains a higher predictive accuracy than the individual models being hybridised. Additionally, our approach outperforms a non-adaptive hybrid model that uses static model weights.

The contributions of this paper are threefold: (1) a generic user- and item-aware weighting scheme for linearly combining predictive user models; (2) two approaches for adaptive online estimation of the model parameters that are specific to the target user; and (3) experimental evidence from an evaluation with a real-world dataset from the museum domain, which shows that our approach to model hybridisation improves predictive accuracy.

This paper is structured as follows. Section 2 discusses related research. Our user- and item-aware weighting scheme for combining predictive user models is presented in Sect. 3, including a discussion of efficient algorithms for learning the weight vectors. We introduce our application scenario in Sect. 4, followed by our evaluation in Sect. 5, and conclusions in Sect. 6.

2   Related Research

Many researchers have investigated hybrid approaches for combining predictive user models [1, 2]. For example, Lekakos and Giaglis [3] utilise lifestyle data to address sparsity-related limitations of collaborative filtering algorithms by *switching* between models. Another prominent hybridisation technique is to linearly combine models in an *ensemble* fashion, where predictions are computed as a weighted average of the predictions generated by the individual models [8]. In contrast to this paper, such research has generally used static model weights, thus making the implicit assumption that the individual models perform uniformly across the user and item space. For example, Mobasher *et al.* [4] use a static weighting scheme to linearly combine similarities that are derived from semantic knowledge about items with item-to-item similarities that are computed collaboratively from the items' ratings. Claypool *et al.* [5] combine content-based and collaborative filters for recommending online newspaper articles using per-user per-item weights, but do not explain the specifics of their approach.

We apply the proposed weighting scheme to a real-world scenario from the museum domain, extending our previous research on non-intrusive statistical user modelling techniques for predicting a visitor's interest in exhibits [9]. Other research projects that investigate techniques for personalising the museum experience include *PEACH* [10] for content presentation, and *CHIP* [11] for exhibit recommendations based on explicit user input.

3   A User-and Item-Aware Weighting Scheme for Combining Predictive User Models

Previous research on linearly combining predictive user models has often used static model weights when computing a weighted average of the predictions generated by the individual models, thus assuming that the individual models perform uniformly across the user and item space [2]. However, generally, the performance of a model varies depending on the user and item for which predictions are generated, with different models being more (or less) well-suited for certain user/item combinations. Instead of the static non-adaptive model weights, one might consider employing model weights that are specific to a given user/item combination. This approach would result in a large number of model weights, e. g., $m \times n$ model weights when combining two predictive user models (where m is the number of users, and n is the number of items). More importantly, such model weights cannot be estimated using observed rating data, as the ratings pertaining to the target user/item combinations are not available when learning the weights.

To address this problem, we propose to map both users and items to a joint latent weight/factor space, where user-item interactions are modelled as inner products (similarly to latent factor models for recommender systems, e. g., [6]). Specifically, we propose to compute the model weights as inner products of user- and item-specific weight vectors (these vectors can be learnt from observed rating patterns, Sects. 3.2 and 3.3). This approach enables the hybrid model to give more (or less) weight to certain individual models depending on the current user and item (Sect. 3.1), thus making the weighting scheme user- and item-aware. In addition, our weighting scheme is *model-independent* in the sense that it is not restricted to particular kinds of predictive user models (except for the requirement that there must be a consistent interpretation of the predictions across the individual models).

3.1   Weighting Scheme Specification

When combining two predictive user models, we compute the scalar model weight w_{ui} as the inner product of a user specific weight vector w_u and an item-specific weight vector w_i, i. e., $w_{ui} = w_u \cdot w_i = \sum_{k=1}^{K} w_{u,k} \, w_{i,k} \in [0,1]$, where u denotes a user, i denotes an item, and K is the number of elements of each weight vector (the user- and item-specific weight vectors have the same number of elements, so that their inner product is defined). The resultant model weight w_{ui} is then used to calculate a hybrid weighted-average prediction $\hat{r}_{ui}$ of user u's rating r_{ui} of item i as follows:

$$\hat{r}_{ui} = w_{ui} \, \tilde{r}_{ui}^{(1)} + (1 - w_{ui}) \, \tilde{r}_{ui}^{(2)}, \tag{1}$$

where $\tilde{r}_{ui}^{(1)}$ and $\tilde{r}_{ui}^{(2)}$ are the rating predictions for user u and item i generated by user models M_1 and M_2 respectively. To ensure that $w_{ui} = w_u \cdot w_i \in [0,1]$, we restrict the elements $w_{u,k}$ and $w_{i,k}$ of w_u and w_i respectively to the interval $[0, 1/\sqrt{K}]$ for all $k = 1, \ldots, K$ (and for all users u and items i).[1]

---

[1] The optimal value of K can be determined by minimising an error measure of choice. Without loss of generality, we use $K = 1$ for our evaluation (Sect. 5). This simplifies the user- and item-specific weight vectors w_u and w_i to scalar weights.

By computing w_{ui} as the inner product of user- and item-specific weight vectors, the hybrid model generates weighted rating predictions in a user- and item-adapted fashion, exploiting potential user- and item-specific benefits of each individual model. The user and item weight vectors have a clear interpretation in the sense that larger weights characterise a preference for model M_1 (and weights close to 0 represent a preference for model M_2). Specifically, users and items can be characterised as "being more like" model M_1 or model M_2 for each of the elements of their weight vectors (which in turn represent the model preference for each dimension of the joint latent weight/factor space). Our weighting scheme can easily be extended to combine more than two predictive models (provided the sum of the model weights is 1).

3.2   Offline Weight Vector Learning

The above weighting scheme has $Km + Kn$ model parameters (this corresponds to the total number of elements of all weight vectors w_u and w_i), which can be learnt from observed rating patterns. We propose to determine the optimal weight vectors w_u and w_i by minimising the predictive error of the hybrid model using a cross validation approach, where rating predictions are computed using Equation 1. This learning procedure is independent of the objective function (i.e., the error measure). We minimise a variant of the *mean absolute error (MAE)*, called MAE_{avg} (Sect. 5.2), which is specific to our application scenario, using a standard derivative-free algorithm for constrained optimisation problems in combination with line search for the singleton dimensions (we use the Matlab function `fmincon`).

This procedure requires some ratings r_{ai} of a target user a to be included in the training data, so that an estimate of the target user's weight vector w_a is generated. However, often (e.g., in the museum domain), few or no ratings are available for the target user a when learning the weight vectors *offline*, thus making accurate offline estimation of w_a difficult. This problem is addressed in Sect. 3.3.

3.3   Online Estimation of the Current User's Weight Vector

We propose the following two *online* estimation approaches for adaptively estimating a current user a's weight vector w_a as ratings become available: (1) a cross validation approach to learn w_a from user a's available ratings, and (2) a nearest-neighbour collaborative approach, which estimates w_a from the weight vectors of other like-minded users. For both approaches, a prediction $\hat{r}_{ai}$ of a target user's item rating r_{ai} is calculated using Equation 1 once w_a is estimated.

A Cross Validation (CV) Approach. The *CV* approach determines a current user a's optimal weight vector $\hat{w}_a$ by minimising the MAE with respect to the current user a's available ratings (Sect. 5.1). The MAE in turn is approximated using leave-one-out cross validation. That is (for all possible weight vectors w_a), we calculate the MAE as follows.[2] First, we use Equation 1 to compute a prediction $\hat{r}_{ai}$ for each available

---

[2] The set of all possible weight vectors w_a is generated by varying each of the elements of the weight vector over the range $[0, 1/\sqrt{K}]$.

rating r_{ai} (using the given user weight vector $\boldsymbol{w}_a$, the previously learnt item weight vectors $\boldsymbol{w}_i$, and withholding rating r_{ai}). We then compute the MAE by comparing the resultant predictions $\hat{r}_{ai}$ with the observed ratings r_{ai}. The weight vector $\boldsymbol{w}_a$ that yields the lowest MAE, called $\hat{\boldsymbol{w}}_a$, is considered optimal. When the current user a has rated less than M items, we estimate $\boldsymbol{w}_a$ using a non-adapted default prediction $\boldsymbol{w}_d$ (in our experiments, we use $\boldsymbol{w}_d$ with elements $w_{d,k} = 1/\sqrt{K}$ for all $k = 1, \dots, K$, which gives maximum weight to model M_1).

A Nearest-Neighbour (NN) Collaborative Approach. The *NN* approach estimates a current user a's weight vector from the weight vectors of other similar users, making the assumption that their weight vectors accurately reflect the current user a's weight vector. Specifically, we compute $\tilde{\boldsymbol{w}}_a$, a personalised prediction of $\boldsymbol{w}_a$, as the similarity-weighted average of the other users' previously learnt weight vectors $\boldsymbol{w}_u$ (learnt as described in Sect. 3.2), i.e.,

$$\tilde{\boldsymbol{w}}_a = \frac{\sum_{u \in N(a)} sim(a, u)\, \boldsymbol{w}_u}{\sum_{u \in N(a)} sim(a, u)},$$

where $N(a)$ is the set of nearest neighbours, and $sim(a, u)$ is the similarity between users a and u. We calculate $sim(a, u)$ using Pearson's correlation coefficient on the rating vectors of users a and u. The current user a's set of nearest neighbours $N(a)$ is constructed by selecting up to K_{NN} users that are most similar to the current user a. These users are selected from those who have at least a certain minimum number C of co-rated items with user a, and whose similarity score $sim(a, u)$ is above a certain non-negative threshold S (these calculations are independent of the current item i).

Having calculated $\tilde{\boldsymbol{w}}_a$, we apply *shrinkage to the mean* to compute a shrunken personalised prediction $\hat{\boldsymbol{w}}_a$ of $\boldsymbol{w}_a$. This regularises our similarity-weighted personalised prediction $\tilde{\boldsymbol{w}}_a$ by linearly combining it with a default prediction, i.e.,

$$\hat{\boldsymbol{w}}_a = \boldsymbol{w}_d + \omega\left(\tilde{\boldsymbol{w}}_a - \boldsymbol{w}_d\right),$$

where $\boldsymbol{w}_d$ is a default prediction (as for the *CV* approach, we use $\boldsymbol{w}_d$ with elements $w_{d,k} = 1/\sqrt{K}$ for all $k = 1, \dots, K$), and $\omega \in [0, 1]$ is the shrinkage weight. When the set of nearest neighbours is empty (i.e., a similarity-weighted prediction $\tilde{\boldsymbol{w}}_a$ is not possible) or the current user a has rated less than M items, we estimate $\boldsymbol{w}_a$ using simply the default prediction $\boldsymbol{w}_d$.

In summary, the *NN* approach for estimating a current user's weight vector $\boldsymbol{w}_a$ has the following adjustable parameters: (1) the minimum number of rated items M (similarity-weighted prediction), (2) the minimum number of co-rated items C (nearest neighbour), (3) the minimum similarity S (nearest neighbour), (4) the maximum number of nearest neighbours K_{NN}, and (5) shrinkage weight ω.

4   Application Scenario

To evaluate our weighting scheme, we apply it to a real-world scenario from the museum domain. Specifically, we combine two variants of a Gaussian *Spatial Process*

Model (SPM) [7] which predicts a visitor's interest in museum exhibits, and evaluate the resultant hybrid model with a real-world dataset of museum visits. In the museum domain, no information regarding a visitor's interests in exhibits is generally available at the beginning of a visit. This fits well with the ability of our hybrid models to adaptively estimate a current visitor's weight vector as observations become available with the progression of a visit.

Our application scenario is motivated by the need to automatically recommend exhibits to museum visitors based on non-intrusive observations of their movements in the physical space. Employing recommender systems in this scenario is challenging, as predictions differ from recommendations (we do not want to recommend exhibits that visitors are going to see anyway). This challenge will be addressed by (1) predicting a visitor's interests in exhibits, e. g., using *SPM*, (2) calculating a prediction of a visitor's pathway through the museum [12], and (3) combining these models to recommend personally interesting exhibits that may be overlooked if the predicted pathway is followed. The focus of this paper is the interest prediction step.

4.1 Dataset

Our dataset of visitor pathways was obtained by manually tracking visitors at Melbourne Museum (Melbourne, Australia) from April to June 2008, using a custom-made tracking tool running on laptop computers [9]. In total, we recorded 158 pathways of first-time adult visitors travelling on their own, in the form of time-annotated sequences of visited exhibit areas.[3] Hence, although obtained manually, this dataset provides information of the type that may be automatically inferred from sensors. The resultant dataset (described in detail in [9]) contains 8327 viewing durations at the 126 exhibit areas of Melbourne Museum, yielding an average of 52.7 exhibit areas per visitor (41.8% of the exhibit areas). Hence, on average 58.2% of the exhibit areas were not viewed by a visitor, indicating a potential for pointing a visitor to relevant but unvisited exhibits.

In the museum domain, non-intrusive observations of a visitor's viewing behaviour (such as the viewing durations provided by our dataset of visitor pathways) can be used as an indirect measure of interest, as viewing time correlates positively with preference and interest. As previously argued [9], we use log viewing time instead of raw viewing time. This is because a log-transformation of the viewing times generates approximately normal exhibit-specific viewing time distributions, which have appealing analytical properties. Additionally, this transformation fits well with the idea that for high viewing times, an increase in viewing time indicates a smaller increase in the modelled interest than a similar increase in the context of low viewing times (we view a visitor's log viewing times at exhibits as implicit exhibit ratings).

4.2 Hybridised Individual Models

In our previous research, we used theory from spatial statistics to develop a Gaussian *Spatial Process Model (SPM)*, which we employed to predict a visitor's interests in

---

[3] Prior to collecting the data, we grouped the individual exhibits of Melbourne Museum into 126 semantically coherent and spatially confined *exhibit areas*.

exhibits (measured by means of log viewing times) [7]. The use of spatial processes requires a measure of distance between items in addition to users' ratings (to enable a functional specification of the correlation structure between the items). In this paper, we combine two variants of *SPM*: (1) *SPM-PD* [9], which utilises *Physical Walking Distance (PD)* to measure item distances (in a museum, walking distances between exhibits are meaningful, as the museum space is carefully themed by curatorial staff such that semantically related exhibits are in physical proximity); and (2) *SPM-SD*, which uses *Semantic Distance (SD)* derived from keyword-based exhibit representations. The *SD* measure was developed using bags of keywords that are representative of exhibits, which in turn were derived from manual exhibit annotations made by four independent annotators [13]. This approach is based on the assumption that subjective manual annotations comprising keywords associated with exhibits can be used to represent the content of the exhibits. The *SD* measure was implemented by calculating semantic exhibit-to-exhibit similarity using the cosine similarity between the exhibit-specific bags of keywords. The resultant similarity values from within the interval $[0, 1]$ were transformed into distances by performing an inverse linear mapping onto values in $[0, 1]$ (a similarity value of 0 yields a distance of 1, and a similarity of 1 yields a distance of 0).

5 Evaluation

This section evaluates our user- and item-aware weighting scheme with the Melbourne Museum dataset (Sect. 4.1). We hybridise *SPM-PD* and *SPM-SD* (Sect. 4.2), yielding the hybrid models *SPM-HY-CV* and *SPM-HY-NN* (*SPM-HY-CV* denotes the hybrid model which adaptively estimates a current user's weight vector using *Cross Validation (CV)*, and *SPM-HY-NN* uses our *Nearest-Neighbour (NN)* collaborative approach). The experimental setup is described in Sect. 5.1, and our results are discussed in Sect. 5.2.[4]

5.1 Experimental Setup

To evaluate the performance of *SPM-HY-CV* and *SPM-HY-NN*, we implemented one additional hybrid model: *SPM-HY-SW*. *SPM-HY-SW*, which we use as a baseline, employs a static model weight $w \in [0, 1]$ for linearly combining the predictions of *SPM-PD* and *SPM-SD* (the optimal value of w can be determined by minimising an error measure of choice).

We used leave-one-out cross validation to evaluate the predictive performance of our models. For each fold (i.e., for each of the 158 visitors), we learnt the 157 visitor weights and 126 exhibit weights while withholding the data of the testing visitor (following the learning procedure from Sect. 3.2). In this process, we used fold-specific instances of *SPM-PD* and *SPM-SD* which were trained for each fold prior to the evaluation (i.e., with the data of the testing visitor withheld). In line with our previous research, we performed two types of experiments:

---

[4] Recall that we simplify the weight vectors $\boldsymbol{w}_u$ and $\boldsymbol{w}_i$ to scalar weights w_u and w_i for our evaluation (i.e., $K = 1$). This was done to enable timely completion of the experiments.

- **Individual Exhibit (IE).** *IE* evaluates predictive performance for a single exhibit. For each observed visitor-exhibit pair (u, i), we first removed the log viewing time r_{ui} from the vector of a visitor u's log viewing durations. We then computed a prediction $\hat{r}_{ui}$ from the other observations. This experiment is lenient in the sense that all available observations except the observation for exhibit i are kept in a visitor's log viewing time vector.

- **Progressive Visit (PV).** *PV* evaluates performance as a museum visit progresses, i. e., as the number of viewed exhibit areas increases. For each visitor, we started with an empty visit, and iteratively added each viewed exhibit area to the visit history, together with its log viewing time. At each visit stage, we predicted the log viewing times of all yet unvisited exhibit areas.

For both experiments, we measured predictive accuracy using the *mean absolute error (MAE)* with respect to log viewing times as follows:

$$\text{MAE} = \frac{1}{\sum_{u \in U} |I_u|} \sum_{u \in U} \sum_{i \in I_u} |\hat{r}_{ui} - r_{ui}|,$$

where U is the set of all visitors, and I_u denotes a visitor u's set of exhibit areas for which predictions were computed. For *IE*, we calculated the total MAE for all valid visitor-exhibit pairs; and for *PV*, we computed the MAE for the yet unvisited exhibit areas for all visitors at each time fraction of a visit (to account for different visit lengths, we normalised all visits to a length of 1 prior to the *PV* experiment, and considered fractions of a visit).

5.2   Results

Weight Learning.   For visualisation purposes, we learnt the visitor and exhibit weights with the complete dataset of 8327 log viewing times. Figure 1 summarises the results of this training process. The plot shows the 158 visitor weights and 126 exhibit weights in increasing order of weight magnitude, with the vertical black line separating visitor and exhibit weights. The horizontal black line marks $1/\sqrt{2}$. When assigned to w_u and w_i, this value yields a hybrid model that gives equal weight to the predictions of *SPM-PD* and *SPM-SD* (if $w_u, w_i = 1/\sqrt{2}$, then $w_{ui} = w_u \times w_i = 1/2$). For both visitors and exhibits, the majority of learnt weights is larger than $1/\sqrt{2}$ (57.0% of the visitor weights w_u, and 76.2% of the exhibit weights w_i). This means that for the majority of visitors and exhibits, more weight is given to *SPM-PD* than to *SPM-SD* (in the notation of Sect. 3.1, *SPM-PD* is model M_1, and *SPM-SD* is model M_2). Further, more weight is given to *SPM-PD* for just over half of the observed visitor/exhibit combinations (54.6% of the model weights w_{ui} are larger than $1/2$).[5]

The visitor and exhibit weights can be interpreted as follows. For visitors with large visitor weights w_u, the observations at spatially close exhibits are better predictors of interest than the observations at exhibits that are semantically similar, whereas for visitors with small weights w_u, the observations at semantically similar exhibits are better predictors than the observations at spatially close exhibits. Similarly, depending on whether

---

[5] Individually, *SPM-PD* attains a higher predictive accuracy than *SPM-SD*.

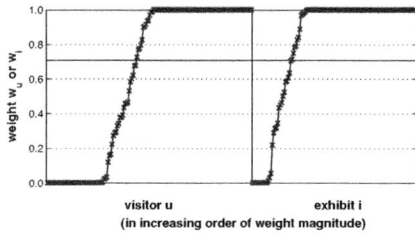

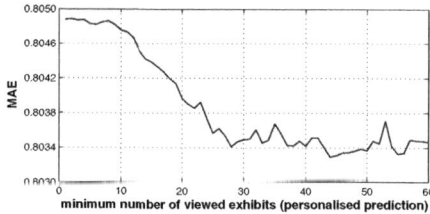

Fig. 1. *SPM-HY* – Weight estimates obtained with the complete Melbourne Museum dataset

Fig. 2. *SPM-HY-CV* – Varying M, the minimum number of viewed exhibits for computing a personalised prediction

exhibit weight w_i is small or large, exhibit i's viewing times are more correlated with those at semantically similar exhibits than with those at spatially close exhibits, and vice versa (visitor and exhibit weights are interpreted differently due to *SPM*'s different view of visitors and exhibits). From a user modelling perspective, one could thus infer from a visitor weight's magnitude whether the visitor is guided more by the spatial layout of the museum or by the semantic content of the exhibits.

Configuration Assessment (*IE* and *PV* Experiments). *SPM-HY-SW*, *SPM-HY-CV* and *SPM-HY-NN* are configurable approaches. Hence, using leave-one-out cross validation, we tested thousands of configurations to assess the influence of the different parameters on the performance of these models, and to determine the best-performing variants. Specifically, the models were assessed by comparing the MAE_{avg} scores of the various configurations (the MAE_{avg} score averages the results of the *IE* and *PV* experiments). For *SPM-HY-SW*, we varied the static model weight w. The minimum MAE_{avg} is achieved for $w = 0.81$, which means that the optimal non-adaptive hybrid model gives about four times as much weight to the predictions of *SPM-PD* than to those of *SPM-SD*. For *SPM-HY-CV*, we varied the current visitor's minimum number M of viewed exhibits for computing a personalised prediction. Figure 2 depicts *SPM-HY-CV*'s MAE_{avg} for $M = 1, \ldots, 60$. The minimum MAE_{avg} is achieved for $M = 44$. For *SPM-HY-NN*, we tested thousands of model configurations. The configuration that achieves the minimum MAE_{avg} is $\{M = 39, C = 39, S = 0.10, K_{NN} = 3, \omega = 0.65\}$ (the symbols are explained at the end of Sect. 3.3). We omit the results of a sensitivity analysis of *SPM-HY-NN*'s parameters due to space limitations.

Table 1 summarises the results for the *IE* experiment, where the highest-scoring configurations of *SPM-HY-CV* and *SPM-HY-NN* outperform both individual models *SPM-PD* and *SPM-SD*. Further, both hybrid models attain a higher predictive accuracy than *SPM-HY-SW*. Specifically, *SPM-HY-CV* achieves an MAE of 0.7520 (standard error 0.0066), and *SPM-HY-NN* attains an MAE of 0.7507 (standard error 0.0066). These values are statistically significantly lower than the MAEs of *SPM-PD*, *SPM-SD* and *SPM-HY-SW* (which are 0.7548, 0.7680 and 0.7539 respectively, $p < 0.05$).[6] By contrast, *SPM-HY-SW* does not statistically significantly outperform the individual

---

[6] The statistical tests performed are one-tailed paired t-tests (significance level $\alpha = 0.05$).

Table 1. Model performance for the *IE* experiment (MAE)

	MAE	Stderr
Spatial Process Model using *SD* (SPM-SD)	0.7680	0.0067
Spatial Process Model using *PD* (SPM-PD)	0.7548	0.0066
Non-Adaptive Hybrid Spatial Process Model using static model weights (SPM-HY-SW)	0.7539	0.0066
Adaptive Hybrid Spatial Process Model using cross validation (SPM-HY-CV)	**0.7520**	**0.0066**
Adaptive Hybrid Spatial Process Model using nearest neighbours (SPM-HY-NN)	**0.7507**	**0.0066**

models (p > 0.05). This means that using static model weights is not sufficient. Additionally, the performance difference between the two hybrid models *SPM-HY-CV* and *SPM-HY-NN* is not statistically significant at the significance level $\alpha = 0.05$ (i.e., p > 0.05), which means that for the *IE* experiment, *SPM-HY-CV* and *SPM-HY-NN* perform similarly.

Figure 3 depicts the performance of the highest-scoring configurations of the hybrid models compared to *SPM-PD* and *SPM-SD* for the *PV* experiment. Our results show that individually, *SPM-PD* statistically significantly outperforms *SPM-SD* for almost 70% of a visit (hence, *SPM-PD* is the better-performing individual model). Further, *SPM-HY-SW* performs better than *SPM-PD* and *SPM-SD* for 24.3% and 87.3% of a visit respectively, while *SPM-PD* and *SPM-SD* outperform *SPM-HY-SW* slightly for 5.2% and 0.3% of a visit respectively. This means that non-adaptive *SPM-HY-SW* is only marginally better than *SPM-PD* (as seen in Fig. 3, the models have a largely identical performance). By contrast, *SPM-HY-CV* and *SPM-HY-NN* attain a statistically significantly higher predictive accuracy than *SPM-PD* and *SPM-SD* for at least 58% of a visit (mostly in the second half), with the performance of the hybrid models steadily improving relative to *SPM-PD* with the progression of a visit. Specifically, *SPM-HY-NN* outperforms *SPM-PD* for 58.2% of a visit, and *SPM-HY-CV* performs better than *SPM-PD* for 61.9% of a visit (*SPM-HY-NN* and *SPM-HY-CV* outperform *SPM-SD* for 91.6% of a visit, and are never outperformed by *SPM-PD* or *SPM-SD*). Comparing the hybrid models against each other, *SPM-HY-NN* and *SPM-HY-CV* outperform *SPM-HY-SW* for 78.3% and 78.6% of a visit respectively (*SPM-HY-SW* never outperforms *SPM-HY-NN* or *SPM-HY-CV*). Hence, using our user- and item-aware weighting scheme improves predictive accuracy compared to the individual models being hybridised, and importantly, compared to non-adaptive *SPM-HY-SW*. Finally, *SPM-HY-CV* performs at least as well as *SPM-HY-NN*, achieving a statistically significantly higher predictive accuracy than *SPM-HY-NN* for 26.4% of a visit (the models perform identically for most of the first half of a visit, and *SPM-HY-NN* never outperforms *SPM-HY-CV*). That is, the computationally more expensive *CV* approach for estimating a current visitor's weight performs slightly better than the *NN* collaborative approach.

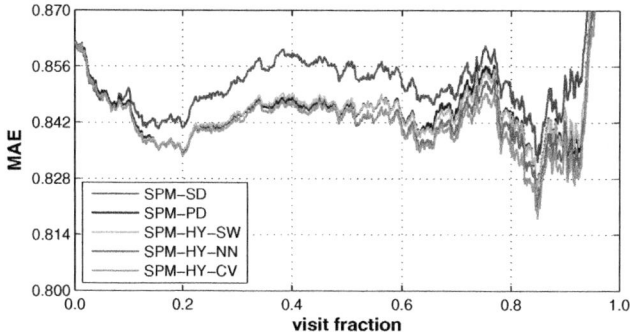

Fig. 3. Model performance for the *PV* experiment (MAE)

As seen in Fig. 3, the performance of all models drops at the end of a visit. This phenomenon (which was also observed in [9]) may be explained by the increased influence of outliers on the MAE towards the end of a visit, as the number of predictions is reduced with the progression of a visit (this hypothesis is supported by a widening of the standard error bands for all models towards the end of a visit). Outliers might occur more frequently in the final stages of a visit, because visitors' personal interests play a reduced role at that point. Rather, visitors are increasingly influenced by factors such as receptive saturation and increased awareness of limited remaining time (such factors are not yet considered by our models).

6 Conclusions and Future Work

This paper presented a user- and item-aware weighting scheme for linearly combining the predictions of two user models. Additionally, we proposed two online approaches for estimating a target user's weight vector as observations become available: (1) a *Cross Validation (CV)* approach, and (2) a *Nearest-Neighbour (NN)* collaborative approach. Focusing on the museum domain, we provided experimental evidence that our approach to model hybridisation (i.e., user- and item-aware model combination) improves predictive accuracy. In particular, we combined two interest-based Gaussian *Spatial Process Models SPM-PD* and *SPM-SD*, and evaluated the resultant hybrid models *SPM-HY-CV* and *SPM-HY-NN* with a real-world dataset of museum visits. We showed that the hybrid models outperform the component models *SPM-PD* and *SPM-SD*. Additionally, both hybrid models attain a higher predictive accuracy than a non-adaptive hybrid model called *SPM-HY-SW*, which uses static model weights. Our results also indicate that for our hybridisation approach, the learnt visitor and exhibit weights (and hence, the model weights) vary greatly among the visitors and exhibits. This validates our use of visitor- and exhibit-specific weights instead of static model weights. Particularly important for the museum domain are our results for the realistic *Progressive Visit* experiment, where *SPM-HY-CV* outperforms *SPM-PD*, *SPM-SD* and *SPM-HY-SW* for 61.9%, 91.6% and 78.6% of a visit respectively.

In the future, we plan to train our weighting scheme for weight vectors of length ≥ 2, and explore hybrid models which employ more than two component models. We also intend to apply our approach to other datasets from the recommender systems domain.

Acknowledgements. This research was supported in part by grant DP0770931 from the Australian Research Council. The authors thank Carolyn Meehan and her team from Museum Victoria for fruitful discussions and their support; and David Abramson, Jeff Tan and Blair Bethwaite for their assistance with using their computer cluster.

References

1. Adomavicius, G., Tuzhilin, A.: Toward the next generation of recommender systems: A survey of the state-of-the-art and possible extensions. IEEE Transactions on Knowledge and Data Engineering 17(6), 734–749 (2005)
2. Burke, R.: Hybrid web recommender systems. In: Brusilovsky, P., Kobsa, A., Nejdl, W. (eds.) Adaptive Web 2007. LNCS, vol. 4321, pp. 377–408. Springer, Heidelberg (2007)
3. Lekakos, G., Giaglis, G.M.: A hybrid approach for improving predictive accuracy of collaborative filtering algorithms. User Modeling and User-Adapted Interaction 17(1-2), 5–40 (2007)
4. Mobasher, B., Jin, X., Zhou, Y.: Semantically enhanced collaborative filtering on the web. In: Berendt, B., Hotho, A., Mladenič, D., van Someren, M., Spiliopoulou, M., Stumme, G. (eds.) EWMF 2003. LNCS (LNAI), vol. 3209, pp. 57–76. Springer, Heidelberg (2004)
5. Claypool, M., Gokhale, A., Miranda, T., Murnikov, P., Netes, D., Sartin, M.: Combining content-based and collaborative filters in an online newspaper. In: Proceedings of the SIGIR 1999 Workshop on Recommender Systems (1999)
6. Koren, Y., Bell, R., Volinsky, C.: Matrix factorization techniques for recommender systems. IEEE Computer 42(8), 30–37 (2009)
7. Bohnert, F., Schmidt, D.F., Zukerman, I.: Spatial processes for recommender systems. In: Proceedings of the 21st International Joint Conference on Artificial Intelligence (IJCAI-09), pp. 2022–2027 (2009)
8. Polikar, R.: Ensemble based systems in decision making. IEEE Circuits and Systems 6(3), 21–45 (2006)
9. Bohnert, F., Zukerman, I.: Non-intrusive personalisation of the museum experience. In: Houben, G.-J., McCalla, G., Pianesi, F., Zancanaro, M. (eds.) UMAP 2009. LNCS, vol. 5535, pp. 197–209. Springer, Heidelberg (2009)
10. Stock, O., Zancanaro, M., Busetta, P., Callaway, C., Krüger, A., Kruppa, M., Kuflik, T., Not, E., Rocchi, C.: Adaptive, intelligent presentation of information for the museum visitor in PEACH. User Modeling and User-Adapted Interaction 18(3), 257–304 (2007)
11. Wang, Y., Aroyo, L., Stash, N., Sambeek, R., Schuurmans, Y., Schreiber, G., Gorgels, P.: Cultivating personalized museum tours online and on-site. Interdisciplinary Science Reviews 34(2), 141–156 (2009)
12. Bohnert, F., Zukerman, I.: Personalised pathway prediction. In: De Bra, P., Kobsa, A., Chin, D. (eds.) UMAP 2010. LNCS, vol. 6075, pp. 363–368. Springer, Heidelberg (2010)
13. Bohnert, F., Zukerman, I.: Using keyword-based approaches to adaptively predict interest in museum exhibits. In: Nicholson, A., Li, X. (eds.) AI 2009. LNCS, vol. 5866, pp. 656–665. Springer, Heidelberg (2009)

PersonisJ: Mobile, Client-Side User Modelling

Simon Gerber, Michael Fry, Judy Kay, Bob Kummerfeld,
Glen Pink, and Rainer Wasinger

School of Information Technologies, University of Sydney, Sydney, NSW 2006,
Australia
{sger6218,mike,judy,bob,gpin7031,wasinger}@it.usyd.edu.au

Abstract. The increasing trend towards powerful mobile phones opens
many possibilities for valuable personalised services to be available on
the phone. Client-side personalisation for these services has important
benefits when connectivity to the cloud is restricted or unavailable. The
user may also find it desirable when they prefer that their user model be
kept only on their phone and under their own control, rather than under the control of the cloud-based service provider. This paper describes
PersonisJ, a user modelling framework that can support client-side personalisation on the Android phone platform. We discuss the particular
challenges in creating a user modelling framework for this platform. We
have evaluated PersonisJ at two levels: we have created a demonstrator
application that delivers a personalised museum tour based on client-side
personalisation; we also report on evaluations of its scalability. Contributions of this paper are the description of the architecture, the implementation, and the evaluation of a user modelling framework for client-side
personalisation on mobile phones.

1 Introduction

Personalisation has the potential to offer many benefits, particularly in reducing
information overload by enabling a person to be more efficient in finding the
information they need or want. Personalised systems can also be valuable in an
active role, alerting the user to useful information. But there is a tension between
such personalisation and privacy; the user model that drives personalisation is
based upon the user's personal information. Moreover, there is evidence of considerable community concern about the proper protection of such information,
for example [1].

One way to address such concerns is to perform the personalisation at the
client-side, with the user's model stored on their own system. This is in contrast to the widespread *server-side* personalisation. Consider, for example, an
e-commerce website such as 'www.amazon.com', where customers must register
with the site in order to shop. The site can log every action they take while
they are logged in, such as the items they view, add to their shopping carts and
ultimately buy. This is used to create a user-profile which is held on the server.
The website's owners are in control of this user model and the way it is used.

P. De Bra, A. Kobsa, and D. Chin (Eds.): UMAP 2010, LNCS 6075, pp. 111–122, 2010.

In client-side personalisation, the user model is controlled by the user and the personalised applications that also run on their machine, under their control.

We now consider the issue of mobile personalisation. Mobile phones are providing an increasingly important interface for people to access information. There are currently over 3.5 billion mobile phone subscribers [2] and there is an increasing trend for these to have data subscriptions (for example, 40% of mobile users in Japan have a data plan). Interestingly, studies of how people are actually using their mobiles reveal that many use their phones for internet access, even while in their own homes or with another computer nearby [3].

At present, personalisation of the information delivered to mobile phones is typically performed at the server-side, by services in the *cloud*. This has been a necessity due to the limited computational power and memory of mobile phones. However, with widely-available consumer phones becoming increasingly powerful, it is becoming feasible to support client-side personalisation for these devices.

To support mobile personalisation on phones, we need new tools to support the creation of new applications. PersonisJ is one such tool, providing a framework for developing context-aware, personalised applications on a mobile phone. It can support reuse of user modelling information by arbitrary personalised applications running on the device. PersonisJ is unlike other personalisation shells or context-aware frameworks in that it treats the mobile device as a platform, rather than simply as an actor in a larger framework [4]. It also has to operate under very different constraints from previous user modelling frameworks, because it must take account of the power constraints for programs running on a mobile phone.

The next section reviews related work. We then describe the architecture and implementation of PersonisJ followed by our validation of it by demonstrating its use in the MuseumGuide application and our evaluations of its scalability. Finally, we discuss the implications of the work and future directions.

2   Related Work

The relatively recent emergence of powerful mobile phones has created new possibilities for mobile personalisation and the associated needs for personalisation. For example, studies point to the need for personalisation of mobile search interfaces [5]. There has been considerable exploration of mobile personalisation for a range of contexts and types of application. For example, personalisation of information available has been based on the user's social context [6] and location [7]. In e-commerce, personalisation has been widely deployed, with an increasing role for mobile, m-commerce [8]. Some interesting forms include systems that enable retailers to push recommendations to the mobile customer [9] and to offer both personalised product details and in-store customer advice [10]. Another important class of applications, mobile guides [11], can cover roles as diverse as museum guides, navigation systems and shopping assistants [12]. For example, the PEACH [13] system delivered personalised information about the art in a museum on PDAs.

Early research in mobile personalisation has been dominated by a view of the mobile phone as the client of (and portal to) a powerful server which was responsible for the personalisation, often restricted to a particular space such as a university or hospital [14]. In the mobile personalisation work described above, the architecture of the systems places personalisation at the server side. We have not found reports of mobile client-side personalisation, or even mobile applications that reuse frameworks for client-side personalisation.

One of the barriers for client side personalisation is the lack of a framework for the user modelling. A recent review of generic user modelling systems [4] points to the considerable work on such frameworks for server side personalisation.

PersonisJ is strongly influenced by the PersonisAD context-aware modelling framework [15]. Distinctive features of this modelling framework are: the same mechanisms model users as well as devices and places; it supports distributed user modelling, particularly important for pervasive computing applications; it provides *scrutable* modelling, meaning that it was designed, from its foundations, to support a user's scrutiny of their user model and the way that it is used. It is also able to perform lightweight user modelling, making it a promising foundation for the phone where power consumption is a major concern.

We now describe key elements from PersonisAD that are important for PersonisJ. PersonisAD represents a model as an hierarchy of *contexts* which can contain *components*. It is based on the *accretion/resolution* representation. Applications interact with PersonisAD via three primitive operations. The first looks up a Model for a particular person, device or place. The application can then can use a *tell* operation to supply evidence, and an *ask* operation to request the value of a component. This value is dynamically determined at the time of the *ask* based on a two part process. First, an *evidence filter* selects just the evidence allowed for the application which performed the *ask*. Then a *resolver* interprets the set of evidence. For example, a playlist application might use a resolver for a person's favourite genre from a list of evidence that includes the songs they have most recently played. Notably, flexibility and power in the reasoning comes from the availability of a range of evidence filters and resolvers. Some aspects of PersonisAD are not suitable for use on a mobile device. Notably, it is a distributed server application and *always on* so that clients can make TCP/IP connections. While the essence of PersonisAD gives a conceptual foundation for this work, the demands of creating a framework for client side personalisation on a mobile phone have meant that we created PersonisJ from scratch and independently from it. Unlike PersonisAD, written in Python, PersonisJ is written in Java and runs as a native application on Android.

3   Architecture and Implementation

We now describe the PersonisJ framework. We begin with the conceptual level, which has much in common with PersonisAD as described above. Then we present the high level architecture. The actual implementation was on the Open

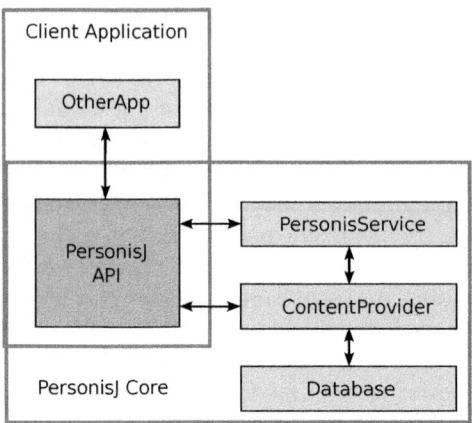

Fig. 1. PersonisJ Architecture

Handset Alliance's Android platform[1]. Pure implementation details are relegated to footnotes.

At a conceptual level, PersonisJ represents user models as an hierarchical structure of *contexts*, which can contain the *components* to be modelled. For example, it may have a context for the user's visits to museums and within this it may have components modelling the museums they prefer. Each component accretes evidence. This is essentially a value with metadata indicating how and when it was received. The metadata supports flexible evidence filtering and resolution and is a basis for supporting scrutability. The hierarchy of contexts and components constitute an *ontology* and a sub-tree within the hierarchy is a *partial ontology*.

Figure 1 shows the key modules of the PersonisJ architecture. The *PersonisJ API* enables an arbitrary application to access the user model, albeit only after the application has been granted read and/or write access to this model. *PersonisJ Core* provides the *API*, *Database*, *ContentProvider* and *PersonisService*. We now describe how each of these have been designed to represent the model and to secure access management.

The *Database*[2] has tables, with rows for each context, component and piece of evidence in the model. The *Database* is not accessible outside the *PersonisJ Core*. Instead, access is mediated by the *ContentProvider*, which specifies a unique "content URI" for content it exposes to the client application.

When a client interacts with the model, via the *PersonisJ API*, it must use a generic *ContentResolver* to act upon data, using content URIs. It is critical that client applications cannot directly modify anything within the PersonisJ database; for this reason, custom security permissions have been created for PersonisJ (READ_CONTENT, WRITE_CONTENT, and TELL_PERMISSION). These permissions are not intended to restrict unauthorised access; rather, it

---

[1] Android, http://www.android.com

[2] SQLite, on the Android platform.

forces applications to explicitly declare how they wish to interact with PersonisJ. All applications intending to access the PersonisJ database must state the permissions they require upfront[3]. To avoid direct third party application access to the database, the WRITE_CONTENT permission has been given the Android protection level of "signature", meaning that only applications signed with the same certificate as the PersonisJ code have direct write access to PersonisJ content. This has the effect that only two of the three above defined permissions are publicly available; third-party applications intending to read from the database use the READ_CONTENT permission (which corresponds to the *ask* operation) and those intending to contribute to the database do so via the PersonsisService module using the TELL_PERMISSION (corresponding to the *tell* operation).

To allow client applications to interact with PersonisJ, *PersonisJ Core* provides the *PersonisService* in Fig. 1. Because this resides within the same application as the *ContentProvider*, it can freely write to the PersonisJ *Database*.

The *PersonisService* is also in charge of handling imports. It is not possible to expose the ability to create new contexts and components via the *ContentProvider* without also exposing the ability to modify or delete them. This problem is solved by allowing client applications to pass the *PersonisService* a description of the partial ontology they require encoded in JavaScript Object Notation (JSON). The *PersonisService* then creates any necessary contexts and components.

PersonisService has one other responsibility. Any time it processes a *tell* operation it will broadcast a message[4] indicating which component was changed. It will then walk back up the context tree to the root, broadcasting notifications for each parent context in turn. This allows client applications to 'listen in', and discover if a particular component has received new evidence or, more generally, if any component beneath a particular context has changed.

The ontology is normally specified by applications. However, location is so fundamental that PersonisJ provides a predefined *location monitor* context in the phone model, with components for the co-ordinates of each location value. When turned on, this uses the phone's GPS to record any change in location as Evidence in the PersonisJ model.[5]

PersonisJ provides an Application Programming Interface [API] that applications can use to interact with the system. The *ask* operation returns a single value for a component. An important aspect of PersonisJ is that each component may have a list of evidence. In order to resolve multiple pieces of evidence into a single value the PersonisJ API provides a *Resolver* interface, which interprets a

---

[3] Security permissions are enforced by the Android OS, and the user is made aware of the required permissions when installing the application.

[4] An *Intent* in Android.

[5] The location monitor utilises the location API provided by Android. To conserve battery power, it does not turn on the GPS of its own accord. Instead, it hooks in to its operation whenever another program turns it on. The user can disable, or re-enable, the location monitor. While the GPS is disabled the location monitor does not reside in memory and uses no additional battery.

set of evidence, returning a single value. The API includes a selection of 'default' resolvers for numerical values, booleans, strings and dates. The actual resolution of values is executed within the client application, enabling them to provide their own *Resolver* implementations. The evidence passed to the *Resolver* can optionally be passed through an *EvidenceFilter*, which chooses the pieces of evidence to be used by the *Resolver*.

The API also exposes the ability to import and export partial ontologies. It provides a method, which can nominate any context, to gain the exported ontology. This operation occurs within the calling process. Optionally, a URL can be provided to the export function. After encoding into JSON, the generated string will be passed to the PersonisService which, in a background thread, will transmit the data to the specified URL via an HTTP POST. The import function works in a similar fashion, to upload the partial ontology defining a new part of the model with *PersonisService* performing this operation as external applications do not have write access to PersonisJ itself.

4   Evaluation

This section reports the two approaches we have used to validate the PersonisJ framework. First, we used it to create a personalised client-side application called MuseumGuide, which is able to notify a user of nearby museums and then download content for any museum that the user is keen to visit. Then we conducted scalability evaluations.

4.1   MuseumGuide Application

Consider the following scenario:

Alice and Bob, with their young family, are on a driving holiday in Sydney. Their phone has a model of the family's entertainment preferences including: low cost; suitable for children; kids are interested in ancient Egypt. The MuseumGuide, running on Alice's phone and aware of their location, sends an alert that they are near the Nicholson Museum. They decide to go to the museum and, on arrival, download a personalised museum tour based on a detailed model of the family's interests.

We now describe our implementation of MuseumGuide, making use of the PersonisJ framework to model a family's entertainment interests, as outlined in the scenario, and their more detailed interest model.

Figure 2 shows the MuseumGuide architecture, including the PersonisJ API module described earlier on (see Fig. 1), providing the application with access to the PersonisJ database. It operates as a client-side application, and like other Android applications requires the user to confirm when they are installing the application that they are happy with granting it the requested permissions. In this case, it requires access to the Internet, to the PersonisJ *ask* and *tell* operators, as well as several other permissions.

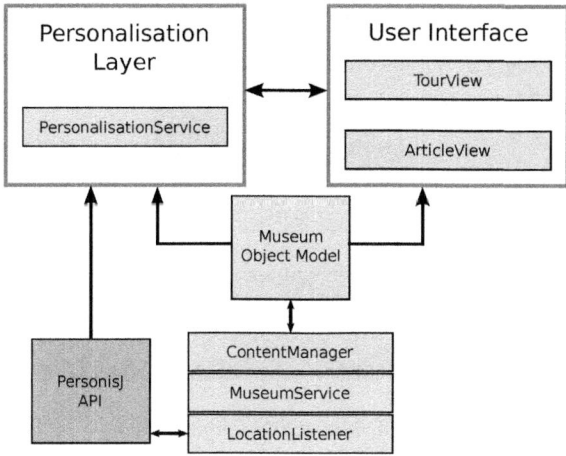

Fig. 2. MuseumGuide architecture

MuseumGuide uses a BroadcastReceiver, named `LocationListener`, to listen for component changes broadcast by PersonisJ. If the Location component changes, the alert is passed to `MuseumService`. This compares the new location against a stored list of museums. It sends a notification as shown on the left in Fig. 3. When viewed it starts an Activity that checks to see if there is content that can be downloaded (or updated) for the museum in question. If so, a prompt is displayed asking the user to confirm whether it should go ahead with the download (middle of Fig. 3).

The content is downloaded from a URL that is generated based upon the name of the museum. Once loaded, the background service invokes the `Content-Manager` which imports it to MuseumGuide and then sends a second notification (right screen in Fig. 3). Responding to this notification brings MuseumGuide into the foreground.

At this point the content is ready for viewing. A `PersonalisationService` is situated, architecturally, between the ContentProvider and its UI. This service is responsible for personalising the content before it is displayed. Figure 4 shows one such personalised article. In this example, text content that has been adapted for a young child. We used content for the Nicholson Museum, located on the University of Sydney campus, taken from an existing museum guide [16].

The PersonalisationService exists as a separate Service to enforce a separation between the content, the user-interface and the personalisation of the content. The service takes a museum and article identifier as input and returns personalised output, based on a simple personalisation algorithm based on the age of the user.

The above description illustrated how PersonisJ enables applications to register for updates in a manner that is no more complicated than requesting an update from existing Android system services. Once the logic for resolving a value from evidence has been encapsulated within a Resolver class it requires only a single *ask* to retrieve it.

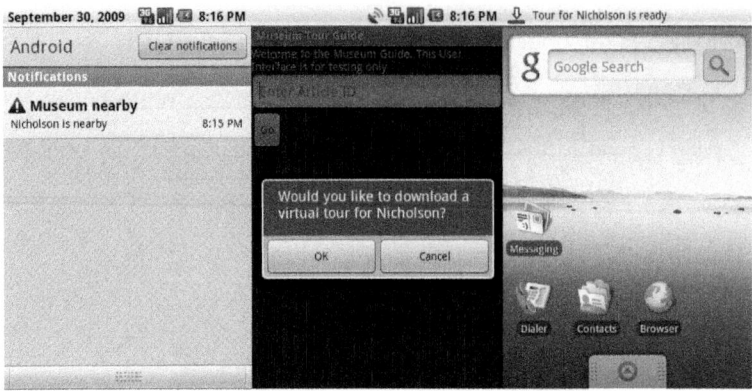

Fig. 3. MuseumGuide interface

In this way, PersonisJ makes it possible to create a context-aware application, which achieves its personalisation by accessing the user model via calls to a high-level API. We note that obtaining location updates from Android directly requires only slightly less code. However, PersonisJ provides a higher level of abstraction.

4.2 Scalability Evaluation

Before beginning our evaluation of PersonisJ we note that the DalvikVM[6] does not provide any 'Just in Time' [JIT] compilation. This has three important consequences. Firstly, it means managed code will always run slower on Android than its native equivalent. Secondly, it means that method level optimisations are important, as one cannot rely on minor inefficiencies being 'optimised away'. Finally, the lack of JIT compilation means we can be relatively naive about our performance testing, knowing that the code we write is, more or less, the code that Android executes.

Performance was measured using the profiler built into the Android framework, which has a resolution of micro seconds. The results are shown in Table 1. The critical column is the final column which shows the relative performance for the tested actions. This was calculated as the ratios of the average time per call normalised to one, then rounded to the nearest ten. The 'time' and 'average time' columns were included for completeness, but should not be relied upon as a measure of real-world performance.

Tests 1 to 4 involve actions selected from the sample actions in Table 1. This gives a baseline for PersonisJ performance against some simple operations. Tests 5, 6 and 7 reflect some basic PersonisJ API operations. Test 5 is the *tell* operation. We would expect it to be a relatively quick API call as it does not do much work itself, but rather encapsulates a call through to the PersonisService class in the core framework. Test 6 is the corresponding *ask* operation. Test 7 is

---

[6] The virtual machine used in Android.

Fig. 4. Example of a personalised article

an example of an API call that would be used to drill down through an ontology. Finally, Test 8 retrieves the same information from the PersonisJ database as Test 7, but does so using a ContentProvider directly and is optimised to use only one query instead of two.

These results indicate that PersonisJ's fastest operation is clearly the *tell* operation. This is expected, as a *tell* operation is asynchronous. The *ask* operation, in contrast, is a lot heavier and can be expected to differ significantly based on the supplied parameters. The *ask* operation tested in Test 6 was as lightweight as possible. Only the latest piece of evidence was examined and a default String resolver was used. The String resolver performs no additional computation, as all evidence is stored internally as a String. Even so, this simple *ask* took approximately three times longer than a *tell*. Test 7 is an example of a fairly common class of method. The ability to retrieve a model, context and Component by name is part of the required setup any client application must perform before it can even start to call *ask* or *tell*. However, once acquired, these methods do not usually need to be called again until the next time the application is killed and then relaunched. Test 8 acquires the same data as Test 7, but does so using only a single query. Its performance is roughly half that of Test 7. From this, we can determine that when only a few columns are involved it is much more efficient to return all columns in the initial query. This translates into an easy optimisation that can be applied to all PersonisJ calls that require ContentResolver queries.

We can also see from the results that the PersonisJ operations are the same order of magnitude as inflating a trivial user interface, with just one button, from XML (Test 4). Any user interface of practical use would far take longer, We can see that an *ask* on the model (test 6) takes of the order of 0.1 seconds.

Table 1. Performance Tests: The result of 200 calls to various Android and PersonisJ methods, profiled using 'traceview'. Results are sorted by relative performance. An asterisk * before the test number indicates a test involving PersonisJ.

Test Description	#	Time (ms)	Time/Call	Relative Performance
String.length()	1	7.606	0.038	1
HashMapIterator.next()	2	70.822	0.354	10
HashMap.put()	3	142.474	0.712	20
PersonisContext.tell()	*5	5763.410	28.82	760
LayoutInflator.inflate()	4	13904.226	69.521	1830
'Raw' context query	*8	14719.000	73.595	1930
PersonisContext.ask()	*6	24472.835	122.356	3220
PersonisContext.getChildContext()	*7	28371.960	141.860	3730

We can therefore conclude that acquiring complex information from PersonisJ would have not add a noticeable delay to the loading of a typical interface screen.

The PersonisJ API takes up 44.7kb when converted to dex format. This further compresses to 20.414kb in a .jar file. The PersonisJ application is 37.24kb as an APK package. It takes up 108kb of space on the phone when uncompressed. These sizes are negligible when compared to most Android applications. The PersonisJ Core Framework has essentially no UI and no other packaged resources, such as images. This makes it smaller, when first installed, than all but the simplest of applications. However, PersonisJ does create a database which will grow continuously over time.

The data requirements of PersonisJ are not dictated by PersonisJ itself, but rather the client applications. A full analysis of disk space requirements is therefore neither possible nor particularly helpful. An empty PersonisJ database is 10KB (10240 bytes).

In summary, the time performance of PersonisJ is adequate for the tasks required of it. Simple operations and resolvers that do not require much evidence are fast enough to be performed in the main UI thread without affecting application performance. More complex resolvers and contextual reasoning should be performed in a background thread, but are unlikely to take longer to run than setting up a typical UI screen. With respect to space performance, the size of the PersonisJ API is modest. The PersonisJ database, however, could grow too large over a period of only a few months. Future versions of PersonisJ will address this, with options to move the database to other storage media, prune old evidence, or back up old data to secure, personal storage over a network.

5   Conclusions

The goal of PersonisJ was to provide a personalisation framework that could support *client-side* personalisation on a *mobile phone*. This was motivated by two important potential benefits. First, client-side personalisation enables the phone to deliver a personalised service even when the phone is not connected to a data network. Second, it stores the user model on a device that is *controlled* by the user. We have shown the effectiveness of the PersonisJ framework in its capability to support the demonstrator application MuseumGuide. This made use of a model which illustrates some of the breadth of user modelling power of PersonisJ. The demonstrator used of the model for: location to determine the nearby museums; phone owner's *preferences* for inexpensive and educational entertainment which caused a recommendation for the free Nicolson museum; their *interest* in ancient Egypt which also affected recommendation of Nicolson; age of user of MuseumGuide, which affected presentation of museum information. We also showed that PersonisJ runs efficiently enough that a typical phone interface screen will load in essentially the same time, whether there is personalisation or not. We have shown that the space demands are modest for short term user models; however, for a long term user model, we still need to create mechanisms for archiving or removing parts of the model.

With the user model restricted to the mobile phone, the privacy of that model depends upon the effectiveness of the associated security model. As we have described in this paper, the PersonisJ architecture was carefully designed to address security issues. Notably, PersonisJ mediates all accesses to the model. (For the details of the implementation, see [17]).

Another element of security relates to the behaviour of the *applications* that a user loads onto their phone. This is outside the scope of this paper. However, it is clearly a critical issue. This is why we have conducted parallel work on a security framework [18] which enables the user to control what an arbitrary application is permitted to do. For example, the user can limit an application to have no communication outside the phone. Or it may simply restrict the application from exporting any information from the phone. This is essential if a user is to download an arbitrary application, such as a personalised museum tour, since it ensures that the application can provide personalisation, based on the user model on the phone, but cannot send any information outside the phone. Our MuseumGuide application operated within an environment controlled by the security environment.

Client-side personalisation provides an important foundation for life-long user modelling, in which the user is able to create, edit, reuse, and extend their user model throughout their digital life experiences. We have described PersonisJ, a user modelling framework that can support client-side personalisation on the Android phone platform. Contributions of this work are: the first architecture for a user modelling framework for client-side personalisation on mobile phones; and its validation in terms of a demonstrator application and scalability tests.

References

1. Shilton, K.: Four billion little brothers?: Privacy, mobile phones, and ubiquitous data collection. Queue 7(7), 40–47 (2009)
2. Church, K., Smyth, B.: Understanding the intent behind mobile information needs. In: Proceedings of the 13th international conference on Intelligent user interfaces, Sanibel Island, Florida, USA, pp. 247–256. ACM, New York (2008)

3. Nylander, S., Lundquist, T., Brannstrom, A.: At home and with computer access: why and where people use cell phones to access the internet. In: Proceedings of the 27th international conference on Human factors in computing systems, Boston, MA, USA, pp. 1639–1642. ACM, New York (2009)
4. Kobsa, A.: Generic User Modeling Systems, 136–154 (2007)
5. Church, K., Smyth, B.: Who, what, where & when: a new approach to mobile search. In: Proceedings of the 13th international conference on Intelligent user interfaces, Gran Canaria, Spain, pp. 309–312. ACM, New York (2008)
6. Kjeldskov, J., Paay, J.: Public Pervasive Computing: Making the Invisible Visible. Computer 39(9), 60 (2006)
7. Bilandzic, M., Foth, M., De Luca, A.: Cityflocks: designing social navigation for urban mobile information systems. In: DIS '08: Proceedings of the 7th ACM conference on Designing interactive systems, pp. 174–183. ACM, New York (2008)
8. Goy, A., Ardissono, L., Petrone, G.: Personalization in e-commerce applications, 485–520 (2007)
9. Kurkovsky, S., Harihar, K.: Using ubiquitous computing in interactive mobile marketing. Personal Ubiquitous Comput. 10(4), 227–240 (2006)
10. Li, J., Ari, I., Jain, J., Karp, A.H., Dekhil, M.: Mobile in-store personalized services. In: ICWS '09: Proceedings of the 2009 IEEE International Conference on Web Services, Washington, DC, USA, pp. 727–734. IEEE Computer Society, Los Alamitos (2009)
11. Kruger, A., Baus, J., Heckmann, D., Kruppa, M., Wasinger, R..: Adaptive Mobile Guides, 521–549 (2007)
12. Krüger, A., Baus, J., Heckmann, D., Kruppa, M., Wasinger, R.: Adaptive Mobile Guides. In: Brusilovsky, P., Kobsa, A., Nejdl, W. (eds.) Adaptive Web 2007. LNCS, vol. 4321, pp. 521–549. Springer, Heidelberg (2007)
13. Stock, O., Zancanaro, M., Busetta, P., Callaway, C., Kruger, A., Kruppa, M., Kuflik, T., Not, E., Rocchi, C.: Adaptive, intelligent presentation of information for the museum visitor in PEACH. User Modeling and User-Adapted Interaction 17(3), 257–304 (2007)
14. Chen, G., Kotz, D.: A survey of Context-Aware mobile computing research. Technical report, Dartmouth College (2000)
15. Assad, M., Carmichael, D., Kay, J., Kummerfeld, B.: PersonisAD: distributed, active, scrutable model framework for Context-Aware services. In: LaMarca, A., Langheinrich, M., Truong, K.N. (eds.) Pervasive 2007. LNCS, vol. 4480, pp. 55–72. Springer, Heidelberg (2007)
16. Czarkowski, M.: A Scrutable Adaptive Hypertext. PhD, University of Sydney (March 2006)
17. Gerber, S.: PersonisJ: A Platform for Context-Aware, Client-Side, Mobile Personalisation. PhD thesis, Unversity of Sydney (2009)
18. Pink, G.A.: Safe Execution of Dynamically Loaded Code on Mobile Devices. PhD thesis, Unversity of Sydney (2009)

Twitter, Sensors and UI: Robust Context Modeling for Interruption Management

Justin Tang and Donald J. Patterson

University Of California, Irvine, USA
justinwktang@yahoo.com,
djp3@ics.uci.edu

Abstract. In this paper, we present the results of a two-month field study of fifteen people using a software tool designed to model changes in a user's availability. The software uses status update messages, as well as sensors, to detect changes in context. When changes are identified using the Kullback-Leibler Divergence metric, users are prompted to broadcast their current context to their social networks. The user interface method by which the alert is delivered is evaluated in order to minimize the impact on the user's workflow. By carefully coupling both algorithms and user interfaces, interruptions made by the software tool can be made valuable to the user.

1 Introduction and Related Work

Working with online tools makes vast networks of information and social resources available, but reciprocally exposes users to the reach of both software and people. Interruptions inevitably follow and have become a hallmark of modern information work while in the office [1, 2], while working nomadically [3], and while mobile [4].

"Interruptions", while generally perceived as negative, play an important role in the accomplishment of work and don't always have a correspondingly negative impact. Depending on their context, interruptions can be helpful or harmful and interruptees often express ambivalence toward their disruption and value [5]. An interruption from an assistant, for example, while disruptive at the moment, may provide critical information that will prevent wasted effort, enable effective response to time critical situations and reduce the need for future interruptions.

Interruptions are disruptive because they act in such a way as to break one's concentrated creative energy and give rise to feelings of loss of control [6]. Examples include receiving a phone call, having a supervisor enter a work space or having an instant message window appear. However, another class of interruptions is internally generated. Examples of these include suddenly remembering a forgotten task, a self-recognition of fatigue, or an unexpected insight into a forgotten problem. Internal interruptions are easier to accommodate fluidly however, because unlike a supervisor standing in front of you, the degree of response is under the control of the interruptee.

P. De Bra, A. Kobsa, and D. Chin (Eds.): UMAP 2010, LNCS 6075, pp. 123–134, 2010.
© Springer-Verlag Berlin Heidelberg 2010

Recognizing technology's role in *creating* opportunities for interruption, many researchers have undertaken studies that attempt to evaluate technology's capability to help *manage* them as well. Initially such studies were directed at busy office managers and explored the potential for exposing interruptibility cues to colleagues, for example based on sensors in the office [7, 8] or changes in workflow on the desktop [9, 10]. As cell phones have grown in capabilities and prominence, researchers have expanded this sensor-based approach to address mobile individuals as well. Much of this work has been done to advance the methodological concerns associated with the experience sampling method, which invokes automated interruptions in user experience studies [11, 12], but has broad applicability in other domains.

This paper builds on these sensor-based approaches for determining interruptibility to support mobile laptop users. Our goal is to create software that will interrupt the user, not for conducting a user study, but for reminding the user to report their current context to their social network (for example through the status update service Twitter[1]). We hypothesize that such software can be effective because it has access to local sensor information, and will provide value to users by reminding them to inform their social network of changes to their availability before they are blindly interrupted from afar.

In this paper we report on a field study in which we compare the effectiveness of various approaches to accomplish this goal. We make three novel contributions. First, we leverage the user's current Twitter status as a *virtual sensor* in conjunction with existing sensors as input to a context modelling algorithm. Second we use a novel algorithm based on *Kullback-Leibler Divergence* (KLD) to detect changes in context. Third, we study the response of users to a variety of user interface (UI) techniques in order to identify properties of UI interruptions that make them able to be handled like an internal interruption by a user.

2 Background

Researchers have studied how instant messaging (IM) awareness cues have been appropriated by users as a means for inviting and discouraging interactions [13]. As with many social networking systems, IM users maintain buddy-lists in which they can report a custom status message that is broadcast to all of their contacts. The UIs of IM clients are designed in such a way that status messages can be viewed just prior to interrupting a remote user with a message. Users leverage this flexibility to, among other things, express their mood, promote issues and causes and *describe their current context* [14, 15]. Many of the practices around custom IM status messages have also been adopted in social networking systems such as Twitter and Facebook[2].

In previous studies we hypothesized that this last type of status message could be accurately guessed by modeling users' past behavior in light of sensor readings from the environment. Because of the complexity and privacy concerns related

---

[1] http://www.twitter.com

[2] http://www.facebook.com

Fig. 1. Nomatic*IM steady state

to accurately representing a user [16] we avoid automatically broadcasting status from a predefined ontology, opting instead to make it easy for a user to manually update their status in their own words. To test this hypothesis we built a status message broadcast tool called Nomatic*IM [4]. This tool gives users a single point of entry for updating their status across a wide variety of IM and web services. In previous studies we observed that this single-point-of-entry strategy was effective and helpful in assisting users in managing social contact [13]. However, effectiveness of, and satisfaction with the tool were closely correlated to how frequently the status messages were updated.

Previously, to keep status updates accurate, simple hand-crafted rules were used to recognize changes in a user's context [17]. Rules based on elapsed time were the most effective, but contrary to our hypothesis, rules based on sensors proved to be ineffective because of the complexity of the sensed environment. For example, although a change in IP address might be assumed to be associated with a change in location, and subsequently a change in availability, in practice, IP addresses change for a wide variety of reasons not associated with mobility. Temporary network disconnections, automatic switching between co-located WiFi access points, and DNS credential refreshing all caused fluctuations in IP addresses. This motivated the more robust treatment of sensor data described in this paper.

3   User Interface

Because of the many ways in which people use status updates that are not related to availability context, we developed the user interface of Nomatic*IM to guide the user in making the type of status updates that are relevant for our study.

The steady state Nomatic*IM interface is shown in Fig. 1. In this window the user's current status is displayed while the software collects readings from sensors in the background. There are buttons that allow a user to initiate a change in status, rebroadcast the current status (reaffirm), clear the current status, or manage their preferences.

When users want to change their status, the window is extended to present three fields in which a sentence can be constructed consisting of a location, an activity, and a free form field. The user can type in new data or use historic data that is presented in drop down lists as shown in Fig. 2. When the user is satisfied and accepts the current status by clicking the update button, Nomatic*IM makes a current reading of the sensors, pairs them with the current status message, stores them locally in order to later form a user model and then broadcasts the status message out to user-configured services such as Skype, AIM, Twitter, Facebook, etc.

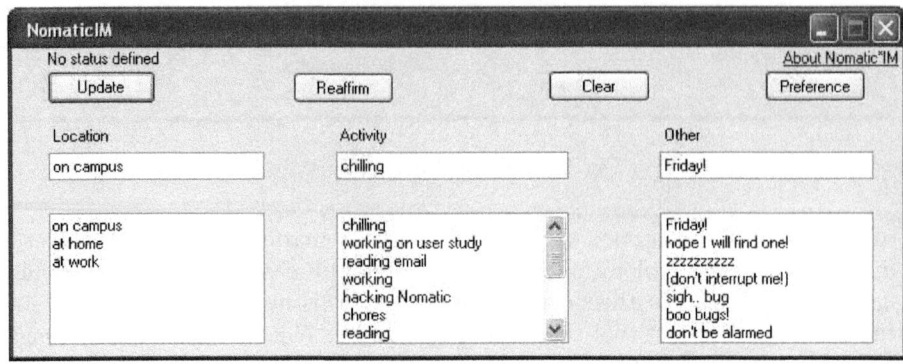

Fig. 2. Nomatic*IM status entry dialog

In addition to UI elements for the user-initiated flow of events, we also implemented and subsequently studied the acceptance of five interruption techniques to alert the user when the system identified that a context change had occurred.

1. POPUP-WINDOW-UI, which brings the window from Fig. 2 to the forefront
2. CURSOR-CHANGE-UI, which changes the mouse pointer to a custom icon until the user updates their status (see Fig. 3 left).
3. FADING-WINDOW-UI, which creates a popup window of varying opacity depending on the urgency of the interruption (see Fig. 3 center).
4. SYSTRAY-BALLOON-UI, which activates a small alert in the bottom right system tray (see Fig. 3 right).
5. AUDIO-INTERRUPT-UI, which plays a short audio chime.

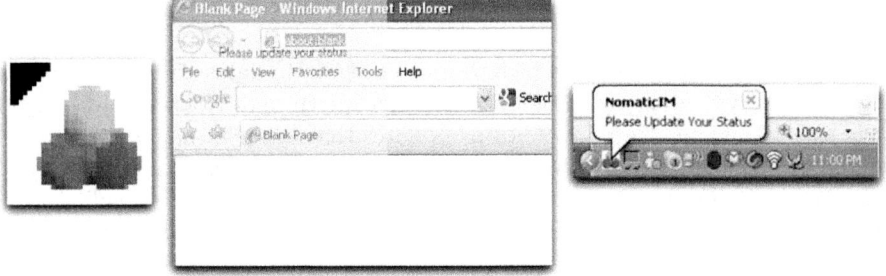

Fig. 3. CURSOR-CHANGE-UI (left), FADING-WINDOW-UI overlaid on a web browser window (center), SYSTRAY-BALLOON-UI (right)

4   Robust Context Change Detection

To robustly detect a change in context, we developed an alternative to the previously studied rule-based strategy. It is based on the the Kullback-Leibler Divergence (KLD), a non-negative numeric measure of how similar two multinomial

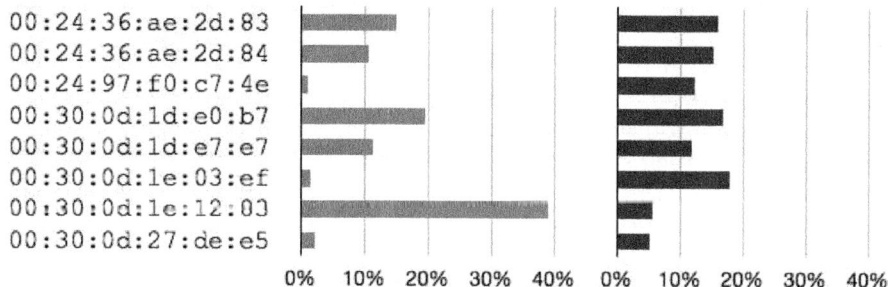

Fig. 4. Example probability distributions for the WiFi MAC address sensor when at the place, "at home", $X^{WiFi\text{"}at\_home\text{"}}$ and $Y^{WiFi\text{"}at\_home\text{"}}$

probability distributions are. A value of 0 indicates exact similarity. Formally, KLD measures the number of extra bits required to encode a distribution, Y, using the information in another distribution, X. Informally, KLD can be used to measure the similarity between two normalized histograms. It is defined as follows:

$$D_{\mathrm{KL}}(X\|Y) = \sum_i X(i) \log_2 \frac{X(i)}{Y(i)}$$

While using Nomatic*IM, our participants built up a history of contextual status messages and sensor readings. The sensor readings were collected both when the context message was changed, but also periodically in the background. During this study, sensor readings were collected every 60 seconds. A wide variety of sensors were captured including current time, network parameters, battery status, display configuration, WiFi parameters, UI activity and active processes. Depending on the hardware available additional sensors such as light and sound levels, accelerometer readings, and location were collected as well. With the exception of the status message, no semantic content (e.g., keystrokes, IM content, browser URLs) was collected.

In order to detect a change in context, a collection of probability distributions were constructed using the historic sensor readings that had been observed by the system when the user was reporting their status. Separate distributions were calculated for each of the three status field components. When users changed their status message, $X^{S_{place}}, X^{S_{activity}}, X^{S_{other}}$ were selected from the collection of historic probability distributions based on the three status fields. $X^{S_{place}}$ subsequently contained the probability $P(S = i|place = p)$, where i ranges over the possible values of the sensors and p is the current value of "place" (or "activity" or "other"). A second set of three distributions were created and augmented every time a new sensor reading was taken in the background. These became the distributions, $Y^{S_{place}}, Y^{S_{activity}}, Y^{S_{other}}$. Each of the three distributions differ from the others because they use the subsets of sensors that are relevant for modeling place, activity and other.

The change detection algorithm is based on a comparison of the historic distribution of sensors, X, compared with the currently observed sensors, Y, given the

most recently entered status message. The comparison is captured in the values
$D_{\mathrm{KL}}(X^{S_{place}}\|Y^{S_{place}})$, $D_{\mathrm{KL}}(X^{S_{activity}}\|Y^{S_{activity}})$, $D_{\mathrm{KL}}(X^{S_{other}}\|Y^{S_{other}})$.

Figure 4 shows an example of these distributions. The probability distribution
on the left shows an example of the distribution of WiFi MAC addresses that
have ever been historically collected for a user when reporting being "at home" in
the place field of their status. This is the distribution, $X^{S_{place}}$. The figure on the
right is the distribution over the currently observed WiFi MAC addresses since
the last update when the user entered "at home" as their current location. This
is $Y^{S_{place}}$. Our algorithm monitors $D_{\mathrm{KL}}(X\|Y)$ and after collecting five minutes
worth of samples, alerts when the metric begins to rise. An increasing value
is associated with diverging distributions between the sensor values historically
expected and currently observed.

5   Methodology

We conducted a within-subjects experiment across two variables to compare the
techniques that we developed. The first variable was the UI element that was
used to alert the user. It took on one of the five conditions corresponding to the
techniques specified in Sect. 3.

The second variable was the context change detection algorithm. This variable
had six conditions corresponding to five rules plus our new KLD technique.

1. WIFI-MAC-CHANGE-R: This rule asserted a context change when a user's
   laptop connected to a new WiFi access point as determined by the MAC
   address.
2. LOCAL-IP-CHANGE-R: This rule asserted a context change if the local network IP address changed on the primary adaptor.
3. REMOTE-IP-CHANGE-R: This rule detected if the IP address of the user's
   laptop changed as viewed from the internet (the "remote" IP address).
4. STALE-R: This rule was activated if the user did not change their status for
   2 hours.
5. START-UP-R: This rule activated if the user did not change their status
   within 3 minutes after starting the software.
6. KLD-R: This technique detected a change the first time that a user's current
   distribution of WiFi access point MAC addresses diverged from the historic
   distribution for the same place status or when the current distribution of
   active processes diverged from the historic distribution of active processes
   for the current activity status. All techniques were prevented from activating
   within five minutes of the last change.

Each time the user entered a new status or focussed on our tool following an
interruption, we randomly selected a new pair of conditions to test. When the
FADING-WINDOW-UI was selected with the KLD-R alert technique, the window's opacity was set based on the KLD measure such that as the two distributions diverged more, the window became more opaque. When a rule based alert
was paired with FADING-WINDOW-UI the opacity was set to 75%.

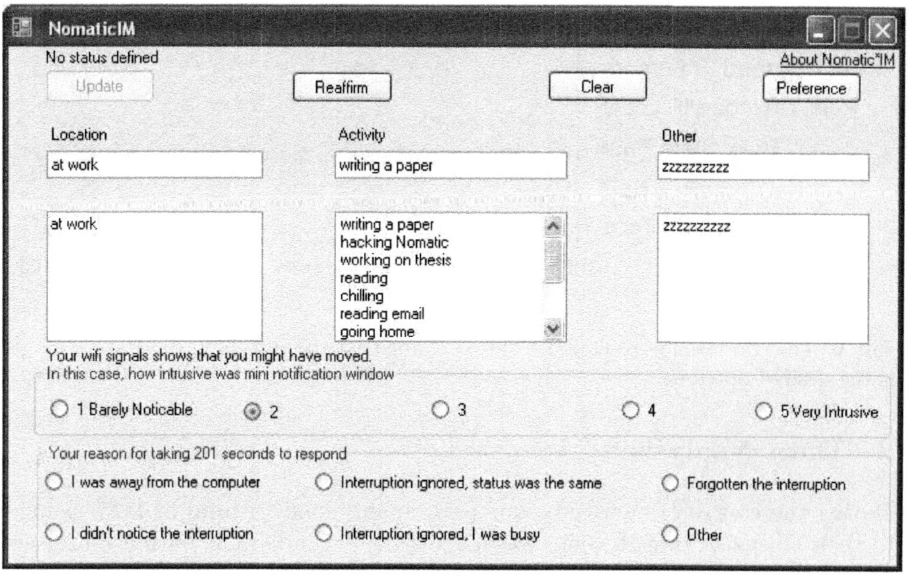

Fig. 5. UI following a long delay

Additionally when Nomatic*IM initiated an alert, the status change window was augmented with a description of why the system initiated an alert and a question asking the user to rate how "intrusive" this alert was on a 5-point Likert scale. To ensure that the data that we collected didn't have cofounding temporal factors, we monitored the time for the user to respond to an alert. If the user took more than three minutes to respond to the interruption, the user was asked a second question about why it took so long to respond, the choices included: "I was away from the computer", "I didn't notice the interruption", "Interruption ignored, my status was the same", "Interruption ignored, I was busy", and "Other" which allowed for a free-form response (see Fig. 5).

Fifteen participants were recruited from the University of California, Irvine community via email advertisement and mailing lists, website advertisements, and flyers. Participants were also recruited by in-person invitations. We enrolled adults who use Windows XP or Windows Vista laptops as well as Skype, Facebook, and/or Twitter, and who self-reported using their laptop in more than 2 physical locations per day via WiFi. Upon enrollment, participants were provided with a copy of Nomatic*IM, and given step-by-step instructions via an online video on installation and usage. Participants were instructed to use their computer normally, with the exception of setting their status via Nomatic*IM. Participants were compensated $1.00 for each day that they entered a status up to a maximum of $42.00. The study lasted a total of two-months.

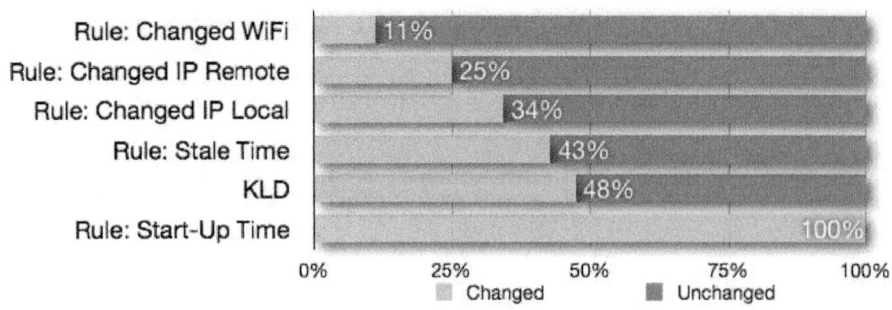

Fig. 6. The effectiveness of various context change detection techniques. 100% changed is the desired outcome.

6   Data Analysis

During the course of our study, our participants made a total of 1157 changes to their status. Although there was a great deal of individual variance in users, some broad trends could be observed. We were interested in looking at the relative effectiveness of the context change detection algorithms, how different UI techniques affected status updating behavior and overall user satisfaction with the techniques.

Status Change. Figure 6 compares different ways of detecting context change and whether or not users ultimately changed their status in response to the alert. A change in status is our ground truth. We are not interested in knowing whether the user's status is "true" in some understandable way (as is typically the case in the activity recognition literature), but rather whether our system successfully supported the user in keeping their status up to date. We would like a technique that only alerts users when they want to change their status. An algorithm that is too aggressive in suggesting that a user's context has changed, requiring a status update, risks disrupting the user too often. An algorithm that is too conservative fails to provide value.

Unsurprisingly, START-UP-R was 100% effective in prompting a status change. This rule coincides with a user starting the program which is likely prompted by a self-initiated need to broadcast a status or the initial boot process of the laptop. The least effective context detection algorithms were sensor-based rules. WIFI-MAC-CHANGE-R was only successful 11% of the time, followed by rules detecting changes in IP address. REMOTE-IP-CHANGE-R was accurate 25% of the time and LOCAL-IP-CHANGE-R was accurate 34% of the time. Simply suggesting a change after a period of time had elapsed was effective 43% of the time, which likely captured the scenario when a user awakened their laptop from a sleep state. KLD-R managed to outperform all the non-trivial rules with a 48% success rate and addresses the situations meant to be captured by the 3 least successful rules. All of the results in Fig. 6 showed a statistically significant difference from KLD-R ($p < 0.05$).

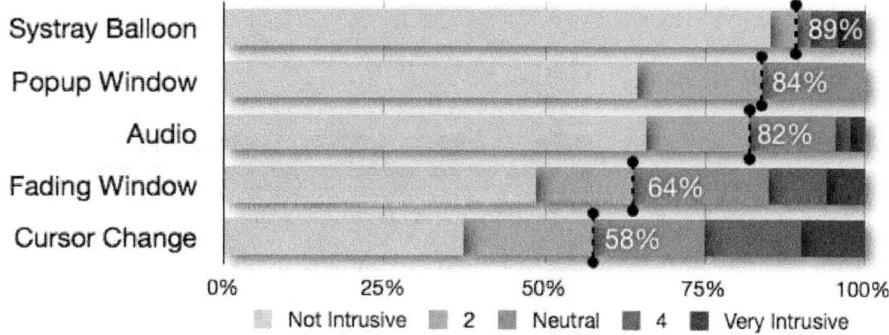

Fig. 7. User reported intrusiveness levels of various UI interruption methods

Intrusiveness Level. An aggregate analysis of user evaluation of different user interface interruption techniques is visualized in Fig. 7. This figure lists each of the 5 different techniques that were used to get the attention of the user when one of the context change detection algorithms alerted. The percentage of Likert scale ratings is shown as horizontal bars totalling 100%.

Treating a rating of 1 or 2 (low intrusiveness) as being acceptable, our data shows that the SYSTRAY-BALLOON-UI technique was acceptable 89% of the time, followed by the POPUP-WINDOW-UI 84% of the time, the AUDIO-INTERRUPT-UI 82% of the time, the FADING-WINDOW-UI 64% of the time and the CURSOR-CHANGE-UI 58% of the time. The difference in preference between the AUDIO-INTERRUPT-UI and the FADING-WINDOW-UI techniques was statistically significant ($p < 0.05$).

Long Response. During the study we evaluated the length of time it took before our participants would respond to the UI alerts. Our goal was to attempt to appropriately credit changes to status that were caused by our alerts, as opposed to changes in status that were made by a user who did not see an alert and later self-initiated. The reasons for the long alerts are shown in Fig. 8. Across the board, the most common reason for the prolonged response is that the users did not notice the interruption. This suggests that mobile users may not be as focussed on their laptops as we had assumed. This point is particularly highlighted by the responses to POPUP-WINDOW-UI. 55% of the responses were that the user did not notice the interruption. Considering that this particular interruption would take up a majority of the real estate on screen, and tying this data with the fact that POPUP-WINDOW-UI is scored as *not intrusive* 68% of the time, suggests that our participants were not completely focused on their laptops while they were running.

When a user did notice an alert, the decision to change their status was not statistically significantly correlated with the UI technique used to alert the user. This is a reasonable outcome and suggests that no UI technique caused the user to want to broadcast a different status just by its use. We were able to track a no-change-but-noticed case because a user can notice an alert, switch focus

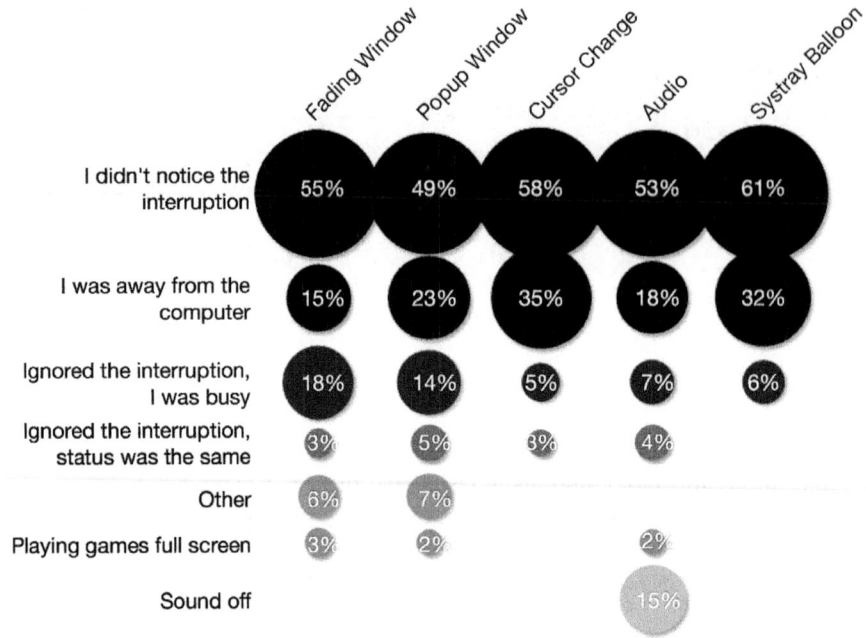

Fig. 8. Why participants took a long time to respond to UI alerts. Each column is normalized to reflect relative importance of each reason given a UI technique.

to our tool and explicitly tell our tool to rebroadcast the same status using a button for this purpose. This clears any persistent alerting mechanism, and our users did this on many occasions even for non-persistent alerts.

7   Conclusion

At the beginning of this study, we identified differences between internal and external interruptions and suggested that users would be more satisfied with external interruptions that could be handled more like an internal interruption. This was the motivation for creating the FADING-WINDOW-UI and CURSOR-CHANGE-UI elements. They are both small and persistent but easily ignorable.

Empirically, however, we observed that FADING-WINDOW-UI and CURSOR-CHANGE-UI were the most likely UI techniques to be rated as intrusive. One explanation for this result is that since these techniques were developed for this study, they were novel to our participants. Encountering these UI interruptions may have seemed more intrusive simply because they were unfamiliar experiences.

Although CURSOR-CHANGE-UI was scored as the most intrusive UI technique, it was one of the most *effective*, as only 8% of the time was the interruption method ignored. However, SYSTRAY-BALLOON-UI also had a similar low probability of being ignored (6%), but was evaluated as being the least intrusive. One explanation for this difference could be that non-persistent interruption techniques can be just as effective as persistent techniques, while being much less irritating. More support for this position comes from the AUDIO-INTERRUPT-UI, which plays its chime only once and was rated as non-intrusive. These alerts may be more easily mentally shifted from being an external interruption to being an internal interruption and thus able to be attended to at a natural break in the user's workflow. A competing hypothesis is that non-persistent interruption methods may also be perceived as less intrusive because users are away from their computer, or the sound is off, when the interruption occurs. More evaluation is required to isolate this effect.

This user study of context modeling and interruption techniques demonstrates that using the Kullback-Leibler Divergence metric to model context with inputs of both traditional sensors and Twitter-like status messages as *virtual sensors* can be more effective than hand-coded rules based on traditional sensors alone. The data suggest that an optimized context change detection would use a hybrid of rules based on elapsed time coupled with robust evaluations of the sensed environment and user entered status updates.

Our motivation was to create a software application that broadcasts context for social interruption management. As such it was important that the interruptions done by our software were as non-intrusive as possible, so that, like a good human assistant, the user sees interruptions by the software as an overall benefit. We demonstrated that although context change can be modelled with some accuracy, that is not the complete picture. User satisfaction is also heavily influenced by the UI technique and timing that is used to present alerts to the user. Our data showed that effective techniques for getting users' attention are not necessarily the ones that they prefer. Our participants preferred non-persistent, familiar UI alerts that we hypothesize can be more easily internalized until a natural break in the workflow permits them to be attended to.

References

1. González, V.M., Mark, G.: Constant, constant, multi-tasking craziness: Managing multiple working spheres. In: Dykstra-Erickson, E., Tscheligi, M. (eds.) CHI, pp. 113–120. ACM, New York (2004)
2. Rouncefield, M., Hughes, J.A., Rodden, T., Viller, S.: Working with "constant interruption": CSCW and the small office. In: CSCW, pp. 275–286 (1994)
3. Su, N.M., Mark, G.: Designing for nomadic work. In: van der Schijff, J., Marsden, G. (eds.) Conference on Designing Interactive Systems, pp. 305–314. ACM, New York (2008)
4. Patterson, D.J., Ding, X., Noack, N.: Nomatic: Location by, for, and of crowds. In: Hazas, M., Krumm, J., Strang, T. (eds.) LoCA 2006. LNCS, vol. 3987, pp. 186–203. Springer, Heidelberg (2006)
5. Hudson, J.M., Christensen, J., Kellogg, W.A., Erickson, T.: "I'd be overwhelmed, but it's just one more thing to do": Availability and interruption in research management. In: CHI '02: Proceedings of the SIGCHI conference on Human factors in computing systems, pp. 97–104. ACM, New York (2002)

6. Csikszentmihalyi, M.: Flow: The Psychology of Optimal Experience. Harper Perennial (March 1991)
7. Horvitz, E., Koch, P., Apacible, J.: Busybody: creating and fielding personalized models of the cost of interruption. In: Herbsleb, J.D., Olson, G.M. (eds.) CSCW, pp. 507–510. ACM, New York (2004)
8. Fogarty, J., Hudson, S.E., Atkeson, C.G., Avrahami, D., Forlizzi, J., Kiesler, S., Lee, J.C., Yang, J.: Predicting human interruptibility with sensors. ACM Trans. Comput.-Hum. Interact. 12(1), 119–146 (2005)
9. Iqbal, S.T., Bailey, B.P.: Understanding and developing models for detecting and differentiating breakpoints during interactive tasks. In: Rosson, M.B., Gilmore, D.J. (eds.) CHI, pp. 697–706. ACM, New York (2007)
10. Shen, J., Irvine, J., Bao, X., Goodman, M., Kolibaba, S., Tran, A., Carl, F., Kirschner, B., Stumpf, S., Dietterich, T.G.: Detecting and correcting user activity switches: algorithms and interfaces. In: Conati, C., Bauer, M., Oliver, N., Weld, D.S. (eds.) IUI, pp. 117–126. ACM, New York (2009)
11. Ho, J., Intille, S.S.: Using context-aware computing to reduce the perceived burden of interruptions from mobile devices. In: van der Veer, G.C., Gale, C. (eds.) CHI, pp. 909–918. ACM, New York (2005)
12. Kapoor, A., Horvitz, E.: Experience sampling for building predictive user models: a comparative study. In: CHI, April 2008, pp. 657–666 (2008)
13. Ding, X., Patterson, D.J.: Status on display: a field trial of Nomatic*Viz. In: Wagner, I., Tellioğlu, H., Balka, E., Simone, C., Ciolfi, L. (eds.) ECSCW 2009. Computer Science, pp. 303–322. Springer, London (2009)
14. Smale, S., Greenberg, S.: Broadcasting information via display names in instant messaging. In: GROUP '05: Proc. of the 2005 Intl ACM SIGGROUP Conference on Supporting group work, pp. 89–98. ACM, New York (2005)
15. Cheverst, K., Dix, A., Fitton, D., Rouncefield, M., Graham, C.: Exploring awareness related messaging through two situated-display-based systems. Hum.-Comput. Interact. 22(1), 173–220 (2007)
16. Dourish, P.: What we talk about when we talk about context. Personal and Ubiquitous Computing 8(1), 19–30 (2004)
17. Patterson, D.J., Ding, X., Kaufman, S.J., Liu, K., Zaldivar, A.: An ecosystem for learning and using sensor-driven IM messages. IEEE Pervasive Computing 8(4), 42–49 (2009)

Ranking Feature Sets for Emotion Models Used in Classroom Based Intelligent Tutoring Systems

David G. Cooper[1], Kasia Muldner[2], Ivon Arroyo[1],
Beverly Park Woolf[1], and Winslow Burleson[2]

[1] University of Massachusetts, Department of Computer Science,
140 Governors Drive, Amherst MA 01003, USA
dcooper@cs.umass.edu
[2] Arizona State University, School of Computing and Informatics,
Tempe AZ 85287, USA

Abstract. Recent progress has been made by using sensors with Intelligent Tutoring Systems in classrooms in order to predict the affective state of students users. If tutors are able to interpret sensor data with new students based on past experience, rather than having to be individually trained, then this will enable tutor developers to evaluate various methods of adapting to each student's affective state using consistent predictions. In the past, our classifiers have predicted student emotions with an accuracy between 78% and 87%. However, it is still unclear which sensors are best, and the educational technology community needs to know this to develop better than baseline classifiers, e.g. ones that use only frequency of emotional occurrence to predict affective state. This paper suggests a method to clarify classifier ranking for the purpose of affective models. The method begins with a careful collection of a training and testing set, each from a separate population, and concludes with a non-parametric ranking of the trained classifiers on the testing set. We illustrate this method with classifiers trained on data collected in the Fall of 2008 and tested on data collected in the Spring of 2009. Our results show that the classifiers for some affective states are significantly better than the baseline model; a validation analysis showed that some but not all classifier rankings generalize to new settings. Overall, our analysis suggests that though there is some benefit gained from simple linear classifiers, more advanced methods or better features may be needed for better classification performance.

1 Introduction

Student affect plays a key role in determining learning outcomes from instructional situations [1, 2]. For instance, learning is enhanced when empathy or support is present [3, 4]. While human tutors naturally recognize and respond to affect [5, 6], doing so is quite challenging for Intelligent Tutoring Systems (ITS), in part due to the lack of directly-observable information on a student's affect. A promising avenue for increasing model bandwidth, i.e., the quality and degree of information available to a student model, in terms of affect recognition

P. De Bra, A. Kobsa, and D. Chin (Eds.): UMAP 2010, LNCS 6075, pp. 135–146, 2010.

is sensing devices that capture information on students' physiological responses as they interact with adaptive systems. With the advent of inexpensive sensor technology, we have been able to deploy such sensing systems and use their output to infer information on student affect. Specifically, in the Fall of 2008 we performed a number of experiments in the classrooms of schools in both Western Massachusetts and Arizona, with a total of just under 100 students. In each experiment, students were queried about four emotional states (*confident, interested, frustrated,* and *excited*), providing the standard for validating our models. The study data was used to construct a number of linear classifiers for each emotional state, as we reported in [7]. The best classifiers for a given emotion obtained accuracies between 78% and 87% according to a leave-one-student-out cross-validation.

While these results are promising, it is important to validate the classifiers and verify that their performance generalizes to a new and/or larger population. This is particularly the case for our data, obtained from a classroom setting which involves a higher degree of noise and other distractions than standard controlled laboratory experiments. One aspect of validation involves verifying that our classifiers perform better than the baseline classifier (i.e., one that always outputs yes if the labels are yes most of the time, or no if the labels are no most of the time). In addition to validating our classifier performance, we also wanted to investigate if and how the sensors (or subsets of sensors) improved model performance over using only features from the tutor data (e.g. the number of hints requested). With an understanding of how each combination of sensor and tutor features predicts a given emotion, we can recommend which sensors to use for emotion recognition, and we can also rank the classifiers so that if some sensor data is unavailable, for instance due to an error, a comparable (or the next best) sensor set can be selected.

Thus, in this paper, we report on how we realized these objectives by utilizing a large data set for validation from experiments that we conducted in the Spring of 2009 with over 500 students. Our results show that our method is successful on three of our four target emotions: for each success, at least one linear classifier performs better than the baseline classifier and generalizes to a new and larger population.

We begin by presenting the related work in Sect. 2, and then describing in Sect. 3 the setup and apparatus of the experiments used to collect the data. Section 4 outlines the method for constructing and validating the student emotion classifiers. Section 5 describes the comparison of classifiers. Section 6 summarizes the results, discusses the design of affective interventions based on the classifier output, and suggests future work on improving the classifiers.

2   Related Work

The results of a feature selection competition in 2004 suggest that feature selection can be very useful for improving classifiers [8]. In addition to using simple correlation coefficients as criteria for selection (as stepwise linear regression does), treed methods, wrapper and embedded methods have been used for feature selection. [9] compares features of a number of individual sensors used for detecting affective state with an ITS, but does not compare disparate sensors, nor are multiple sensors used in conjunction in a classifier. In this paper we use a method from [10] to compare and rank the different feature sets used in the linear classifiers as a way of ranking our features selected by stepwise linear regression.

There are a number of adaptive systems in existence that use real-time information about a student in order to address the student's affective state. Recent work includes [11], which discusses the use of electromyogram (EMG) data to improve an affective model in an educational game. This work does careful collection, cross-validation, and uses a pairwise t-test (a parametric test) for ranking the classifiers. [12] aimed to predict learners' affective states (boredom, flow/engagement, confusion, and frustration) by monitoring variations in the cohesiveness of tutorial dialogues during interactions with an ITS with conversational dialogs; here, both student self reports and independent judges were used to identify emotional states. The study compared the correlation between self-reports and independent judges, and used tutor and dialogue features automatically classify emotion with accuracies between 68% and 78%.

Other work, such as [13, 14], does not incorporate any sensor data to construct affective models. [13] uses Dynamic Bayesian Networks and Dynamic Decision Models specified by an expert to determine and respond to each student's affective state, while [14] uses self-reports to determine affective state and focuses on how affective feedback changes the student's experience. This work does use cross-validation and a parametric ranking for classifiers, but does not do a feature comparison or a validation with a separate population.

Much of this past research has focused on constructing models based on a fixed set of sensors or solely on expert knowledge. In contrast, our research compares the utility of different sensors as well as sensor and tutor interaction features in a variety of empirically-based models. Another difference relates to the source of the data: Since our data is obtained from actual schools rather than the laboratory, the ecological validity of our results is strengthened. Our features are ranked using non-parametric procedures and take an extra step of validating on a separate population in order to address the additional artifacts created by a classroom setting.

3 Data Collection: Sensors with Wayang Outpost in the Classroom

3.1 Setup

In the Fall of 2008 and the Spring of 2009 the geometry tutor Wayang Outpost was deployed with a set of sensors into real classroom environments [7, 15, 16]. The set of sensors included: a mouse that captured degree of pressure placed on its various points, a bracelet that measured skin conductance of the wrist, a chair that sensed the level of pressure on the chair back and seat, and a camera supplemented with software for facial emotion recognition.

These four sensors collected data on students' physiological responses while students worked with Wayang Outpost. Each student's physiological data and interactions with the tutor were logged. Subsequently, the interaction and sensor data were time-aligned and converted into tutor and sensor features, as described in [7]. At intervals of five minutes in the Fall, and three minutes in the Spring, students were presented with an emotional query about one of four affective states (*confident, interested, frustrated,* or *excited*) selected from a uniform random distribution. The queries were presented as shown in Fig. 1; to respond, students selected from the options shown in Table 1. The sensor and tutor features were used as predictors for the levels of the self-reported affective states.

Fig. 1. An example of the Emotion query. Table 1 below has the values for each <> enclosed word, except for (<*Name*>), which is the name of the student.

Table 1. The mapping of tags to text in Fig. 1 above

<emotion>	<Left>	<Right>
confident	I feel anxious	I feel very confident
interested	I am bored	I am very interested
frustrated	Not frustrated at all	Very frustrated
excited	I'm enjoying this a lot	This is not fun

The Fall 2008 data collection involved 93 students using the Wayang Tutor. Of the 93 students 85 of them had at least one working sensor connected to them while using the tutor. Students used the tutor as part of a class, and class sizes ranged from three to twenty-five students with one teacher in the classroom and between one and three experimenters. The students had between two and five sessions with Wayang Outpost, based on teacher preference and availability of the student. The student ages were 15-16, 18-22, and 22-24. These data were used as our training set.

The Spring 2009 data collection involved over 500 students using the Wayang Tutor. 304 of the students were connected to at least one working sensor. The Spring collection differed from the Fall collection as follows: (1) The students in the Spring were from different schools; (2) The ages of Spring students were 13-14, and 15-16; (3) The camera sensor in the Spring had upgraded software. The Spring data was used purely for validation of the Fall Data.

3.2   Tutor and Sensor Features

We considered nine tutor features and forty sensor features as potential predictors for the emotion classifiers (see Table 2). The forty sensor features are based on four ways of summarizing ten specific features: the mean, the standard deviation, the min value, and the max value over the course of a problem. Since the sensor and tutor logging happens asynchronously, their data are interpolated in a piecewise constant fashion with the constraint that only data from the past is used to predict missing sensor or tutor values. The tutor logs when a problem is opened and closed, creating boundaries for summarizing the interpolated sensor data (i.e. to compute each feature, we use data over the span of a single problem). When there is an emotional query after a problem, the result becomes the affective state label for that problem. For each student and for each emotion there are between two and five affective-state labels. For more detail on the full specification of these features see [7].

Table 2. Features used for each problem that includes an affective state label in order to train the emotion classifiers (features are shown in abbreviated form). The nine tutor features are shown on the left and the ten sensor features are on the right. Features used in a classifier that is significantly better than the baseline ($p < 0.05$) are in **bold**.

Tutor feature	Definition	Sensor feature	Definition
Solv. on 1st	1st attempt correct	Agreeing	
Sec. to 1st	time to 1st attempt	Concentrating	
Sec. to solv.	time to a correct	Thinking	camera mental states
# incorrect	responses	**Interested**	
# hints	requested	Unsure	
LC	learning companion	**Mouse**	sum of pressure
Group	which LC (Jake, Jane, or none)	**Sit Forward**	
		Seat change	movement in chair
Time in session	same day	Back change	
Time in tutor	all days	Skin conductance	value from wrist

4   Method

The current standards for evaluating affective classifiers do not address our need to rank classifiers for the purpose of actionable affect detection. Though each individual step in our method has been established and tested, the combination of these steps yields a more robust test for the classifiers constructed. The use of our classifiers in a classroom environment necessitates our method described in the rest of this section and summarized in Table 3.

4.1   Collection

The data collection described in Sect. 3 is the first step in our methodology for building affect classifiers. The key parts of the data collection are that the emotion labels are made at the time of the experience, and the training and validations sets are taken from distinct populations using the same basic setup, allowing the validation results to be more likely to generalize. Here, the Fall collection is our training data set and the Spring collection is our validation data set.

Table 3. Our affect detection method summarized

1. Data Collection
– in situ self-reports of emotion
– training and validation sets from different population
2. Feature Selection
– remove central self-report values
– use step-wise linear regression to select features and train classifiers
3. Cross-validation (leave-one-student-out)
– compute the mean accuracy, sensitivity, and specificity per student
4. Classifier ranking
– parametric and nonparametric ranking using $p < 0.05$
5. Validation
– run steps 3 and 4 on validation set using classifiers from step 2

4.2 Predictor Selection

Once the data were collected and summarized as described in Sect. 3.2, we used the entire set of labeled training data to create a subset of predictors using a combination of tutor and sensor features. For each combination of features, a subset of the data set that was not missing data for the features was selected. Then stepwise linear regression was performed in R to select the 'best' subset of features from those available. The subset of features was stored as a formula for use in training the classifiers and performing cross validation.

4.3 Cross Validation

For each set of features determined by the feature selection, we performed leave-one-student-out cross-validation on linear classifiers for each affective state. During the cross-validation, we calculated the mean accuracy, sensitivity, and specificity for each test student. We also performed the same cross-validation on a linear classifier with a constant model, which we used as our baseline. This step differs from [7] in two ways: 1) The mean was taken across each test student instead of across tests. 2) We calculated sensitivity and specificity in addition to accuracy.

Though the cross-validation described above provides a general indication of the performance of each classifier, the information is not sufficient to enable appropriate pedagogical action selection by an ITS for *new* populations of students. Thus, we validated that the classifiers are generalizable and so can be used with a new population without having to be retrained. We also ranked the classifiers according to how sensors and features impact accuracy, allowing us to make informed decisions about sensor selection (e.g. if some sensors become unavailable, to select the next best alternative).

4.4 Classifier Ranking

A number of alternative techniques exist for classifier comparison. One is to use classifier accuracy, which identifies the overall performance of a classifier, but does not express accuracy on positive vs. negative instances. To do so, the following two measures can be used: (1) sensitivity, also referred to as the true positive rate, which provides information about the accuracy of a positive response; (2) specificity, the true negative rate, which provides information about the accuracy of negative responses.

Since the purpose of our classifiers is to help an ITS make decisions of how to appropriately respond to student emotion, one approach would be to only make a decision when there is confidence in the prediction. So, if one classifier has very good sensitivity relative to the baseline, then the ITS would act when the classifier reports a positive result. Similarly, if a classifier has a very good specificity relative to the baseline, then the ITS would act when the classifier reports a negative result.

In order to compare our classifiers' accuracy, sensitivity, and specificity for each affective state, we first performed a one-way analysis of variance (ANOVA), with classifier as the independent variable and either accuracy, sensitivity, or specificity as the dependent variable. When there was a significant difference between classifiers, we performed Tukey's HSD test to rank the differences in the means.

There is some question about the soundness of the ANOVA and Tukey's HSD test for these comparisons because the design is not balanced (not every student had all sensors available), and the responses are not normally distributed. So, in addition to the ANOVA, a Kruskal-Wallace test was performed; when there was a significant difference between classifiers, a Nonparametric Multiple Comparison Procedure (NPMC) for an unbalanced one-way layout was performed, as described in [10].

We conducted both parametric and non-parametric tests because the parametric tests are known to be robust to violations of the assumptions, so performing both was a way to verify the findings. Here, for all tests, we only report results with significant differences.

4.5   Validation with Follow-on Data

As mentioned above, we used the Spring data set to validate the classifiers trained on the Fall Data set (the Spring data set was not used to inform any of the training). The validation consisted of the following three steps. First, for each feature set selected by the feature selection step, a linear classifier was created using the entire subset described in Sect. 4.2. Second, each classifier was tested on the relevant subset of data from the Spring data set. Third, the accuracy, sensitivity, and specificity values and rankings were compared to the cross-validated values and rankings to determine how the classifiers generalized to a new and larger population.

5   Results

The classifier sets were designed to compare the performance of (1) a classifier using just tutor features vs. (2) one using features from one sensor in addition to the tutor features vs. (3) a classifier using all of the available features. The collection, feature selection, and cross validation results from the training data (Fall 2008) are described in [7]; however, a couple of important details are needed here. First, although the feature selection has the option of using both tutor data and other sensor data, sometimes it only selected tutor data. Table 4 shows the results of the feature selection. Second, we extended the cross-validation results to include sensitivity and specificity. Third, we modified the grain size, in that the samples in this work are on a per student rather than per test basis. The ranking and validation results are discussed below.

Table 4. These are the results of the feature selection. The baseline classifier for each emotion is just a linear model trained on a constant. The classifier names are the concatenation of an abbreviated emotion and the contributing sensor features. If there are no sensor features, then Tutor comes after the emotion, and when there is more than one classifier with the same feature set a letter is added to disambiguate the names. Names in **bold** are for classifiers that performed significantly better than the baseline for that emotion in at least one way.

Classifier name	Features
confBaseline	constant
confTutorA	Solv. on 1st + Hints Seen
confTutorM	# Incorrect + Solv. on 1st + Session
confSeat	# Incorrect + Solv. on 1st + sitForward Std Dev.
intBaseline	constant
intMouse	Group + # Hints + mouse Std Dev + mouse Max
intCamera	Group + # Hints + interestedMin
excBaseline	constant
excTutor	Group + # Incorrect
excCamera	interested Mean + # Incorrect
excCameraSeat	netSeatChangeMean + interestedMin + sitForwardMean

5.1   Classifier Ranking

Accuracy had a significant main effect on both the *interested* and *excited* affective states, but not for the *confident* and *frustrated* states. For the *interested* state, the classifier using the mouse and tutor features is significantly better than the baseline with a mean of 83.56% vs. 42.42%, according to both Tukey's HSD and NPMC tests. For the *excited* state, the classifiers with the tutor features were significantly better than the baseline with a mean of 73.62% vs. 46.31%.

As far as sensitivity is concerned, there is a significant main effect for *confident*, *interested*, and *excited* affective states using both parametric and nonparametric tests. However for *confident*, no classifier performed better than the baseline. For *interested*, both the camera and tutor, and mouse and tutor features were better than the baseline. For *excited*, the camera with seat sensors, camera sensors, and tutor only performed better than the baseline.

For specificity, there is only a significant main effect for *confident*, with TutorA, TutorM, and Seat classifiers performing better than the baseline. The details of these results are shown in Table 5.

Table 5. Classifier ranking using cross-validation data ($p < 0.05$)

Confident	Tukey HSD	NPMC
Specificity	$(confTutorA \sim confTutorM \sim confSeat) > confBaseline$	$(confTutorA \sim confTutorM) > confBaseline$
Interested	Tukey HSD	NPMC
Accuracy	$intMouse > intBaseline$	$intMouse > intBaseline$
Sensitivity	$(intCamera \sim intMouse) > intBaseline$	$(intCamera, intMouse) > intBaseline$
Excited	Tukey HSD	NPMC
Accuracy	$excTutor > excBaseline$	$excTutor > excBaseline$
Sensitivity	$(excTutor \sim excCamera \sim excCameraSeat) > excBaseline$	$(excTutor \sim excCamera \sim excCameraSeat) > excBaseline$

Given these results, our findings suggest that the tutor could generate interventions more reliably when it detects interest and excitement. If the tutor wanted to intervene when the student is *interested*, then using the mouse and tutor features or the camera and tutor features would be most appropriate. If the tutor wanted to intervene when the student is *excited* then either the camera with seat features, camera features, or tutor features classifier would all be appropriate.

It may be more relevant to intervene when a student is not *interested* or not *excited*, or not *confident*. Our results do not provide information on which features to use to predict low interest or low excitement, but to detect lack of confidence, we could use either the TutorA, TutorM, or Seat features trained on *confident*. The corresponding features are shown in Table 4.

5.2   Validation with Follow-on Data

In order to verify that our classifier ranking generalizes to new data sets, we tested the classifiers by training them with all of the Fall data and testing them with the Spring data. Performance results of the significantly ranked classifiers from the cross-validation done above are compared to the validation set and shown in Table 6. Since the data are from an entirely separate population, it is likely that the overall performance will degrade somewhat; however, if each classifier's performance is similar, then that will provide evidence that the classifiers should be preferred as they were ranked during the cross-validation phase.

When comparing mean accuracy for the training vs. test sets, there is a general drop in accuracy of between 2% and 15%, though in some cases, there is a much larger difference of up to 37%. The larger differences suggest that some of the features do not generalize well to new populations.

Table 6. This shows validation results of all classifiers that performed better than the baseline classifier during training. All values are the mean value per student. Fall specifies the training set based on the leave-one-student-out cross-validation, and Spring specifies the results of the classifiers trained on the training set (Fall Data), and tested on the validation set (Spring Data). Values in **bold** are significantly better ($p < 0.05$) than the baseline.

model	Accuracy		Sensitivity		Specificity	
	Fall	Spring	Fall	Spring	Fall	Spring
confBaseline	65.06%	62.58%	72.22%	76.13%	55.56%	44.14%
confTutorA	70.49%	65.49%	47.07%	46.04%	**90.43%**	**84.88%**
confTutorM	68.64%	67.53%	52.31%	52.26%	**82.41%**	**80.68%**
confSeat	65.70%	67.13%	54.63%	60.17%	**79.26%**	**70.32%**
intBaseline	42.42%	78.30%	0.00%	0.00%	81.82%	100.00%
intMouse	**83.56%**	63.34%	**29.73%**	5.09%	90.54%	81.60%
intCamera	69.44%	57.65%	**52.08%**	**12.11%**	64.58%	68.53%
excBaseline	46.31%	74.31%	0.00%	0.00%	96.15%	100.00%
excTutor	**73.62%**	62.99%	**36.54%**	**12.45%**	87.88%	77.28%
excCamera	66.33%	51.53%	**38.67%**	**28.39%**	72.00%	52.24%
excCameraSeat	70.67%	43.34%	**32.00%**	**15.97%**	83.00%	54.07%

   Results of ranking the classifiers on the validation data are shown in Table 7. Note that the accuracy rankings no longer hold, and the mouse classifier for the *interested* affective state is no longer significantly better than the baseline.

Table 7. Classifier ranking using validation data from the Spring of 2009. All differences indicated by '>' are significant with $p < 0.01$.

Confident	Tukey HSD	NPMC
Specificity	$(confCameraA \sim confTutorA \sim$ $confTutorM) > (confSeat \sim$ $confTutorW) > confBasline$ $confCameraB > confTutorW >$ $confBaseline$	$(confCameraA \sim confTutorA \sim$ $confTutorM) > (confSeat \sim$ $confTutorW) > confBasline$ $confCameraB > confTutorW >$ $confBaseline$
Interested	Tukey HSD	NPMC
Sensitivity	$intCamera > intBaseline$	$intCamera > intBaseline$
Excited	Tukey HSD	NPMC
Sensitivity	$((excCamera > excTutor) \sim$ $excCameraSeat) > excBaseline$	$excCamera > excCameraSeat >$ $excTutor > excBaseline$

6   Discussion

In this paper we describe a method for discovering actionable affective classifiers for Intelligent Tutoring Systems (ITS). Though the method was used with specific sensors, features, ITS and classifiers based on linear models, each of these could conceivably be swapped out for another system.

Our results identify a clear ranking for three classifiers designed to detect low student confidence, one classifier to detect interest, and three classifiers for detecting excitement. For not *confident*, two different sets of tutor only features performed better than the tutor and seat features, so it is unlikely that there would be a time that we would use the classifier with the seat sensor.

Now that we have actionable classifiers for three affective states, our ITS will be able to leverage the results to make a decision. For instance, the ITS could intervene whenever the classifier detects low student confidence, in order to help the student gain self efficacy. This intervention will have to also take into account other emotions detected, e.g., the detection of high excitement and/or high interest may change the type of intervention that is most appropriate.

Future work will involve implementing these various affect-based interventions, and evaluating their impact on student learning, affect and motivation. We also plan to explore how we can design classifiers for affect recognition that perform better than the baseline for the subset of affective states that our classifiers performed poorly on. One approach for doing so that we plan to implement is to identify more complex features based on the sensor data than those currently used. A more complete set of affective classifiers will likely improve the ITS interventions. For example, if we had a classifier that had good sensitivity for confidence, then that classifier could be used to stop interventions relating to low confidence.

Acknowledgments. We thank Sharon Edwards and Sarah English for their coordination of the school studies. This research was funded by awards from the National Science Foundation, 0705554, IIS/HCC *Affective Learning Companions: Modeling and Supporting Emotion During Teaching*, Woolf and Burleson (PIs) with Arroyo, Barto, and Fisher and the U.S. Department of Education to Woolf, B. P. (PI) with Arroyo, Maloy and the Center for Applied Special Technology (CAST), *Teaching Every Student: Using Intelligent Tutoring and Universal Design To Customize The Mathematics Curriculum*. Any opinions, findings, conclusions or recommendations expressed in this material are those of the authors and do not necessarily reflect the views of the funding agencies.

References

1. Beebe, S.A., Ivy, D.K.: Explaining student learning: An emotion model (1994)
2. Kort, B., Reilly, R., Picard, R.: An affective model of interplay between emotions and learning: reengineering educational pedagogy-building a learning companion. In: IEEE International Conference on Advanced Learning Technologies, Proceedings, pp. 43–46 (2001)
3. Graham, S., Weiner, B.: Theories and principles of motivation. In: Berliner, D., Calfee, R. (eds.) Handbook of Educational Psychology, vol. 4, pp. 63–84. Macmillan, New York (1996)
4. Zimmerman, B.J.: Self-efficacy: An essential motive to learn. Contemporary Educational Psychology 25, 82–91 (2000)

5. Lepper, M.R., Woolverton, M., Mumme, D.L., Gurtner, J.L.: Technology in education. In: Motivational techniques of expert human tutors: Lessons for the design of computer-based tutors, pp. 75–105. Lawrence Erlbaum Associates, Inc., Mahwah (1993)

6. Derry, S.J., Potts, M.K.: How tutors characterize students: a study of personal constructs in tutoring. In: ICLS '96: Proceedings of the 1996 international conference on Learning sciences, International Society of the Learning Sciences, pp. 368–373 (1996)

7. Cooper, D.G., Arroyo, I., Woolf, B.P., Muldner, K., Burleson, W., Christopherson, R.: Sensors model student self concept in the classroom. In: Houben, G.-J., McCalla, G., Pianesi, F., Zancanaro, M. (eds.) UMAP 2009. LNCS, vol. 5535, pp. 30–41. Springer, Heidelberg (2009)

8. Guyon, I., Gunn, S., Ben-Hur, A., Dror, G.: Result analysis of the nips 2003 feature selection challenge. In: Advances in Neural Information Processing Systems (2004)

9. D'Mello, S., Picard, R.W., Graesser, A.: Toward an affect-sensitive autotutor. IEEE Intelligent Systems 22(4), 53–61 (2007)

10. Munzel, U., Hothorn, L.A.: A unified approach to simultaneous rank test procedures in the unbalanced one-way layout. Biometrical Journal 43(5), 553–569 (2001)

11. Conati, C., Maclaren, H.: Modeling user affect from causes and effects. In: Houben, G.-J., McCalla, G., Pianesi, F., Zancanaro, M. (eds.) UMAP 2009. LNCS, vol. 5535, pp. 4–15. Springer, Heidelberg (2009)

12. D'Mello, S.K., Craig, S.D., Graesser, A.C.: Multimethod assessment of affective experience and expression during deep learning. Int. J. Learn. Technol. 4(3/4), 165–187 (2009)

13. Hernandez, Y., Arroyo-Figueroa, G., Sucar, L.: Evaluating a probabilistic model for affective behavior in an intelligent tutoring system, pp. 408–412 (July 2008)

14. Robison, J., McQuiggan, S., Lester, J.: Evaluating the consequences of affective feedback in intelligent tutoring systems, pp. 1–6 (2009)

15. Arroyo, I., Cooper, D.G., Burleson, W., Woolf, B.P., Muldner, K., Christopherson, R.: Emotion sensors go to school. In: Dimitrova, V., Mizoguchi, R., du Boulay, B., Graesser, A.C. (eds.) AIED, vol. 200, pp. 17–24. IOS Press, Amsterdam (2009)

16. Arroyo, I., Woolf, B.P., Royer, J.M., Tai, M.: Affective gendered learning companions. In: Dimitrova, V., Mizoguchi, R., du Boulay, B., Graesser, A.C. (eds.) AIED, vol. 200, pp. 41–48. IOS Press, Amsterdam (2009)

Inducing Effective Pedagogical Strategies Using Learning Context Features

Min Chi[1], Kurt VanLehn[2], Diane Litman[3], and Pamela Jordan[4]

[1] Machine Learning Department, Carnegie Mellon University, PA, 15213 USA
minchi@cs.cmu.edu

[2] School of Computing and Informatics, Arizona State University, AZ, 85287 USA
Kurt.Vanlehn@asu.edu

[3] Department of Computer Science, University of Pittsburgh, PA, 15260 USA
litman@cs.pitt.edu

[4] Department of Biomedical Informatics, University of Pittsburgh,
Pittsburgh, PA 15260
pjordan@pitt.edu

Abstract. Effective pedagogical strategies are important for e-learning environments. While it is assumed that an effective learning environment should craft and adapt its actions to the user's needs, it is often not clear how to do so. In this paper, we used a Natural Language Tutoring System named Cordillera and applied Reinforcement Learning (RL) to induce pedagogical strategies directly from pre-existing human user interaction corpora. 50 features were explored to model the learning context. Of these features, domain-oriented and system performance features were the most influential while user performance and background features were rarely selected. The induced pedagogical strategies were then evaluated on real users and results were compared with pre-existing human user interaction corpora. Overall, our results show that RL is a feasible approach to induce effective, adaptive pedagogical strategies by using a relatively small training corpus. Moreover, we believe that our approach can be used to develop other adaptive and personalized learning environments.

1 Introduction

Natural Language (NL) Tutoring Systems are a form of Intelligent Tutoring Systems (ITSs) that use natural dialogue for instructional purposes such as helping students to learn a subject by engaging in a natural language conversation. Why2-Atlas and Why2-AutoTutor [1], for example, are NL tutoring systems that teach students conceptual physics. One central component of NL Tutoring Systems is the dialogue manager, which uses *dialogue strategies* to decide what action to take at each point during the tutorial dialogue. For tutoring systems, dialogue strategies are also referred to as pedagogical strategies.

It is commonly believed that an effective tutoring system would craft and adapt its actions to the students' needs based upon their current knowledge

P. De Bra, A. Kobsa, and D. Chin (Eds.): UMAP 2010, LNCS 6075, pp. 147–158, 2010.

level, general aptitude, and other salient features [2]. However, most pedagogical strategies for ITSs are encoded as hand-coded rules that seek to implement cognitive and/or pedagogical theories. Typically, the theories are considerably more general than the specific decisions that designers must make, which makes it difficult to tell if a specific pedagogical strategy is consistent with the theory. Moreover, it is often not easy to empirically evaluate these decisions because the overall effectiveness of the system depends on many factors, such as the usability of the system, how easily the dialogues are understood, and so on. Ideally, several versions of a system are created, each employing a different pedagogical strategy. Data is then collected with human subjects interacting with these different versions of the system and the results are compared. Due to the high cost of experiments, only a handful of strategies are typically explored. Yet, many such other reasonable ones are still possible.

In recent years, work on the design of NL non-tutoring Dialogue Systems has involved an increasing number of data-driven methodologies. Among these, Reinforcement Learning (RL) has been widely applied [3]. RL is a machine learning method that centers on the maximization of expected rewards. It has many features well-suited to the problem of designing the dialogue manager such as unobservable states, delayed rewards, and so on. Its primary advantage is its ability to compute an optimal policy within a much larger search space, using a relatively small training corpus. In this work, rather than implementing pedagogical strategies drawn from human experts or theories, we applied RL to derive pedagogical strategies using pre-existing interactivity data.

While most previous work on using RL to train non-tutoring dialogue systems has been successful [3], whether it can be used to improve the effectiveness of NL tutoring systems is still an open question. One major source of uncertainty comes from the fact that the rewards used in RL are much more delayed in NL tutoring systems than those in non-tutoring dialogue systems. Much of this work in NL non-tutoring Dialogue Systems is focused on systems that obtain information or search databases such as querying bus schedules [4]. For example, in non-tutoring Systems like the train scheduler, the interaction time is often less than 20 minutes, and the number of interactions within user-dialogue systems is generally less than 20 turns [3]. In the training corpora reported here, the time is roughly 4-9 hours and the number of interactions is about 280 turns. More immediate rewards are more effective than more delayed rewards for RL induction. This is because the issue of assigning credit for a decision, attributing responsibility to the relevant decision is substantially easier in the former case. The more we delay rewards, the more difficult it becomes to identify the decision(s) responsible for our success or failure. Additionally, to train an RL model, a large amount of data is generally needed. In this work, we use human data only instead of data from simulators as in applying RL in non-tutoring dialogue systems. This is because the cause of human learning is still an open question and thus it would be difficult to accurately simulate students' responses to the tutor and simulate how students would learn. Given the high cost of collecting human data, we were more likely to encounter the issue of data sparsity.

For RL, as with all machine learning tasks, success is dependent upon an effective state representation or state model. An effective state representation should be an accurate and compact model of the learning context. Compared with non-tutoring Dialogue Systems, where success is primarily a function of communication efficiency, communication efficiency is only one of the factors determining whether a student learns well from an NL tutoring system. Moreover, the other factors are not well understood, so to be conservative, states need to contain features for anything that is likely to affect learning. Hence, state models for RL applications to tutoring systems tend to be much larger than state models for non-tutoring applications. Unfortunately, as states increase in size and complexity, we risk making the learning problem intractable or the decision space too large to sample effectively. In order to obtain an effective state model that both minimizes state size while retaining sufficient relevant information about the learning context, we began with a large set of features to which we applied a series of feature-selection methods in order to reduce them to a tractable subset. Before describing our approach in detail, we will briefly describe the two types of tutorial decisions covered by the induced pedagogical policies.

2   Two Types of Tutorial Decisions

Among the many tutorial decisions that must be made, we focus on two types of decisions, Elicit/Tell (ET) and Justify/Skip-Justify (JS). The ET decision asks "should the tutor *elicit* the next problem-solving step from the student, or should he or she *tell* the student the next step directly?". For example, when the next step is to select a principle to apply and the target principle is the "definition of Kinetic Energy', the tutor can choose to elicit this from the student by asking the question, "Which principle will help you calculate the rock's kinetic energy at T0?" By contrast, the tutor can elect to tell the student the step by stating, "To calculate the rock's kinetic energy at T0, let's apply the definition of Kinetic Energy." The JS decision asks "should the tutor include a *justify* for a step just taken or or not". For example, after deciding to use the "definition of Kinetic Energy", the tutor can choose to ask the student why the principle is applicable or to skip to ask. There is no widespread consensus on how or when any of these actions should be taken [5–8]. This is why our research objective is to derive policies for them from empirical data.

3   Applying RL to Induce Pedagogical Strategies

Previous research on using RL to improve non-tutoring dialogue systems (e.g. [9]) has typically used Markov Decision Processes (MDPs) [10] to model dialogue data. The central idea behind this approach is to transform the problem of inducing effective pedagogical strategies into computing an optimal policy for an agent that is choosing actions in an MDP. An MDP formally corresponds to a 4-tuple (S, A, T, R), in which: $S = \{S_1, \cdots, S_n\}$ is a state space; $A = \{A_1, \cdots, A_m\}$ is an action space represented by a set of action variables; T :

$S \times A \times S \rightarrow [0, 1]$ is a set of transition probabilities $P(S_j|S_i, A_k)$, which is the probability that the model would transition from state S_i to state S_j after the agent takes action A_k; $R : S \times A \times S \rightarrow R$ assigns rewards to state transitions. Finally, $\pi : S \rightarrow A$ is defined as a policy, which determines which action the agent should take in each state in order to maximize the expected reward.

The set of possible actions, A, is small and well-defined. In our application, we have $A = \{Elicit, Tell\}$ for inducing pedagogical strategies on ET decisions and $A = \{Justify, Skip - Justify\}$ for inducing those on JS decisions. The set of possible states, S, however is not well-defined in advance and can potentially be astronomically large if we include everything that could possibly influence the effectiveness of a tutorial action. In this study, we assumed that S is the Cartesian product of a set of state features $F = \{F_1, \cdots, F_p\}$ and our challenge now becomes finding a set of features F to model the state or learning context compactly and yet effectively. Features must be operational, in that there is some way to determine their value prior to just before each tutor action in the dialogue. For instance, one operational feature would be a count of the number of words uttered by the student since the last tutor turn.

Each student-system interaction dialogue d can be viewed as a trajectory in the chosen state space determined by the system actions and student responses:

$$S_1 \xrightarrow{A_1, R_1} S_2 \xrightarrow{A_2, R_2} \cdots S_n \xrightarrow{A_n, R_n}$$

Here $S_i \xrightarrow{A_i, R_i} S_{i+1}$ indicated that at the i_{th} turn in the tutorial dialogue d, the system was in state S_i, executed action A_i, received reward R_i, and then transferred into state S_{i+1}. Because our primary interest is to improve students' learning, we used Normalized Learning Gain (NLG) as the reward because it measures students' gain *irrespective of their incoming competence*. The NLG is defined as: $NLG = \frac{posttest - pretest}{1 - pretest}$. Here *posttest* and *pretest* refer to the students' test scores before and after the training respectively; and 1 is the maximum score. Given that a student's NLG will not be available until the entire tutorial dialogue is completed, only terminal dialogue states have non-zero rewards. Thus for a tutorial dialogue d, $R_1 \cdots$, R_{n-1} are all equal to 0 and only the final reward equal to the student's $NLG \times 100$, which is in the range of $(-\infty, 100]$.

Once the MDP structure $\{S, A, R\}$ has been defined, the transition probabilities T are estimated from the training corpus, which is the collection of dialogues, as: $T = \{p(S_j|S_i, A_k)\}_{i,j=1,\cdots,n}^{k=1,\cdots,m}$. More specifically, $p(S_j|S_i, A_k)$ is calculated by taking the number of times that the dialogue is in state S_i, the tutor took action A_k, and the dialogue was next in state S_j divided by the number of times the dialogue was in S_i and the tutor took A_k. The reliability of these estimates clearly depends upon the size and structure of the training data. Once a complete MDP is constructed, a dynamic programming approach can be used to learn the optimal control policy π^* and here we used the toolkit developed by Tetreault and Litman [11]. The rest of this section presents a few critical details of the process, but many others must be omitted to save space.

3.1   Knowledge Component (KC) Based Pedagogical Strategies

In the learning literature, it is commonly assumed that relevant knowledge in domains such as math and science is structured as a set of independent but co-occurring Knowledge Components (KCs) and that KC's are learned independently. A KC is "a generalization of everyday terms like concept, principle, fact, or skill, and cognitive science terms like schema, production rule, misconception, or facet" [12]. For the purposes of ITSs, these are the atomic units of knowledge.

The domain selected for this project is a subset of the physics work-energy domain, which is characterized by eight primary KCs. For instance, one KC is the definition of kinetic energy ($KE = \frac{1}{2} * m * v^2$) and another is the definition of gravitational potential energy ($GPE = m * g * h$). It is assumed that a tutorial dialogue about one KC (e.g., kinetic energy) will have no impact on the student's understanding of any other KC (e.g, of potential energy). This is an idealization, but it has served ITS developers well for many decades, and is a fundamental assumption of many cognitive models [13, 14].

When dealing with a specific KC, the expectation is that the tutor's best policy for teaching that KC (e.g., when to Elicit vs, when to Tell) would be based upon the student's mastery of the KC in question, its intrinsic difficulty, and other relevant, but not necessarily known, factors specific to that KC. In other words, an optimal policy for one KC might not be optimal for another. Therefore, one assumption made in this paper is that inducing pedagogical policies specific to each KC would be more effective than inducing an overall KC-general policy. In order to learn a policy for each KC, we annotated our tutoring dialogues and action decisions with the KCs covered by each action. For each KC, the final kappa was ≥ 0.77, which is fairly high given the complexity of the task. Additionally, a domain expert also mapped the pre-/post test problems to the sets of relevant KCs. This resulted in a KC-specific NLG score for each student. Thus, for the decision of when to Elicit vs. Tell about the definition of kinetic energy KC_{20}, we consider all and only the dialogue about that KC and consider only the learning gains on that KC.

Given these independence assumptions, the overall problem of inducing a policy for ET decisions and a policy for JS decisions is decomposed into 8 sub-problems of each kind, one per KC. Among the eight KCs, KC_1 does not arise in any JS decisions and thus only an ET policy was induced for it. For each of the remaining seven KCs, a pairs of policies, one ET policy and one JS policy, were induced. So we induced 15 KC-based NormGain policies. During the tutoring process, there were some decision steps that did not involve any of the eight primary KCs. For them, two KC-general policies, an ET policy and a JS policy, were induced. To sum, a total of 17 NormGain policies were induced in this study.

3.2   Training Corpora

In order to apply RL to induce pedagogical strategies and evaluate the induced strategies, we used Cordillera. Cordillera is a NL tutoring system that teaches introductory physics[12]. To reduce confounds due to imperfect NL understanding, the NL understanding module was replaced with human wizards whose only task is to match students' answers to the closest response from a list of potential responses and they *cannot* make the tutorial decisions. As the first step, we developed an initial version of Cordillera, called random-Cordillera on which both ET and JS decisions on it were made randomly. 64 college students were then trained on random-Cordillera in 2007 and the collected training data is called the Exploratory corpus.

From the Exploratory corpus, we tried our first round of policy induction. It is done by first defining 17 state features and then used some sort of greedy-like procedure to search for a small subset of it as the state representation. For the reward functions, we had dichotomized the NLGs scores so that there were only two levels of reward and thus the derived policies were named DichGain policies. We next tested our hypothesis that these RL-induced policies would improve the effectiveness of a tutoring system. The version of Cordillera that implemented the DichGain policies was named DichGain-Cordillera. Except following the policies (random vs. DichGain), the remaining components of Cordillera, including the GUI interface, the same training problems, and the tutorial scripts, were left untouched. DichGain-Cordillera's effectiveness was tested by training a new group of 37 college students in 2008. Results showed that although the DichGain policies generated significantly different patterns of tutorial decisions than the random policy, no significant difference was found between the two groups on the pretest, posttest, or the NLGs.

3.3   Inducing NormGain Strategies

Although the previous experiment seemingly failed to confirm our hypothesis, it did generate more training data. We now have three training corpora: the Exploratory corpus in 2007, the DichGain corpus in 2008, and a combined training corpus dataset consisting of the 101 dialogues from both the Exploratory and the DichGain corpora. This time we started with a larger set of possible state features. We included 50 features based upon six categories of features considered by previous research [15–17] to be relevant. They include not only student's performance and background related features such as student's overall performance but also domain-oriented and system behavior related features. Moreover, we explored more domain -general methods of searching the power set of the 50 features and instead of dichotomizing learning gains as rewards, we used the $NLG \times 100$ directly. Based on the reward function, the induced policies are named normalized Gain (NormGain) policies in the following.

Figure 1 shows an example of a learned NormGain policy on KC_{20}, "Definition of Kinetic Enegy", for JS decisions. The policy involves five features. They are:

TimeInSession: The total time spent in the current session. This feature reflects a student's fatigue level.

nKCs: The number of times the present KC has occurred in the current dialogue. This feature reflects the students' familiarity with the current KC.

pctElicit: The percentage of ET decisions turned out to be elicit during the dialogue. This feature reflects how active a student is overall.

stuAverageWords: The average number of words per student turn. This reflects the student's level of activity and verbosity.

stuAverageConceptSession: The ratio of the number of the student's turns which involves at least one physics concept to all the student turns in this session. This feature reflects how often the student's answers involved at least one physics concepts since the start of the training.

[**Feature:**]

    TimeInSession: $[0, 3040.80) \rightarrow 0$; $[3040.8, \infty] \rightarrow 1$

    nKCs: $[0, 66) \rightarrow 0$; $[66, \infty] \rightarrow 1$

    pctElicit: $[0, 0.49) \rightarrow 0$; $[0.49, 1) \rightarrow 1$

    stuAverageWords: $[0, 4.18) \rightarrow 0$; $[4.18, \infty] \rightarrow 1$

    stuAverageConceptSession: $[0, 0.29) \rightarrow 0$; $[0.29, 1] \rightarrow 1$

[**Policy:**]

  Justify:

    0:0:0:0:0 0:0:1:1:0 0:1:0:0:1 0:0:1:0:0 0:1:0:1:1 0:1:1:0:0 0:1:1:0:1 0:1:1:1:0

    0:1:1:1:1 1:0:0:0:0 1:0:0:1:0 1:0:1:0:0 1:0:1:0:1 1:0:1:1:0 1:0:1:1:1 1:1:0:0:1

    1:1:1:0:0 1:1:1:0:1 1:1:1:1:0 1:1:1:1:1

  Skip-Justify:

    0:0:0:0:1 0:0:0:1:0 0:0:0:1:1 0:0:1:0:1 0:0:1:1:1 0:1:0:0:0 0:1:0:1:0 1:0:0:0:1

    1:0:0:1:1 1:1:0:0:0 1:1:0:1:0 1:1:0:1:1

Fig. 1. An NormGain Policy on KC_{20} For JS Decisions

MDP generally requires discrete features and thus all the continous features need to be discretized. Figure refFig.ExampleNormGainPolicy describes how each of the five features was discretized. For example, for TimeInSession, if its value is above 3040.80 sec (50.68 min), it is 1 otherwise, it is 0. There were a total of 32 rules learned: in 20 situations the tutor should execute the justification step, in the other 12 situations the tutor should skip. For example, 0:0:0:0:0 is listed as the first situation under the [Justify], it means that when the student has spend less than 50.68 min in this session, the occurrence of KC_{20} in the student's dialogue history is less than 66, the student has got less than 49% of elicit in the past, the average number of words in student's entries is less than 4.18 words, and the percentage of times times that the student mention a physics concept in his/her turn is less than 29%, then the tutor should execute the justification. As you can see, the RL induced policies are very subtle and adaptive to the learning context and they are not like most of the tutorial tactics derived from analyzing human tutorial dialogues.

The resulting 17 NormGain policies were implemented back into Cordillera yielding a new version of the system, named NormGain-Cordillera. In order to test our hypothesis that RL can be used to improve tutoring systems, we tested the effectiveness of NormGain-Cordillera on a new group of students as described in the next section. The section is written as if one large experiment was done with 3 conditions, when in fact the 3 groups of students were run sequentially, as described above.

4   Methods

The purpose of this experiment is to compare the learning gains of students using random-Cordillera, DichGain-Cordillera and NormGain-Cordillera respectively. All participants were required to have basic knowledge of high-school algebra, no experience with college-level physics, and were paid for their time. Each participant took between six and fourteen hours (3-7 sessions) to finish the study in a period of two to three weeks. Each session typically lasted about two hours.

The domain selected here is Physics work-energy domain as covered in a first-year college physics course. The eight primary KCs were: the weight law (KC1), definition of work (KC14), Definition of Kinetic Energy (KC20), Gravitational Potential Energy (KC21), Spring Potential Energy (KC22), Total Mechanical Energy (KC24), Conservation of Total Mechanical Energy (KC27), and Change of Total Mechanical Energy (KC28).

All three groups experienced the identical procedure and materials. More specifically, participants all completed a background survey; read a textbook covering the target domain knowledge; took a pretest; solved the same seven training problems in the same order on Cordillera; and finally took a posttest. The pretest and posttest were identical.

Only three salient differences existed across the three groups:

1. The Exploratory group with a population of 64 was recruited in 2007; the DichGain group with a population of 37 was recruited in 2008; and the NormGain group with a population of 29 was recruited in 2009.
2. Random-Cordillera made random decisions and the DichGain-Cordillera and NormGain-Cordillera followed the induced DichGain and NormGain policies respectively.
3. A group of six human wizards were used by the Exploratory and DichGain groups; but only one of six wizards were involved in the NormGain group.

4.1   Grading

All tests were graded by a single experienced grader who did not know which student belonged to which group. For all identified relevant KCs in a test question, a KC-based score for each KC application was given. We assigned an overall competence to a student by the sum of these KC-based scores and normalizing to a [0,1] interval. We also tried other methods of computing an overall score, and this did not affect the pattern of results discussed below.

5   Results

The primary goal reported below is twofold: first, to test whether our improved RL methodology and software produced more effective pedagogical strategies than either random policies or the policies used by the DichGain group; and second, to determine the features selected in the state models in the NormGain policies.

5.1   Learning Results

A one-way ANOVA showed that there were no significant differences among the three groups on overall training time: $F(2, 122) = 1.831$, $p = .17$. After solving seven training problems on Cordillera, all three groups scored significantly higher in the posttest than pretest: $F(1, 126) = 10.40$, $p = 0.002$ for the Exploratory group, $F(1, 72) = 7.20$, $p = 0.009$ for the DichGain group, and $F(1, 56) = 32.62$, $p = 0.000$ for the NormGain group respectively. The results suggested that the basic practices and problems, domain exposure, and interactivity of Cordillera might cause students to learn even from tutors with non-optimal pedagogical skills.

A one-way ANOVA was used for comparing the learning performance differences among the three groups. While no significant pre-test score differences were found: $F(2, 127) = 0.53$, $p = 0.59$, there were significant differences among the three groups on both post-test scores and NLG scores: $F(2, 127) = 5.16$, $p = .007$ and $F(2, 127) = 7.57$, $p = 0.001$ respectively. Figure 2 compares the three groups on the pre-test, post-test, and NLG scores. Moreover, a t-test comparison showed that the NormGain group out-performed the DichGain on both post-test scores and NLG scores: $t(64) = 3.28$, $p = .002$, $d^1 = 0.82$ and $t(64) = 3.68$, $p = 0.000$, $d = 0.95$ respectively. Similar results were found between the NormGain and Exploratory groups: $t(91) = 2.76$, $p = .007$, $d = 0.63$ on post-test, and $t(91) = 3.61$, $p = 0.000$, $d = 0.84$ on NLG scores respectively.

To summarize, the comparison among the three groups shows that the NormGain group significantly outperformed both the Exploratory and DichGain groups. These results were consistent both for the post-test scores and the NLGs and the effect sizes were large by Cohen's d criteria.

5.2   Feature Choices in INDUCED POLICIES

Only 30 out of 50 defined features occurred among the 17 NormGain policies. Among them, the most frequent feature appeared seven times. Four features appeared in more than three induced policies and they are:

StepDifficulty (7 Occurrences): Which encodes a step's difficulty level and its value is roughly estimated from the Combined Corpus based on the percentage of answers that were correct on the step.

ConceptToWordRatio (5 Occurences): Which represents the ratio of the physics concepts to words in the tutor's dialogue.

---

[1] Cohen's d, which is defined as the mean learning gain of the experimental group minus the mean learning gain of the control group, divided by the groups' pooled standard deviation.

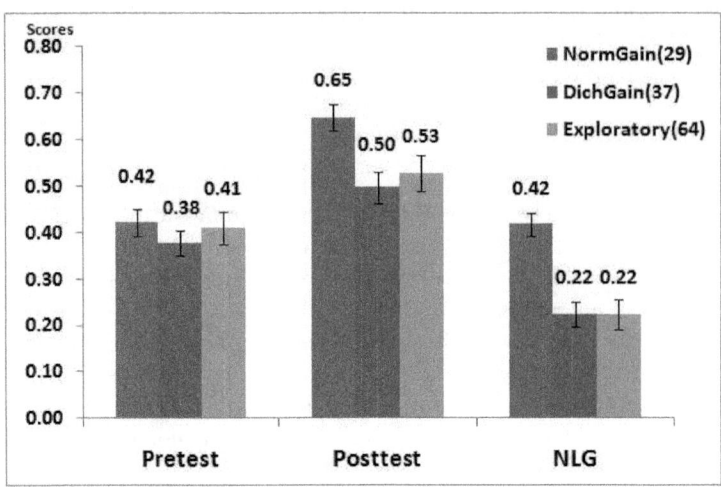

Fig. 2. Compare Three Groups Learning Performance under Overall Grading

NumberTellsSinceElicit (5 Occurences): Which represents the number of tells the student has received since the last elicit.

TimeBetweenDecisions (4 Occurences): Which represents the time since the last tutorial decision was made on the current KC.

While StepDifficulty can be seen as domain-oriented feature, the remaining three features are all the system-behavior related features. The high occurrence of StepDifficulty in the NormGain policies is not very surprising because it has been widely believed that difficulty level is an important factor for the system to behave adaptively and effectively. The frequent involvement of System-behavior related features in the induced policy maybe because these features might reflect student's general aptitude, the activeness of their knowledge on a specific KC, and so on. For example, NumberTellsSinceElicit reflects how interactive a student has been recently and TimeBetweenDecisions reflect how active a student's knowledge on the current KC is. When TimeBetweenDecisions is high, it means that the tutor has not mentioned the KC recently so the student's knowledge on the current KC may be still or forgotten.

Much to our surprise, the features related to the students' overall or recent performance (e.g., error rate) and background (e.g., MSAT, VSAT, gender, pretest score) appeared the least or none in the NormGain policies. Although space does not permit a detailed discussion of the prevalence of features, it appears to be a mixture of easily anticipated dependencies (e.g., step difficulty) and a few surprises (why doesn't error rate matter?).

6   Conclusions

We presented a general data-driven method that can be used to improve NL tutoring system over time. We built and improved a large NL tutoring system using our methodology, and showed that RL is able to effectively search a very large continous space of dialogue policies (After discretized, the space is $\geq 2^{50}$ in size) using a relatively small amount of training dialogue data (64 subjects in Exploratory group and 37 in the DichGain group). A post-hoc comparison showed that our learned policy outperformed both sets of training policies in terms of learning performance. This success supports the hypothesis that RL-induced rules are effective and that the approach taken in this project was a feasible one. However, inducing effective tutorial tactics was not trivial. The DichGain tutorial tactics did not seem to be more effective than the random decisions in Random-Cordillera. A number of factors were changed in deriving NormGain policies from the process of inducing DichGain policies. These included the feature choices, the choice of training corpora, feature selection methods, and so on. So it is still not clear which factor or factors caused a change in effectiveness.

Although the discussion of induced features has been cursory, it nonetheless appears that the learning context features that make the most difference for determining when to Tell vs. Elicit and when to Justify vs. Skip-Justify are not always the ones that one would first think of given current theories of learning and tutoring. For instance, it is widely believed that effective tutors adapt their behavior to the individual student knowledge level. However, such feature did not appear in the NormGain policies. Indeed, individualized tutoring is considered a Grand Challenge by the National Academy of Engineering. However, such features appeared to play little role in the effective tutorial policies induced from our data. Overall, our results suggested that when building an accurate learning context model, adding domain-oriented and the system behavior related features would be beneficial.

Acknowledgments. NSF (#0325054) supported this work and NSF (#SBE-0836012) supported its publication. We thank Collin Lynch and the reviewers for helpful comments.

References

1. VanLehn, K., Jordan, P.W., Rosé, C.P., Bhembe, D., et al.: The architecture of why2-atlas: A coach for qualitative physics essay writing. In: Cerri, S.A., Gouardéres, G., Paraguaçu, F. (eds.) ITS 2002. LNCS, vol. 2363, pp. 158–167. Springer, Heidelberg (2002)
2. Chi, M.T.H., Siler, S.A., Jeong, H., Yamauchi, T., Hausmann, R.G.: Learning from human tutoring. Cognitive Science 25, 471–533 (2001)
3. Singh, S.P., Litman, D.J., Kearns, M.J., Walker, M.A.: Optimizing dialogue management with reinforcement learning: Experiments with the njfun system. J. Artif. Intell. Res. (JAIR) 16, 105–133 (2002)
4. Raux, A., Langner, B., Bohus, D., Black, A.W., Eskenazi, M.: Let's go public! taking a spoken dialog system to the real world. In: Proceedings of Interspeech, Eurospeech (2005)
5. Collins, A., Brown, J.S., Newman, S.E.: Cognitive apprenticeship: Teaching the craft of reading, writing and mathematics. In: Resnick, L.B. (ed.) Knowing, learning and instruction: Essays in honor of Robert Glaser, pp. 453–494. Lawrence Erlbaum Associates, Hillsdale (1989)

6. Chi, M.T.H., de Leeuw, N., Chiu, M.H., LaVancher, C.: Eliciting self-explanations improves understanding. Cognitive Science 18(3), 439–477 (1994)

7. Conati, C., VanLehn, K.: Toward computer-based support of meta-cognitive skills: a computational framework to coach self-explanation. International Journal of Artificial Intelligence in Education 11, 398–415 (2000)

8. Katz, S., O'Donnell, G., Kay, H.: An approach to analyzing the role and structure of reflective dialogue. International Journal of Artificial Intelligence and Education 11, 320–343 (2000)

9. Singh, S.P., Kearns, M.J., Litman, D.J., Walker, M.A.: Reinforcement learning for spoken dialogue systems. In: Solla, S.A., Leen, T.K., Müller, K.R. (eds.) NIPS, pp. 956–962. The MIT Press, Cambridge (1999)

10. Sutton, R.S., Barto, A.G.: Reinforcement Learning. MIT Press Bradford Books, Cambridge (1998)

11. Tetreault, J.R., Litman, D.J.: A reinforcement learning approach to evaluating state representations in spoken dialogue systems. Speech Communication 50(8-9), 683–696 (2008)

12. VanLehn, K., Jordan, P.W., Litman, D.: Developing pedagogically effective tutorial dialogue tactics: Experiments and a testbed. In: Proceedings of SLaTE Workshop on Speech and Language Technology in Education ISCA Tutorial and Research Workshop (2007)

13. Anderson, J.R.: The architecture of cognition. Harvard University Press, Cambridge (1983)

14. Newell, A. (ed.): Unified Theories of Cognition. Harvard University Press, Cambridge (1994); Reprint edition

15. Moore, J.D., Porayska-Pomsta, K., Varges, S., Zinn, C.: Generating tutorial feedback with affect. In: Barr, V., Markov, Z. (eds.) FLAIRS Conference. AAAI Press, Menlo Park (2004)

16. Beck, J., Woolf, B.P., Beal, C.R.: Advisor: A machine learning architecture for intelligent tutor construction. In: AAAI/IAAI, pp. 552–557. AAAI Press / The MIT Press (2000)

17. Forbes-Riley, K., Litman, D.J., Purandare, A., Rotaru, M., Tetreault, J.R.: Comparing linguistic features for modeling learning in computer tutoring. In: Luckin, R., Koedinger, K.R., Greer, J.E. (eds.) AIED. Frontiers in Artificial Intelligence and Applications, vol. 158, pp. 270–277. IOS Press, Amsterdam (2007)

"Yes!": Using Tutor and Sensor Data to Predict Moments of Delight during Instructional Activities

Kasia Muldner, Winslow Burleson, and Kurt VanLehn

Arizona State University
{Katarzyna.Muldner,Winslow.Burleson,Kurt.VanLehn}@asu.edu

Abstract. A long standing challenge for intelligent tutoring system (ITS) designers and educators alike is how to encourage students to take pleasure and interest in learning activities. In this paper, we present findings from a user study involving students interacting with an ITS, focusing on when students express excitement, what we dub "yes!" moments. These findings include an empirically-based user model that relies on both interaction and physiological sensor features to predict "yes!" events; here we describe this model, its validation, and initial indicators of its importance for understanding and fostering student interest.

Keywords: interest, motivation, empirically-based model, sensing devices.

1 Introduction

In some cultures, the classic "yes!" gesture is to clench the fist of one's dominant arm, jerk the arm downward and exclaim "yes!" - everyone understands this as an expression of triumphal victory. When we noticed this behavior among students using our physics tutoring system, we began to wonder about it. For instance, what causes a "yes!" during tutoring? Is the "yes!" behavior a desirable outcome in itself or is it also associated with other desirable outcomes?

Because we are interested in building affective learning companions, we are also interested in how a companion could use students' "yes!" behavior for its own ends, such as increased bonding with the student. This requires, however, that the companion can detect "yes!" behaviors in real time. This paper reports our progress on addressing these issues and questions, including:

1. *Is the "yes!" behavior a desirable outcome for a tutoring system or associated with one?* We argue from the literature that it is both.
2. *What causes "yes!" events and how can we increase their frequency?* We compare "yes!" episodes with ones where a "yes!" could have occurred but did not. This descriptive analysis sets the stage for future work on what could cause an increase in "yes!" events.
3. *How can "yes!" events by used by tutors, learning companions or other agents?* We present a review of the literature that suggests some possibilities.

P. De Bra, A. Kobsa, and D. Chin (Eds.): UMAP 2010, LNCS 6075, pp. 159–170, 2010.
© Springer-Verlag Berlin Heidelberg 2010

4. *Can a "yes!" event be detected more accurately than a baseline approach?* We developed a regression model based on sensor and tutor log data analysis that has high accuracy.

The rest of this introduction contains literature reviews that address points 1 and 3, and a review of related work on affect detection (point 4).

1.1 The Likely Role of "yes!" in Learning and Interest

As we describe in Sect. 3, we view "yes!" as a class of brief expressions of (possibly highly exuberant) positive affect. Positive affect has been linked to increased personal interest [1, 2], which is in turn associated with a facilitative effect on cognitive functioning [3], and improved performance on creative problem solving and other tasks [4], persevering in the face of failure, investing time when it is needed and engaging in mindful and creative processing (for a review see [5]). Although there is work in the psychology community on how interest develops and is maintained (e.g., [6, 7]), to date there does not yet exist sufficient work on these topics to understand the role of positive affect in general and of "yes!" events in particular, so calls for additional research are common (e.g., [8]).

We should point out, however, that while positive affect could itself be considered a desirable property during tutoring, it has not always shown strong correlations with learning [9]. For instance, doing unchallenging problems may make students happy but may not cause learning. However, the "yes!" expression of positive affect may well be correlated with learning, because as we show later, "yes!" occurs only after the student has been challenged, and challenge fosters learning [10].

1.2 How Can "yes!" Events Be Used during Tutoring and Learning?

In general, about 50% of human tutor interventions relate to student affect [11], highlighting the importance of addressing affect in pedagogical interactions. As far as addressing "yes!" events, work on the impact of tutorial feedback provides some direction regarding how "yes!" detection can be valuable to a tutoring system for generating subsequent responses. For instance, praise needs to be delivered at the right moment, e.g., be perceived as representative of effort and sincere, to be effective [12], and so a "yes!" event may be exactly the right time for an agent to give praise.

If "yes!" events do predict increased learning, interest and motivation, then they can be used as proximal rewards for reinforcement learning of agent policies. For instance, Min Chi et al. [13] found that a tutorial agent's policies could be learned given a distal reward, namely, a students' learning gains at the end of six hours of tutoring. It seems likely that even better policies could be learned if the rewards occurred more frequently. That is, if a "yes!" event occurs, then perhaps the most recent dialogue moves by the agent could be credited and reinforced.

1.3 Related Work on Detecting Brief Affective States

Affect recognition has been steadily gaining prominence in the user modeling community, motivated by the key role of affect in various interactions. Like us, some researchers have proposed models for identifying a single emotion. For instance, Kappor et al. [14] rely on a sensor framework, incorporating a mouse, posture chair, video camera and skin conductance bracelet, to recognize frustration. McQuiggan and Lester [15] describe a data-driven architecture called CARE for learning models of empathy from human social interactions. In contrast to Kappor's and our work, CARE only uses situational data as predictors for empathy assessment. Others have focused on identifying a set of emotions. Cooper et al.'s [16] four linear regression models each predict an emotion (frustration, interest, excitement, confusion). Like our work, these models are built from a combination of tutor and sensor log data, although only we explore the utility of eye tracking and student reasoning data. D'Mello et al. [17] use dialog and posture features to identify four affective states (boredom, flow, confusion, and frustration). In Conati's model, [18] a set of six emotions are assessed (joy/regret, admiration/reproach, pride/shame) from tutor log data, subsequently refined to include one sensor modality, namely an EEG [19].

While there is some work on modeling users with eye tracker information, most of it has focused on how attention shifts predict focus (e.g., [20]), or how pupillary response predicts cognitive load [21]. This latter work is inspired by findings in psychology showing that pupillary response is increased by cognitive load [22]; likewise, affect also increases pupillary size [23]. However, results from experiments less tightly controlled than traditional psychology ones have been mixed, with many failing to find the anticipated link between pupillary response and state of interest (e.g., [24]). In the past we investigated how *only* pupillary response distinguishes different types of affect [25], and did not propose a model based on our results. In contrast, here we present a model that relies on a broad range of features across both interaction and sensor data to predict "yes!" moments. In doing so, we provide insight into the utility of pupillary information for predicting "yes!" events.

In short, although others have investigated predicting positive affective states, including joy [26], engagement [17] and excitement [16], our work distinguishes itself in several ways. First, we identify a novel set of features unique to "yes!", including time on task, degree of reasoning and pupillary response. A more important difference relates to our methodology. A fundamental challenge in inferring affect from data is finding the appropriate gold standard against which to compare a model's predictions. A common approach is to elicit affect information by explicitly querying users [16, 26]. This approach has the potential to be disruptive, thus resulting in inaccurate affect labels; it can also miss salient moments of interest (i.e., when affect is actually occurring). Another common approach relies on using human coders to identify affect in users [17], a technique that also suffers from limitations since human coder performance can be variable [17]. In contrast, we rely on talk-aloud for obtaining affective labels. Doing so has the potential to avoid the above pitfalls, because it is a form of naturally occurring data that has been shown to not interfere with the task at hand [27]. Talk-aloud is also used in [28], although there, only conversational cues are considered as affect predictors, while we use an array of tutor and sensor features.

2 Obtaining Data on "yes!" Moments

We obtained data on "yes!" moments from a previous user study we conducted [25], which involved students interacting with an intelligent tutoring system (ITS) for introductory Newtonian physics. This ITS, referred to as the Example Analogy (EA)-Coach [29], provides support to students during problem solving in the presence of worked-out examples. To solve problems with the EA-Coach, students use the problem window (Fig. 1, left) to draw free body diagrams and type equations; students are free to enter steps in any order and/or skip steps. For each solution entry, the EA-Coach responds with immediate feedback for correctness, realized by coloring entries red or green, indicating correct vs. incorrect entries. Instead of providing hints, for instance on instructional material, the EA-Coach makes examples available to students (accessed with the "GetExample" button); these are displayed in the example window (Fig. 1, right). The system relies on a decision-theoretic approach to tailor the choice of example to a student's needs by considering problem/example similarity, a student's knowledge and reasoning capabilities (see [29] for details).

The study involved 15 participants, all Arizona State University students, who either were taking or had taken an introductory-level physics course. Each participant solved two physics problems with the EA-Coach of the type shown in Fig. 1; each problem solution involved about 15 steps (for further study details, see [25]). We used a variety of data collection techniques. First, the EA-Coach logged all interface actions. Second, we used talk aloud protocol [27]: we asked students to verbalize their thoughts; all sessions were taped and subsequently transcribed. Third, a sensor logger captured students' physiological responses from four sensing devices (see Fig. 2): (1) a *posture chair pad* measured position shifts (the pad included three pressure points on the chair seat and three on the back); (2) a *skin-conductance (SC) bracelet* captured skin conductance; (3) a *pressure mouse* measured the pressure exerted on the mouse (via six "pressure points"); (4) an *eye tracker* captured pupillary responses (the tracker was an integrated model that appeared as a regular computer screen).

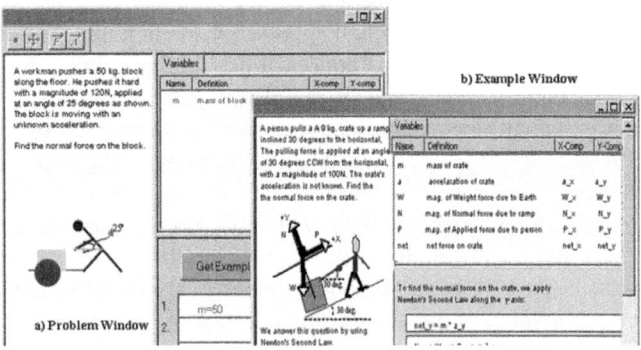

Fig. 1. EA-Coach problem and example windows

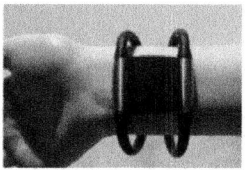

Fig. 2. Affective sensors (from left to right): posture chair, skin conductance (SC) bracelet, pressure mouse, Tobii eye tracker

3 Data Pre-processing

As our "gold standard" for "yes!" moments during the study, we relied on the verbal protocol data. Since a "yes moment" corresponds to excitement and/or positive affect, the transcripts were coded by the first author to identify such instances. As a starting point, we used data from an earlier affect coding [25], reanalyzing the codes to identify "yes!". We identified 68 "yes!" moments; all but one were directly associated with subjects generating a correct solution step and were expressed directly after doing so (recall that the EA-Coach provided immediate feedback for correctness so students were aware of the status of their entries). The one "yes!" that was not associated with a solution step occurred when a participant was reading the second problem statement after having already successfully solved the first problem.

While some of the "yes!" events were expressed in a very effusive manner (*"yes! I'm smart!", "oh yay!"*), others were more subdued (*"I got it right and that makes me feel good"*). In general, we found that when participants expressed a "yes!", it varied in terms of tone, expression, etc. Because we found it very difficult to disambiguate between the various forms of positive affect related to a "yes!", we decided to keep all instances in the analysis without trying to further distinguish between them.

We also had data on how subjects were reasoning during the study, obtained from an earlier coding [25] that included information on various types of reasoning, e.g., whether students were self-explaining by deriving physics principles, drawing comparisons between the problem/example, and/or expressing *some* form of cognitive processing (for examples, see [25]). For the purposes of this study, we collapsed the various types of reasoning into a single *"reasoning"* code, because as a starting point we were interested in how reasoning was related to "yes!" events.

Data Features. To analyze what events predict "yes!" moments, we identified a set of features we believed could be relevant. Note that the list presented here is not meant to be exhaustive, but rather to provide a starting point for understanding predictors of excitement/positive affect in instructional situations. First, we identified *interaction data* features we obtained from the EA-Coach logger corresponding to events in the tutor's interface, as follows:

- *Time*: The amount of time taken to generate a correct solution step (as described in Sect. 4, we focus on correct solution entries);
- *NumAttempts*: The number of attempts required to generate a correct solution step;

- *NumReasoning*: The number of "reasoning" utterances a student expressed in the process of generating a solution step;
- *Type of step*: The type of solution step (e.g., a force, an axis, an equation).

Second, we identified *sensor* features that we obtained from the sensor logger:

- *Pupillary response*: The mean change in pupil dilation around a point of interest (described in Sect. 4). For instance, if the point of interest is when a student generates a solution step, then mean change = (*mean pupil size over time span T directly following the step*) - (*mean pupil size over time span T directly preceding the step*). We set the threshold $T=2$ seconds, since this comparable to that used in other related work involving analysis of pupillary response (e.g., [30]).
- *Skin Conductance (SC)* response: The mean change in SC response around a point of interest (calculated as for pupillary response). We set the threshold $T=2$ seconds, based on the timeframe containing a SC response [14].
- *Mouse response*: The mean change in mouse pressure before and after an event of interest, using the method in [16] (where the mean pressure was obtained by summing over the pressure points, dividing by a constant and finding the mean). We set the threshold $T=10$ seconds, because this sensor does not measure instantaneous responses (like SC and pupillary response) but rather longer scale transitions in behavior.
- *Chair*: The number of "*sitForward*" events, when a student leaned forward prior to generating a solution step, calculated by obtaining the pressure on the seat back via the formula in [16]. Here, we used a threshold $T=10$ seconds, as for the mouse.

4   Results

In order to understand predictors of "yes!" in instructional activities, we compared "yes!" moments to other instances when students obtained a correct solution step but did not generate a "yes!". Since the "yes!" moments directly followed the generation of a correct solution step, we felt this would be the most appropriate comparison; this gave us 67 "yes!" instances[1] and 218 other events. As a final pre-processing step, for each logged correct step we extracted the above-described features, merging across the different log files (transcript, EA-Coach, sensor) to produce a single file.

Our hypotheses were that students would only express a "yes!" if they invested some effort into generating the solution step, and that there would be physiological manifestations of "yes!" that differed from other correct entries. To analyze whether these hypotheses were correct we carried out several types of analysis.

4.1   The Unique Nature of "yes!"

As a starting point, we wanted to determine if "yes!" moments differed from other correct entries (referred to as *other* below) in terms of the features listed above. Thus, we compared data on these two types of entries for our set of features through

---

[1] There was one exception where a student expressed "yes!" when reading an example; given our scheme, we did not consider this one data point in our analysis.

univariate ANOVA. As far as the *interaction* features are concerned, we found that students took significantly longer to generate a correct solution step corresponding to a "yes!" than other correct entries (on average, 206 sec. vs. 54 sec.; $F(1,297) = 77.27$; $p < 0.001$). Students also generated significantly more attempts for "yes!" entries, as compared to other correct entries (on average, 5.1 vs. 1.7; $F(1,297) = 40.47$, $p < 0.001$), and expressed significantly more reasoning episodes for "yes!" (on average, 1.34 vs. 0.58 $F(1,283) = 11.614$, $p = 0.001$). Our data was too sparse to analyze whether type of step had an effect.

As far as the sensor features are concerned, students had a significantly larger pupillary response for a correct solution step associated with a "yes!", as compared to other correct entries (on average, .043mm vs. -.037mm; $F(1,271)=8.422$, $p=0.004$). Skin conductance response had a marginal effect on "yes!" as compared to other entries (.000388μS vs. -.0000422μS, $F(1,291)=3.257$, $p=0.07$), suggesting a higher level of arousal for "yes!". Likewise, students had significantly fewer *sitForward* events before a "yes!", as compared to other entries (6.4 vs. 10.8; $F(1,296)=4.63$, $p=0.032$). One possibility for why this was the case is that students were more focused for "yes" entries and so were fidgeting less. We did not find "yes!" to have a significant effect on mouse response.

4.2 An Empirically-Based Model for Predicting "yes!"

The above analysis showed that "yes!" moments are uniquely distinguishable. To develop a user model, however, we need to understand how the various features predict "yes!" events. Thus, we conducted regression analysis. Because we have a nominal dependent variable ("yes!" vs. other), we used a logistic regression. A key consideration behind our choice of modeling technique was our data set size: while acceptable for modeling with logistic regression, where the rule of thumb is at least 20 data points per independent variable, it was not large enough for some other machine learning techniques, e.g., support vector machine. Of the applicable techniques, regression was chosen based on prior research showing its suitability for classifying affect ([16, 31]); [31] found that regression yielded the highest affect classification accuracy over other machine learning methods.

We begin by presenting the baseline model, one that always predicts the most likely event (here, lack of a "yes!"). Given the base rates of the two decision options and no other information, the best strategy is to predict that each step is not a "yes!". This model achieves 76% accuracy (# of correctly classified events / total # of events), but obviously completely misses our goal of predicting when "yes!" occurs (i.e., never predicts "yes!", and so has a true positive value of 0%, see Table 1, top).

Using the *Step* method, we then added our features to the logistic regression model[2]. The resulting model containing *time, numReasoning, pupil response, SC response* and *chair* was significantly better from the baseline model ($p<0.001$, see Table 1, top). Below, we will analyze the contribution of some of our features to the model's accuracy, but first we examine the full model accuracy.

---

[2] Because increasing the number of predictors decreases experimental power, we omitted *numTries* from this analysis, as it was redundant due to its high correlation with time; we also omitted *mouse response* since it did not significantly distinguish "yes!" from other entries.

Table 1. Logistic regression "yes!" models (TP=Sensitivity, TN=Specificity; Acc= TP + TN / N)

	Overall Logistic Regression Equation	TP	TN	Acc.
Baseline model	-2.206	0	100	76
Full model **	-2.206+time*.008+ numReasoning*.309 + pupilResponse*1.68 +SC*126.4+chair*-.019	60.3	87.2	81.4
Time*+numReasoning*	-2.437 + time*.01 + numReasoning*.345	55.2	89.2	81.6
Time*+pupilResponse*	-2.157+time*.009+pupilResponse*2.082	54.2	85.0	78.4
Time*+SC	-2.308 + time*0.01+ SC*128.97	57.6	89.9	82.6
Time*+Chair	-2.084 +time*0.01 + chair*-.017	56.7	88.3	81.2

** *Significantly better than baseline model, p<0.05*
* *Each feature significantly improves model fit over previous model (i.e., model 1=baseline, model 2=time, model 3= time+2ⁿᵈ feature), p<0.05*

The output of a logistic regression equation is a probability that a given event belongs to a particular class. In order to use the model for prediction, it is therefore necessary to have a decision rule: if the probability of an event is greater or equal to some threshold then we will predict that event will take place (and not take place otherwise). To choose the optimal threshold, we built a Receiver Operating Characteristic (ROC) curve (Fig. 3). The ROC curve is a standard technique used in machine learning to evaluate the extent to which a classifier can successfully distinguish between data points (episodes correctly classified as positive, or *true positives*) and noise (episodes incorrectly classified as positive, or *false positives*), given a choice of different thresholds. Figure 3 shows the ROC curve we obtained for our "yes!" models, where each point on the curve represents a model with a different threshold value. As is standard practice, we chose as our final threshold the point on the curve that corresponds to a reasonable tradeoff between too many false positives vs. too few true positives (P=0.26, labeled by a cross on the curve in Fig. 3).

When reporting classifier accuracy, it is standard to provide *sensitivity* (true positives) and *specificity* (true negatives), since these are more informative then overall accuracy (true positives + true negatives / total number of instances). Our classifier is significantly better than the baseline model (p < 0.05) and obtained a sensitivity of 60.3%, a specificity of 87.2% (and overall accuracy of 81.4% - see Table 1, top). Thus, this classifier correctly identifies 60% of "yes" moments, without incorrectly classifying *other* entries as "yes!" for 87% of the time.

Model Validation. To validate the above model, we conducted a leave-one-out cross validation. Specifically, we trained the classifier using *N-1* data points and tested on the remaining data point, repeating this process *N* times (where *N* is equal to the number of samples, 269 full samples, i.e., without any missing data points that were the result of, for instance, the eye tracker failing to find a valid pupil reading). The validation showed that our model accuracy does not degrade substantially (i.e., sensitivity=55.2%, specificity = 87.1%, accuracy = 79.5%).

Parsimonious Models. We wanted to explore what kind of model fit we could obtain with a subset of our features, which helps to make an informed decision as to which sensors to use if not all are available. Thus, we ran a series of regressions using *time* as the tutor variable (as this variable was highly significant in our regression model)

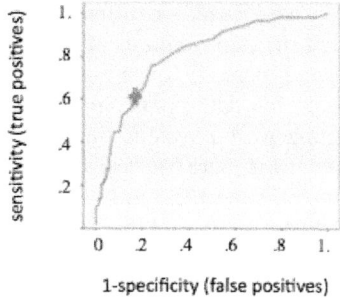

Fig. 3. ROC curve for various decision rule thresholds

and one of the other features. As Table 1 illustrates, we obtained reasonable results in terms of sensitivity and specificity with these reduced models, although only *reasoning* and *pupil response* resulted in significantly better models over a model that only included time (*SC response* and *chair* both improved the model fit, but this did not reach significance, i.e., p=.151 for the *SC response* and p=.153 for the *chair*).

5 Discussion and Future Work

In this paper, we reported on our analysis of moments of excitement and positive affect during instructional activities, which we refer to as "yes!" events. We found that "yes!" always followed a correct solution step, but conversely, a correct step was not always followed by a "yes!". In particular, students were significantly more likely to express a "yes!" after investing more time, generating more attempts, and expressing more reasoning episodes, as compared to correct entries for which corresponding enthusiasm was not expressed. Note that in addition to verbal expression of "yes!", another indication of arousal related to these events was provided by the pupil dilation and skin conductance data. These findings imply that students experience excitement and/or positive affect in tutoring situations when they have invested effort into the process and that effort pays off (i.e., correct solution is obtained). It is possible, however, that students express "yes!" not because they invested thoughtful, deliberate processing but because they guessed and/or arrived at the solution by luck. Our analysis does provide some indication that this is not the case, as students engaged in significantly more reasoning (captured by the "reasoning" code that included self-explanation, a form of deep processing) prior to "yes!". This does not guarantee every student behavior related a "yes!" is an instances of "deep" reasoning – in the future, we plan to delve deeper into this issue of mindful processing and "yes!".

To the best our knowledge, ours is the first work to propose a model for affect recognition incorporating pupillary response data. Although in contrast to the other low-cost sensors we used, eye tracking technology is more expensive, it is becoming more and more accessible, and so investigating its utility for user modeling is important. In a prior study [25], we also found a significant difference in pupil size between affective responses, but there are four key differences between that study and the present. First, in [25] we analyzed how pupillary response differs between positive and negative affect, without developing a model based on this data. Second, here we focus on "yes!" while in [25], we focused on differences between four affective states. Third, in [25] we normalized the pupil data using Z scores – while this approach is sometimes used (e.g., [19]) and increases experimental power, it requires subtracting the overall signal mean from each data point. Since this mean can only be obtained after a user finishes interacting with a system, the findings are difficult to apply for real-time user models. In contrast, here we use the raw signal values, making our findings more applicable to real-time modeling. Fourth, our feature set includes an array of sensors and tutor features, while in [25], we analyzed only pupillary data.

Overall, the tutor and sensor features resulted in a model that predicted "yes!" with 60% sensitivity and 87% specificity, a significant improvement over the baseline model. We also analyzed how using subsets of features impacts model fit: although the model incorporating the full set of features allowed the best trade-off between sensitivity and specificity, using a subset of features also resulted in models with reasonable fit. For instance, a model that includes only information on time and reasoning performs quite well – this may be useful if a system already has the tools to capture reasoning style (e.g., as in [32]) but sensors are not available. As far as the sensor features are concerned, when we explored parsimonious models, each sensor improved model fit over the time-only model. However, this improvement was only reliable, as reported by the p value, for the pupillary response feature. Compared to the pupil-based model, the models incorporating the other sensors resulted in higher specificity and/or specificity. These results, however, have to be interpreted with caution, since they approached but did not reach significance. This may be due to our modest sample size, and so more data is needed to confirm these sensors' utility.

While there is room for improvement, our model is a first step in providing information on "yes!" moments, which in turn can be used for tailoring pedagogical scaffolding to foster interest. For this purpose, it is key that the classifier not misclassify too many other entries as "yes!" (i.e., has high specificity), while still identifying *some* "yes!" moments, as is the case for our classifier. Given our limited sample size and particular instructional context, however, more work is needed to validate and generalize our findings.

Returning to our original four questions, we summarize the progress made so far and directions for future work.

1. *Are "yes!" events desirable outcomes or associated with desirable outcomes?* We argue that it is both. We now know that "yes!" occurs after students appear to have overcome a challenge related to generating a solution step, as indicated by time spent and number of tries produced. Since challenge fosters interest, this suggests that "yes!" events may be suitable as a predictor of increased learning, interest and motivation, something we plan to explore in future studies.

2. *What causes "yes!" events and how can we increase their frequency?* We now know that "yes!" events occur after a challenge is overcome with an example as the only the aid from the tutor. This is consistent with Lepper's advice of keeping the student optimally challenged [10].

3. *How can "yes!" events be useful to tutors, learning companions and other agents?* We offer some suggestions based on theory, but this remains to be empirically explored.

4. *Can "yes!" events be detected more accurately than a baseline approach?* Yes!

Acknowledgements. The authors thank the anonymous reviewers for their helpful suggestions and David Cooper, who contributed the sensor logging software. This research was funded the National Science Foundation, including the following grants: (1) IIS/HCC *Affective Learning Companions: Modeling and supporting emotion during learning* (#0705883); (2) *Deeper Modeling via Affective Meta-tutoring* (DRL-0910221) and (3) *Pittsburgh Science of Learning Center* (SBE-0836012).

References

1. Reeve, J.: The Interest-Enjoyment Distinction in Interest Motivation. Motivation and Emotion 13, 83–103 (1989)
2. Isen, A., Reeve, J.: The Influence of Positive Affect on Intrinsic and Extrinsic Motivation: Facilitating Enjoyment of Play, Responsible Work Behavior, and Self-Control. Motivation and Emotion 29(4), 297–325 (2005)
3. Hidi, S.: Interest and its Contribution as a Mental Resource for Learning. Rev. of Ed. Research 60(4), 549–571 (1990)
4. Isen, A., Daubman, K., Nowicki, G.: Positive Affect Facilitates Creative Problem Solving. J. of Personality and Social Psychology 52, 1122–1131 (1987)
5. Lepper, M.: Motivational Considerations in the Study of Instruction. Cognition and Instruction 5(4), 289–309 (1988)
6. Hidi, S., Renninger, A.: The Four-Phase Model of Interest Development. Educational Psychologist 41(2), 559–575 (2006)
7. Deci, E., Koestner, R., Ryan, R.: Extrinsic Rewards and Intrinsic Motivation in Education: Reconsidered Again. Rev. of Ed. Research 71, 1–27 (2001)
8. Baker, R., D'Mello, S., Rodrigo, M., Graesser, A.: Better to Be Frustrated Than Bored: The Incidence, Persistence, and Impact of Learners' Cognitive-Affective States During Interactions with Three Different Computer-Based Learning Environments. Int. J. of Human-Computer Studies (in press)
9. Boyer, K., Phillips, R., Wallis, M., Vouk, M., Lester, J.: Balancing Cognitive and Motivational Scaffolding in Tutorial Dialogue. In: Woolf, B.P., Aïmeur, E., Nkambou, R., Lajoie, S. (eds.) ITS 2008. LNCS, vol. 5091, pp. 239–249. Springer, Heidelberg (2008)
10. Lepper, M., Malone, T.: Intrinsic Motivation and Instructional Effectiveness in Computer-Based Education. In: Snow, R., Farr, M. (eds.) Aptitude, Learning and Instruction, vol. 3, pp. 255–296. Erlbaum, Hillsdale (1987)
11. Lepper, M., Woolverton, M., Mumme, D., Gurtner, J.: Motivational Techniques of Expert Human Tutors: Lessons for the Design of Computer-Based Tutors. In: Lajoie, S., Derry, S. (eds.) Computers as Cognitive Tools, pp. 75–105. Lawrence Erlbaum Associates, Hillisdale (1993)
12. Henderlong, J., Lepper, M.: The Effects of Praise on Children's Intrinsic Motivation: A Synthesis and Review. Psychological Bulletin 128(5), 774–795 (2002)
13. Chi, M., VanLehn, K., Litman, D., Jordan, P.: Inducing Effective Pedagogical Strategies Using Learning Context Features. In: Proc. of the 18th Int. Conference on User Modeling, Adaptation and Personalization (in press)
14. Kapoor, A., Burleson, W., Picard, R.: Automatic Prediction of Frustration. Int. J. of Human-Computer Studies 65(8), 724–736 (2007)
15. McQuiggan, S., Lester, J.: Diagnosing Self-Efficacy in Intelligent Tutoring Systems: An Empirical Study. In: Ikeda, M., Ashley, K.D., Chan, T.-W. (eds.) ITS 2006. LNCS, vol. 4053, pp. 565–574. Springer, Heidelberg (2006)

16. Cooper, D., Arroyo, I., Woolf, B., Muldner, K., Burleson, W., Christopherson, R.: Sensors Model Student Self Concept in the Classroom. In: Houben, G.-J., McCalla, G., Pianesi, F., Zancanaro, M. (eds.) UMAP 2009. LNCS, vol. 5535, pp. 30–41. Springer, Heidelberg (2009)

17. D'Mello, S., Graesser, A.: Mind and Body: Dialogue and Posture for Affect Detection in Learning Environments. In: Luckin, R., Koedinger, K., Greer, J. (eds.) Proc. of the 13th Int. Conference on Artificial Intelligence in Education, pp. 161–168. IOS Press, Amsterdam (2007)

18. Conati, C., Zhou, X.: Modeling Students' Emotions from Cognitive Appraisal in Educational Games. In: Cerri, S.A., Gouardéres, G., Paraguaçu, F. (eds.) ITS 2002. LNCS, vol. 2363, pp. 944–954. Springer, Heidelberg (2002)

19. Conati, C., Maclaren, H.: Modeling User Affect from Causes and Effects. In: Houben, G.-J., McCalla, G., Pianesi, F., Zancanaro, M. (eds.) UMAP 2009. LNCS, vol. 5535, pp. 4–15. Springer, Heidelberg (2009)

20. Conati, C., Merten, C.: Eye-Tracking for User Modeling in Exploratory Learning Environments: An Empirical Evaluation. Know. Based Systems 20(6), 557–574 (2007)

21. Iqbal, S., Zheng, X., Bailey, B.: Task-Evoked Pupillary Response to Mental Workload in Human-Computer Interaction. In: Dykstra, E., Tscheligi, M. (eds.) Proc. of the ACM Conference on Human Factors in Computing Systems, pp. 1477–1480. ACM, NY (2004)

22. Marshall, S.: Identifying Cognitive State from Eye Metrics. Aviation, Space and Environmental Medicine 78, 165–175 (2007)

23. Vo, M., Jacobs, A., Kuchinke, L., Hofmann, M., Conrad, M., Schacht, A., Hutzler, F.: The Coupling of Emotion and Cognition in the Eye: Introducing the Pupil Old/New Effect. Psychophysiology 45(1), 130–140 (2008)

24. Schultheis, H., Jameson, A.: Assessing Cognitive Load in Adaptive Hypermedia Systems: Physiological and Behavioral Methods. In: De Bra, P.M.E., Nejdl, W. (eds.) AH 2004. LNCS, vol. 3137, pp. 225–234. Springer, Heidelberg (2004)

25. Muldner, K., Christopherson, R., Atkinson, R., Burleson, W.: Investigating the Utility of Eye-Tracking Information on Affect and Reasoning for User Modeling. In: Houben, G.-J., McCalla, G., Pianesi, F., Zancanaro, M. (eds.) UMAP 2009. LNCS, vol. 5535, pp. 138–149. Springer, Heidelberg (2009)

26. Conati, C., Maclaren, H.: Empirically Building and Evaluating a Probabilistic Model of User Affect. User Modeling and User-Adapted Interaction (in press)

27. Ericsson, K., Simon, H.: Verbal Reports as Data. Psych. Rev. 87(3), 215–250 (1980)

28. D'Mello, S., Craig, S., Sullins, J., Graesser, A.: Predicting Affective States Expressed through an Emote-Aloud Procedure from Autotutor's Mixed-Initiative Dialogue. Int. J. of Artificial Intelligence in Education 16(1), 3–28 (2006)

29. Muldner, K., Conati, C.: Evaluating a Decision-Theoretic Approach to Tailored Example Selection. In: Veloso, M. (ed.) Proc. of 20th Int. Joint Conference on Artificial Intelligence, pp. 483–489. AAAI Press, Menlo Park (2007)

30. Van Gerven, P., Paas, F., Van Merrienboer, J., Schmidt, H.: Memory Load and the Cognitive Pupillary Response in Aging. Psychophysiology 41(2), 167–174 (2001)

31. D'Mello, S., Picard, R., Graesser, A.: Towards an Affect Sensitive Auto Tutor. IEEE Intelligent Systems 22(4), 53–61 (2007)

32. Conati, C., VanLehn, K.: Toward Computer-Based Support of Meta-Cognitive Skills: A Computational Framework to Coach Self-Explanation. Int. J. of Artificial Intelligence in Education 11, 389–415 (2000)

A Personalized Graph-Based Document Ranking Model Using a Semantic User Profile

Mariam Daoud, Lynda Tamine, and Mohand Boughanem

IRIT, Paul Sabatier University,
118 Route Narbonne F-31062, Toulouse cedex 9, France
{daoud,tamine,bougha}@irit.fr

Abstract. The overload of the information available on the web, held with the diversity of the user information needs and the ambiguity of their queries have led the researchers to develop personalized search tools that return only documents that meet the user profile representing his main interests and needs. We present in this paper a personalized document ranking model based on an extended graph-based distance measure that exploits a semantic user profile derived from a predefined web ontology (ODP). The measure is based on combining Minimum Common Supergraph (MCS) and Maximum Common Subgraph (mcs) between graphs representing respectively the document and the user profile. We extend this measure in order to take into account a semantic recovery between the document and the user profile through common concepts and cross links connecting the two graphs. Results show the effectiveness of our personalized graph-based ranking model compared to Yahoo[1] search results.

1 Introduction

A major limitation of most existing search engines is that they are based on a content-based query-document matching pattern. Retrieving the most relevant documents for short queries in a large scale document collection, where the user needs are diverse, is a limitation of traditional IR strategies. Indeed, these latter consider that the query is the only key that represents the user information need. Personalized search aims at tackling this problem by considering the user profile that describes the main user interests and preferences, in the search process. The main challenging task in the field is how to represent, infer, and exploit the user profile so as to improve the search performance. User profile models are arranged from very simple representations to complex representations based on semantic resources used to describe the user interests with a rich variety of interrelations among them. The user profile representation model could be based on a bag of words [1], graph of terms [2, 3] defined usually by term co-occurrence, or conceptual representation based on a list of concepts [4, 5] or an instance of the ontology [6, 7]. Improving the search accuracy is then achieved by using

---

[1] http://www.yahoo.com

P. De Bra, A. Kobsa, and D. Chin (Eds.): UMAP 2010, LNCS 6075, pp. 171–182, 2010.

the user profile in a personalized document ranking such as query reformulation techniques, query-document matching models or result re-ranking.

We present in this paper a personalized graph-based document ranking model using a semantic user profile. We have already proposed in previous works [8, 9] a semantic representation of the user profile based on a graph of semantically related concepts issued from a predefined web ontology, namely the ODP ontology[2]. Personalization is based on reranking the search results by calculating for each retrieved document, a personalized score using the cosine similarity measure between the document and the top weighted concepts of the user profile. In this paper, we focus on a personalized document ranking model that represents both the document and the user profile in a graph-based model and uses a graph-based distance measure to calculate a document-profile semantic recovery. It is based on a semantic extension of a graph-based distance measure combining Maximum common subgraph (mcs) and minimum common super-graph (MCS). We extend this measure in order to increase the personalized score of the most related documents to the user profile by taking into account not only common concepts but also cross links connecting the two graphs.

The rest of this paper is organized as follows. In Sect. 2, we present related works in personalized document ranking models and then highlight our contribution. In Sect. 3, our personalized document ranking model is detailed. In Sect. 4, we present the experimental evaluation and results by comparing the performance of our personalized search ranking to Yahoo ranking results. In the last section, we present our conclusion and plan for future work.

2   Related Work

Personalized document ranking is usually achieved by integrating the user profile in the query reformulation process [4, 2], query-document matching [10] or document ranking [7, 6, 5].

Most of query reformulation techniques are based on adding or reweighting terms using *Rocchio* algorithm [4] or by using rewriting rules [2] depending on the user profile representation. In [4], the query is matched with the most appropriate pair of concepts in the user profile; the first concept of the pair is the most similar to the user query, it is used to add or enhance the weight of relevant terms while the second one is the less similar one used to eliminate non relevant terms in the search. In [2], a more elaborated user profile based on terms connected by different edge relations (negation, substitution, etc.) allows to add, eliminate or substitute the query terms with relevant terms.

Using the user profile in the query-document matching model consists of calculating the relevance score of the document relatively to both the user query and the user profile. A bayesian model integrating a document relevance score function is proposed in [10] in order to increase the score of the document when the document vocabulary matches the user profile one.

---

[2] http://www.dmoz.org

Personalization based on result re-ranking consists of combining the content-based document score with the personalized document score [7, 6, 5] or with the personalized *PageRank* of the document [11]. The personalized score is computed in [7, 6] using the cosine similarity measure between each returned document and the most similar concepts of the user profile. In [5], personalization consists of combining personalized categorization and result re-ranking using a voting-based merging scheme. Personalized result-reranking based on *PageRank* is described in [11]. The user selects his preferred pages from a set of hub pages and one personalized PageRank vector is computed for each user interest used to redirect the returned web pages to the preferred ones.

In most of other related works based on an ontological user profile, the document is represented by a vector of weighted terms and the personalized score of the document is computed according to a term-based similarity measure (cosine) [7, 6] between the document and some concepts of the user profile. The main distinctive feature of our work is to use a semantic document-profile matching model that represents both the user profile and the document by graphs derived from the ODP ontology and computes a personalized relevance score of the document based on an extended semantic graph-based distance measure.

3   A Personalized Graph-Based Document Ranking Model

In our approach, we make use of a semantic user profile model proposed in our previous work [8], which holds the user interest built across a search session. This latter is defined by a sequence of queries related to the same user information need. The user profile is represented as a graph of interrelated concepts of the ODP ontology [3] considered as a highly expressive ground to describe a semantic user profile model. It is built by mapping the user's documents of interests on the ODP and selecting the highly weighted group of concepts that are semantically linked with different edge types in the ontology. Formally, the user profile is represented by a hierarchical (tree) component composed of "is-a" links, and a non hierarchical component composed of cross links of different types predefined in the ontology. It is defined as a directed graph G=(V,E) where:

- V is a set of weighted nodes, representing concepts of interest,
- E is a set of edges between nodes in V, partitioned into three subsets T, S and R, such that: T corresponds to edges made of "is-a" links, S corresponds to edges made of "symbolic" cross links and R corresponds to edges made of "related" cross links.

We exploit the user profile in a result reranking process using a personalized graph-based document ranking model. The model consists of calculating a personalized relevance score of a document with respect to the user profile according to a graph-based distance measure. This latter is based on a semantic extension

---

[3] http://www.dmoz.org

of the combined distance measure using minimum common supergraph (MCS) and maximum common subgraph (mcs). We review in this section the most common used graph-based distance measures and then present our extended semantic graph-based measure used in the personalized document ranking model.

3.1   Background: Graph-Based Distance Measures

The most common known graph-based distance measures use the maximum common subgraph (mcs) [12], the minimum common supergraph (MCS) [12], the combined measure using MCS and mcs [13] or the edit distance [14].

The maximum common subgraph (mcs) of two graphs g_1 and g_2, is a subgraph g of both g_1 and g_2 based on common concepts and has among all the subgraphs, the maximum number of nodes [15]. The graph-based distance measure based on the subgraph mcs [12] is given by the following formula:

$$d(g_1, g_2) = 1 - \frac{|mcs(g_1, g_2)|}{max(|g_1|, |g_2|)} \tag{1}$$

$|g_1|$ (resp. $|g_2|$) is the the number of nodes in g_1 (resp. in g_2). This formula gives lower distance for graphs having large mcs.

The minimum common supergraph (MCS) of two graphs, g_1 and g_2, is a graph g that contains both g_1 and g_2 as subgraphs and that has the minimum number of nodes and edges [15]. The distance measure based on only the supergraph (MCS) [15] is given in the following formula:

$$d(g_1, g_2) = 1 - \frac{|g_1| + |g_2| - |MCS(g_1, g_2)|}{max(|g_1|, |g_1|)} \tag{2}$$

The distance measure based on combining MCS and mcs is given in [13] as follows:

$$d_{MMCS}(g_1, g_2) = |MCS(g_1, g_2)| - |mcs(g_1, g_2)| \tag{3}$$

According to this measure, the distance between graphs is lower when the size of the supergraph is smaller and the size of the subgraph is larger.

An alternative of the distance measure based on mcs is the edit distance [14]. It expresses the shortest sequence of edit operations that transform a graph g_1 into another graph g_2. An edit operation is either the deletion, insertion or substitution applied to nodes and edges. This measure is defined as follows:

$$d(g_1, g_2) = min \{C(\xi)\} \tag{4}$$

where ξ is the sequence of edit operations to transform graph g_1 into graph g_2.

3.2   A Semantic Graph-Based Matching Model

We propose a personalized document ranking model based on a graph-based distance measure for calculating the personalized score of a document. This latter reflects a semantic recovery with the user profile based on the following hypothesis: "*a document is ranked higher if it recovers the maximum of concepts of the user profile at both specific and general levels*". Based on this hypothesis, personalization is achieved as follows:

- Map each retrieved document's content d_k represented by a term-based vector on the ODP ontology using the cosine similarity measure and extract the document's graph by connecting the top 20 weighted concepts using different edge types of the ontology.
- calculate the personalized score $S_p(d_k, G_u)$ of each retrieved document d_k using an extended semantic graph-based distance measure combining MCS and mcs with respect to the reference ontology (ODP),
- re-rerank the search results according to the final score $S_f(d_k)$ of each retrieved document d_k calculated by combining, its initial score S_i returned by the system using a content-based ranking model and its personalized score S_p as follows:

$$S_f(d_k) = \gamma * S_i(q, d_k) + (1 - \gamma) * S_p(d_k, G_u) \qquad (5)$$

$0 \leq \gamma \leq 1$. When γ has a value of 1, personalized score is not given any weight. If γ has a value of 0, the original score is ignored and pure personalized score is considered.

In the next sections, we present our motivations behind using a graph-based distance measure and the way of calculating the personalized score of a document using the semantic extension of this measure over the ODP ontology.

Toward a Semantic Graph-Based Measure. We propose a graph-based distance measure based on the combination of MCS and mcs. Our choice is based on the following reasons:

- The Minimum Common Supergraph (MCS) allows measuring the similarity between the document and the user profile at general levels even if the graph do not have common concepts. Indeed, the supergraph of two graphs is smaller when the root nodes of both graphs are close in the ontology.
- The Maximum Common Subgraph (mcs) allows measuring the similarity between the document and the user profile at specific levels. Indeed, the subgraph is larger when the two graphs have common concepts at specific levels, and consequently this enlarge the subgraph with common concepts at general levels.

However, this measure can't deal with approximate matching between the document and the user profile. Indeed, this measure give similar distance for documents that have cross links with the user profile and others that dont have cross links neither common concepts. This is due to the exact recovery assumption using the common concepts to build the subgraph. That's why we argue to extend semantically the subgraph of two graphs by taking into account relative document-profile semantic recovery. Our intuition at this level is to consider two types of recovery in the semantic distance measure:

1. *Exact recovery*: refers to an exact similarity between the document and the user profile. It is calculated by the number of common concepts.
2. *Relative recovery*: refers to a relative similarity between the document and the user profile. It is calculated by the number of common related concepts connecting the two graphs, in other terms, those linked with cross links.

Calculating the Personalized Document Rank. We calculate the personalized document rank by extending first each of the document's graph and the user-profile's graph and then by computing the distance measure based on the combination of MCS and mcs of the two extended graphs.

- *A semantic extension of the mcs:* Formally, let g_1 and g_2 the graphs representing respectively the user profile and the document. As shown in Fig. 1, the set of concepts of graph g_2 connected to graph g_1 with cross links represent the extension of graph g_1 (in Fig. 1, concepts c_{11} and c_{14} forms the extension of graph g_1). Formally, we define the extended graph g_1^{2*} of graph g_1 with respect to g_2, as follows:

$$g_1^{2*} = g_1 \cup \{c_i \in g_2 / \exists c_j \in g_1 \wedge e_{ij} \in S \cup R\} \tag{6}$$

e_{ij} is the edge linking concept c_i to concept c_j, $S \cup R$ is the set of symbolic and related concepts of the ODP ontology (cf. Sect. 3). We create the extended graph g_2^{1*} by the same manner as the graph g_1^{2*}. We obtain two extended graphs g_1^{2*} and g_2^{1*} that will be used in the personalized document ranking model.

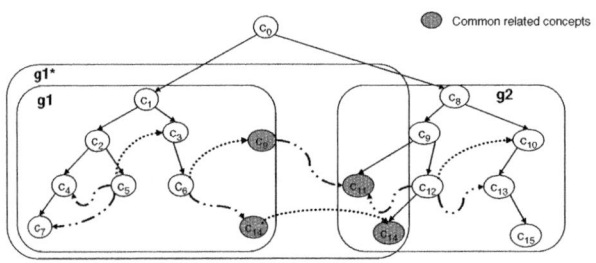

Fig. 1. A semantic extended graph through cross links

- *Calculating the personalized relevance score:* We use the extended graphs g_1^{2*} and g_2^{1*} to calculate the personalized relevance score of the document based on a semantic distance combining the MCS and mcs between the document and the user profile. The mcs of the two extended graphs contains initially common concepts called mcs_{cc} and the activated concepts issued from the graph extension called mcs_{ca} that are either the concepts linking the two graphs through cross links or the inner concepts linking the concepts of the subgraph together.

In order to distinguish the role of the related concepts connected through the cross links relatively to the direct common concepts between graphs, we used a decay factor f_{ca}. This factor is calculated automatically based on the following assumption: *"the number of activated concepts must be reduced as more as we have symbolic or related edges connecting the graphs"*. f_{ca} is given as follows: $f_{ca} = \frac{L_R}{1+L_R}$, L_R is the set of cross edges linking concepts of the two graphs. Finally, in order to take into account the graph size difference between the documents compared to the user profile, we normalized the semantic distance measure between graphs by dividing over MCS as follows:

$$d(g_1^{2*}, g_2^{1*}) = \frac{\left|MCS(g_1^{2*}, g_2^{1*})\right| - \left(\left|mcs_{cc}(g_1^{2*}, g_2^{1*})\right| + f_{ca} * \left|mcs_{ca}(g_1^{2*}, g_2^{1*})\right|\right)}{\left|MCS(g_1^{2*}, g_2^{1*})\right|}$$

(7)

4   Experimental Evaluation

In this section, we present an experimental evaluation undertaken through a user study in order to compare the effectiveness of our personalized search in front of Yahoo search, and to evaluate the impact of different document-profile similarity measures.

4.1   Dataset

We exploited a search log of a commercial web search engine, namely *Exalead*, and we extracted the search history of 10 users collected along three months. As our approach is based on personalizing search across sessions defined by a sequence of related queries, we have selected 25 user search sessions for all the 10 users as follows:

- Each session contains three queries related to the same user information need and are submitted by the same user in a chronological order.
- Each query has at least one clicked document as it is the only source of evidence to build the user profile in a search session. We consider obviously that a document is relevant if it is clicked by the user.

In order to test the personalized search effectiveness along the user search session, we have divided the query set per session into a training query set to learn the user profile and a testing query set to evaluate the retrieval performance. The first two queries of each search session are part of the training query set, which contains a total of 50 queries. The last formulated user query of each session is part of the testing query set, which contains a total of 25 queries. Testing query terms vary between 1 and 4 and the user intent behind these queries is mostly informational (*"Risques auditifs"*) or transactional (*"Le bourg d'oisans hotel"*).

The document collection consists of collecting the top 50 results retrieved from the publicly available Yahoo API[4] for each testing query. In our evaluation setting, these documents are used only for reranking the search results using

---

[4] `search.cpan.org/perldoc?Yahoo::Search`

the user profile. In order to evaluate the retrieval effectiveness, the relevance assessments for the testing queries were given through a user study. To do, 5 computer science students of our lab were presented with the set of top 50 results retrieved from Yahoo. Each participant was considered the user who has formulated the query and asked to judge whether each document was relevant or not according to the subject of a subset of testing queries.

4.2   Evaluation Protocol

The evaluation protocol consists of a training step and a testing step.

1. *Training step:* This step consists of learning the user profiles for each testing query using the clicked documents of the corresponding training queries belonging to the same user.
2. *Testing step:* This step consists of evaluating the personalized retrieval effectiveness for each testing query using the user profile compared to the baseline search performed by Yahoo search using only the testing query. Personalized search is based on reranking the top 50 results of Yahoo for each testing query using the appropriate user profile and by combining for each document its original rank (sorted by Yahoo) and its personalized rank calculated using our graph-based distance measure. We use the precision at top 10 and top 20 documents (P@10, P@20) as an evaluation metric which measures the system performance for documents that are most viewed. We pool together the queries and judgments of all the ten users, so that the evaluation result will be an average over the whole testing queries.

4.3   Experimental Results

In this section, we present the evaluation results by comparing our personalized retrieval effectiveness to the baseline search. Results concern the following objectives: (1) Evaluating the effect of the combination parameter γ on the retrieval effectiveness, (2) Evaluating our personalized search to the baseline search performed by Yahoo and to different similarity measures used to calculate the personalized score of the document.

Effect of γ Combination Parameter on the Retrieval Performance. In this experiment, we study the effect of combining the original document's rank of Yahoo (corresponding to the original document score in formula 5) and the personalized document rank on the retrieval effectiveness using a combination parameter γ (formula 3). Figure 2 shows the improvement of our personalized search with varying γ in the interval [0 1]. Results show that the best performance is obtained when γ is 0, i.e., when the original search engine rankings are ignored altogether. This is likely due to the fact that all the results on the top 50 match the query well and thus the distinguishing feature is how well they match the user profile.

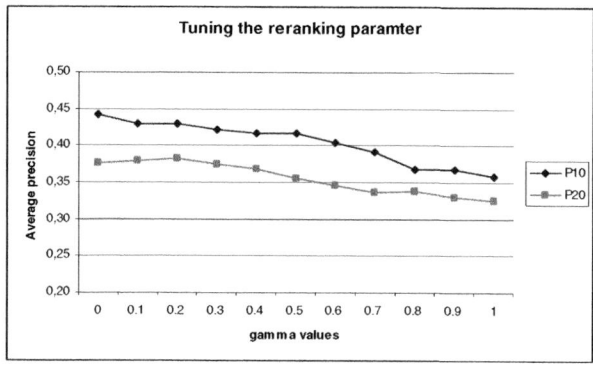

Fig. 2. Effect of γ on the final rank

Evaluating the Personalized Ranking Model Effectiveness. In this experiment, we compare our personalized retrieval effectiveness to the baseline search and to other different measures. These measures are (1) the basic combined measure using MCS and mcs, normalized without using a decay factor (formula 3), (2) the distance measure based on mcs (formula 1), (3) the distance measure using MCS (formula 2) and (4) the cosine similarity measure proposed in our previous works [8, 9]. This latter calculates the personalized score of the document using the most highly weighted concepts of the user profile. In order to set a reliable comparison between the different measures, we have conducted a preliminary experiment in order to identify the best number of concepts used to calculate the personalized score of the document according to the cosine similarity measure. We outline that earlier experiments [8] have shown the effectiveness of personalizing search using the basic cosine similarity measure compared to a typical search and also by comparison to the personalized search approach described in [6].

(A) Effet of the number of concepts used in the cosine similarity measure on the retrieval performance. According to our previous work [8], the personalized score of the document using the cosine similarity measure is computed as follows:

$$S_p(d_k, G_u) = \frac{1}{h} \cdot \sum_{j=1..h} score(c_j) * \cos(\overrightarrow{d_k}, \overrightarrow{c_j})$$ (8)

where $\overrightarrow{d_k}$ and $\overrightarrow{c_j}$ are term-based vectors representing respectively document d_k and concept c_j, $score(c_j)$ is the weight of concept c_j in the user profile G_u and h is the number of concepts considered in the personalized search.

In this experiment, we varied the number of concepts used to calculate the personalized score of the document using the cosine similarity measure. Figure 3 shows P@10 and P@20 of the personalized search. We can see that the best improvement $(17, 24\%$ at P@10 and $8, 07\%$ at P@20) is obtained when using 7 concepts of the user profile. We retain this value for comparing the retrieval effectiveness using the cosine measure with other graph-based measures.

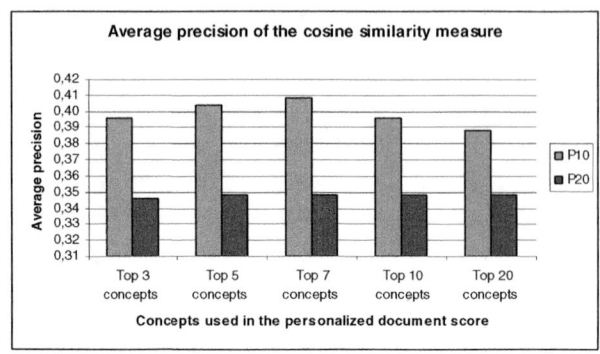

Fig. 3. Effect of the number of concepts using the Cosine measure

(B) Evaluating the effect of the personalized graph-based ranking. In this experiment, we compare the effectiveness of our personalized search to Yahoo search results as well as to personalized search results performed using the measures mentioned above. All graph-based measures calculate the distance between the extended graphs of both the document and the user profile. We recall that for each testing query, the personalized search is based on reranking the top 50 results in descending order according to the graph-based distance measure and ascending order according to the cosine similarity measure.

Table 1 shows the improvement of each measure compared to Yahoo search. Results show a significant improvement of our personalized search at both P@10 and P@20. This indicates the effectiveness of ranking semantically the documents with respect to the user profile using our semantic graph-based distance measure. Indeed, our measure gives higher ranks for documents that are semantically related to the user profile by bringing them to the top 10 and the top 20 results presented to the user. We can confirm also that information from the clicked web pages of training queries can be used to provide an effective personalized search. In order to get a more detailed understanding of the effects of our personalized search, we examined the results on a query by query basis at P@10. For the 25 testing queries reranked using our graph-based measure, 12 (48%) showed an improvement, 8 (32%) were unchanged, and only 5 (20%) were negatively impacted. Thus, the personalized reranking helped 2 times as many queries as it hurts. This is probably due to the difference of clarity degree between the queries measured by the number of query terms.

By comparison to the basic combined distance measure, the decay factor f_{ca} of our measure has a positive effect of increasing the P@20 documents. This proves that the direct common concepts should be given more weights than common related ones. Using the distance measure based only on the subgraph (mcs) or the supergraph (MCS) gives lower improvement compared to the combined measure especially at P@10. This proves that combining semantic distance at general and specific levels allows performing better personalized search improvement. The cosine similarity measure performs a reasonable improvement at P@10 documents, which proves that the user profile is effective even using only

Table 1. Comparison of different document-profile similarity measures

	P@10	P@20
Yahoo search	0,3480	0,3220
Our semantic extended measure ($MCS + mcs + f_{ca}$)	0,4280	0,3660
Improvement	**22,99%**	**13,66%**
Classic combination measure ($MCS + mcs$)	0,4280	0,3620
Improvement	22,99%	12,42%
mcs	0,4160	0,3660
Improvement	19,54%	13,66%
MCS	0,3960	0,3600
Improvement	13,79%	11,80%
Cosine	0,4080	0,3480
Improvement	17,24%	8,07%

few concepts. On the other hand, it performs the lowest improvement at P@20 documents which proves the effectiveness of using a semantic distance measure between the document and the user profile.

5   Conclusion and Outlook

We presented in this paper an approach for personalizing search using a conceptual graph-based user profile. Personalization is achieved by reranking the search results based on a graph-based distance measure combining MCS and mcs and by considering cross links between graphs. Our experimental evaluation is carried out using real user queries issued from *Exalead* web search log. Results show that our model achieves higher performances compared to Yahoo search results and to other graph-based distance measures. In future work, we plan to improve the accuracy of both the document and the user profile graph-based representations and study the effect of tuning the importance of related concepts between graphs relatively to direct common concepts on the retrieval performance.

Acknowledgments. This work was in part supported by OSEO under the Quaero program.

References

1. Gowan, J.: A multiple model approach to personalised information access. Master thesis in computer science, University of College Dublin (2003)
2. Koutrika, G., Ioannidis, Y.: A unified user profile framework for query disambiguation and personalization. In: Proceedings of the workshop on New Technologies for Personalized Information Access, pp. 44–53 (2005)
3. Micarelli, A., Sciarrone, F.: Anatomy and empirical evaluation of an adaptive web-based information filtering system. User Modeling and User-Adapted Interaction 14(2-3), 159–200 (2004)

4. Sieg, A., Mobasher, B., Burke, R., Prabu, G., Lytinen, S.: Using concept hierarchies to enhance user queries in web-based information retrieval. In: AIA '04: Proceedings of the international Conference on Artificial Intelligence and Applications, Innsbruck, Austria, pp. 114–124 (2004)
5. Liu, F., Yu, C., Meng, W.: Personalized web search for improving retrieval effectiveness. IEEE Transactions on Knowledge and Data Engineering 16(1), 28–40 (2004)
6. Gauch, S., Chaffee, J., Pretschner, A.: Ontology-based personalized search and browsing. Web Intelligence and Agent Systems 1(3-4), 219–234 (2003)
7. Sieg, A., Mobasher, B., Burke, R.: Web search personalization with ontological user profiles. In: CIKM '07: Proceedings of the sixteenth ACM Conference on information and knowledge management, pp. 525–534. ACM, New York (2007)
8. Daoud, M., Tamine, L., Boughanem, M.: Towards a graph based user profile modeling for a session-based personalized search. Knowledge and Information Systems 21(3), 365–398 (2009)
9. Daoud, M., Tamine-Lechani, L., Boughanem, M., Chebaro, B.: A session based personalized search using an ontological user profile. In: SAC '09: Proceedings of the 2009 ACM symposium on Applied Computing, pp. 1732–1736. ACM, New York (2009)
10. Tamine-Lechani, L., Boughanem, M., Zemirli, N.: Personalized document ranking: exploiting evidence from multiple user interests for profiling and retrieval. Digital Information Management 6(5), 354–365 (2008)
11. Jeh, G., Widom, J.: Scaling personalized web search. In: WWW '03: Proceedings of the 12th international conference on World Wide Web, pp. 271–279. ACM, New York (2003)
12. Levi, G.: A note on the derivation of maximal common subgraphs of two directed or undirected graphs. Calcolo 9(4), 341–354 (1973)
13. Fernández, M.L., Valiente, G.: A graph distance metric combining maximum common subgraph and minimum common supergraph. Pattern Recognition Letters 22(6-7), 753–758 (2001)
14. El-Sonbaty, Y., Ismail, M.A.: A new error-correcting distance for attributed relational graph problems. In: Amin, A., Pudil, P., Ferri, F., Iñesta, J.M. (eds.) SPR 2000 and SSPR 2000. LNCS, vol. 1876, pp. 266–276. Springer, Heidelberg (2000)
15. Bunke, H., Jiang, X., Kandel, A.: On the minimum common supergraph of two graphs. Computing 65(1), 13–25 (2000)

Interaction and Personalization of Criteria in Recommender Systems

Shawn R. Wolfe[1,2] and Yi Zhang[1]

[1] School of Engineering, University of California Santa Cruz,
Santa Cruz CA 95064, USA
[2] NASA Ames Research Center, Moffett Field CA 94035, USA
{srwolfe,yiz}@soe.ucsc.edu

Abstract. A user's informational need and preferences can be modeled by criteria, which in turn can be used to prioritize candidate results and produce a ranked list. We examine the use of such a criteria-based user model separately in two representative recommendation tasks: news article recommendations and product recommendations. We ask the following: are there nonlinear *interactions* among the criteria; and should the models be *personalized*? We assume that that user ratings on each criterion are available, and use machine learning to infer a user model that combines these multiple ratings into a single overall rating. We found that the ratings of different criteria have a nonlinear interaction in some cases, for example, article novelty and subject relevance often interact. We also found that these interactions vary from user to user.

Keywords: information filtering, multiple criteria, nonlinear models.

1 Introduction

Choosing one or more items among many candidates often requires an evaluation on multiple criteria. For instance, in space science, investigations may involve selecting observations on the basis of measurement type, resolution, range, location and format. In other cases, it may be necessary to trade off competing interests. For example, an air traffic flow manager may need to balance needs of individual airlines while maintaining safety and equity and minimizing overall delay. Proactively, an improved understanding of the involved criteria for a particular user need could also lead to better marketing and product development opportunities.

Criteria-based models, which capture multiple, potentially competing aspects of a user's need, have been developed and used in operations research [1]. These criteria-based models differ from feature-based models in that criteria are inherently subjective and may not be directly observable (however, in this study they are provided as input to the problem). A previous study showed that using a linear combination of multiple criteria to model the user's need can improve information retrieval results [2]; though we restrict ourselves to recommender systems in this study, we presume our results should extend to other information retrieval settings as well. This paper extends previous work by going beyond

P. De Bra, A. Kobsa, and D. Chin (Eds.): UMAP 2010, LNCS 6075, pp. 183–194, 2010.
© Springer-Verlag Berlin Heidelberg 2010

a linear combination to model the interactions among the criteria. Specifically, we seek answers to the following questions:

1. Is there evidence that some criteria interact in the decision/rating process?
2. If so, are there discernible patterns to these interactions?
3. Given interactions, are these interactions consistent across users?

To answer these questions, we perform our study within the context of two very different recommendation tasks: news article recommendations — a representative task for adaptive filtering; and product recommendations (for flat panel televisions) — a representative task for collaborative filtering. We expect certain interactions might exist. For example, low ratings on certain criteria might negate higher ratings in other criteria. Is respect for an article's author still important when the article is not of the desired topic? Is the durability of a television a factor if it has a poor picture? Or it may be that certain high ratings limit the impact of other criteria. For instance, would readability be as important among articles with the same breaking news? It is these sorts of interactions that we are searching for. On the two tasks, we test for the presence of interactions by comparing the root mean squared error (RMSE) of learned linear and nonlinear user models for predicting the overall item rating or recommendation.

The rest of the paper is organized as follows. In Sect. 2, we review related work. In Sect. 3, we describe our recommender datasets. In Sect. 4, we detail our approach to represent criteria interactions and to select the best model. We present our experimental results in Sect. 5 and our conclusions in Sect. 6.

2   Related Work

In information retrieval, the limited adoption of criteria-based user models has been mostly restricted to enhancing standard relevance-based models with novelty. Researchers have studied criteria such as information-novelty for search [3], summarization [4], filtering [5] and topic detection and tracking [6]. Prior research on a user's perception/criteria have found that a wide range of factors (such as personal knowledge, topicality, quality, novelty, recency, authority and author qualitatively) affect human judgments of relevance [7–9].

Most of the research in the information retrieval community that uses multiple criteria has been in information filtering. Manouselis and Costopoulou categorize 37 recommender systems that implicitly use some multi-criteria aspect in their operation [10]. These systems primarily use only the weighted sum (i.e., linear combination) model. Of the information filtering systems we are aware of, PENG [11] is the most similar in application to our experimental domain. PENG is a multi-criteria news bulletin filtering system that utilizes several criteria, including content, coverage, reliability, novelty and timeliness.

Learning user models based on multiple criteria (as opposed to content alone) is not common in information retrieval. Naïve Bayesian classifiers were used to learn content-based user profiles for movie search [12]. A more complicated scheme was used to predict whether a user would watch television programs [13], first by building a model of what genres a user likes, and then classifying each show based on its genres by means of a support vector machine. DIVA [14] uses a somewhat similar approach to recommend movies, using the C5.0 algorithm to classify each movie based on its metadata.

Outside the information retrieval community, general additive independence models have gained some popularity, and are akin to our current approach. One method for estimating generalized additive independent utility functions is to treat criteria as random variables and use Bayesian techniques to estimate them [15]. This same utility decomposition concept was later applied to the multi-issue negotiation task, by representing the utility of a buyer in a utility graph [16].

3 Datasets

We used two recommendation datasets for our research. Each dataset had four criteria and one overall rating defined. The range of these ratings are different for different criteria, as the data were originally collected for other research. For consistency, we have rescaled all ratings to have minimum and maximum values of 0 and 1, respectively. After this rescaling, the ratings were either binary (0 or 1) or five-valued (0.0, 0.2, 0.4, 0.6, 0.8 or 1.0). For both data sets, we restrict ourselves to user-item pairs with complete ratings (i.e., any items with missing ratings were excluded from our study).

3.1 News Recommendation

Our news recommendation data were provided by the University of California, Santa Cruz and Carnegie Mellon University [17]. The data were previously collected in a user study performed on the Yow-now news filtering system. Yow-now was an information filtering systems that delivered news articles to users from various RSS feeds. Approximately twenty-five users used the Yow-now system for about a month, reading news for at least one hour each day, rating approximately 9000 articles in all, with an average of 383 articles rated per user (with a standard deviation of 252.8). This allowed us to explore creating personalized user models with the Yow-now dataset.

The users rated each article according to the following four criteria:

Authoritative: How authoritative the article appeared (binary).
Novel: The novelty of the article (five-valued).
Readable: The ease of reading the article (binary).
Relevant: The degree to which the article was relevant to the general subject category of the article (five-valued).

The overall user rating of the article was given on a five-point scale.

3.2 Product Recommendation

Our product recommendation data came from a crawl of the Epinions.com review site. Our dataset is restricted to flat panel television reviews. Approximately 1100 users reviewed 1200 items, with an average of 1 review per user (with a standard deviation of 0.29). With such a small number of reviews per user, it was clearly not possible to build personalized user models with this dataset.

The users rated each product according to the following four criteria:

Sound: The sound quality of the television (five-valued).
Ease of Use: Ease of use of the various features and menus (five-valued).
Picture Quality: All visual aspects of the television's picture (five-valued).
Durability: Durability of the television set (five-valued).

The overall user rating of the article was given on a five-point scale.

4   Approach

To test for interactions among criteria in the final decision/rating process, we compared the performance of two sets of models on a rating prediction task. The first model is a linear combination of ratings on the criteria, which makes the assumption that the criteria do not interact in the user decision process. The second set of models are nonlinear combinations that explicitly represent interactions among pairs of criteria, assuming that such interactions occur in in the user decision process. Both models take the user's item rating on each criterion as input, and output a prediction of the item's overall rating.

In our experiment, we first used machine learning to estimate the model parameters from training data. We then compared the prediction accuracy of the two sets of models on testing data. If the nonlinear model performed better, then we would have expected similar results in practice under conditions comparable to our study. On the other hand, if there were no such interactions in practice, the nonlinear model should have performed no better than the linear model. As mentioned earlier, we used RMSE as our evaluation measure, as is commonly done for recommender systems.

4.1   Lower Bound of Root Mean Squared Error

Although not necessary to determine if interactions among criteria exist, we defined a lower bound on RMSE to give our findings context. Users were not entirely self-consistent when rating items, occasionally providing different overall ratings on items that were otherwise rated identically. Such differences may have been due to some random variability in their ratings (from difficulty in estimating or user changes over time), or may also have been due to other factors, such as the coarseness of the ratings or from other criteria excluded from the study.

If the true probability of the overall rating conditioned on the criteria ratings were known, it would be possible to create a classifier that makes the optimal decision and hence achieves the Bayes error rate (the overall minimum error rate). As we do not know this probability, we define an similar oracle that makes the optimal decision based on the empirical distribution, measured from the entire dataset (*both* training and test sets). The optimal prediction that minimizes

RMSE is the mean overall rating from all identically rated items (*including* the item to be predicted). We stress that this provides the lower bound in the limit, but is not a learning method and may not be achievable in practice.

4.2   Linear Model

The linear model is simply a linear combination over the ratings for each criterion; the independent variables are the ratings on the criteria, plus a bias term, and the dependent variable is the overall rating. If it was possible to select the best nonlinear model in every case, the RMSE of the linear model would serve as a upper bound on RMSE, as the linear model is a special case of nonlinear models described below. However, due to overfitting, it is possible to select a nonlinear model that is suboptimal and worse than the linear model. The RMSE achieved by the linear model is our baseline and a failure to improve upon it would indicate a lack of evidence for the criteria interactions the nonlinear model tries to capture. The linear model is simply:

$$P_L = \sum_{i=1}^{m} w_i v_i \tag{1}$$

where P_L is the predicted overall rating, v_i is the item rating on the ith criterion, and w_i are the coefficients to be learned.

4.3   Nonlinear Model

The general class of nonlinear models allows for any consistent prediction of overall rating based on the ratings on each criterion. However, this introduced too many possible models to effectively choose from, given the small amount of data, and exacerbated by inconsistency in the overall ratings (as noted earlier). Therefore, we limited ourselves to interactions between pairs of criteria. Observing interactions on this restricted set would be sufficient to show that criteria interactions existed, though we may not have found the optimal nonlinear model. Conversely, a failure to observe interactions would not have indicated that interactions do not exist, as the interaction may have been on several criteria.

We modeled interactions among pairs of criteria by creating derived binary features that correspond to specific ratings on criteria in a linear combination:

$$P_{ab} = \sum_{i=1}^{m} w_i v_i + c_{ab} \sum_{x \in A}^{m} \sum_{y \in B}^{m} I((v_a = x), (v_b = y)) \tag{2}$$

where P_{ab} is the predicted overall rating, a and b are the selected criteria pair, A is the set of possible values for criterion a, B is the set of possible values for criterion b, I is an indicator function that returns 1 when the arguments hold, 0 otherwise, v_i is the item rating on the ith criterion, and w_i and c_{ab} are the coefficients to be learned. Note that the first summation is simply Equation 1, and the second summation is simply a linear combination over a new set of (derived) features. In other words, we have created new binary features for each possible pair of ratings on criteria a and b. For example, when combining *authority* and *readability* (two binary criteria), $2 * 2 = 4$ new binary features are created; when combining *authority* and *novelty* (a binary and a five-valued criteria), $2 * 5 = 10$ new binary features are created. One can think of these induced binary features as correction factors, and as such, any nonlinear combination involving only these two features can be represented.

Since both datasets have four criteria, this gives us $C_4^2 = 6$ pairs of criteria to choose from. We also added a seventh nonlinear form (*all-pairs*) which uses all six pairwise combinations. We are further aided by the fact our criteria are discrete and take on a small set of values; for our data, the number of pair values for a criteria pair ranges from four to twenty-five. Table 1 shows the number of unique pairs of ratings observed for each criteria pair; a binary feature is created for each unique pair of ratings.

Table 1. Number of unique ratings possible when combining pairs of criteria

Yow-now	Authority	Novelty	Readability	(Subject) Relevance
Authority	n/a	10	4	10
Novelty		n/a	10	25
Readability			n/a	10
(Subject) Relevance				n/a

Epinions.com	Sound	Ease of Use	Picture Quality	Durability
Sound	n/a	24	22	24
Ease of Use		n/a	25	24
Picture Quality			n/a	24
Durability				n/a

4.4   Regularization

Since both sets of models take a linear form (as we have represented the nonlinear form as a linear model on a new feature space, described above), we may use linear regression to find model parameters that minimize RMSE on the training data. However, our goal is to minimize RMSE on the unseen testing data, not the training data, and given the small training set size, some form of regularization is needed to avoid overfitting. This is particularly important for the more complex (originally nonlinear) model, as the increased complexity can lead to an overly specific model that fits more of the noise in the data. We use Tikhonov regularization, a special case of L2-norm regularization or ridge regression. The analytical solution to the minimize RMSE with regularization is:

$$\mathbf{W} = (\lambda \mathbf{I} + \mathbf{X}^T \mathbf{X})^{-1} (\lambda \mathbf{W}_0 + \mathbf{X}^T \mathbf{Y}) \tag{3}$$

where an exponent of T indicates matrix transposition, λ controls the amount of regularization, $\mathbf{I}$ is the identify matrix, $\mathbf{X}$ is the instance matrix, $\mathbf{Y}$ is the vector of target values, $\mathbf{W}_0$ is the regularization vector we specify and $\mathbf{W}$ is the vector of coefficients we seek. Larger values of λ causes the solution to be closer to $\mathbf{W}_0$. We also added a constant term to our ratings representation to account for any bias in the overall rating.

For the linear model, we biased towards the following regularization vector:

$$\mathbf{W}_0 = \begin{bmatrix} 0.0 \; 0.25 \; 0.25 \; 0.25 \; 0.25 \end{bmatrix} \tag{4}$$

where the first position is the constant bias term and the remaining terms are the coefficients for the four criteria. We chose W_o such that all criteria would be weighted evenly, and the minimum (maximum) overall rating would be predicted when the minimum (maximum) rating was given on each criterion.

For the nonlinear models, we biased the model against interactions between criteria. The first five terms of the nonlinear regularization vector are the same as in the linear case, followed by zeros for each unique criteria pair value:

$$\mathbf{W}_0 = \begin{bmatrix} 0.0 \; 0.25 \; 0.25 \; 0.25 \; 0.25 \; 0.0 \; .... \; 0.0 \end{bmatrix} \tag{5}$$

Since the number of criteria pairs varies, the size of $\mathbf{W}_0$ also varies.

4.5   Tuning and Model Selection

The λ term in equation 3 controls the tradeoff between coefficients that minimize RMSE on the training set, and coefficients that are closer to the regularization vector ($\mathbf{W}_0$) described above. Higher values of λ moves the solution closer to the regularization vector, while allowing for higher RMSE; lower values of λ do the opposite. We automatically tuned the value of λ with ten-fold cross-validation on the training set alone. For a candidate value of λ and for each fold of the training data, we used the other 90% of the training data to learn the coefficients (using Equation 3); we used these coefficients to predict the overall ratings and record the RMSE. Starting with $\lambda = 0$, we tried successfully higher values of λ until the mean RMSE (i.e., the average over all ten folds) consistently increases. We then tried values of λ between the best two observed until no further reduction in RMSE is found. We did this in parallel for all seven nonlinear models, as well as the linear model, for eight models in all. From these eight models, we selected the one with the lowest mean RMSE across all the folds with the best corresponding value of λ. Note that we could select the linear model as the best model; we would run this model for comparison purposes in any case. Finally, the final coefficients were learned from the entire training set (i.e., no cross-validation) using this chosen model and value of λ.

5   Experimental Results

We tested for interactions among criteria by contrasting the observed RMSE of our criteria interaction models with that of the linear model. To decrease the possibility for random misleading effects, we ran the experiment 1000 times (i.e.,

Table 2. Non-personalized models results over 1000 trials

Method	RMSE Mean	RMSE Std. Dev.	RMSE Median	Mean RMSE Reduction	Possible RMSE Reduction Achieved
Yow-now					
$\mathbf{W}_0$ only	0.2507	0.00408	0.2508	-36.99%	-1281.63%
Lower Bound	0.1830	0.00352	0.1829	2.61%	100.00%
Linear	0.1879	0.00362	0.1880	0.00%	0.00%
Nonlinear	0.1853	0.00117	0.1852	1.38%	52.74%
Epinions.com					
$\mathbf{W}_0$ only	0.2225	0.00911	0.2223	-10.09%	-83.27%
Lower Bound	0.1776	0.00665	0.1774	12.12%	100.00%
Linear	0.2021	0.00659	0.2021	0.00%	0.00%
Nonlinear	0.2008	0.00700	0.2005	0.64%	5.31%

1000 trials). The test set was randomly chosen from the full dataset each time, which means different trials will have different training sets and testing sets, and a single item is likely to serve as both training and test data (but in different trials; no testing data is ever included in training data). This is valid because all of our modeling choices (regularization tuning and learning model coefficients) are done solely on the basis of the training data.

Table 2 shows the RMSE results on both datasets without personalization. Four methods are reported: RMSE results using the regularization vector $\mathbf{W}_0$ only (equivalent to setting regularization parameter λ to infinity); the lower bound on RMSE; the learned linear combination; and the learned nonlinear combination. The mean RMSE reduction shows how much the RMSE decreased as a percentage of the RMSE of learned linear model. However, the lower bound is very close to the linear case, so there is not much potential for RMSE reduction. The possible RMSE reduction shows how much of this potential RMSE reduction is achieved; by definition, it is always 100% at the lower bound.

The nonlinear model has a lower RMSE for both datasets, but the difference is very small. This is not surprising as the RMSE for the linear model is quite close to the lower bound. The possible RMSE gives a different picture. For the Yow-now model, over half of the possible RMSE reduction was achieved with the nonlinear model. For the Epinions.com model, much less of the possible RMSE reduction was achieved. The smaller dataset size may have played a role, as less data will tend to produce poorer learned models but also a lower lower bound (because there are less opportunities for inconsistent ratings).

Table 3 shows the results when a separate model is learned for each user (personalized models), as well as the microaverage and macroaverage. Results are generally poorer for users with less data. Performance varies a lot among users: mean RMSE ranges from 0.0888 to 0.2795; mean RMSE reduction ranges from -2.86% to 9.98%; and the percentage of possible RMSE reduction achieved ranges from -21.73% to 55.37%. In fact, a slight majority of users had negative results with respect to the baseline. Comparing Table 3 with Table 2, we can see

Table 3. Personalized Yow-now model results over 1000 trials

User	Articles	Most Sel.	Most Pct	RMSE Mean	RMSE Std. Dev.	RMSE Median	Mean RMSE Reduction	Possible RMSE Reduction Achieved
u51	305	⟨2,4⟩	46%	0.1654	0.01761	0.1654	-1.45%	-18.38%
u56	362	⟨2,4⟩	53%	0.1277	0.01274	0.1272	1.86%	13.56%
u58	569	B	31%	0.2032	0.01204	0.2033	-1.25%	-19.77%
u59	358	C	66%	0.1280	0.00943	0.1281	0.41%	7.32%
u60	138	⟨1,2⟩	48%	0.1488	0.02475	0.1440	-1.23%	-6.45%
u62	161	B	42%	0.1065	0.01161	0.1065	-2.86%	-21.18%
u63	472	⟨2,4⟩	60%	0.1089	0.01522	0.1092	-1.89%	-15.59%
u65	607	⟨3,4⟩	76%	0.1347	0.01537	0.1329	4.53%	34.26%
u66	443	⟨2,4⟩	53%	0.1487	0.01723	0.1472	0.85%	4.72%
u67	590	⟨2,4⟩	96%	0.2344	0.01240	0.2345	5.23%	45.00%
u68	388	B	57%	0.1455	0.00932	0.1453	-1.11%	-21.74%
u69	848	C	82%	0.1772	0.00772	0.1770	1.01%	13.44%
u73	232	B	33%	0.1678	0.01700	0.1677	-0.68%	-5.95%
u74	14	⟨3,4⟩	33%	0.1969	0.07209	0.1810	1.42%	2.93%
u76	603	⟨1,2⟩	58%	0.0888	0.00834	0.0888	-1.14%	-17.68%
u80	218	⟨3,4⟩	55%	0.2795	0.03244	0.2802	0.00%	0.03%
u82	516	C	97%	0.1064	0.01385	0.1062	9.98%	55.37%
u83	1079	⟨2,4⟩	73%	0.0960	0.00473	0.0960	3.06%	38.19%
u84	426	⟨2,4⟩	43%	0.2318	0.01475	0.2321	-0.89%	-12.35%
u87	129	B	68%	0.1740	0.02131	0.1734	-0.99%	-7.49%
u88	54	⟨2,4⟩	56%	0.1704	0.03602	0.1689	-2.02%	-4.51%
u91	367	C	59%	0.2212	0.01646	0.2215	0.46%	3.79%
u92	310	B	21%	0.1557	0.01097	0.1557	-2.74%	-21.73%
micro				0.1539			1.21%	8.42%
macro				0.1616			0.46%	1.99%

that the microaverage over the personalized models is lower than even the lower bound on the non-personalized model. This shows that there was considerable differences among user models, and thus personalization reduced RMSE.

Table 3 also shows the most frequently selected nonlinear model (Most Sel.) for each user. Due to space limit, the criteria are numbered as 1 (authority), 2 (novelty), 3 (readability) and 4 (subject relevance). For example, our method selected the novelty and subject relevance pair (listed as ⟨2,4⟩) for user *u51* in 46% of the trials. Additionally, *B* indicates the basic linear model (no criteria interactions) and *C* indicates the *all-pairs* nonlinear model. The same model was not always selected for the same user on every trial, because it was dependent on the trial's randomly selected training set. Users that showed mostly a linear trend had an increase in RMSE because overfitting occurred when a nonlinear model was selected. Also, users that did not show a consistent preference for a particular form also had an increase in RMSE, for similar reasons.

We observe that a variety of criteria interact in the personal models, and in fact each pair was selected as the best at least once on some trial. However, some

pairs tend to interact more than others. For the non-personalized models, the
all-pairs nonlinear form was always selected for the Yow-now dataset, while the
⟨sound,picture quality⟩ pair was selected 82% of the time for the Epinions.com
dataset. For the personalized Yow-now models, ⟨novelty,relevance⟩ was the most
commonly selected pair, and indeed along with ⟨readability,relevance⟩ and the
all-pairs nonlinear form accounted for all mean reductions in RMSE.

Figure 1 show the mean learned interactions for users *u67* and *u83*, who had
some of the largest RMSE reductions. Though our method consistently selected
⟨novelty,relevance⟩ for both users, the learned interactions were quite different.
The plot for *u67* has a smooth surface, with an upward adjustment for higher
values on either of the criteria while the other criterion remains low. On the
other hand, *u83*'s plot has no such easily interpretable pattern, which was also
true for most users. More research is needed to understand these interactions.

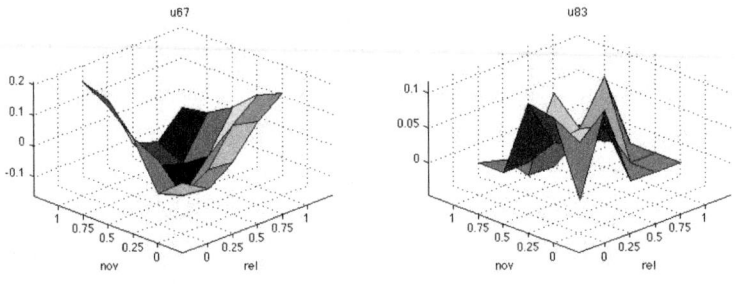

Fig. 1. Interactions of the ⟨novelty,relevance⟩ criteria pair for users *u67* and *u83*

6   Conclusions and Future Work

Our results show that interactions among criteria exist in criteria-based information retrieval models, at least in some cases, as measured by an observed
reduction in RMSE. We observed this reduction in both non-personalized and
personalized models. However, the amount of RMSE reduced by exploiting interactions was slight in the datasets we used; in fact, it often *increased* RMSE,
but the magnitude of the reduction for some users outweighed the increases for
the rest. Personalization was more clearly beneficial. In terms of the interactions
themselves, certain criteria had consistently stronger observed interactions than
others, but we could not discern an interpretable pattern in these interactions.

Despite our use of regularization, overfitting remained a problem, as evidenced
by the occasional *increase* in RMSE over the linear model. This could potentially
be avoided by opting for the linear model when there is insufficient evidence
for interactions (i.e., when the reductions are not consistently observed in the
training data, or not large enough relative to the training set size). This could be
expanded to a Bayesian framework, using prior probabilities to avoid selecting
less probable models when there is not sufficient support in the data. Even
without these improvements, in our experiments we were successful in reducing
the overall mean RMSE by exploiting criteria interactions and personalization.

Acknowledgements. Part of this research is funded by National Science Foundation IIS-0713111 and AFOSR. Any opinions, findings, conclusions or recommendations expressed in this paper are the authors', and do not necessarily reflect those of the sponsors.

References

1. Triantaphyllou, E.: Multi-Criteria Decision Making Methods: A Comparative Study. Kluwer Academic Publishers, Dordrecht (2000)
2. Wolfe, S.R., Zhang, Y.: User-centric multi-criteria information retrieval. In: Allan, J., Aslam, J., Sanderson, M., Zhai, C., Zobel, J. (eds.) SIGIR '09: Proceedings of the 32nd international ACM conference on Research and development in information retrieval, pp. 818–819. ACM, New York (2009)
3. Chen, H., Karger, D.R.: Less is more: probabilistic models for retrieving fewer relevant documents. In: Efthimiadis, E.N., Dumais, S., Hawking, D., Järvelin, K. (eds.) SIGIR '06: Proceedings of the 29th annual international ACM conference on Research and development in information retrieval, pp. 429–436. ACM Press, New York (2006)
4. Carbonell, J., Goldstein, J.: The use of mmr, diversity-based reranking for reordering documents and producing summaries. In: Croft, W.B., Moffat, A., van Rijsbergen, C.J., Wilkinson, R., Zobel, J. (eds.) SIGIR '98: Proceedings of the 21st annual international ACM conference on Research and development in information retrieval, pp. 335–336. ACM, New York (1998)
5. Zhang, Y., Callan, J., Minka, T.: Novelty and redundancy detection in adaptive filtering. In: Järvelin, K., Beaulieu, M., Baeza-Yates, R., Myaeng, S.H. (eds.) SIGIR '02: Proceedings of the 25th annual international ACM conference on Research and development in information retrieval, pp. 81–88. ACM, New York (2002)
6. Allan, J., Wade, C., Bolivar, A.: Retrieval and novelty detection at the sentence level. In: Clarke, C., Cormack, G., Callan, J., Hawking, D., Smeaton, A. (eds.) SIGIR '03: Proceedings of the 26th annual international ACM conference on Research and development in information retrieval, pp. 314–321. ACM, New York (2003)
7. Barry, C.L.: User-defined relevance criteria: an exploratory study. J. Am. Soc. Inf. Sci. 45(3), 149–159 (1994)
8. Maglaughlin, K., Sonnenwald, D.: User perspectives on relevance criteria: a comparison among relevant, partially relevant, and not-relevant judgements. J. Am. Soc. Inf. Sci. Technol. 53(5), 327–342 (2002)
9. Tombros, A., Ruthven, I., Jose, J.M.: How users assess web pages for information seeking. J. Am. Soc. Inf. Sci. Technol. 56(4), 327–344 (2005)
10. Manouselis, N., Costopoulou, C.: Analysis and classification of multi-criteria recommender systems. World Wide Web 10(4), 415–441 (2007)
11. Pasi, G., Bordogna, G., Villa, R.: A multi-criteria content-based filtering system. In: Kraaij, W., de Vries, A.P., Clarke, C.L.A., Fuhr, N., Kando, N. (eds.) SIGIR '07: Proceedings of the 30th annual international ACM conference on Research and development in information retrieval, pp. 775–776. ACM, New York (2007)

12. de Gemmis, M., Semeraro, G., Lops, P., Basile, P.: A retrieval model for personalized searching relying on content-based user profiles. In: Mobasher, B., Anand, S.S., Kobsa, A., Jannach, D. (eds.) 6th AAAI Workshop on Intelligent Techniques for Web Personalization and Recommender Systems (ITWP 2008), pp. 1–9 (2008)
13. Pognačnik, M., Tasič, J., Košir, A.: Optimization of multi-attribute user modeling approach. International Journal of Electronics and Communications 58(4), 402–412 (2004)
14. Nguyen, H., Haddawy, P.: The decision-theoretic video advisor. In: Kautz, H. (ed.) AAAI-98 Workshop on Recommender Systems, pp. 77–80 (1998)
15. Chajewska, U., Koller, D.: Utilities as random variables: Density estimation and structure discovery. In: Boutilier, C., Goldszmidt, M. (eds.) Proceedings of the 16th Conference on Uncertainty in Artificial Intelligence, pp. 63–71 (2000)
16. Robu, V., Somefun, D.J.A., La Poutré, J.A.: Modeling complex multi-issue negotiations using utility graphs. In: Pechoucek, M., Steiner, D., Thompson, S. (eds.) AAMAS '05: Proceedings of the fourth international joint conference on Autonomous agents and multiagent systems, pp. 280–287. ACM, New York (2005)
17. Zhang, Y.: Yow user study data: Implicit and explicit feedback for news recommendation, http://www.soe.ucsc.edu/~yiz/papers/data/YOWStudy

Collaborative Inference of Sentiments from Texts

Yanir Seroussi, Ingrid Zukerman, and Fabian Bohnert

Faculty of Information Technology, Monash University
Clayton, Victoria 3800, Australia
firstname.lastname@infotech.monash.edu.au

Abstract. Sentiment analysis deals with inferring people's sentiments and opinions from texts. An important aspect of sentiment analysis is polarity classification, which consists of inferring a document's polarity – the overall sentiment conveyed by the text – in the form of a numerical rating. In contrast to existing approaches to polarity classification, we propose to take the authors of the documents into account. Specifically, we present a nearest-neighbour collaborative approach that utilises novel models of user similarity. Our evaluation shows that our approach improves on state-of-the-art performance, and yields insights regarding datasets for which such an improvement is achievable.

1 Introduction

Polarity classification is one of the earliest tasks attempted in the sentiment analysis field [1]. The binary case consists of classifying a piece of text as either positive or negative. Less attention has been paid to *Multi-way Polarity Classification (MPC)*, i.e., inferring the "star rating" of texts on a scale of more than two values [2].

A key challenge in MPC is that ratings on a non-binary scale are more open to interpretation than binary ratings (e.g., the difference between a rating of 6 and 7 on a 10-point scale is less clear cut than the difference between "good" and "bad"), meaning that every user has a different "feel" for the rating scale. This challenge was noted in [2], but not dealt with directly. In this paper, we address this challenge by considering users when performing MPC, rather than relying solely on standalone texts. We do this by introducing a nearest-neighbour collaborative MPC framework: we train user-specific classifiers, and consider user similarity to combine the outputs of the classifiers to yield an estimate of a text's polarity rating. This approach decreases the error in cross-user MPC compared to user-blind methods, while requiring less computational resources.

This paper investigates several models of user similarity built on the basis of users' textual input, and compares their impact on MPC performance. These models are employed to address two main challenges: (1) modeling user similarity when the item/rating matrix is sparse; and (2) modeling user similarity for new users that have not submitted ratings or reviews (but have written other texts). When the item/rating matrix is sparse, we show that basing similarity on all the reviews by the users reduces the error compared to basing similarity only on reviews for co-reviewed items (Sect. 6.1). When no ratings or reviews are available, we show that message board posts can be used to model user similarity (Sect. 6.2).

P. De Bra, A. Kobsa, and D. Chin (Eds.): UMAP 2010, LNCS 6075, pp. 195–206, 2010.

This paper is organised as follows. Related work is surveyed in Sect. 2. Polarity classification methods and user similarity models are presented in Sect. 3 and 4 respectively. Our dataset is described in Sect. 5. Section 6 presents the results of our evaluation, and Sect. 7 discusses our conclusions and plans for future work.

2   Related Work

In recent years, the sentiment analysis task of *polarity classification* has received much attention [1]. Polarity scales can be either binary (e.g., positive/negative) or multi-way (e.g., star ratings). Binary polarity classification has been an active research area since the early days of sentiment analysis. *Multi-way Polarity Classification* (*MPC*) was attempted in several domains, e.g., restaurant reviews [3], movie reviews [2], and customer feedback [4]. Results vary depending on the domain and size of the texts. Unsurprisingly, MPC results are inferior to those obtained for binary classification.

Several researchers found that authorship *affects* performance in sentiment analysis [2, 5]. Pang and Lee [2] found that a classifier trained on film reviews by one user and tested on reviews by a different user is likely to perform poorly. Lin *et al.* [5] obtained similar results when classifying pro-Palestinian and pro-Israeli articles, half of them written by two editors and the other half by various guest writers. Our work takes these insights one step further, in that we harness user similarity to *improve* performance.

The inspiration for considering users in MPC comes from recommender systems, specifically from *Collaborative Filtering* (*CF*), which employs a *target user*'s previous ratings and ratings submitted by similar users to predict the ratings that the target user will give to unrated items [6]. MPC resembles CF in that both produce ratings. However, there are two fundamental differences: (1) in MPC, the ratings are obtained from classifiers that take as input a user's textual review of an unrated item, whereas CF relies on the user's ratings of other items; and (2) CF systems generally require some ratings by the target user, while MPC systems infer the polarity (numeric rating) of a target user's text even when no ratings by this user are available.

Despite the differences between MPC and CF, they have two problems in common: (1) sparsity of the item/rating matrix, where rating prediction is based on a relatively small number of known ratings, compared to the number of possible item/rating pairs; and (2) the new user problem – predicting ratings for users who supplied few or no ratings [6]. Our text-based measures address these problems by reducing the dependency on rated items to calculate similarity between users (Sect. 6.1), and by decreasing the need for ratings and opinion-bearing texts (Sect. 6.2).

3   Polarity Classification Methods

In this section, we describe several methods for MPC. In the following descriptions, the *target user* is the author of the reviews to be classified (i.e., reviews for which the polarity is unknown), and the *training users* are the authors of reviews for which the polarity labels are known. The target user may be a new user, for whom few or no labeled reviews are available. All methods rely on classifiers that are trained on labeled reviews and output the classification of unlabeled reviews.

Pang and Lee [2] focused on training a single classifier on labeled reviews. They used training data from a single user (*SCSU – Single Classifier, Single User*) or from multiple users (*SCMU – Single Classifier, Multiple Users*). SCSU is similar to content-based recommender systems [6], as it is based only on the target user's past ratings and reviews. In our experiments [7], SCSU was found to perform best on reviews by the user on whom it was trained, but it requires many reviews to achieve acceptable performance. In addition, SCSU is unsuitable when there are few or no labeled reviews by the target user.

SCMU addresses SCSU's problem of target users with few reviews, since it does not rely solely on labeled reviews from the target user. However, SCMU's classification performance is worse than that of SCSU, as differences between the training users make it hard for the classifier to generalise [7]. Moreover, training an SCMU classifier on all the available data is infeasible in a system with many users and reviews. To address this problem, one could randomly sample a subset of the available reviews and use them for training the classifier, but this does not result in satisfactory performance [7].

Our method, *Multiple Classifiers, Multiple Users* (*MCMU*), addresses these problems by training a separate classifier for each training user and combining the normalised ratings inferred by the classifiers into a single rating. Specifically, we use Equation 1 to estimate $\hat{r}_{q_a}$ (the polarity classification of review q by the target user a):

$$\hat{r}_{q_a} = \mu_a + \sigma_a \sum_{u \in U} \left(\frac{w_u}{\sum_{u \in U} w_u} \right) \left(\frac{\hat{r}_{u,q_a} - \mu_u}{\sigma_u} \right) \tag{1}$$

where U is the set of training users, μ_x and σ_x are user x's rating mean and standard deviation, $\hat{r}_{u,q_a}$ is the rating inferred by user u's classifier for review q_a, and w_u is the weight of user u's classifier. The weights are calculated using the methods introduced in Sect. 4, and normalised by dividing each weight by the sum of the weights.

Equation 1 can be used when no ratings by the target user are available (Sect. 6.2). In this case, the training users' mean and standard deviation are used as an estimation of μ_a and σ_a. When used for CF, Equation 1 yielded better performance than a simple weighted average of ratings [8], because it considers user-specific rating biases. We obtained this result in our MPC experiments [7], and hence use Equation 1 in this paper.

4   User Similarity Models

This section introduces several methods for modeling user similarity. These methods yield a similarity score $sim(a, u)$ for users a and u, which is then incorporated into the MCMU classifier ensemble described in Sect. 3. That is, given a target user a, the weights used in Equation 1 are $w_u = f(sim(a, u))$, where $f(x)$ is a transformation function that ensures non-negative weights. We chose $f(x) = e^x$ in order to give more weight to similar users.[1]

In this setup, all available users are given weights, but it can easily be modified to consider only users above a similarity threshold s (Sect. 6.1):

$$w_u = \begin{cases} f(sim(a, u)) & sim(a, u) > s \\ 0 & \text{otherwise} \end{cases} \tag{2}$$

---

[1] We experimented with several functions, with $f(x) = e^x$ yielding the best results [7].

Table 1. Similarity Measures Taxonomy

	Rating-based (Sect. 4.1)	Text-based (Sect. 4.2)
All items	AIR	AIT, AIP
Co-reviewed items	CRR	CRT, CRP

Table 1 groups the similarity models based on the type of information they use and the sources for this information.[2] Rating-based methods rely only on ratings in order to measure similarity between users, while text-based methods employ only the users' documents. Measures based on all items calculate general statistics on the entire set of user reviews or documents, while measures based on co-reviewed items perform a pairwise comparison of the reviews for items reviewed by two users.

We expect measures based on co-reviewed items to be more informative than measures based on all items. This is because the former take into account the actual items reviewed, while the latter may be computed on the basis of users who have only a few or no items in common. However, measures based on co-reviewed items may require more labeled reviews, as the size of the set of co-reviewed items should be sufficiently large to give meaningful similarity values. They may also underperform when the item/rating matrix is sparse (i.e., many items are reviewed by only a few users).

4.1 Rating-Based Models

All Item Ratings (AIR). Let Q_x denote the set of all reviews written by user x and $Q_{x,r}$ denote x's reviews with rating r ($r \in \{1, 2, \ldots, 10\}$). The similarity between users a and u is one minus the *Hellinger distance* between their rating distributions:

$$sim(a, u) = 1 - \sqrt{\frac{1}{2} \sum_{r=1}^{10} \left(\sqrt{\frac{|Q_{a,r}|}{|Q_a|}} - \sqrt{\frac{|Q_{u,r}|}{|Q_u|}} \right)^2} \in [0, 1] \qquad (3)$$

This similarity measure accounts for the relative positivity or negativity of the users. For instance, if one user mostly gives low ratings and another mostly high ratings, they are considered dissimilar. For this measure to be meaningful, we need a sufficiently large sample of ratings for the two users, so that it accurately represents the overall rating distribution. However, no textual analysis is required to calculate this measure, and thus its computation is faster than that of text-based models (Sect. 4.2).

Co-reviewed Ratings (CRR). Basing user similarity on co-reviewed item ratings is common in CF, with the two most popular approaches employing *Pearson correlation* or *cosine similarity* to compare the rating vectors of co-reviewed items [6]. In our experiments, we found that the former yields better performance than the latter [7], and thus we use the Pearson correlation coefficient in this paper:

$$sim(a, u) = \frac{\sum_{i \in I_{a,u}} (r_{a,i} - \bar{r}_a)(r_{u,i} - \bar{r}_u)}{\sqrt{\sum_{i \in I_{a,u}} (r_{a,i} - \bar{r}_a)^2 \sum_{i \in I_{a,u}} (r_{u,i} - \bar{r}_u)^2}} \in [-1, 1] \qquad (4)$$

---

[2] All the methods except CRR, which is commonly used in CF, were devised by us.

where $I_{a,u} = I_a \cap I_u$ is the set of items co-reviewed by users a and u, and I_x is the set of items reviewed by user x. User x's rating for item i is $r_{x,i}$, and $\bar{r}_x$ denotes the mean of user x's ratings for the items in $I_{a,u}$.

4.2 Text-Based Models

All Item Terms (AIT). We employ the *Jaccard coefficient* of the sets of terms used in two documents to measure the similarity between them:

$$J(d_1, d_2) = \frac{|T(d_1) \cap T(d_2)|}{|T(d_1) \cup T(d_2)|} \in [0, 1] \tag{5}$$

where $T(d)$ is the set of terms that appear in document d. We chose the Jaccard coefficient, rather than cosine similarity of tf-idf vectors, because our experiments showed that the former yields better results than the latter [7]. This is in line with the results reported by Pang *et al.* [9], who found that unigram presence performs better than frequency when classifying textual polarity.

Equation 5 can be modified to consider only certain types of terms. For example, instead of all the terms, $T(d)$ may include only adjectives or nouns. The type of terms included in $T(d)$ is determined experimentally in Sect. 6.1.

We define the AIT similarity between two users a and u as the Jaccard coefficient of the documents (not necessarily reviews) written by these users:

$$sim(a, u) = J(d_a, d_u) \in [0, 1] \tag{6}$$

where d_x is the concatenation of the documents written by user x.

Co-reviewed Terms (CRT). This measure considers only the reviews of co-reviewed items (unlike AIT, which considers all the documents written by the users). Thus, we define CRT as the mean of the Jaccard coefficients (Equation 5) of every review pair:

$$sim(a, u) = \sum_{i \in I_{a,u}} \frac{J(q_{u,i}, q_{a,i})}{|I_{a,u}|} \in [0, 1] \tag{7}$$

where $q_{x,i}$ is user x's review of item i.

All Item PSPs (AIP). *Positive Sentence Percentage (PSP)* was defined by Pang and Lee [2] as the percentage of positive sentences out of the subjective sentences in a review. To detect the positive sentences, they trained a Naive Bayes (NB) classifier on the *Sentence Polarity Dataset* (v1.0). When used to model *review* similarity, PSP outperformed term-based methods for multi-way polarity classification [2].

Here we introduce a *user* similarity measure based on PSP. This measure replaces ratings with PSPs, thereby obviating the need for explicit ratings. In contrast to Pang and Lee, we define PSP as the percentage of positive sentences among *all* the sentences in a document (rather than just subjective sentences). This generalises the PSP definition to include objective texts, such as message board posts. Additionally, we use a Support Vector Machine (SVM) trained on the Sentence Polarity Dataset, because the classification accuracy of the SVM implementation we use was found to be higher than that of the NB classifier (running a 10-fold cross validation on this dataset).

AIP is defined in a similar way to AIR (Sect. 4.1), as one minus the Hellinger distance between the PSP distributions of users a and u:

$$sim(a, u) = 1 - \sqrt{\frac{1}{2}\sum_{l=1}^{L}\left(\sqrt{p_{a,l}} - \sqrt{p_{u,l}}\right)^2} \in [0, 1] \qquad (8)$$

where L is the number of bins (determined experimentally), and $p_{x,l}$ is defined as:

$$p_{x,l} = \frac{|\{q \in Q_x : l - 1 \le L \times psp(q) < l\}|}{|Q_x|}, \quad l \in \{1, 2, ..., L\} \qquad (9)$$

where Q_x is the set of user x's reviews, and $psp(q)$ is the PSP of review q. The last element $p_{x,L}$ is calculated using $L \times psp(q) \le L$ (instead of $< L$) to include reviews with $psp(q) = 1$.

Co-reviewed PSPs (CRP). Like CRR, this model is based on co-reviewed items (Sect. 4.1), but it does not require explicit ratings. Instead of ratings, it uses the Pearson correlation coefficient of PSPs, yielding the following similarity measure:

$$sim(a, u) = \frac{\sum_{i \in I_{a,u}} (psp(q_{a,i}) - \overline{psp}_a)(psp(q_{u,i}) - \overline{psp}_u)}{\sqrt{\sum_{i \in I_{a,u}} (psp(q_{a,i}) - \overline{psp}_a)^2 \sum_{i \in I_{a,u}} (psp(q_{u,i}) - \overline{psp}_u)^2}} \in [-1, 1] \qquad (10)$$

where $\overline{psp}_x$ denotes the mean of user x's PSPs for the items in $I_{a,u}$.

5 Dataset

We created the *Prolific IMDb Users* dataset by collecting data from the *Internet Movie Database (IMDb)* at www.imdb.com in May 2009.[3] This dataset contains 184 users with at least 500 movie reviews per user. However, not all users have a large number of *labeled* reviews, as IMDb users may choose not to assign a rating to their reviews. In addition to movie reviews, users may write message board posts. IMDb message boards are mostly movie-related, but some are about television, music and other topics.

In the experiments presented in this paper, we use a subset of the Prolific IMDb Users dataset, called *IMDb62*, which contains reviews of 62 prolific users (1000 reviews per user, 62,000 reviews in total), and includes only reviews with ratings (all ratings are on a 1–10 star scale). Each user's reviews were obtained using proportional sampling without replacement (i.e., for each user, the 1000 reviews have the same rating frequencies as their complete set of reviews). It is worth noting that in our evaluation we do not assume that every user has submitted 1000 labeled reviews. In fact, we show that our methods yield improved performance compared to the baselines even when the number of prolific users is small (Sect. 6.3).

We ensured that each item in IMDb62 is reviewed only once by each user. Reviews for items with multiple reviews by the same user were discarded to reduce ambiguity. In addition, explicit ratings were automatically filtered out from the review texts, e.g., "5/10" was removed from texts such as "this movie deserves 5/10".

---

[3] We could not use the *Sentiment Scale Dataset* (v1.0) [2], as it includes movie reviews by only four users – too few to support experiments regarding the impact of authorship on sentiment.

Table 2. IMDb62 Dataset Properties

Users:	62	Words per review mean:	300
Labeled reviews:	62000	Words per review standard deviation (stddev):	198
Reviewed items:	29116	Message board posts:	17560
Items with only one review:	18322	Number of users with no posts:	11
Item/rating matrix sparsity:	96.57%	Posts per user mean (for users with posts):	344
		Posts per user stddev (for users with posts):	743

IMDb62 also includes all the message boards posts for each user. Some users did not submit any posts, while others wrote hundreds to thousands of posts. No sub-sampling of message board posts was performed, because posts do not have assigned ratings (unlike reviews), and thus proportional sampling is impossible.

Table 2 displays some statistics for the IMDb62 dataset. Notable properties of the dataset are the large percentage of items that were reviewed by only one user (around 63%), and the high standard deviation of the review word count. These properties make cross-user MPC a challenging task.

As discussed in Sect. 1, one challenge in MPC is that different users may have different interpretations of the rating scale, e.g., two users may express a similar opinion of an item, but assign it a different rating [2]. Further, users select the items they review, and therefore they might choose to submit only reviews with extreme ratings. These characteristics are visible in IMDb62, which displays a large variability of rating distributions. For example, it contains users with more than 40% 10-star ratings and almost no 1–4 star ratings, while others have most of their ratings in the 1–5 star range.

6   Evaluation

In this section, we evaluate the models introduced in Sect. 4 by running experiments on the IMDb62 dataset (Sect. 5). For all experiments, we perform leave-one-out cross validation, training on at most 61 users and testing on the remaining user. We report the *mean absolute error (MAE)* across all classified reviews: $\text{MAE} = \sum_{q \in Q} |r_q - \hat{r}_q|/|Q|$, where Q is the set of reviews to classify, r_q is the actual rating of review q, and $\hat{r}_q$ is the rating inferred using Equation 1.[4] We chose this measure because the ratings for the reviews in the dataset are given on an *ordinal* 10-point scale, and using MAE (rather than classification accuracy or ROC curves) gives different weights to different classification errors: misclassifying a 10-star review as a 1-star review is different from misclassifying it as a 9-star review. Statistically significant differences in MAE are reported when $p < 0.05$ according to a paired two-tailed t-test.

To infer ratings from reviews, we used Support Vector Regression as implemented in Weka 3.6.0 (www.cs.waikato.ac.nz/ml/weka), with default settings (using different settings did not improve performance in preliminary experiments). This algorithm was found to outperform other algorithms when tested on the IMDb62 dataset [7].

---

[4] Our methods return integer ratings, not star fractions. That is, on a 10-star scale the only possible values are 1–10. This was found to reduce the MAE [7].

The features used by the algorithm are unigrams extracted from the review texts. AIT and CRT require part-of-speech tagging (Sect. 4.2), which was done using OpenNLP 1.4.3 (opennlp.sourceforge.net) with the default English language models.

6.1   Similarity Modeling Experiment

Experimental Setup. In this section, we evaluate our approach for inferring sentiment from text, focusing on the similarity measures introduced in Sect. 4. We use these measures to give different weights to the training users, and compare the resulting MAEs to those obtained using equal weights and SCSU (Sect. 3).

To calculate the similarity between a target user and the training users, we need a certain amount of reviews by the target user. Thus, we split the target user's reviews into two sets by sampling uniformly without replacement: the first set, for which the ratings are known, is used for similarity calculation; and the second set, with unknown ratings, is used to test the classifier. We consider different set sizes of labeled reviews by the target user (5, 10, 20, 50, 100, 200, ..., 950), and repeat this process 50 times for each set size. The same sets are used to train and test the SCSU baseline.

We experiment with training user selection by setting a threshold for user similarity as specified in Equation 2 (if thresholding filters out all the users, then all classifiers are given the same weight). The optimal threshold is expected to be user-specific, and thus it is learned separately for each target user from their labeled reviews. This is done as follows. We vary the threshold over the interval $[0, 1]$,[5] and classify the *labeled* reviews of the target user. The threshold yielding the lowest MAE on the labeled reviews is used for classifying the *unlabeled* reviews of the target user.

In addition to the similarity threshold, we set a threshold on the size of the set of co-reviewed items for the similarity measures that depend on co-reviewed items. This threshold is set dynamically for each target user in a similar manner to the similarity threshold. Our results show that applying the set size threshold reduces the MAE [7].

Results. Figure 1 displays the results of this experiment.[6] The number of bins L for AIP (Equation 9) is set to 100, and $T(d)$ consists of all nouns for AIT, and all unigrams for CRT (Equation 5). These options yield the lowest MAE from those we tested (for 100 labeled reviews or less): for L, we experimented with 10, 20, 50, 100, 150, 200, 500, 1000; for $T(d)$ we considered all unigrams, all unigrams except stop words, open word classes (adjectives, adverbs, nouns and verbs), and combinations of these classes [7].

As seen in Fig. 1, EQW – obtained by assigning equal weights to the training models in MCMU – outperforms SCSU when up to 100 reviews are available. Additionally, the best similarity measure, AIT, outperforms EQW for every number of labeled target user reviews, and performs better than SCSU for up to 200 reviews. This is an encouraging result, since in general users submit a relatively small number of reviews (e.g., as seen

---

[5] This removes negatively correlated users from the set of neighbours for CRR and CRP (Sect. 4). The other similarity measures only produce values within the [0,1] range.

[6] All the differences are statistically significant except for CRT vs. CRR for 200 labeled reviews by the target user, and AIR vs. AIP and CRP vs. CRR for 950 labeled reviews.

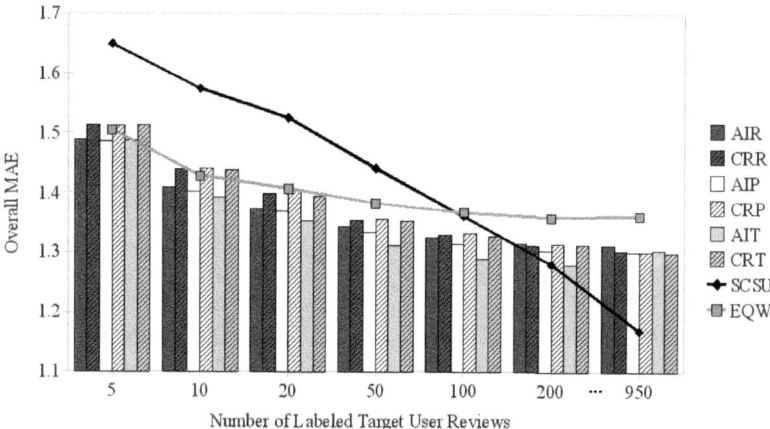

Fig. 1. Similarity Modeling Experiment Results

in www.imdb.com/title/tt0068646 in January 2010, 90% of the 1420 reviews for the movie "The Godfather" were submitted by users with less than 200 reviews).[7]

Another interesting result is that similarity measures based on co-reviewed items generally require many labeled reviews by the target user to achieve comparable performance to that of the measures which are based on all items (200 for CRR vs. AIR, and 950 for CRP vs. AIP and CRT vs. AIT). This may be attributed to the sparsity of our dataset, which results in small sets of co-reviewed items. Note that the inferior performance of the co-reviewed items measures is not due to a fallback to equal weights, as the results for these measures differ from the results obtained by EQW for almost all target users when more than 5 labeled reviews are available (when 5 reviews are available, equal weights are used only for about 20% of the target users).

6.2   Similarity Modeling without Reviews or Ratings

Experimental Setup. One advantage of modeling user similarity based on texts is that no explicit ratings are required. Texts are generally easier to obtain than ratings, since users commonly communicate textually, e.g., using emails, instant messaging or message boards. In addition to labeled reviews, our dataset includes message board posts (Sect. 5). In this section, we consider the case where no target user ratings or reviews are available, and thus we model user similarity based *only* on message board posts. In this case, the only relevant similarity models are AIT and AIP (Sect. 4). Since we have no labeled reviews by the target user, we cannot set the similarity threshold dynamically for each user. Thus, we set a global threshold for all users.

Results. The results of this experiment are displayed in Table 3 (all the differences are statistically significant). Modeling user similarity based on message board posts yields

---

[7] MCMU reaches optimal performance for 200–600 labeled reviews, at which point the system should switch to SCSU. In the future, we will train the system to switch automatically to SCSU when it performs better than MCMU.

Table 3. Message Board Posts Similarity Experiment Results

Similarity Measure	Optimal Threshold	MAE
EQW	—	1.51
AIT (all unigrams)	0.02	1.50
AIP	0.34	**1.49**

a lower MAE than EQW, but the margin is smaller than when labeled reviews by the target user are available (as seen in Fig. 1, AIT's lowest MAE is 1.28 for 200 reviews, and for EQW it is 1.36). One reason for the smaller improvement in MAE is that there are 11 users with no message board posts, in which case their similarity to other users is 0. Another possible reason is that we use a global similarity threshold. If we take the mean of the MAEs obtained using the optimal threshold for each individual target user, we get an MAE of 1.45 for AIT (all unigrams) and 1.32 for AIP. This shows that modeling user similarity based on message board posts can be beneficial when no other information is available (e.g., for a completely new user).

AIP outperforms AIT in this experiment, unlike most cases in the experiment from Sect. 6.1. However, as seen in Fig. 1, AIP yielded the lowest MAE for 5 labeled reviews by the target user. The combined results of both experiments indicate that AIP is preferable when little information is available about the target user, while AIT performs better when more information is available. The reason for this might be that AIT compares the vocabulary of users and thus requires a relatively large number of documents to produce reliable results, while AIP compares users' positivity (in the form of PSP distribution), which may be accurately represented by fewer documents.

6.3 Learning from a Few Training Users

Experimental Setup. In some scenarios, we may not have many prolific users to train on. Thus, we evaluated our methods for different numbers of training users. This was done by uniformly sampling without replacement a subset of the training users, and running the experiment from Sect. 6.1. This process was repeated 10 times for each subset (we experimented with 5, 10, 20, 30, 40 and 50 training users). We experimented with the best performing methods from Sect. 6.1, one for each cell in Table 1: AIR, CRR, AIT and CRT. The resulting MAE is compared to the MAE yielded by SCSU and EQW.

Results. Figure 2 shows the results of this experiment for AIT and EQW with 5, 10, 20 and 61 training users (the trend for the other similarity models resembles that of AIT, and the results for 30, 40 and 50 training users resemble those obtained for 61 users – these results are omitted from the graph for clarity of exposition).[8] As seen in Fig. 2, AIT outperforms EQW even when a small number of training users is considered. The difference between EQW and AIT increases as more training users are added, up to the point where all 61 users are considered. This is not surprising, as our thresholding

---

[8] All the differences are statistically significant except for EQW vs. AIT for 5 training users and 10 labeled reviews by the target user.

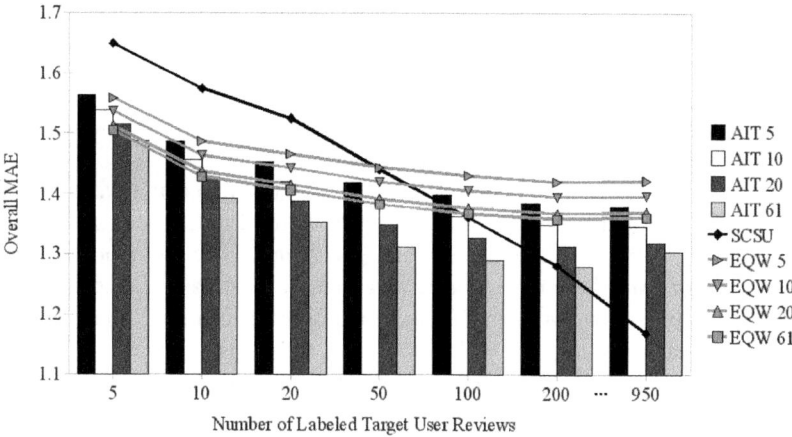

Fig. 2. Training Users Experiment Results

mechanism selects the best training users for each target user (since the selected threshold depends on the labeled target user reviews), and it is expected to perform better when a larger selection of users is available.

Another trend that is demonstrated in Fig. 2 is that the performance of AIT improves compared to SCSU as the number of training users increases. Further, when up to 50 labeled reviews are available for the target user, AIT outperforms SCSU for 5 or more training users; for 100 labeled reviews, 10 or more training users are required for AIT to outperform SCSU; and for 200 labeled reviews, all 61 prolific users are needed to match the performance of SCSU. These results show that our collaborative approach outperforms SCSU in common situations where target users have not submitted many reviews and only a few prolific training users are available.

7 Conclusions and Future Work

This paper introduced a collaborative approach to MPC, and our experimental results show the merits of this approach. As demonstrated in Sect. 6.1, the approach of training on a single user (SCSU) yields a high MAE in the common scenario where this user has not submitted enough reviews. By contrast, even the simple switch to a user-based ensemble of classifiers results in a reduced MAE, and modeling user similarity decreases the MAE even further.

The similarity modeling experiment (Sect. 6.1) showed that when the item/rating matrix is sparse, basing user similarity on *all* the users' texts or ratings yields better performance than basing user similarity only on reviews or ratings of co-reviewed items. This result was reinforced in Sect. 6.3, where our approach improved on the SCSU baseline even when only a few prolific training users were considered. Thus, our collaborative approach and similarity models are likely to apply to many real-life datasets, which tend to be sparse, as long as some prolific users are available.

As shown in Sect. 6.2, the new user problem is addressed by applying our text-based models of user similarity to forms of texts other than rated reviews. This allows us to measure similarity based on activities that are more commonly performed than writing reviews. We envision a system where users can benefit from personalisation without having to rate or review items. This requires further testing on various types of texts, but our results show that this is a promising direction.

As mentioned in Sect. 2, our approach to MPC has much in common with CF. Thus, we conjecture that text-based user similarity can be used in CF – an area where similarity is traditionally calculated based solely on ratings. The fact that our text-based measures outperformed the rating-based measures in most cases (Sect. 6.1) indicates that using text-based similarity in CF can result in improved performance, especially for sparse datasets and new users. In the future, we plan to develop additional text-based similarity measures, and apply our measures to recommender systems.

References

1. Pang, B., Lee, L.: Opinion mining and sentiment analysis. Foundations and Trends in Information Retrieval 2(1-2), 1–135 (2008)
2. Pang, B., Lee, L.: Seeing stars: Exploiting class relationships for sentiment categorization with respect to rating scales. In: Proceedings of the 43rd Annual Meeting of the Association for Computational Linguistics (ACL), Ann Arbor, Michigan, pp. 115–124 (2005)
3. Snyder, B., Barzilay, R.: Multiple aspect ranking using the Good Grief algorithm. In: Human Language Technologies 2007: The Conference of the North American Chapter of the Association for Computational Linguistics; Proceedings of the Main Conference (HLT/NAACL), Rochester, New York, pp. 300–307 (2007)
4. Gamon, M.: Sentiment classification on customer feedback data: noisy data, large feature vectors, and the role of linguistic analysis. In: Proceedings of the 20th International Conference on Computational Linguistics (COLING), Geneva, Switzerland, pp. 841–847 (2004)
5. Lin, W., Wilson, T., Wiebe, J., Hauptmann, A.: Which side are you on? Identifying perspectives at the document and sentence levels. In: Proceedings of the 10th Conference on Natural Language Learning (CoNLL), New York, pp. 109–116 (2006)
6. Adomavicius, G., Tuzhilin, A.: Toward the next generation of recommender systems: A survey of the state-of-the-art and possible extensions. IEEE Transactions on Knowledge and Data Engineering 17(6), 734–749 (2005)
7. Seroussi, Y., Zukerman, I., Bohnert, F.: A user-based approach to multi-way polarity classification. Technical Report 2010/253, Faculty of Information Technology, Monash University, Clayton, Victoria 3800, Australia (2010)
8. Herlocker, J.L., Konstan, J.A., Borchers, A., Riedl, J.: An algorithmic framework for performing collaborative filtering. In: SIGIR '99: Proceedings of the 22nd Annual International ACM SIGIR Conference on Research and Development in Information Retrieval, Berkeley, California, pp. 230–237 (1999)
9. Pang, B., Lee, L., Vaithyanathan, S.: Thumbs up? Sentiment classification using machine learning techniques. In: Proceedings of the 2002 Conference on Empirical Methods in Natural Language Processing (EMNLP), Philadelphia, Pennsylvania, pp. 79–86 (2002)

User Modelling for Exclusion and Anomaly Detection: A Behavioural Intrusion Detection System

Grant Pannell and Helen Ashman

WebTech and Security Lab, University of South Australia
{grant.pannell,helen.ashman}@unisa.edu.au

Abstract. User models are generally created to personalise information or share user experiences among like-minded individuals. An individual's characteristics are compared to those of some canonical user type, and the user included in various user groups accordingly. Those user groups might be defined according to academic ability or recreational interests, but the aim is to include the user in relevant groups where appropriate. The user model described here operates on the principle of exclusion, not inclusion, and its purpose is to detect atypical behaviour, seeing if a user falls outside a category, rather than inside one. That is, it performs anomaly detection against either an individual user model or a typical user model. Such a principle can be usefully applied in many ways, such as early detection of illness, or discovering students with learning issues. In this paper, we apply the anomaly detection principle to the detection of intruders on a computer system masquerading as real users, by comparing the behaviour of the intruder with the expected behaviour of the user as characterised by their user model. This behaviour is captured in characteristics such as typing habits, Web page usage and application usage. An experimental intrusion detection system (IDS) was built with user models reflecting these characteristics, and it was found that comparison with a small number of key characteristics from a user model can very quickly detect anomalies and thus identify an intruder.

Keywords: user model, exclusion, anomaly detection, behavioural IDS.

1 Introduction

1.1 How User Models Are Applied

User modelling is frequently a part of applications where some element of personalisation is required in information delivery. Commercial applications such as recommender systems and educational delivery systems are two areas where user models are now a mainstream technology. In particular, adaptive hypermedia systems have focused on delivery of educational materials for many years [4], but there is now a greater awareness of the potential application areas of user models, such as Cultural heritage, Health care, Assistive technologies and so on.

Interestingly, these application areas generally implement user models with the intention of being able to classify a user into a particular group, with the hope that the aggregated experiences of other members will be able to assist that user. The

P. De Bra, A. Kobsa, and D. Chin (Eds.): UMAP 2010, LNCS 6075, pp. 207–218, 2010.
© Springer-Verlag Berlin Heidelberg 2010

purpose of the user modelling is to *include* the user into some group of like-minded others. In recommender systems in particular, the user model allows the inclusion of an individual into a larger group based on their resemblance to the majority of the group.

The approach in this paper departs significantly from this. We propose an alternative use of user models which is much less common in the literature, namely to model a user with the intention of determining if they do not belong in some group, namely, the aim is to *exclude* users or at least to identify non-members. In many cases, a user is assumed by default to belong to a certain group, such as the group of permitted users, or the group of healthy users. Where such implicit judgements are made about users, it can be helpful to identify when the assumption is wrong. Examples include identifying where users are no longer part of a group of healthy people (i.e. identifying illness or stress), or identifying when students do not belong in a student group (i.e. a person masquerading as another student in order to achieve better results by proxy). In particular, the work reported here is motivated by a need to exclude unauthorised users who may have bypassed the barrier security methods (such as passwords) and have entered a computer system.

We propose the use of anomaly detection over user models, where anomalies indicate that the user is behaving in an uncharacteristic way. This might be a user who normally opens word-processing files now creating Unix executables, or someone with error-prone typing suddenly showing proficiency. To perform anomaly detection successfully, it is necessary to create a user model that reflects characteristics of the individual, and to perform statistical analyses that identify anomalies.

This work is similar to and extends the user-recognition work that statistically analysed Unix command line data in order to characterise and subsequently recognise an individual [9]. Anomaly detection is a natural use of such automatically-recognised users and Unix command line data has been successfully used for around 15 years in intrusion detection systems, although the data being analysed is not purely user commands but also includes keystrokes, session times and lengths and resource usage (see Sect. 1.2). This paper also extends that of prior IDS research on user-specific anomaly detection, as it performs analyses over multiple characteristics, not just command line data, but also over GUI-based characteristics such as Web page accesses and game playing. We also consider different classes of users, categorised according to their proficiency (basic computing skills versus deep technical knowledge), purpose of use (work versus leisure) and direct versus indirect user characteristics (inferred data such as typical CPU and memory use versus direct user data such as keystrokes). Interestingly, we found that the most useful characteristics for anomaly detection (and hence for user recognition more generally) were user-specific ones that directly reflect some user feature, such as typing habits and Web page usage, more so than application-specific features which only indirectly reflect user activity.

Another aspect of this work is that instead of comparing the user's current behaviour against a stereotypical user model or aggregated user model, we compare a user's model against their own past behaviour. That is, the user model is compared over time, rather than over a user group. This alternative approach is starting to appear in medical uses, such as detecting degradation in a user's skills over time [17].

1.2 Overview of Intrusion Detection Systems

The main focus of this paper is on the application of user models to anomaly detection. However since the principle is demonstrated in an intrusion detection system (IDS), we briefly overview intrusion detection systems here, focusing mostly on IDSs which capture and use information about users.

There are two main principles behind intrusion detection systems, the first being *rule-based systems*, where rules explicitly describe disallowed activities (e.g. use of the "su" command, or any access within a honeypot area), and *anomaly-based systems*, where specific behaviours are not prescribed but activity significantly outside the norm triggers an alert. These latter are generally statistical systems where a range of acceptable usages is characterised, either by statistical analysis of typical activity, or by defining a "canonical" activity, such as the number of accesses from a given IP in a time frame, or the number of some types of accesses (e.g. "ping" commands).

Intrusion detection based on user models is very much at an experimental stage, but generally can be characterised as being a sort of statistical, anomaly-based IDS except that the statistics are not calculated on the network traffic but instead are calculated on user behaviour. This approach implements a *canonical* user model that represents the trusted user (either a typical user or an identified individual), and a *sessional* user model, representing the activity and behaviour of the user who is currently operating under the appropriate user ID, which is compared to the trusted model. Should there be a significant discrepancy between the canonical user model and the sessional user model, an alert is generated. The canonical user model is of course persistent, while the sessional user model is only sessional (although pertinent data is retained for evidence or for statistical analysis).

In intrusion detection, user models can be applied in different ways:

- *"Role"-based canonical modelling*: An IDS may not map a user's activities to a specific individual's user model, but rather compare the sessional user model against a "typical" user model. If a user's activity falls outside the "typical" behaviour, an alert is generated [15]. However this is not especially useful in an environment where there are no "typical" users or working behaviours.
- *Personal user models*: We aim to detect intrusions by comparing the behaviour of a user currently active with their own user model of past behaviour. It will compare the user's behaviour in such things as common typing errors, frequently-used commands and applications, normal times and duration of connections, command sequences, and so on. This is the application area described in this paper that demonstrates the usefulness of anomaly detection over user models.

Anderson [1] outlines the idea of audit trails, where activity logs are taken from machines and manually analysed by activity security officers to find unauthorized access to files and other resources. He introduces the idea of using an automated surveillance system that looks out for characteristics such as session logs, durations, program usage, device usage, and file usage. The proposed surveillance system used basic statistics, such as averages, standard deviations, maxima and minima, to determine abnormal usage of the machine.

Denning [5] furthers Anderson's work by formally defining the structure of an IDS that is capable of automated audit trail analysis and adds other characteristics such as CPU and memory usage. This focuses on the IDS being statistical rather than rule-based. Lunt [11] extends Denning's structure by combining the rule-based system with statistical methods. One of the first implementations of a behavioural intrusion detection system, based on Denning's work, analysed daily audit files to find anomalous activity against user behaviour models and specific constraints [15].

User profiling for intrusion detection is established in the Unix systems and many have profiled users with command line data [2, 8, 16]. Characteristics used include commands, session times and resource usage, and the data analysis algorithm.

Very few user-profiling IDSs have combined the characteristics to improve performance although it is done quite often in network IDSs and it has been shown that the performance of a network IDS can be increased by using multiple, differently implemented, network IDSs [10]. Other experiments with using OR and AND operations with multiple systems improve both detection and false-positive rates [7]. One of the contributions of this paper is to implement a behavioural IDS that combines numerous characteristics in the same way that network IDSs do.

Adapting the IDS to a GUI-based system influences the user characteristics that are modelled. For example, capturing data from the mouse and from the way the user interacts with their windows now becomes possible. Implementations of a behavioural system built on GUI-based systems seem to be rare. One example bases the system on Windows 2000, using data from the operating system's performance monitor to create user profiles, and it is claimed that this system obtains a 95% detection rate and a low false-positive rate of "less than one alert per day" [14].

The physical attributes of the user can be characterised such as mouse movements [13] and the delays between keys typed on a keyboard [3]. It is feasible to use key delays and typing patterns to determine whether the user is cognitively or physically stressed [17]. This application clearly shows the potential benefits of anomaly detection over user models within health applications.

There have also been other profiling IDSs, which however are not user-focused, such as profiling network traffic from a host to determine whether it has been compromised [12] and profiling the system event log to determine the execution flow of an application to ensure that it is not being misused or exploited [6].

In summary, behavioural intrusion detection systems have so far seen little work in the combination of user characteristics, especially where data can be collected from GUI-based sources. The work reported in this paper considers intrusion detection applied over combined user characteristics and for individual users of many different types, such as office worker, advanced worker, game-playing and so on.

2   The User Model and the IDS

2.1   Characteristics Stored in the User Model

The system profiles a user using multiple characteristics, and combines different statistical methods such as standard deviations, averages, and limits, and then uses a score-based system to determine whether the combination of characteristics triggers an intrusion based on previous user data. The profiling engine retrieves characteristics at an interval of 30 seconds to ensure that new anomalies are detected quickly.

The system profiles users around the following characteristics:

- Applications running - The applications that are running on a machine would allow profiling of a user to determine the default or typical applications they use. For example, a user may exclusively use a specific set of office applications. Any new applications opened by the user could indicate anomalistic behaviour;
- Number of windows open - The number of windows currently open also can determine one user from another, depending on their style of use;
- Performance of running applications - Performance of running applications, such as measured by CPU usage and memory usage could determine how the applications are being used. For example, an abnormally high CPU usage in a database application could mean an intruder is extracting data;
- Keystroke analysis - Keystroke analysis includes such characteristics as speed, combination of keys and pauses between key presses; and
- Websites viewed - The websites viewed characteristic looks at web browser history to determine if new sites have entered the profile.

2.2 Analytical Algorithms Performed over User Characteristics

The *CPU usage* and *memory usage* characteristic engines were a collection of three different algorithms and a scoring system. The algorithms used included a Standard deviation analysis, a Rolling average, and Upper and lower limits. The standard deviation analysis stored the previous 120 values collected by the system, i.e. the past 60 minutes of data with an interval of 30-second collections. An anomaly was detected if a new value was three deviations from the mean. The rolling average was simply a cumulative average that took the previous usage value, added it to the new value, and divided it by two. This gives an overall view of the process for its entire lifetime, compared to the standard deviation algorithm, which only provides a view over the past hour. The final algorithm used, sliding limit, would only be changed during the learning phase of the system. This allows the system to learn the upper and lower limits of the process, and since it is assumed that the profile is perfectly trained, these values should show when the process is acting abnormally.

After these three algorithms had analysed the incoming data, a scoring system was used to validate any detected anomaly. If more than one algorithm triggered, or if the same algorithm triggered multiple times, it is less likely to be a false-positive. The scoring system essentially attempts to smooth out false-positives given by the simple algorithms. Each algorithm was assigned a point value: Standard Deviation given 0.5 points, Rolling Average getting 1 point and Sliding Limit getting 2 points. The points would accumulate over three collections and once the 3-point limit was breached, the system would trigger an anomaly, otherwise the accumulated points would reset to 0.

The *number of processes* characteristic looked at two different sets of user data related to number of processes, namely the variance of number of processes, over the hour and the number of new processes to the user's profile. The variance of number of processes per hour was calculated the same as the standard deviation algorithm used for CPU usage and memory usage characteristics. The number of processes are collected every 30 seconds over the past hour and the mean and standard deviation are calculated. If the current number of processes were 2.5 deviations away from the mean, the number of running processes would be anomalistic. The number of processes characteristic also had to determine if the process was new to the profile, and if so, the system should theoretically mark as anomalous all new processes. In a real-world situation this may be desirable. However, it is dependent on the actual user type, e.g. a power user frequently installs new applications.

The *number of windows* characteristic algorithm works exactly like the number of processes algorithm except that only the variance of the number of windows per hour is used and the system does not look for new windows to add to the profile.

The *websites viewed* characteristic algorithm also works in this way but looks at the number of new sites that had been added to the browser's history, per hour.

Finally, the *keystroke analysis* algorithm adapted the ideas from Bergadano et al. [3], using digraphs to capture a user's typing pattern. If a user were to type "Digraphs", the system would store key pairs: "Di", "ig", "gr", "ra", "ap", "ph" and "hs" with the delays between each of the keys. To determine if a typed digraph was anomalistic, the algorithm would determine the standard deviation and mean then check if the new delay value was more than 2.5 standard deviations from the mean.

2.3 System Architecture

The architecture is based upon a generic intrusion detection system. The data collection engine gathers data from the performance monitor, Windows API, browser history locations and from keystrokes, at an interval of every 30 seconds. It then sends data for each user characteristic to its own analysis engine. The analysis engine contains two types of detection: rule-based and anomaly. Both types analyse the data to determine if the behaviour falls within the normal range of the user's profile, according to the algorithms described above (see Sect. 2.2). If any anomaly is significant, the user's activity for that characteristic in that 30-second time slice is deemed to be unauthorised.

Each characteristic engine then sends its results to the data-mining engine which collects results and determines what action to take. The engine uses a score-based threshold system, determined during testing, and past results to lower false-positives. This may mean that multiple alarms will need to be triggered or the same alarm triggered many times for the system to take action. The data-mining engine then sends its results to the alert/action engine that will perform actions specified by the user depending on if the system has determined authorised or unauthorised behaviour.

3 Design of Experiment Validating Behavioural Anomaly Detection

Eleven users tested the system, running the IDS in the background for approximately 10 days. The systems that were used had a range of different operating systems and uses. Each user categorised their machine for its primary use as follows:

No.	Machine main use
3	Web Browsing - Primarily surfing the web
1	Gaming - Primarily playing 3D games
3	Office/University Work - Primarily using Microsoft Office-like tools
0	Entertainment - Primarily using a media player to watch or listen to media
4	Power User - A combination of the above plus arbitrary other applications

Data was gathered in two modes: learning, and intrusion detection. In learning mode, the system treated all user actions as normal and added them to the profile, i.e. the learning phase was assumed to be clean of intrusions. However while in learning mode, especially initially, most of the user actions were anomalous compared to the user model which was necessarily incomplete at this stage, so the reduction in anomalies (i.e. false-positives) detected was used as a measure of how successfully the system was able to create the user's profile of normal behaviour, especially for each profiled characteristic. The system was placed in learning mode for 10 days (28880 30-second collections), after which it was switched to detection mode, where the system treated anomalies as intrusions and would create an appropriate alert.

The experiment then aimed to generate anomalies against the user profile by stressing each characteristic in two ways, firstly by deliberately challenging each specific characteristic such as by changing a keystroke pattern, visiting unusual websites or running unusually CPU-intensive processes. The second testing method was done by putting a new user in front of the machine, effectively masquerading as the legitimate user. This allowed us to measure how long the system took to detect a different user and what characteristics were best for this.

4 Results and Discussion

4.1 In the Learning Phase – Separate Characteristics

In this section, we consider how frequently anomalies were detected for each user characteristic, focusing on those reflecting the user's behaviour directly, and then how the data mining engine that combines them all performed during this phase.

The assumption was made that during the learning phase, the machines were not exposed to intrusion and that all user activity was valid. However it was useful to generate a measure of how well and how quickly the system learned the user's typical behaviours. We measured to what extent the user profile had converged on a typical collection of values for the given characteristics by counting the number of "false-positives" during the learning period, i.e. the number of times some user behaviour appeared to be anomalous, according to the as-yet-incomplete user model. Because the learning phase was not exposed to intruders, the apparent anomalies were not actually anomalies but were known to be false-positives. The number of false-positives fell as the user model became more representative of normal user activity.

All characteristics showed a steadily-decreasing learning curve, where the number of false-positives would decrease over time. This shows that the IDS was successfully able to learn the user's activities during the learning period. In fact, for some classes of user it would be feasible to reduce the learning time since the number of false-positives later in the learning period became negligible and the system was not learning anything "new" by continuing to observe the user.

The *CPU usage characteristic*, overall produced a significant number of false-positives compared to other characteristics. Nevertheless, the system showed a learning curve that resulted in fewer than 50 false-positives per two days near the end of the learning period. The power user machines produced the highest false-positive rate of 6.73% during the learning period, showing that power users have a wider range of typical usages and that perhaps ten days is insufficient for learning the typical behaviour of a power user. The lowest CPU usage false-positive rates during the learning phase come from the web browsing and office machines, ranging between 0.6% and 0.8%, indicating a lower level of variance in such activity. The CPU usage characteristic may not be the most suitable for anomaly detection, given its ongoing false-positive rate and inapplicability across all user types. It also does not directly reflect user personal characteristics except very coarsely, such as when a user is imposing an abnormal load on the processor.

The *memory usage characteristic* produced fewer false-positives per day during the learning period, but also showed a shallower learning curve. This suggests that the system was not as successful in profiling the memory usage of processes on the machine. By the end of the learning period, the memory usage characteristic would output approximately 30 false-positives per two days. Again, the highest false-positive rates come from the power user machines at 1.18%. Like the CPU usage characteristic, this characteristic may be less useful for anomaly detection. It also does not directly reflect the user's personal characteristics except very coarsely, and may be prone to error when very large files are opened.

The *websites viewed* and *number of windows* characteristics were more promising. They all produced the fewest false-positives with rates of less than 0.05%. While these characteristics were not triggered as often, they show a steady decrease over learning period. The false-positives for the websites viewed characteristic would occasionally increase over time then continue to decrease. This might be explained by the user initially viewing a usual set of sites, and then changing their pattern by browsing to new sites. The system increases its alert level during the pattern change over and then decreases as it learns the new behaviour.

The characteristic that provided the most rapid learning, and perhaps the one most promising for profiling a user, is *keystroke usage*. This is likely due to the physical relation it has to the user. Again, the highest false-positive rates come from the power user machines at approximately 1.89% although the system would output fewer than 50 alerts per two days at the end of the learning period.

The gaming machine, in keystroke usage results, has a high false-positive rate of 3.84% with false-positives equally spread over collection periods, meaning the system could not determine a consistent typing pattern. This may be because when gaming, the user is reacting to what they see on the screen and the delay between keys depends more on the game, i.e. user activity is not autonomous but rather interactive. In contrast, an office machine user tends to generate a more consistent typing pattern.

Aside from the gaming machine, the keystroke analysis was one of the more successful characteristics for profiling quickly with the majority of users shown to have an easily-characterised typing style. The false-positive results for the keystroke usage characteristic for the different user classes is summarised in Fig. 1.

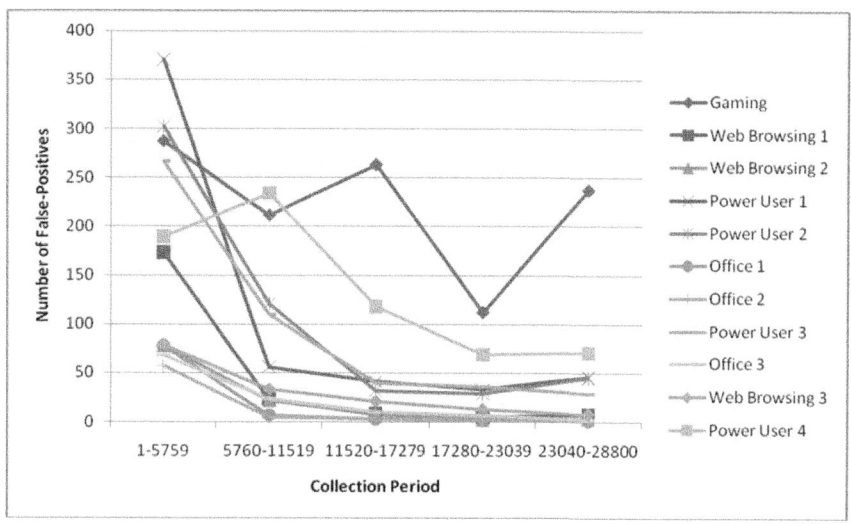

Fig. 1. False-positive rates over the learning period for keystroke usage

4.2 In the Learning Phase – Combined Characteristics

Finally, we look at the combination of outputs by the *data mining engine* using a scoring system which combines the anomalies detected by the separate characteristics.

Once the scoring system is applied to combine the characteristics via the data mining engine, the learning curve starts to flatten and the system begins to output the same number of false-positives per day. This is not necessarily good as it is likely that the system will continue to output the same level of false-positives in the future. However, the flat curve shows a consistent false-positive rate and with improved algorithms, so this rate could possibly be decreased. The false-positive rates during learning mode peak at 0.62% for a power user machine, representing fewer than 10 false-positives per day. This indicates that a combination of characteristics is significantly better than any single characteristic. Figure 2 shows the false-positive rate during the learning period for combining characteristics using the scoring system.

These results indicate that the system can learn typical user behaviour in a reasonable time by observing normal use during a dedicated learning. The most effective characteristics are those that directly reflect user behaviour, such as keystrokes, websites visited and number of windows opened. These are governed by the user's own physical attributes (keystrokes) or their own choices (opening windows or selecting websites to view) and are easier to characterise than those only indirectly related to the user, which may be affected by other causes other than the user.

In the scoring system, the number of false-positives was much lower than the false-positive rate for some of the individual characteristics, showing that the scoring system was able to correct for false-positives from individual characteristics.

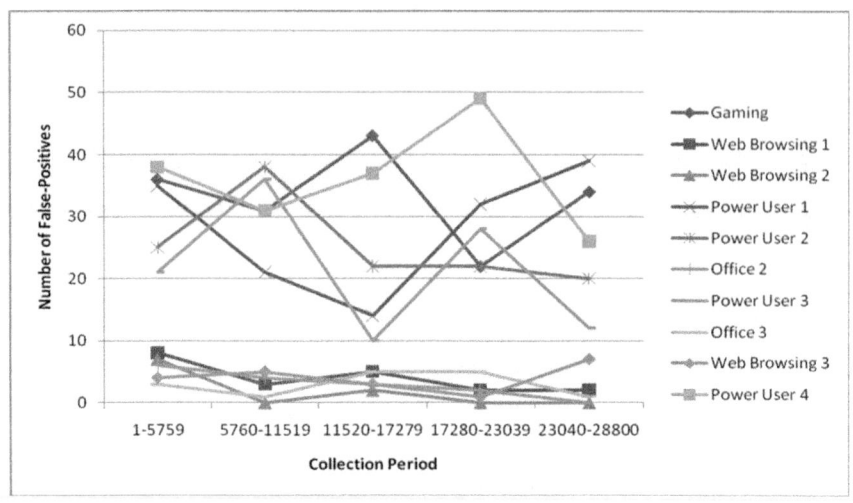

Fig. 2. False-positive rate during the learning period using scoring system

4.3 Anomaly Detection Phase

Once the learning phase was completed, the system was switched to detection mode where instead of subsuming the detected new behaviour into the user model, the system triggers an alert when an intrusion occurs.

Anomaly detection was tested by stressing each characteristic with an abnormal usage, such as opening many windows. It was also tested by placing a new user in front of the machine and waiting for the system to trigger with their usage pattern.

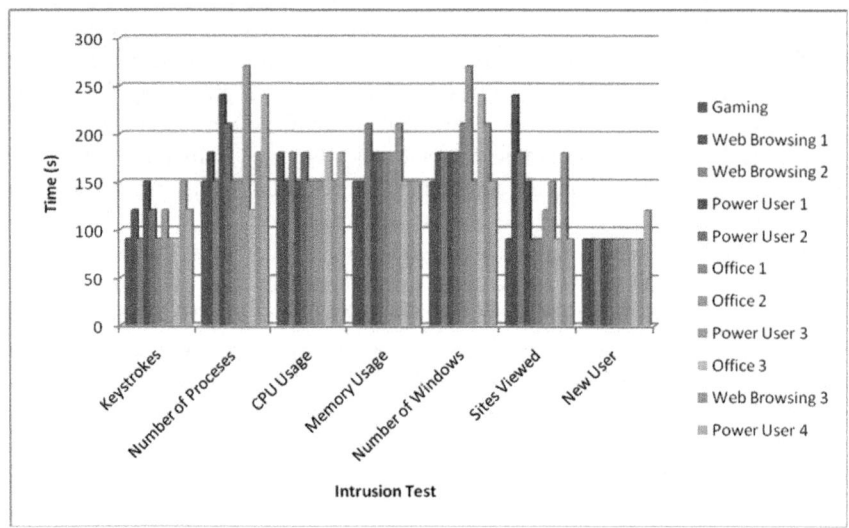

Fig. 3. Average time to detect intrusions on all machine usage types

The keystroke analysis engine triggered the system most quickly, generally in less than 120 seconds (4 cycles of data collection). The more application-related characteristics (number of processes, CPU usage and memory usage) all had higher detection times, with an average of 180 seconds. Placing a new user in front of the machine triggered alerts in the shortest time (90 seconds), for almost all machine usage types. Figure 3 shows the average time taken to detect intrusions on all machine usage types.

The most useful characteristics are those that directly reflect user behaviour, keystrokes and sites viewed. The number of windows characteristic, also user-specific, is the slowest,. The less direct characteristics were the least efficient for identifying anomalies, requiring up to 10 cycles of data collection.

The ability to detect genuine intrusions within three data collections shows that a combination of characteristics is more effective than any individual characteristic, even when the scoring system that combines the characteristics is not optimised.

5   Conclusions and Future Work

This paper reports on the facility of user models for anomaly detection, and that a combination of user characteristics achieves the most rapid detection. The more personal characteristics were the most efficient for anomaly detection. However there are still many improvements possible, such as more personal features in the user model. This could include mouse use and n-gram analysis of writing styles for commonly-used words and phrases. It may also be possible to create a database to categorize websites and applications. e.g. to track the primary text editor used by the user. We aim to optimise the scoring system and learning algorithms to achieve detection in a single collection. More complex algorithms, such as a genetic algorithm, could increase the performance of the system. It may also be valuable to use anomaly detection over other user models to see if the user is behaving abnormally, e.g. fluctuations in student achievement may indicate a false user.

References

1. Anderson, J.: Computer Security Threat Monitoring and Surveillance. James P. Anderson Co., Fort Washington (1980)
2. Balajinath, B., Raghavan, S.V.: Intrusion detection through learning behavior model. Computer Communications 24(12), 1202–1212 (2001)
3. Bergadano, F., Gunetti, D., Picardi, C.: Identity verification through dynamic keystroke analysis. Intelligent Data Analysis 7(5), 469–496 (2003)
4. Brusilovsky, P.: Methods and techniques of Adaptive Hypermedia. User Modeling and User Adapted Interaction 6(2-3), 87–129 (1995)
5. Denning, D.E.: An Intrusion-Detection Model. IEEE Transactions on Software Engineering 13(2), 222–232 (1987)
6. Forrest, S., Hofmeyr, S.A., Somayaji, A., Longstaff, T.A.: A sense of self for Unix processes. In: Proc. 1996 IEEE Symposium on Security and Privacy, pp. 120–128. IEEE Computer Society, Washington (1996)

7. Gu, G., Cardenas, A.A., Lee, K.: Principled reasoning and practical applications of alert fusion in intrusion detection systems. In: Proc. ASIACCS '08, pp. 136–147. ACM, New York (2008)
8. Gunetti, D., Ruffo, G.: Intrusion Detection through Behavioral Data. In: Hand, D.J., Kok, J.N., R. Berthold, M. (eds.) IDA 1999. LNCS, vol. 1642, pp. 383–394. Springer, Heidelberg (1999)
9. Iglesias, J.A., Ledezma, A., Sanchis, A.: Creating User Profiles From a Command-Line Interface: A Statistical Approach. In: Houben, G.-J., McCalla, G., Pianesi, F., Zancanaro, M. (eds.) UMAP 2009. LNCS, vol. 5535, pp. 90–101. Springer, Heidelberg (2009)
10. Julisch, K., Dacier, M.: Mining intrusion detection alarms for actionable knowledge. In: Proc. 8th ACM SIGKDD Int. Conf. on Knowledge discovery and data mining, pp. 366–375. ACM, New York (2002)
11. Lunt, T.F.: Real-time intrusion detection. In: COMPCON Spring '89. 34th IEEE Computer Society Int. Conference: Intellectual Leverage, Digest of Papers, pp. 348–353. IEEE Press, Washington (1989)
12. Mazzariello, C., Oliviero, F.: An Autonomic Intrusion Detection System Based on Behavioral Network Engineering. In: Proc. INFOCOM 2006, pp. 1–2. IEEE Press, Washington (2006)
13. Pusara, M., Brodley, C.E.: User re-authentication via mouse movements. In: ACM workshop on Visualization and data mining for computer security, pp. 1–8. ACM, New York (2004)
14. Shavlik, J., Shavlik, M.: Selection, combination, and evaluation of effective software sensors for detecting abnormal computer usage. In: Proc. 10th ACM SIGKDD, pp. 276–285. ACM, New York (2004)
15. Smaha, S.E.: Haystack: an intrusion detection system. In: 4th ACSAC, pp. 37–44. IEEE Press, Washington (1988)
16. Tan, K.: The application of neural networks to UNIX computer security. In: IEEE International Conference on Neural Networks, Proc., vol. 1, pp. 476–481. IEEE Press, Washington (1995)
17. Vizer, L.M., Zhou, L., Sears, A.: Automated stress detection using keystroke and linguistic features: An exploratory study. IJHCS 67(10), 870–886 (2009)

IntrospectiveViews: An Interface for Scrutinizing Semantic User Models

Fedor Bakalov[1], Birgitta König-Ries[1], Andreas Nauerz[2], and Martin Welsch[2]

[1] Friedrich Schiller University of Jena
{fedor.bakalov,birgitta.koenig-ries}@uni-jena.de
[2] IBM Deutschland Research and Development GmbH
{andreas.nauerz,martin.welsch}@de.ibm.com

Abstract. User models are a key component for user-adaptive systems. They represent information about users such as interests, expertise, goals, traits, etc. This information is used to achieve various adaptation effects, e.g., recommending relevant documents or products. To ensure acceptance by users, these models need to be scrutable, i.e., users must be able to view and alter them to understand and if necessary correct the assumptions the system makes about the user. However, in most existing systems, this goal is not met. In this paper, we introduce Introspective-Views, an interface that enables the user to view and edit her user model. Furthermore, we present the results of a formative evaluation that show the importance users give in general to different aspects of scrutable user models and also substantiate our claim that IntrospectiveViews is an appropriate realization of an interface to such models.

1 Introduction

Adaptive Web systems are the systems that tailor their content, appearance, and behavior to the needs of individual users or groups of users. Such systems are being developed as an answer to the overwhelming and steadily growing amount of information available in typical Web systems. An adaptive Web portal, e.g., could be a portal that places on its front page links to the resources that are relevant to a user based on her interests, expertise and/or current context. The basis for such adaptation effects is a user model containing information about users, such as their interests, expertise, goals, traits, etc. [1]. In many adaptive systems the user model is considered as purely internal system information, hence it is partially or completely hidden from the user. This results in a number of grave usability problems and may well result in the user not accepting the system. For instance, it violates two of Nielsen's ten usability principles [2]. Hiding user models occludes the *system status* and hinders *control* on the adaptation, which might lead to errors, e.g. issuing irrelevant recommendations.

In order to avoid the above mentioned problems, user models need to be scrutable. This means, the user needs to be able to view and adapt the information contained in her user model [3]. Jameson [4] argued that allowing inspection and parametrization of user models are important measures to achieve

P. De Bra, A. Kobsa, and D. Chin (Eds.): UMAP 2010, LNCS 6075, pp. 219–230, 2010.

predictability, transparency, and controllability of an adaptive system. According to Cook and Kay [5], the user needs to be able to understand the provenance of information in her user model, e.g., the user needs to understand why the system believes she is interested in a certain topic. Finally, Orwant [6] argued that scrutability is an essential step towards establishing trust between the user and an adaptive system. Section 2 provides a short overview of the previous research related to visualization of scrutable user models.

In Sect. 3 we introduce IntrospectiveViews, an interface that visualizes content of user models in a comprehensible way and allows users to inspect and alter them. Through this interface users can see what the system "knows" about them and how this information is used for the adaptation. Moreover, through the interface users can edit that information and control how it is used, hence achieve better adaptation effects and better control on their privacy. An important feature of IntrospectiveViews is its capability of visualizing and managing *semantic* user models. In a semantic user model, information about users is augmented with machine-understandable semantics defined, e.g., in an ontology-based domain model. Such models are more powerful than simpler user models since the system can use the semantics for interest and knowledge propagation, hence increase correctness and completeness of the user model. On the downside, these mechanisms may make adaptation decisions more difficult to understand and result in an even greater need for scrutability. In this paper, we demonstrate application of IntrospectiveViews on the example of semantic user *interest* models as presented in [7]. However, the interface can be also used for visualizing and scrutinizing other features, such as knowledge and goals.

In Sect. 4 we elaborate on a formative evaluation consisting of two parts. In the first part, we determined how important users deem visualization and editing features of a scrutable user model. These results are independent of the concrete implementation, namely IntrospectiveViews, that we evaluated in the second part of the study. The evaluation of the implementation showed that IntrospectiveViews meets the user requirements identified earlier to a large degree and is therefore a very suitable approach to our problem. Finally, Sect. 5 concludes the paper and outlines the directions for our future work.

The main contributions of this paper are twofold. First, it offers insights into user requirements towards scrutable user models. These insights are of use to anyone interested in developing such models. Second, it introduces IntrospectiveViews as an appropriate implementation of such models.

2   Related Work

A number of approaches have been proposed to visualizing scrutable user models. PeerGlass architecture [8] provides a visual method to exploring user models through a Rolodex of model planes, where each plane represents a certain type of user interests, including manually entered interests and automatically induced interests. The um_view interface [5] allows traversing through a user model by expanding the tree of leaves and viewing detailed information about the items in the model. VIUM [9] and its successor SIV [10] are capable of visualizing large user models and enable users to get an overview of the whole model, view a subset of related beliefs, filter items by relevance, and obtain detailed information about the displayed items. STyLE-OLM [11] and Flexi-OLM [12] visualize open learner models using concept graphs and trees.

One of the main distinctive features the interface described in this paper is that it *fully supports all seven tasks for information visualization* postulated by Shneiderman [13], while providing *intuitive and easy-to-use mechanisms for editing* semantic user models. In IntrospectiveViews the user can gain an overview of the entire user model and zoom into a certain part of the model to get a better view on it. It enables the user to filter out unwanted items in order to focus on the relevant ones. Additionally, it provides detailed information about a selected item when needed as well as reveals relationships among the items with respect to a number of attributes. The interface is capable of visualizing the history of changes in the model and allows extracting a certain portion of the model and saving it, so that it can be reused in other systems. Finally, the interface allows the user changing, adding, and deleting items in her user model.

3 IntrospectiveViews for User Interest Model

IntrospectiveViews is a visualization of overlay user models [1] representing user knowledge or interests. To show the features of this visualization, however, we need a concrete example. In this paper, we use the MINERVA User Interest Model developed in our earlier work [7] as this example. We will give a brief introduction to this model before delving into the details of IntrospectiveViews.

3.1 User Model and Modeling Approach

The MINERVA user interest model is implemented as an overlay model: User interests are represented as an overlay of vocabulary defined in the domain knowledge model. User interests are modeled with a hybrid approach that combines an unobtrusive method to capturing interests based on the user browsing history and a manual method allowing the user to specify her interests herself. This section provides a short description of the user interest model and the modeling approach.

We define user interest as a fact indicating that a given user is interested to a certain degree in a certain term. Here, the term is a reference to an ontology instance (e.g., company, location, or person) defined in the *domain knowledge model*, an OWL-based ontology providing machine-understandable semantics of the contained entities. The degree of interest denotes the extent to which the user is interested in a given term. We distinguish three levels of interest: *interested*, *partially interested*, and *not interested*. Also, we model user interests as time dependent features. We assume that a user might be interested in a certain term only for a certain period of time. Thus, the user interest model is represented as a collection of tuples *(U, T, I, V)*, where *U* is the user ID, *T* is the URI of an instance from the domain model, I is the linguistic variable indicating the degree of interest, V is the time period of the interest validity.

Our approach to identifying user interests involves the following processes. First, the terms indicating user interests are collected into the user model either by analyzing the content of visited pages or explicitly entered by users through IntrospectiveViews. Second, the collected terms are semantically enriched by referring them to the corresponding instances in the domain model. Finally, interest degree is determined for every term either by leveraging the term frequency, or semantic relations among the terms, or specified by the user explicitly through the interface.

3.2   IntrospectiveViews

IntrospectiveViews follows Shneiderman's Visual Information Seeking Mantra [13]: "overview first, zoom and filter, then details-on-demand". It offers users an overview over all terms present in their interest model, it allows for zooming into different parts of the model, filtering terms according to different criteria, and it will provide details, for instance additional information on a term or information about how the system determined the user interest in a term, on demand. IntrospectiveViews also supports the further tasks identified by Shneiderman: It allows revealing relationships among the items, supports browsing through the history of the user model, and allows exporting the entire model or parts of it. Let us take a closer look how this is achieved.

IntrospectiveViews, shown in Fig. 1, is implemented as a Java Applet. The interface displays user interests as term labels on a circular surface consisting of a number of colored rings. Positioning of terms on the surface, namely distance from the circle center, is determined by the terms' exact degree of interest. Here it means that the closer a term appears to the center, the higher interest it represents. Font size of terms denotes the term's frequency in the user browsing log: The terms that the user encounters often will appear bigger in relation to the terms that she reads seldom about.

Each ring, distinguished by its color, represents a certain interest group. The color scheme of rings is chosen according to the hot-and-cold metaphor, where hot, represented by red color, denotes interest and cold, represented by blue color, denotes no interest. The colors between red and blue denote partial interest. The border areas of the rings are painted in a gradient color to denote the fuzziness between the groups, i.e., uncertainty of interest degree. User interests can be grouped by their type into circular sectors, which are distinguished by different shades of gray and identified by labels containing the names of types. Such grouping allows users to place together the terms that belong to the same class, e.g., people, companies, locations, and so on.

The user is enabled to zoom in/out the entire collection of interests. By zooming in the user can get a detailed view on terms in a certain area and by zooming out she can switch back to the overview of the entire user model. In a enlarged view, the user can navigate through the collection of terms by dragging the surface in a respective direction. The interface supports a number of filtering

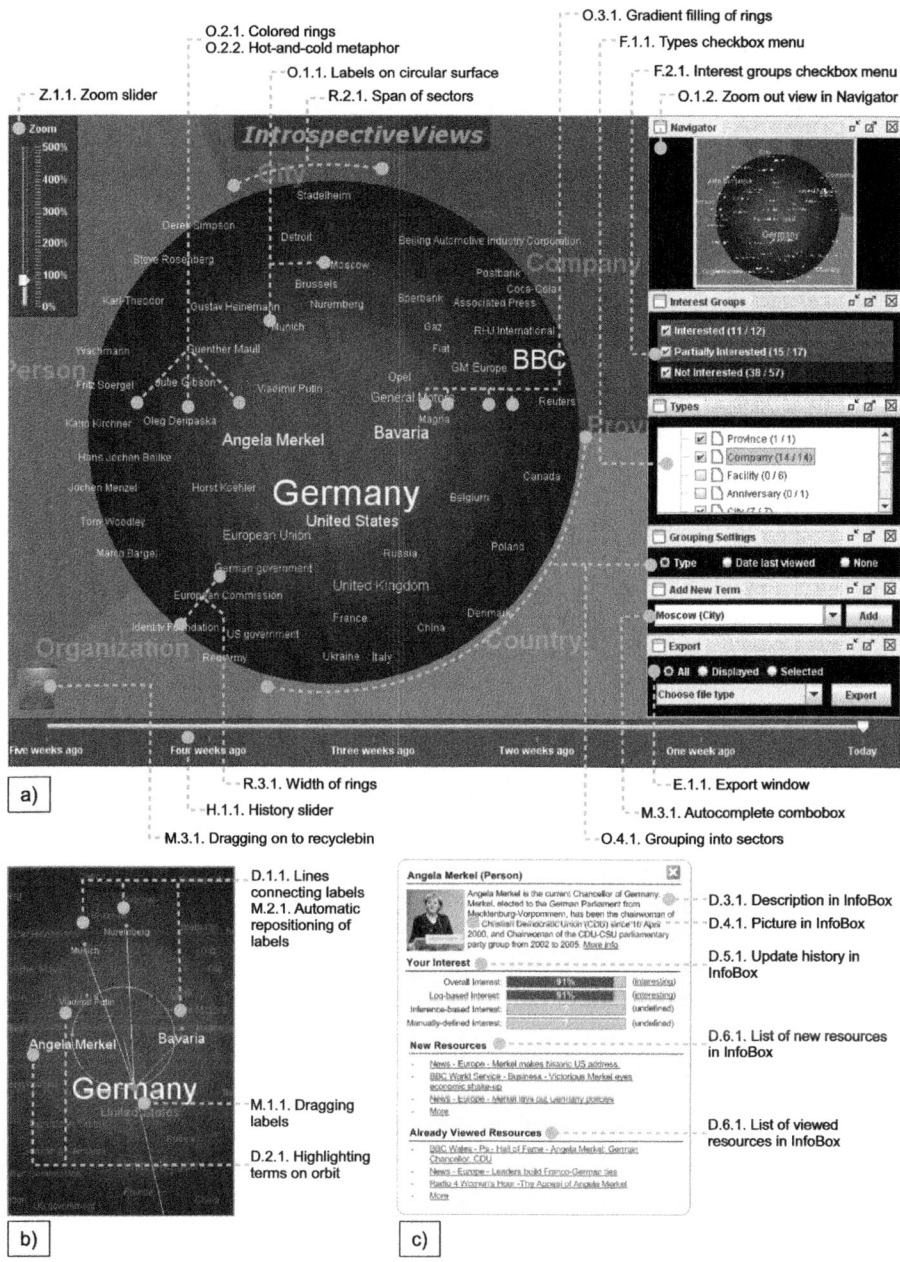

Fig. 1. Screenshots of IntrospectiveViews: (a) provides full view of the interface; (b) displays the terms that are semantically related to the selected term (white lines outgoing from Germany), the terms with the same degree of interest (highlighted on the yellow orbit), and the path along which the term can be dragged (the circle's radius represented as a yellow line); (c) displays the term's InfoBox.

Color screenshots and a screencast are available at
http://www.minerva-portals.de/research/introspective-views/

options. Terms can be filtered by type and by interest group. For instance, the user can display only companies, people, and countries and/or display only the terms that she is interested in.

Also the user can obtain additional information about the terms. For instance, by pressing mouse left button on a term, the user can see its related terms (Fig. 1.b). This allows the user to find out the terms that potentially might be interesting. Right click on a term will display an infobox (Fig. 1.c) containing the term's description and picture, evidence of the user interest (whether the user has read many documents about it, or her interest was propagated from other terms, or she specified her interest manually). Also the infobox displays the list of related documents that the user has not seen yet and the list of related documents that the user has already viewed.

In addition to viewing the contents of user models, the interface enables users to edit them. In order to change interest degree in a certain term, the user can simply drag the term into an appropriate interest group (here represented by one of the rings). For instance, the user has read a number of news articles from BBC, thus this term has appeared in the user model as interesting. However, in fact the user is not really interested in BBC as a company. Using the interface, the user can simply drag the term BBC from the red ring to the blue one. This will trigger a respective update in the user model, i.e., he term will be marked as uninteresting. By dragging a term that has relevant terms, their interest will be automatically updated and their labels will be repositioned accordingly.

Also users are enabled to manually enter new terms into their user models. In the *Add new term* window, the user can simply type a term of her interest and click *Add* button. The new term will be placed in the circle (by default in the red ring). Users can remove terms by dragging them into the recycle bin. In this case, the system will stop tracking such terms and they will not appear on the circle. Respectively, this will affect the adaptation: E.g., the system will stop issuing recommendations on that term. Through the interface the user can get a retrospective view on her user model, i.e., see her interests in the past, by dragging the pointer in the *History slider*. Finally, the interface enables the user to extract a certain portion of her model and save it in a file on the disk.

4   Evaluation and Results

To evaluate usefulness and usability of IntrospectiveViews, we conducted a formative evaluation of the interface with 26 participants, 15 of which were experts in at least one of three areas of interest (user modeling and adaptation, information visualization, and usability), whereas the other 11 were non-expert users.

The aims of the evaluation were twofold. First, when designing Introspective-Views, we followed Shneiderman's recommendations for information visualisation [13]. Our aim was to determine how much importance users actually put on certain features of a scrutable user model: Would they like an overview? How important would it be to filter terms? Do they want to export the model? The findings of this first part of the evaluation are only very weakly coupled to IntrospectiveViews and offer a guideline for anyone developing a scrutable user

Table 1. Questionnaire (elements marked with * are implemented as a mockup)

Overview
O.1. The task of getting an overview of the entire user model is implemented in two ways. First, user interests are represented as labels on a circular surface (***O.1.1***), where the position denotes exact interest degree. Second, a zoomed out copy of the entire model is shown in *Navigator window* (***O.1.2.***).
O.2. The interest groups available in the user model are represented as colored rings (***O.2.1.***), which are painted according to the hot-and-cold metaphor (***O.2.2.***).
O.3. The fuzziness between the interest groups is represented by the gradients filling of border areas of rings (***O.3.1.***).
O.4. To get an overview of the available types of user interests, they can be grouped into circular sectors labeled with the names of types (***O.4.1.***).

Zoom
Z.1. Zooming can be performed by dragging the pointer in *Zoom slider* (***Z.1.1.***) as well as by rotating the mouse wheel (***Z.1.2.***).

Filter
F.1. Interests can be filtered by type by selecting/deselecting corresponding checkboxes in *Display types* window (***F.1.1.***).
F.2. Interests can be filtered by interest group by selecting/deselecting corresponding checkboxes in *Display interest groups* window (***F.2.1.***).

Details-on-demand
D.1. Related terms can be viewed by pressing mouse left button on a term, which will display connecting lines from the selected term to its related terms (***D.1.1.***).
D.2. Terms with the same interest can be viewed on the orbit that is displayed when the user presses mouse left button on a term (***D.2.1.***).
D.3. Textual description of a term can be obtained from the term's InfoBox (***D.3.1.****).
D.4. Graphical depiction of a term can be seen in the term's InfoBox (***D.4.1.****).
D.5. Evidence of user interest in a certain term is represented as the update history in the term's InfoBox (***D.5.1.****).
D.6. Relevant resources that the user has not viewed yet are represented as a list of hyperlinks the term's InfoBox (***D.6.1.****).
D.7. Relevant resources that the user has already viewed are represented as a list of hyperlinks the term's InfoBox (***D.7.1.****).

Relate
R.1. The relation of a term to its frequency in the browsing log is encoded in the term's font size (***R.1.1.***).
R.2. The relation of a term type to its size, i.e., number of user interests of that type, is encoded in the sector's span (***R.2.1.***).
R.3. The relation of an interest group to its importance is encoded in the width of the ring containing the terms of that interest group (***R.3.1.***).

History
H.1. View on a snapshot of the user model in the past can be obtained by setting the pointer in the *History slider* to the desired date (***H.1.1.****).

Extract
E.1. The entire collection of user interests or a selected part of it can be exported into a file on disk by specifying the corresponding export options in the *Export* window and clicking *Export* button (***E.1.1.****).

Table 1. (*continued*)

Modify
M.1. Interest in a term can be increased or decreased by dragging the term's label closer to or further from the circle center (***M.1.1.***).
M.2. Interest in related terms is changed automatically when the user drags a label of term that has related terms (***M.2.1.***).
M.3. New terms of interest can be entered into the user model by typing them in the autocomplete box in *Add New Term* window (***M.3.1.***).
M.4. Terms can be blocked from being tracked and used for adaptation by dragging them on to *Recyclebin* (***M.4.1.***).
M.5. Modify the color scheme of the rings (the task is not implemented).

model. Second, we wanted to evaluate how well IntrospectiveViews met user requirements, i.e., whether this implementation was considered successful and how we could further improve it. To achieve these goals, the participants were first introduced to the interface and the problems that it aims to solve. Then the participants were asked to test the interface on their own and evaluate it using an evaluation questionnaire. In the questionnaire, we listed 24 tasks that the interface can support (Table 1). The tasks were classified into eight categories: *overview, zoom, filter, details on demand, relate, history, extract,* and *modify.* The first seven categories belong to Shneiderman's task by data type taxonomy for information visualizations [13]. The *modify* category was introduced to cover the edit tasks specific to IntrospectiveView. 23 out of 24 tasks are functionally implemented or implemented as mockups. Table 1 describes the eight categories of tasks and their GUI implementation. For each task, the participants were asked to cast two votes. First, how important did they find the task. This importance was rated on a 1-5 scale: from 1-not at all important to 5-very important. Second, how usable is the implementation of the task in IntrospectiveViews. The usability of the GUI implementation was rated on a 1-5 scale: from 1-not at all usable to 5-very usable. Additionally, the participants could leave free-text comments either regarding the tasks, or implementation, or the interface in general.

4.1   Results

Key goals of the evaluation were to assess the usefulness of the interface by examining the importance of the tasks it can support and to evaluate the usability of the GUI elements implementing the tasks. The overall feedback regarding the importance of tasks and the usability of GUI elements was very favorable (see Fig. 2). The participants of both focus groups provided a number of valuable free-text comments. The rest of this section details these results.

Overview. All tasks in the overview category received relatively high importance. Especially interesting was that the task of getting an overview of available types in the user model received the highest importance in its category. The participants of both focus groups confirmed our hypothesis that organizing interests

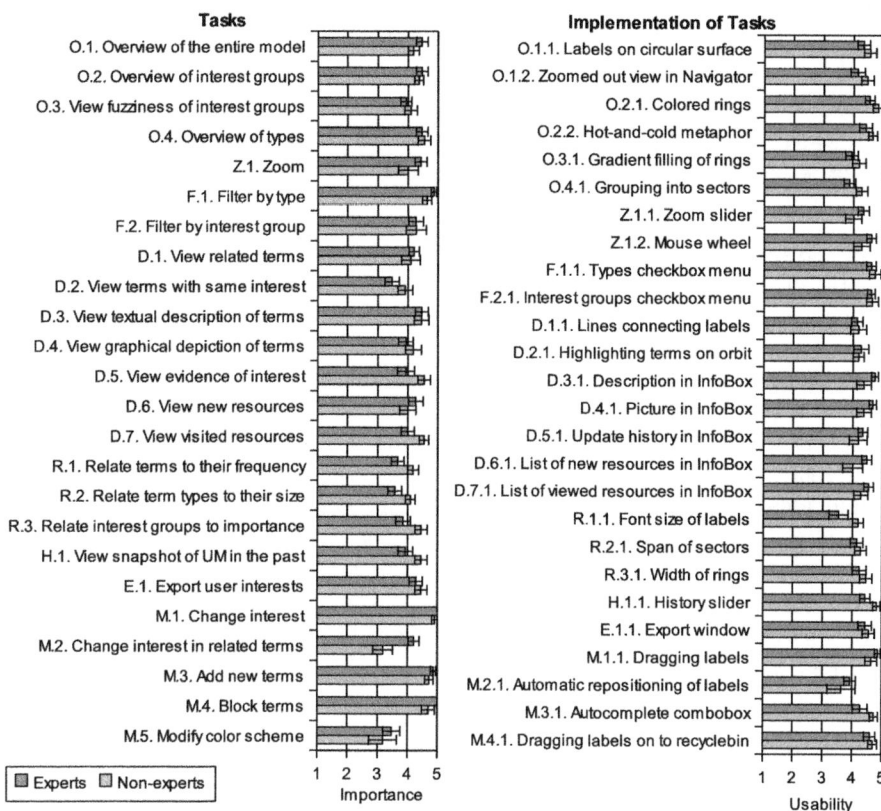

Fig. 2. Rating of importance of tasks (left) and usability of implementation (right)

by type can improve perception of the entire model. With respect to the usability of the implementation, representing user interests on a circular surface consisting of colored rings filled out according to the hot-and-cold metaphor appeared to be very intuitive and comprehensible for both experts and non-experts. However, despite the task of getting an overview of types was rated as very important, its implementation, grouping into sectors painted in different shades of grey, received the lowest usability in its category. In the free-text comments the participants of both groups mentioned that the information about types is very important, hence it must be more prominent. Experts suggested making the sectors visible not only outside of the circle, but also within it.

Zoom. The importance of zoom task was differently rated by experts and non-experts. The majority of experts considered the task as substantially important, whereas the average rating of the task's importance by non-experts appeared to be considerably lower than the experts' one. However, the usability rating of the two methods implementing the zoom task, Zoom slider and rotating the mouse wheel, was very favorable.

Filter. The average rating of both filtering tasks, filter by type and filter by interest group, is above 4, which indicates their high importance. The usability of their implementation is also rated very high. However, there was a useful suggestion from an expert on improving the usability of the filtering features. The expert suggested adding animation effects to smoothen the transition of terms when one of the filtering options is changed, e.g., smooth expansion or shrinking of sectors.

Details-on-demand. Four of the tasks in this category received relatively high importance. These are: viewing related terms, viewing textual description of terms, viewing new resources about a term, and viewing visited resources about a term. However, regarding the last two tasks, a number of respondents mentioned that their importance strongly depends on the application. For instance, one expert highlighted high importance of the task of viewing visited resources in a website that does not have a fixed navigation topology, e.g., YouTube, where it is relatively difficult to find the previously viewed content. The usability of implementation of the tasks in this category was rated good. An interesting observation is the relatively high difference between the rating of experts and non-experts with respect to the implementation of the two tasks of viewing resources about a term. Non-experts rated usability lower then the experts. In the free-text comments some of them suggested having a structure within the list, e.g., grouping by topic.

Relate. The overall importance of the relate tasks is considerably lower then the importance of tasks in other categories, in average between 3-somewhat important and 4-important. With respect to the usability of their implementation, encoding the size of a term type (number of interests of that type) into the sector's span was rated as very intuitive and usable. So was rated the encoding importance of interest groups into the ring's width. However, encoding the term's frequency into the label's font size received relatively low rating of usability. In the free-text comments, many respondents mentioned that it is unintuitive and misleading use of font size. For them, bigger font size means higher interest, which in the current implementation is not the case. It could be that uninteresting terms appear large (see example of BBC explained in Sect. 3), whereas some interesting ones can appear small.

History. The average rating of the task of viewing a snapshot of the user model in the past is above 4-important and the usability of its implementation, history slider, was also rated very high. In the free-text comments, one of the experts suggested adding a bookmarking feature to the slider, which would allow the user to bookmark certain time frames, e.g., when she was interested in Italy, and jump to this frame later on by simply clicking on the bookmark.

Extract. The importance of exporting user model was rated very high. Many respondents considered this task useful and practical if many websites support this format. The implementation, Export window, received good rating of usability.

Modify. Three modify tasks were rated as very important: change interest, add new term, and block terms. The importance of other two tasks, change interest in related terms and modify color scheme, received relatively low importance. Regarding the automatic change of interest in related terms, many respondents mentioned that the relations among the terms captured by the system do not always reflect the relations that influence their personal interest. Except for the task of changing interest in related terms, the usability of implementation of other tasks received very high rating. Most respondents rated the implementation of the interest change task, i.e., dragging labels closer to or further from the circle center, as very usable. An interesting suggestion on improving the usability of adding new terms was made by one of the experts. The expert suggested implementing this feature in a way that new terms could be added in a specific place of the model by making right click on that place and selecting terms from a popped-up menu.

General Comments. A number of participants said that it is important for them to be able to organize the terms in their user models according to their own categories. Some of them would also like to be able to define their own relations among terms. Regarding the overall usability of the interface, a number of experts requested such features as a search box for quickly jumping to a certain term and an undo feature.

5   Conclusions and Future Work

Making user models accessible to the users is a key requirement to the acceptance and success of adaptive systems. In this paper, we have presented Introspective-Views, a novel user interface that allows users to scrutinize user models. Following Shneiderman's recommendations, IntrospectiveViews provides users with an overview of the model as well as possibilities to zoom in, to filter, to obtain details including relationships among concepts and temporal aspects, and to export the model or part of it. Additionally, IntrospectiveViews allows the user to change the model thus actively influencing future adaptations. We have performed a formative evaluation of IntrospectiveViews using the MINERVA user interest model as an example. This evaluation showed on the one hand that the tasks supported by IntrospectiveViews are indeed deemed important by users (and should thus be supported by any scrutable user model) and showed on the other hand that the implementation of these tasks provided by IntrospectiveViews was judged as very usable for nearly all tasks. Additionally, the participants provided us with valuable feedback to further improve IntrospectiveViews.

In our future work, we will incorporate these suggestions into a new version of IntrospectiveViews and evaluate its usability with respect to completing concrete tasks in a real adaptive system. We also plan to develop implementations of other models than the one we used in this paper to further substantiate our claim that it can be widely used. Possible applications range from simply visualizing other aspects of a user model to developing visualizations comparing a person's interest to those of communities she is a member of, to using IntrospectiveViews to visualize the knowledge a student should obtain in a course and to compare it to the knowledge she has at a certain point in time.

Acknowledgements. This research is carried out in the framework of the Minerva project and is supported by IBM Deutschland Research & Development GmbH.

References

1. Brusilovsky, P., Millàn, E.: User models for adaptive hypermedia and adaptive educational systems. In: Brusilovsky, P., Kobsa, A., Nejdl, W. (eds.) Adaptive Web 2007. LNCS, vol. 4321, pp. 3–53. Springer, Heidelberg (2007)
2. Nielsen, J.: Enhancing the explanatory power of usability heuristics. In: Adelson, B., Dumais, S., Olson, J. (eds.) Proc. of the SIGCHI Conf. on Human Factors in Computing Systems, pp. 152–158. ACM, New York (1994)
3. Kay, J.: Scrutable adaptation: Because we can and must. In: Wade, V.P., Ashman, H., Smyth, B. (eds.) AH 2006. LNCS, vol. 4018, pp. 11–19. Springer, Heidelberg (2006)
4. Jameson, A.: Adaptive interfaces and agents. In: Sears, A., Jacko, J. (eds.) The human-computer interaction handbook: Fundamentals, evolving technologies and emerging applications, 2nd edn., pp. 433–458. CRC Press, Boca Raton (2008)
5. Cook, R., Kay, J.: The justified user model: a viewable, explained user model. In: Goodman, B., Kobsa, A., Litman, D. (eds.) Proc. of the 4th Int. Conf. on User Modeling, pp. 145–150 (1994)
6. Orwant, J.: Appraising the user of user models: Interface guidelines. In: Goodman, B., Kobsa, A., Litman, D. (eds.) Proc. of the 4th Int. Conf. on User Modeling, pp. 73–78 (1994)
7. Bakalov, F., König-Ries, B., Nauerz, A., Welsch, M.: A Hybrid Approach to Identifying User Interests in Web Portals. In: Erfurth, C., Eichler, G., Schau, V. (eds.) Proc. of the 9th Int. Conf. on Innovative Internet Community Systems, Bonn. LNI, vol. 148, pp. 123–134. GI (2009)
8. Kliger, J.: Model planes and totem poles: Methods for visualizing user models. Master's thesis, MIT Media Lab. (1995)
9. Apted, T., Kay, J., Lum, A., Uther, J.: Visualisation of ontological inferences for user control of personal web agents. In: Banissi, E., Börner, K., Chen, C., Clapworthy, G., Maple, C., Lobben, A., Moore, C., Roberts, J., Ursyn, A., Zhang, J. (eds.) Proc. of the 7th Int. Conf. on Information Visualization, Washington D.C., pp. 306–313. IEEE Computer Society, Los Alamitos (2003)
10. Kay, J., Lum, A.: Building user models from observations of users accessing multimedia learning objects. In: Nürnberger, A., Detyniecki, M. (eds.) AMR 2003. LNCS, vol. 3094, pp. 36–57. Springer, Heidelberg (2004)
11. Dimitrova, V.: STyLE-OLM: Interactive open learner modelling. International Journal of Artificial Intelligence in Education 17(2), 35–78 (2003)
12. Mabbott, A., Bull, S.: Alternative views on knowledge: Presentation of open learner models. In: Lester, J.C., Vicari, R.M., Paraguaçu, F. (eds.) ITS 2004. LNCS, vol. 3220, pp. 689–698. Springer, Heidelberg (2004)
13. Shneiderman, B.: The eyes have it: A task by data type taxonomy for information visualizations. In: Proc. of the 1996 IEEE Symposium on Visual Languages, Washington D.C., pp. 336–343. IEEE Computer Society, Los Alamitos (1996)

Analyzing Community Knowledge Sharing Behavior

Styliani Kleanthous and Vania Dimitrova

School of Computing, University of Leeds
{stellak,vania}@comp.leeds.ac.uk

Abstract. The effectiveness of personalized support provided to virtual communities depends on what we know about a particular community and in which areas the community may need support. Following organizational psychology theories, we have developed algorithms to automatically detect patterns of knowledge sharing in a closely-knit virtual community, focusing on transactive memory, shared mental models, and cognitive centrality. The automatic detection of problematic areas enables taking decisions about notifications targeted at different community members but aiming at improving the functioning of the community as a whole. The paper presents graph-based algorithms for detecting community knowledge sharing patterns, and illustrates, based on a study with an existing community, how these patterns can be used for community-tailored support.

Keywords: Community Knowledge Sharing, Closely-knit Communities, Graph Mining for Community Modelling.

1 Introduction

Virtual communities (VC) allowing people to gather together and share knowledge are becoming an integral part of today's organizational, educational and business practices. There is a growing interest in developing adaptation and personalization techniques to facilitate the effective knowledge construction and information sharing in virtual communities, aiming at the creation of stimulating and sustainable online environments. The community type (spanning from large, loosely structured to small, closely-knit) underlies the support needed for effective community functioning, and the corresponding adaptation and personalization techniques required. The research presented in this paper considers closely-knit communities that usually exist in relatively well-defined organizational or educational settings, and can share common characteristics with large teams/groups. A number of approaches, such as visualizations, notifications, and community ratings, have been exploited to facilitate community/group awareness, motivate participation, and improve community knowledge sharing [3–5, 13]. However, a key challenge that still remains is how to utilize organization and psychology theories to develop *holistic mechanisms* that support effective community functioning and sustainability [2].

The implication of this argument (raised by researchers in Computer Supported Cooperative Work) to personalization technologies for groups and communities, drives our research to explore *how aspects of organizational theory can be applied to*

P. De Bra, A. Kobsa, and D. Chin (Eds.): UMAP 2010, LNCS 6075, pp. 231–242, 2010.
© Springer-Verlag Berlin Heidelberg 2010

design a novel community-tailored adaptation approach to support effective knowledge sharing and sustainability of a VC. Following [7], we have selected three key team processes applicable to VC, which can be monitored by analyzing community log data[1]. Effective and sustainable communities have a well-developed **Transactive Memory (TM)** system where members are aware how their knowledge relates to the knowledge of others [17]. **Shared Mental Models (SMM)** are also crucial for the effective functioning of a community - members develop a shared understanding of the key processes and the relationships that occur between them [12]. **Cognitive Centrality (CCen)** identifies the members who hold strong relevant expertise and influence the cognitive processes in the community; in effective communities, members gradually become more central and engaged [11].

The first step in our research was the development of ontology-enriched algorithms to mine the log data and extract a *community model* (CM) comprising of: (i) *individual user profiles* (represent the interests of each community member), (ii) a *community context* (defined by a topic-specific ontology); (iii) a model of the *semantic relationships between community members* (graphs that represent derived connections between members), (iv) a list of *cognitively central members* (represents the people who are influential in the community), (v) a list of *popular and peripheral topics* (based on members uploading and downloading and relevance to community context). The algorithms for deriving CM are presented in [10]. Our second step was to apply the community modeling algorithms to tracking data from a real community in order to get an understanding of what was happening in the community, and to identify what support could be provided to improve the functioning of the community following TM, SMM and CCen [8]. The study was used to uncover knowledge sharing behavior patterns that could indicate possible problems with TM, SMM and members' centrality. These patterns were *manually detected* by examining the community model with appropriate visualization tools [8].

This paper presents the third step in our research - automatic detection of community knowledge sharing patterns based on TM, SMM, and CCen. We will define graph-based patterns to *automatically identify* problematic areas in the VC where adaptive support would be required. The main contribution of this paper is:

- Definition of graph-based knowledge sharing patterns related to TM, SMM, CCen;
- Illustration how the patterns can be used to generate community-adapted notifications to selected members aimed at benefiting the community as a whole;
- Application upon archival data from a real community illustrating how the detected patterns could inform what notifications could have been sent to members to improve knowledge sharing and to help the community sustain.

2 Semantic Relationships Model

An important part of this work is the semantic relationships model which defines the relationships and similarities between community members derived by comparing

---

[1] Not all processes important for the effective functioning of a community [7] have been chosen, e.g. trust is excluded as it is not related to cognitive aspects and cannot be monitored by analyizing the generic log data we consider.

resource keywords/tags enriched with related concepts extracted from the ontology presenting the community context. Let a and b denote two community members. We consider the following relationships: $ReadRes(a,b)$ indicates that resources uploaded by b are read by a, and its strength corresponds to the relevance of the resources to the community context; $ReadSim(a,b)$ indicates that a and b have read semantically similar resources; $UploadSim(a,b)$ indicates that a and b have uploaded semantically similar resources; and $InterestSim(a,b)$ represents the similarity of interests between a and b. The algorithms used for extraction of the above relationships have been presented elsewhere [10]. We will introduce here only the main notations needed for defining community knowledge sharing patterns.

For each relationship, a graph is derived capturing the links between people based on that relationship type: $G_{RS}(V_{RS}, E_{RS})$ is the graph derived for $ReadSim$, $G_{US}(V_{US}, E_{US})$ is the $UploadSim$ graph, $G_{IS}(V_{IS}, E_{IS})$ is the $InterestSim$ graph, and $G_{RR}(V_{RR}, E_{RR})$ is the graph for $ReadRes$. $G_{RS}(V_{RS}, E_{RS})$, $G_{US}(V_{US}, E_{US})$ and $G_{IS}(V_{IS}, E_{IS})$ are non-directed graphs of type $G(V, E)$ where v is the set of nodes representing community members and E is the set of edges representing the existence of the corresponding relation between two members (nodes), the strength is calculated by applying the algorithms presented in [10]. An edge is present in a relationship graph only if the weight of that edge is greater than a pre-set threshold value. A neighborhood of a node v, denoted as $N_G(v)$, represents the ego network of v and indicates the members that v has corresponding similarity with.

$G_{RR}(V_{RR}, E_{RR})$ is a directed graph, where the direction of each edge represents that a member (head) has read a resource uploaded by another member (tail). Each node v has out-neighborhood $N_G^+(v): \{x \in V(G): v \to x\}$ representing community members who have downloaded resources uploaded by v, and in-neighborhood $N_G^-(v): \{x \in V(G): x \to v\}$ representing members whose resources v has downloaded.

In addition, the automatic detection of patterns exploits information from the individual user profiles. The user profile of a member a includes: (i) $uRate(a)$ and $dRate(a)$ which denote the upload and download rate of a (see [10] for detail); (ii) $CCen(a)$ which indicates how important the knowledge a holds is for the VC, calculated as the sum of all the relationships a has with any member b; a member might appear to be cognitively central because he/she uploads resources that are read by other members and are relevant to the community context (see [10] for detail).

By analyzing the community relationships model and the individual user profiles, we can identify patterns of knowledge sharing behavior related to TM, SMM, and CCen. The corresponding algorithms are presented in the next section.

3 Detecting Community Knowledge Sharing Patterns

A pattern is important if it can be detected and used in order to provide support to community members. We define seven types of patterns.

P1. Unexplored similarity between community members
Two members have ReadSim with the same members but not among themselves.

Importance: Identifying the above situation and making people aware of their unexplored similarity with others may motivate them to participate more actively, as pointed out in [6]. In addition, helping members understand that they hold complimentary knowledge improves the community TM system [17] and can promote collaboration within the community [7].

Detection: To detect unexploited similarity between a and b, we extract the neighborhoods of both members from G_{RS}. If one of the members does not belong to the other's neighborhood, pattern P1 is discovered: $\left(N_{RS}(v_a) \cap N_{RS}(v_b) \neq \varnothing\right) \wedge \left(v_a \notin N_{RS}(v_b)\right)$

In the same way, P1 is defined for *UploadSim* and *InterestSim* relationships.

P2. Community members may not be aware of their similarity
Two members have ReadSim with the same members and among themselves.

Importance: Community members are not aware of how similar they are in terms of uploading, reading or interests with other members of the community. Detection of this pattern can be used to promote the development of SMM [12] (members will become aware of what others are working on), and enhance TM [17] (members will know who they relate to in the community and how similar they are to others).

Detection: This pattern is detected by extracting the neighborhoods of both members from G_{RS}. If one of the members belongs to the other's neighborhood, pattern P2 is identified: $\left(N_{RS}(v_a) \cap N_{RS}(v_b) \neq \varnothing\right) \wedge \left(v_a \in N_{RS}(v_b)\right)$

In a similar way, P2 is defined for *UploadSim* and *InterestSim*.

P3. Members not benefiting
A member uploads resources but does not download.

Importance: This pattern can be useful to identify members who are not downloading from the community. Support can be provided to those members in order to benefit and make the most of their time in the community.

Detection: Detection of P3 is done by using the upload and download rates of a member: $\left(uRate(a) > 0\right) \wedge \left(dRate(a) = 0\right)$

P4. Members not contributing
Similarly to P3, P4 detects members who *download but do not upload* resources.

P5. Important peripheral members not downloading
Members who do not download and occasionally upload valuable resources.

Importance: We can use this pattern to motivate peripheral members to benefit from the community. Notifying him/her that others are interested in what he/she uploads can motivate that member to start reading resources uploaded by the members he/she has similarity with. This pattern may help to promote collaboration.

Detection: P5 is calculated using the upload and download rates for a and the out-neighborhood in G_{RR} to check that a uploads relevant resources:

$$\left(uRate(a) > 0\right) \wedge \left(dRate(a) = 0\right) \wedge \left(N_{RR}^+(v_a) \neq \varnothing\right)$$

P6. Important peripheral members not uploading
A member appears to download only and has InterestSim *with other members.*

Importance: This pattern can be used to motivate people who are only downloading from the community to start uploading, by showing them how similar their interests are to other members. This can improve the TM system of the community since members will be aware of others' interests [17]. Motivating them to upload to the community may help the community sustain.

Detection: To detect P6, we check a member's upload and download rates and his/her neighborhood in G_{IS}: $\left(uRate(a) = 0\right) \wedge \left(dRate(a) > 0\right) \wedge \left(N_{IS}(v_a) \neq \varnothing\right)$

P7. Unexplored complimentary similarity between members
Two members have UploadSim *but do not have* ReadSim.

Importance: Members who upload similar resources in the community but are not reading similar resources, have similar and complimentary interests but are unaware of this. Making these people aware of their similarity and difference may improve the TM system since members will be able to identify where important knowledge, for them, is located [7]. At the same time, this may improve the building of SMM [12], since members can appreciate how everybody contributes to the community. Awareness where complimentary knowledge is located may encourage collaboration.

Detection: P7 is identified using G_{US} and G_{RS}, and checking that one of the members belongs to the other member's neighborhood in G_{US} but does not belong to the neighborhood of that member in G_{RS}: $\left(v_a \in N_{US}(v_b)\right) \wedge \left(v_a \notin N_{RS}(v_b)\right)$

Table 1 summarizes the importance of each pattern to the community processes. The patterns indicate when and what interference can be made, as discussed next.

Table 1. Summary of how the detection of a pattern can affect the relevant processes

Pattern	Affects
P1: unexploited similarity between members	Collaboration, TM System, SMM
P2: members unaware of their similarity	SMM, TM System
P3: members participating but not benefiting	Improve participation, Sustainability
P4: members not contributing	Improve participation, Sustainability
P5: peripheral members not downloading	SMM, Collaboration
P6: peripheral members not uploading	TM System, Collaboration
P7: unexplored member complementarities	SMM, TM System, Collaboration

4   Generating Community-Tailored Notifications

The purpose of detecting knowledge sharing behavior patterns is to find out when community support is needed, whom to approach and how. Support in this work is designed as *personalized notification messages* that target individuals or groups of members pointing at actions that can have an effect on the VC overall functioning.

Notifications for Cognitively Peripheral Members (CPerM): Studies have shown that acknowledging the uniqueness of peripheral members' expertise may increase their confidence, and thus improve their level of participation and contribution [15]. In addition, CPerM can be motivated to participate by becoming aware of the importance of their unique expertise for the rest of the community [15].

Notifications for Cognitively Central Members (CCenM): CCenM are influential to other VC members due to their status and knowledge. Research showed that less central members are influenced and usually follow the CCenM [8]. Hence, notifications for CCenM should aim at helping members from the periphery to gain confidence and become influential. The participation of CCenM may be motivated by acknowledging their importance to the community [15].

Table 2. Example notifications based on detection of knowledge sharing patterns

ID	Detected situation	Target users	Notification goal	Content template
Na	P3: $\{M_{i_j}\} \in \mathcal{M}$ is downloading and has a < RelationshipType > with $\{M_{i_1}, M_{i_2}, ..., M_{i_n}\}$	$\{M_{i_j}\}$	Develop awareness of how the member relates to others, and help him/her integrate. TM.	*"Share your knowledge with the rest of the community! Start uploading resources.* $\{M_{i_1}, M_{i_2}, ..., M_{i_n}\}$ *have* < RelationshipType > *with you and will benefit from what you share."*
Nb	P6: $\{M_{i_j}\} \in$ CPerM is uploading only; $\{M_{i_j}\}$ has < RelationshipType > with $\{M_{i_1}, M_{i_2}, ..., M_{i_n}\}$	$\{M_{i_j}\}$	Help a CPerM integrate by acknowledging their importance and referring to similar members.	*"$\{M_{i_1}, M_{i_2}, ..., M_{i_n}\}$ are interested in what you are uploading. You may find what they are uploading interesting. Follow the links to navigate through resources these members are uploading"*
Nc	$P_1 \wedge \{M_{i_j}\} \in$ CCenM	$\{M_{i_j}\}$	Inform a CCenM of his importance to the VC, encourage him to continue and suggest he helps them integrate.	*"You are an important member connecting* $\{M_{i_1}, M_{i_2}, ..., M_{i_n}\} \setminus \{M_{i_j}\}$ *Keep the good work and upload more interesting resources. You may suggest resources that these members may read?."*

Notifications to improve TM: When a TM system is developed in a VC, members are able to locate important knowledge to them and identify who the experts in specific areas are [17]. By providing notification messages that include personalized information, we can help individuals in the VC to become aware of what others are working on, who they are similar to and what resources might be of their interest.

Notifications to improve SMM: Understanding what processes are happening in a community, what the VC purpose is, and being aware of the activities that relate members, creates a awareness and develops SMM [12].

For every notification message a standard structure is followed, including: (i) *Detection* – the situation that triggers the notifications e.g. knowledge pattern detected; (ii) *Target users* - the list of community members detected at a given pattern to whom a notification should be sent; (iii) *Goal* –the aim of the notification related to TM, SMM, and CCen; (iv) *Content template* – pattern of the text to be sent. Table 2 gives example notifications, the full list of notifications is given in [9]. The VC members are represented as $\mathcal{M} = \{M_1, M_2, ..., M_n\}$ where n is the total number of members.

Notifications target subsets of members based on detected patterns. Each recognized pattern may trigger different notification messages to different members, according to: the member's status in the VC (if central or peripheral), the member's activities (e.g. downloading/uploading rates), and relations with other members.

5   Evaluation with Existing Community Data

To validate the algorithms for detecting knowledge sharing behavior patterns, we conducted a study with archival data from an existing community. The log data give an inside of what has happened in a *real knowledge sharing community* during a fairly *long period from active functioning to standstill*. The evaluation approach followed is similar to evaluation using simulated data, applicable when large amount of data is needed, data is too expensive to collect, or when people have to be involved and there is no available sample. The major advantage of our evaluation approach is the use of longer term authentic data. Since both authors actively participated in the community and few other members were still available for clarification, we were able to check the appropriateness of patterns recognized involving these members and the suitability of the notifications that could have been generated to community members.

5.1   Study with a BSCW Virtual Community

Community. The VC included 34 members (researchers and doctoral students) from two research groups working on similar research areas, sharing documents and research papers (referred in this paper as resources) with the BSCW system that provides general support for collaboration over the web. Members were using BSCW to create folders, upload and download resources. The groups were based in two European countries, some members knew each other but many had never met.

Data. We collected log data for 15 months using BSCW features allowing every member to see what was happening in the community. At the beginning of the data collection period, the community was quite active and had already functioned for about a year. During Month1 – Month8 of the monitored period, members were uploading and downloading papers. After that, the activities lessened, and during the last few months of the monitored period, Month13 – Month15, there was no uploading and very little downloading. The community declined and stopped.

Procedure. We needed to indicate *when* detected patterns would have been useful and *what* interventions could have been triggered then. A quantitative summary of the community participation identified that the activity minimized rapidly in Month3 and in Month7. Thus, we applied the algorithms on the data collected in Month4, Month5 and Month6 - the months between the two activity drops. The log data was stored in a text file, fully anonymized, and converted to database tables used as input for the community modeling algorithms presented in [10]. Relationships were extracted between 1122 member pairs (considering the relationship between any two members apart from themselves). The community model was used as input to the pattern detection algorithms, implemented in Java following the definitions in Section 3.

We validated 60% of some 90 detected pattern occurrences (from 11 pattern types). A pattern was validated when (a) the detected relationships between members were appropriate (which was checked by looking at the resources members shared); (b) the log data of the follow up behaviour confirmed the pattern, e.g. when it was found that members might have not been aware of their similarity, in their subsequent interactions they indeed did not read papers from each other; when members were available, it was confirmed that they were indeed unaware of the detected similarities; (c) one of the community moderators confirmed that the detected useful patterns indicated situations when intervention could have been made.

The application of the patterns on the data collected uncovered that the community had in general poor TM system and SMM, and collaboration between community members was difficult to achieve. Due to space constraints, we present here only examples from the crucial period (months 4-6) when some intervention could have been beneficial. Detailed description of the study is given in [9].

5.2 Results

The results in Month4, Month5, and Month6 show that each month the VC was coming closer to a halt. The analysis regarded the discovered patterns as relevant. The analyzers (both authors) appeared aware of some patterns, but other patterns showed links that the authors were unaware of and had to validate by examining the resources read/uploaded. We present below examples which illustrate how patterns could have been used to generate notifications to community members.

P1. Unexplored similarity between community members. In Month4 P1 was detected for 10 pairs, while in Month6 this pattern was detected for 22 pairs of members. This situation creates a problem to the VC as it shows that a TM system is not in place, people are not aware of whom they have similarity with and there is a lack of SMM as members do not know what others are working on. See Fig. 1.

In Month4, members M9, M24, M31, and M5 had *ReadSim* with M28 but did not have *ReadSim* among themselves. M28 was found to be the connecting node between these four members (Fig. 1 Left). In Month 5, members M5 and M31 continue in the same situation but members M9 and M24 have stopped contributing or downloading from the community (Fig. 1 Right).

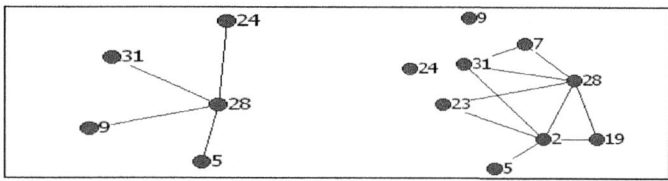

Fig. 1. Left: ReadSim in Month4 - M9, M24, M31 have ReadSim with the same members as M5 but not with M5. Right: ReadSim in Month5 showing members M9 and M24 to be disengaged.

Furthermore, in Month4, members M5 and M9 were detected to have *InterestSim* with M24 and M28 but not among themselves; M23 has *InterestSim* with 28 (Fig. 2). Members M5, M9, M23, M24, and M28 have closely related interests but might have not been aware of their similarity with each other. Consequently, in Month5 and Month6, members M9, M23, and M24 became inactive. Notification Na (Table2) could have been sent in Month 4 to M28, M5, M9, M24, and M31. Notification Nc (Table 2) could have been sent in Month 5 to M28 who is the connection to these members and to M31 detected as cognitively influential member.

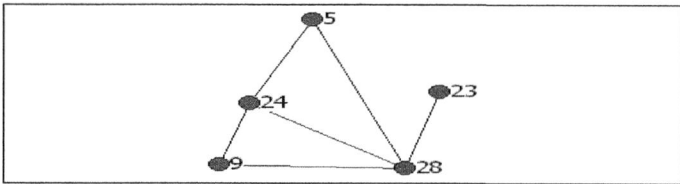

Fig. 2. In Month4 members have InterestSim with the same members but not among themselves

P2. Community members may not be aware of their similarity. Examples of this pattern are members M9, M24 and M31. In Month4, M9 and M24 had *ReadSim* among themselves and with M31, while in Month5 M9 and M24 disengage from the community. Where these members aware of the similarity they had with each other and with member M31 (cognitively central)? Additionally, the pattern detection algorithms found that in Month5 members M2, M5, M7 and M19 had *InterestSim* with each other. In Month6, members M2 and M7 were detected as disengaged from the community and the activity of members M5 and M19 was very low. Notifications making M31 aware of her influence in the VC and the decline in participation of relevant peripheral members might have been helpful to keep those peripheral members engaged in the community.

P3 & P4. Members not benefiting or not contributing. In Month4, 1 member was only uploading and 4 members were only downloading. By Month6, 2 members were only uploading and 9 members were only downloading. The patterns detected in Month6 showed that members began to disengage from the community either because they lost interest or because they could not find information useful for them. Downloading only excessively is a behavior that newcomers develop when they struggle to locate information important for them. For example, in Month6, member M19 downloaded 33 resources, without uploading anything from Month4 to Month6. Members like M19 can be supported by providing them with information of members with similar interests or members who are reading similar resources.

P5. Important peripheral members not downloading. Example for this pattern is member M33 who was detected uploading but not downloading from the community, while other members were interested in what M33 was uploading. In Month 6, this member disengaged from the VC feeling that she had not benefited from this community. The detection of this pattern could have been used to generate appropriate notifications. Na (Table 2) could have been sent to M33 in Month4 and Month 5, making her aware of her importance to the VC. In addition, in Month5, Member M31, who was a cognitively central member, has read resources uploaded by member M33 could have been sent notification Nc (Table2). This might have helped form a link between M33 and M31 to motivate and channel M33's contribution.

P6. Important peripheral members not uploading. An example for this pattern is member M5. In Month4, M5 joined the community and extensively downloaded resources. Based on this, he had *InterestSim* with members M24 and M28. In Month5, M5 was still only downloading and had *InterestSim* with M2, M7 and M19. In Month6, M5 stopped participating. Notification could have been sent to M5 in Month4 helping him identify people in the community or resources of his interest. Notification Nb (Table 2) could have been sent in Month5 pointing to M5 that his knowledge is relevant to the community, helping him discover relevant resources, and encouraging him to contribute to the VC.

P7. Unexplored complimentary similarity between members. Member M13 had *UploadSim* with member M33 but not *ReadSim* with M33. The detection of this pattern could have been used to keep both M33 and M13 aware of their similarity and motivate them to read resources uploaded by each other.

6 Discussion and Conclusion

The study indicates that the approach can be beneficial when an active VC starts to experience problems. The detected patterns can provide a better understanding of these problems and suggest possible interventions. The analysis of what was *automatically detected* in the study corresponds to what was *manually detected* in an earlier study in [10]. Moreover, the automatic analysis discovered patterns *missed by the human analyzer* who was exploring only visualization tools in the earlier study. A careful look into the human-missed problems confirmed their importance to the functioning of the community. Hence, the advantage of the approach presented here is the ability to discover patterns when large log data is collected, and the suggestion of corresponding community-tailored interventions. However, it was pointed out that a snapshot of community behavior may lead to too many notifications and some filtering/prioritization would be required. The use of a large scale data of a community which genuinely experienced problems and indeed did not sustain, allowed us to confirm that the automatic detection would be helpful.

Comparison with related work. Visualization techniques can be employed to present group and community models in a graphical way to help groups function more effectively [16] to motivate community participation [4], and to make members aware of reciprocal relationships [13]. The key limitation of visualization techniques is their *passive influence* on the functioning of the community, e.g. while examining graphical representations members may not be able to see how their contribution could be beneficial for the community as a whole and what activities they can engage in. In contrast, we analyze a community model to *automatically detect* problematic cases which can be used to decide when and how to intervene, offering *support* to improve the knowledge sharing and sustainability of the whole community. Different tools and algorithms have been developed to support people in locating expertise on a specific subject inside small or large groups or VC [14, 18]. In addition to identifying the interests and expertise of community members, we detect possible connections between members which have not been exploited in the community. This is used to encourage cognitively central and peripheral members to engage in interactions beneficial for the VC. The closest to our approach is research on intelligent group/community interventions, e.g. notification [1], feedback[2], or promotion of cognitively central members [3, 5]. The key novelty of our work is that we consider semantic between relationships and suggest community interventions aimed at improving the functioning of the VC as an entity.

Conclusion. To conclude, this paper has described a new approach to identify knowledge sharing behavior patters in a VC driven by processes important for the effective functioning of closely-knit communities. We have shown how these patterns can be detected and used to provide community-tailored support. The examples used are representative of what patterns can be discovered, how they can be automatically detected, and how the detection can be used. This work does not aim at defining an exhaustive list of patterns that can be discovered in a VC. Indeed, patterns can vary from community to community depending on the topic, people and the VC purpose. New patterns can be included as long as they can be defined with appropriate graph characteristics.

The next step of this research project is to apply the whole approach to an active VC and examine the effectiveness of the automatic interventions. We are currently conducting a study with another community which has been running for 3 months and is experiencing problems with forming. A series of notifications are being sent via email to community members, and the changes to community behavior are being analyzed applying [9]. Furthermore, in order to examine the suitability and generality of this approach in larger virtual communities, e.g. social networks or wiki-based VC, we plan to test our approach to improve awareness and participation in learning communities[2] or to help researchers discover connections that may lead to innovation[3].

References

1. Ardissono, L., et al.: Context-aware notification management in an integrated collaborative environment. In: Proceedings of International Workshop on Adaptation and Personalization for Web 2.0 (AP-Web 2.0 2009) at UMAP'09: Trento, Italy (2009)

---

[2] http://awesome.leeds.ac.uk/
[3] http://innovation1.coventry.ac.uk/brain/

2. Baghaei, N., Mitrovic, A.: From modelling domain knowledge to metacognitive skills: Extending a constraint-based tutoring system to support collaboration. In: Conati, C., McCoy, K., Paliouras, G. (eds.) UM 2007. LNCS (LNAI), vol. 4511, pp. 217–227. Springer, Heidelberg (2007)
3. Bretzke, H., Vassileva, J.: Motivating Cooperation on Peer to Peer Networks. In: Brusilovsky, P., Corbett, A.T., de Rosis, F. (eds.) UM 2003. LNCS (LNAI), vol. 2702, pp. 218–227. Springer, Heidelberg (2003)
4. Cheng, R., Vassileva, J.: Design and evaluation of an adaptive incentive mechanism for sustained educational online communities. J. of UMUAI 16(3), 321–348 (2006)
5. Farzan, R., et al.: Spreading the honey: a system for maintaining an online community. In: Proceedings of the ACM GROUP 2009 conference Florida, USA, pp. 31–40. ACM, New York (2009)
6. Harper, M., et al.: Talk amongst yourselves: inviting users to participate in online conversations. In: Proceedings of the 12th Int. Conf. on Intelligent User Interfaces (IUI'07), Honolulu, Hawaii, USA, pp. 62–71. ACM, New York (2007)
7. Ilgen, D.R., et al.: Teams in Organizations: From Input - Process - Output Models to IMOI Models. Annual Review of Psychology (56), 517–543 (2005)
8. Kameda, T., et al.: Centrality in Sociocognitive Networks and Social Influence: An Illustration in a Group Decision-Making Context. Journal of Personality and Social Psychology 73(2), 309 (1997)
9. Kleanthous, S.: Personalised Support for Knowledge Sharing in Virtual Communities. School of Computing University of Leeds, Leeds (expected submission) (May 2010)
10. Kleanthous, S., Dimitrova, V.: Modelling Semantic Relationships and Centrality to Facilitate Community Knowledge Sharing. In: Nejdl, W., Kay, J., Pu, P., Herder, E. (eds.) AH 2008. LNCS, vol. 5149, pp. 123–132. Springer, Heidelberg (2008)
11. Lave, J., Wenger, E.: Situated Learning: Legitimate Peripheral Participation. Cambridge University Press, New York (1991)
12. Mohammed, S., Dumville, B.C.: Team mental models in a team knowledge framework: expanding theory and measurement across disciplinary boundaries. Journal of Organizational Behavior 22(2), 89–106 (2001)
13. Raghavun, K., Vassileva, J.: Visualizing reciprocal and non-reciprocal relationships in an online community. In: AP-Web 2.0 workshop @ UMAP'09, Trento, Italy (2009)
14. Shami, S., et al.: That's what friends are for: facilitating 'who knows what' across group boundaries. In: Proceedings of the ACM 2007 GROUP conference, Florida, USA, pp. 379–382. ACM, New York (2007)
15. Thomas-Hunt, M., et al.: Who's Really Sharing? Effects of Social and Expert Status on Knowledge Exchange Within Groups. Management Science 49(4), 464–477 (2003)
16. Upton, K., Kay, J.: Narcissus: group and individual models to support small group work. In: Houben, G.-J., McCalla, G., Pianesi, F., Zancanaro, M. (eds.) UMAP 2009. LNCS, vol. 5535, pp. 54–65. Springer, Heidelberg (2009)
17. Wegner, D.M.: Transactive Memory: A Contemporary Analysis of the Group Mind. In: Mullen, B., et al. (eds.) Theories of Group Behavior, pp. 185–208. Springer, Heidelberg (1986)
18. Zhang, J., et al.: Expertise networks in online communities: structure and algorithms. In: Int. Conf. on WWW 2007, Alberta, Canada, pp. 221–230. ACM, New York (2007)

A Data-Driven Technique for Misconception Elicitation

Eduardo Guzmán, Ricardo Conejo, and Jaime Gálvez

Dpto. Lenguajes y Ciencias de la Computación. Universidad de Málaga,
Bulevar Louis Pasteur, 35. 29071 Málaga, Spain
{guzman,conejo,jgalvez}@lcc.uma.es

Abstract. When a quantitative student model is constructed, one of the first tasks to perform is to identify the domain concepts assessed. In general, this task is easily done by the domain experts. In addition, the model may include some misconceptions which are also identified by these experts. Identifying these misconceptions is a difficult task, however, and one which requires considerable previous experience with the students. In fact, sometimes it is difficult to relate these misconceptions to the elements in the knowledge diagnostic system which feeds the student model. In this paper we present a data-driven technique which aims to help elicit the domain misconceptions. It also aims to relate these misconceptions with the assessment activities (e.g. exercises, problems or test questions), which assess the subject in question.

Keywords: Student Modeling, Misconception Elicitation, Student Knowledge Diagnosis.

1 Introduction

Intelligent learning environments base their intelligence on the adaptive instruction they supply. The use of student models in such environments has emerged as a consequence of the fact that these systems have to work with incomplete information about the students [1]. A student model represents *who* is being taught, that is, what the student does (or not) know about the domain. Most learning environments construct this model from the student knowledge and the gaps in this knowledge. Using this information, they adapt the teaching process to the student's need. The quality of this adaption strongly depends on the accuracy of the student model. However, inferring the student model (and, in general, any user model) is a very difficult and costly process. Many researchers such as Self [2], have highlighted the intractable nature of this problem. Nevertheless, researchers recognize that although student models may not be highly accurate, and may not be complete from the cognitive perspective, they are indeed useful. In the traditional classroom approach, teachers also use less accurate student models; however, the teaching process is usually effective.

Perhaps the most commonly used strategy in student modeling is overlay modeling. When a learning system is constructed with overlay modeling, one of the first tasks to accomplish is to identify the concepts involved in the domain, and this

P. De Bra, A. Kobsa, and D. Chin (Eds.): UMAP 2010, LNCS 6075, pp. 243–254, 2010.
© Springer-Verlag Berlin Heidelberg 2010

process can be relatively easily performed by a human expert in the domain. However, there are other modeling techniques, such as perturbation models, which also incorporate incorrect knowledge. Even though the inclusion of student errors in their model provides some benefits, it also entails some problems. Specifically, the main problem of this modeling approach is the construction and maintenance of the bug library. The elicitation of this library is a time-consuming task which requires an exhaustive analysis of expert-student interactions and, accordingly, requires an expert with considerable experience. Despite this problem, the research on misconceptions has primarily focused on its diagnosis and remediation more than on its elicitation.

In this paper we present a technique for semi-automatic misconception discovery, primarily for declarative domains. The main goal is to provide the teachers with a collection of potential misconceptions. Subsequently, they have to decide whether or not they are misconceptions. A data-driven procedure is carried out based on the performance of students who have completed assessment activities (such as exercises, problems or test questions) in a certain subject domain.

The paper is structured as follows: The next three sections provide some background to the basis for this work. The first section contains a brief review of student modeling, focusing particularly on how researchers have approached the problem of modeling student error. Section 3 describes the state of the art in misconception modeling. Section 4 briefly looks at the basis of association rule algorithms, which have been used to develop the technique presented here. Then, the misconception inference technique is described in detail. Section 6 is devoted to the description of the experiment we have conducted to explore the performance of our technique. Finally, the conclusions we can extract from this work are outlined and also some of the research lines we plan to follow in the future.

2   Misconception Modeling

Holt et al. [3] reviewed the different types of student models. In their classification they identified the most commonly used *overlay models*, where the student is represented by his/her knowledge of a particular domain. Student behavior is compared to the expert behavior; therefore the student knowledge is a subset of the expert's. Differences between the two are assumed to be gaps in the student knowledge, but are not modeled specifically.

To overcome this problem, in the *perturbation models*, the student knowledge is *not only* a subset of the expert knowledge. In these kinds of models, the student may possess certain different knowledge in terms of both quantity and quality from that of the expert. Additionally, correct student knowledge (overlay modeling) is merged with the representation of faulty knowledge (misconceptions). This approach provides a more sophisticated model than other proposals. The set of misconceptions is usually stored in a *bug library* [4, 5].

The idea of bug libraries has considerable potential, but it also involves some problems [6]: The manual construction of the library is a difficult and time-consuming task; and the resulting library may not cover all the domain misconceptions, i.e., a student may have a misconception which is not included in the library.

3 Related Work

Most authors have focused on misconception diagnosis rather than on its elicitation, leaving this issue to the domain experts. Regarding the elicitation task, most of the research carried out in misconception modeling focuses on procedural errors, rather than declarative ones. The first efforts, in this area, were focused on automatically extending a bug library instead of creating it from scratch. Most notably, we can highlight the two rule-based algorithms INFER* and MALGEN [7], developed for the algebra equation domain and applied to high-school students. The first one creates new rules from incorrect actions taken by a student while solving a problem. MALGEN tries to elicit new rules automatically and without student intervention. Taking the domain rules as a starting point, it attempts to form new problem solving operators representing incorrect states by modifying existing operators. Both approaches required expert intervention to decide whether or not the inferred rule was suitable for the bug library.

ASSERT [6] is an algorithm for building tutoring systems using machine learning techniques to construct student models. It was the first system able to construct bug libraries automatically without needing the active participation of experts. To this end, it used a machine learning technique called *theory refinement* and has been applied to an introductory course of the C++ programming language in a rule-based tutoring system called NEITHER. The technique takes examples of student's behavior as input. If this behavior cannot be explained with the domain rules, the rules are modified and new ones are added to model the misconception.

MEDD [8] is an approach focused on the Prolog language programming domain. It is based on a similarity- and causality-based clustering technique of discrepancies between student programs and a set of reference programs. MEDD defines an error hierarchy containing error classes.

4 Itemsets and Association Rules

Nowadays most organizations have large databases with huge data sets. Nevertheless, because of the sheer amount of data available, important decisions are often taken using the intuition and experience of a human expert, instead of using the rich information stored [9]. In educational environments, the use of data mining techniques is growing [12]. We can find both techniques involving the analysis of the student-computer interaction logs and others studying other forms of educational data [13].

Association rules are an alternative means of extracting useful information from huge databases. They are used to establish the relationship (if X=a then Y=b) among different attribute values (i.e. if attribute X has the value a, the attribute Y will also have the value b). The algorithms which infer association rules take a set of transactions as input, each one labeled with a univocal identifier. Each transaction is formed by a set of attribute-value pairs which occur together.

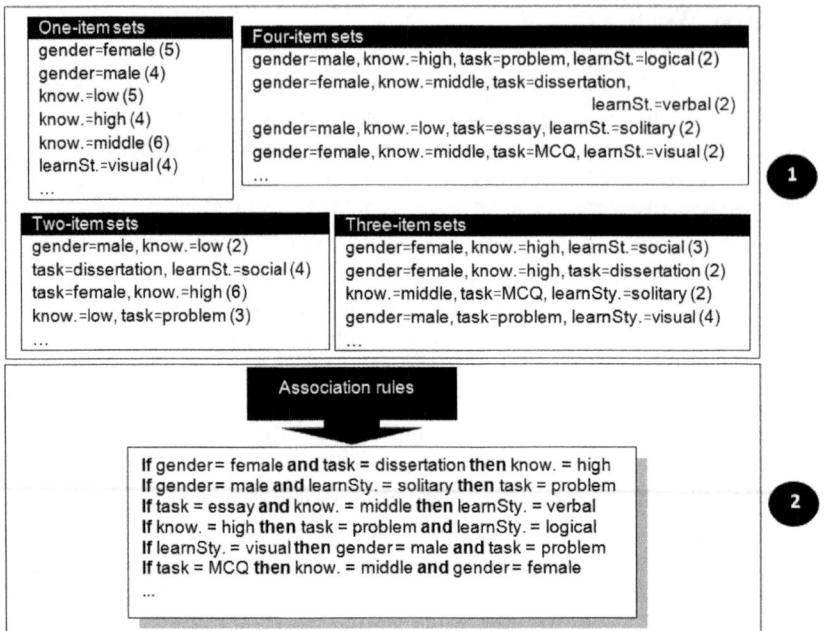

Fig. 1. Results of applying the Apriori method for mining the association rules of an educational dataset example

When inferring the association rules, the algorithms follow mainly two different phases:

1. The *itemset discovery*: In this stage, the most frequent collections of attribute-value pairs are found. Each value-attribute pair is called *item* and the set of items is called *itemset*. For this purpose, a minimum *support* must be defined a priori. The support is the frequency with which a set is found in the transaction dataset.

2. The *association rule inference*: Once the most frequent attribute sets are computed, this information is used as the basis of the association rules. Thus, rules are constructed taking the itemsets as antecedents or consequents. For each rule, its *support* is calculated, i.e. the fraction of transactions satisfying the rule, and its *confidence*, i.e. the number of data instances that the rule predicts correctly. A minimum support is indicated as an algorithm input parameter. At the end, those rules with a support lower than this threshold are discarded.

For instance, let us consider an educational domain example. Let us assume a dataset which contains information about the following four attributes, whose nominal values are included in brackets: `knowledge (low, middle, high)`, `learning style (visual, aural, verbal, physical, logical, social, solitary)`, `task (problem, MCQ, dissertation, essay)` and `gender (male, female)`. The goal is to mine association rules from these data. Figure 1 illustrates the results of applying the Apriori method [10], perhaps the most popular algorithm for mining association rules. This figure shows the results of the two phases of the algorithm. In the first (labeled number one), itemsets were discovered. Several itemsets were found and shown in different categories (see the boxes in the figure), in terms of the item number. Each item set also includes its *support* in brackets. In the example, the algorithm found that the itemset `knowledge=middle` appears six times and the set `gender=male, task=problem, learningStyle=visual` appears four times.

The lower part of Fig. 1 (labeled number two) shows the association rules inferred from the former itemsets. For example, the algorithm found that in the educational datasets, when the attribute `gender` is `male` and the `learning style` is `solitary`, then the `task` is `problem`.

5   A Technique for Misconception Inference

Our hypothesis is that if a student has a misconception, when he/she answers a question involving this misconception, he/she will tend to select those responses which best fit in with his/her incorrect knowledge state. Thus, these wrong responses could provide evidence about their misconception.

Let us consider, for instance, the algebra domain in which a student is solving questions in a test about fractions. If he/she does not know how to add fractions correctly, they may think, for example, that the fraction resulting from adding two fractions has a numerator equal to the sum of the numerators, and that the denominator is also the addition of the denominators. If in a test there are several questions involving the addition of two fractions, and these questions have an option where this addition is computed wrongly in the way that the student misunderstands, then he/she will select this response.

Consequently, if we have information available about student performance in a particular domain, we could try to discover the domain misconceptions by observing the selection patterns of wrong answers to questions, i.e., find out the most frequent associations among incorrect answers. The technique presented in this paper is based on applying an association rule inference algorithm to discover potential misconceptions.

Our technique takes as input the performances of a student population who took one or more tests about the domain whose misconceptions are being researched. Thus, we map the items which feed the association rule algorithms to the test questions, and the values of these items will be the options possible for answering a question. So the algorithm input is composed of student test sessions (equivalent to customer transactions). Each session contains a session identifier, the question posed and the set of choices selected for each question. To identify the misconceptions, we only pay attention to incorrect answers. For this reason, all the correct answers are filtered and blank responses (where allowed) are also discarded. Next, this "filtered" information is processed by the association rule algorithm. Consequently, it will return the most frequent sets of choices selected by the student sample and the frequency value, i.e. the number of times each choice association has been observed. In the post-processing, we order these sequences according to frequency. Our hypothesis is that the most frequent answer associations could correspond to misconceptions.

The next three subsections describe first the prior information needed by the technique and, later, the two stages of this technique which take place before and after the association rule algorithm execution.

5.1 The Evidence Model

The *Evidence-Centered Design* (ECD) proposal is a framework for designing, producing and delivering educational assessments [14]. ECD models incorporate representations of what a student knows and does not know, in terms of the results of his/her interaction performance (evidence) with assessment tasks [15]. In this line, we propose an evidence model which is based on the ECD framework. Our model (see Fig. 2) is composed of three layers: the *concept layer*, the *misconception layer* and the *task model*.

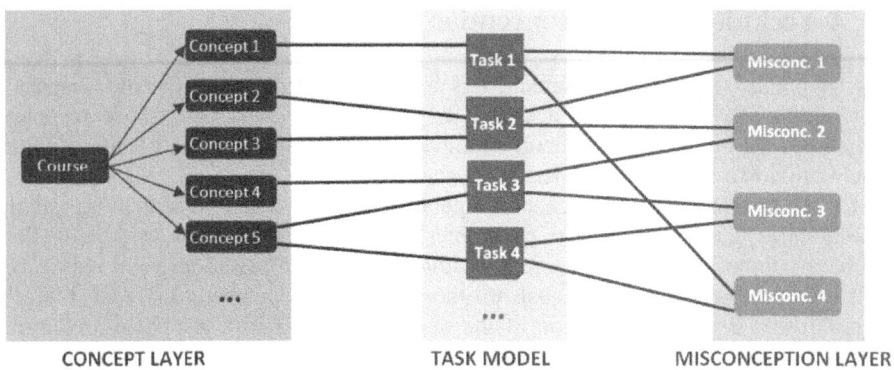

Fig. 2. The evidence model

In traditional teaching, the most popular tendency in the construction of domain models is to structure the contents of a course into parts, which are in turn divided into subparts, and so on. In this way, a hierarchy of variable granularity, called *curriculum* [16], is obtained. Curricula are often represented in intelligent learning environments by semantic networks, i.e., by directed acyclic graphs whose nodes are the pieces originated by the division of the course, and whose arcs represent relationships among the nodes. In the literature, a huge set of proposals exist (e.g., [17]) in which those parts have different names depending on their level in the hierarchy, e.g. topics, concepts, entities, chapters, sections, definitions, etc. Throughout this paper, all the nodes in the hierarchy will be generically called *concepts*. As Reye [18] states, concepts are curriculum elements which represent knowledge pieces or cognitive skills acquired (or not) by students.

From the point of view of student diagnosis, concepts are those elements susceptible to being assessed. However, we will not approach the structure and relationships among these concepts in this paper, since we consider it is not relevant for the work presented here. We only consider that all the concepts involved in the domain are collected in what we have called the *concept layer*.

In addition to this concept layer, our domain modeling approach is completed with a *misconception layer* which will be inferred semi-automatically. These misconceptions could be related to one (or even more) concepts of the former layer through the task model.

The *task model* is formed by assessment activities (e.g. exercises, problems or test questions), i.e., any kind of task which after been solved, could supply evidence about the student knowledge state. In this sense, we impose a restriction: the solutions to the activity reported by the student could be found in a finite number of states. That is, if a student takes a problem, his/her solution must reach a finite number of final states. To simplify, in this paper, we will consider that only one of these states is correct. Thus, the others supply evidence about the student's incorrect knowledge.

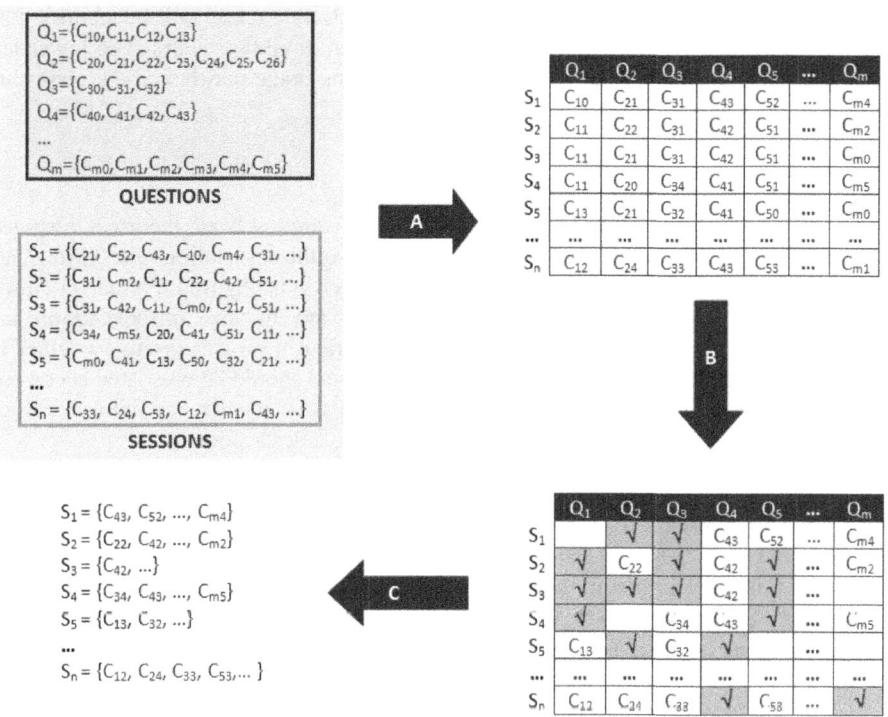

Fig. 3. Preprocessing stage of the Misconception Inference Technique

5.2 Preprocessing Stage

Before starting the inference procedure, a preprocessing stage is needed to prepare the input data. Fig. 3 illustrates graphically this procedure which takes as input the performances of students who took n assessment sessions: S_1, S_2, ..., S_n. Each student session could contain different exercises, all of them included in the following set: Q_1, Q_2, ..., Q_m. Exercises could be posed randomly to the students.

Let us assume, for instance, that these exercises are multiple-choice questions, where a question Q_i has r_i choices C_{i0}, C_{i1}, C_{i2}, C_{i3} and C_{ri}. The first choice C_{i0} represents the blank response; the second, C_{i1}, the correct one; and the rest of them are incorrect answers. For example, in the session S_1 of Fig. 3, the student was posed question Q_1 which he/she left blank, Q_2 and Q_3 which were answered correctly, and finally, Q_4 and Q_m which were answered incorrectly.

The upper left side of Fig. 3 shows the set of questions and sessions before preprocessing. Firstly, we construct a matrix of student sessions versus the questions (transition A). Next, we identify and remove the correct question choices and the blank responses (transition B). Finally, data are prepared to be used by the association rule inference algorithm (transition C). As mentioned before, the input required by this algorithm is composed of transactions. In this case, each transaction represents the incorrect responses of a session. As a result, each transaction will be identified by the session identifier and will be composed of question-answer pairs. The lower left side of Fig. 1 shows the result of this preprocessing stage and hence the algorithm input.

5.3 Post-Processing Stage

The rule association algorithm provides the discovered sets of incorrect response associations (i.e. the itemsets), their support, and also the association rules. After that, the participation of the teachers (i.e. the domain experts) is required. For each itemset, the questions involved and the choices which led to this misconception are highlighted and subsequently, the teachers should reason on and discuss whether or not this itemset maps to a misconception. The itemsets should be presented according to their support, in reverse order. When the teachers identify an itemset which corresponds to a misconception, they should produce a description.

6   A Case Study

The experiment described in this section was carried out with undergraduate students of a Programming Fundamentals course, corresponding to the second semester of the first year of the B.Sc. in Telecommunications at the University of Málaga. The course curriculum is divided into five concepts:

1. *Files:* Main memory and secondary storage. Text and Binary files.
2. *Dynamic Memory:* Physical Management of Dynamic Memory. Pointers. Linked Lists: simple, double, circular.
3. *Recursion:* Concept. Physical Implementation. Use and Examples.
4. *Data Abstraction Introduction:* Data Abstraction and Object Oriented Programming. Basic concepts of Object Oriented Programming. Advanced concepts of Object Oriented Programming. Pointers to objects.
5. *Abstract Data Types (ADT):* Concept of ADT. Stack: Definition, Examples and Implementation. Queue: Definition, Examples and Implementation. Positional List: Definition, Examples and Implementation. Binary Trees: Definition, Examples and Implementation.

The course has around 300 students per academic year (average age, 18 years old) and a high academic failure rate. In 2009, we suggested that the teachers should use this approach to help them identify the most common misconceptions regarding student knowledge in this subject. Thus, the goal of the experiment was: first, to assist the teachers in the process of misconception discovery and, additionally, to explore the relationships among these misconceptions.

6.1 Study Design

To infer the Programming Fundamental course misconceptions, we needed information about the performances of students who had taken this course previously. Accordingly, we used the test and the data from an unpublished experiment we conducted in June of 2007. This experiment consisted of taking the same idea we applied to another course of Artificial Intelligence and Knowledge Engineering during the 2005-06 academic year [19]. We developed a new activity which was to construct a self-assessment test using our web-based adaptive system Siette. Consequently, the students had a drill-and-practice activity to prepare them for the final exam. The main goal of this test was to provide students with an environment to be able to train for the exam from their own homes. We call this type of test *open test*. In [19], empirical evidence suggested that these tests are useful for facilitating the student learning process.

The open test of this experiment had several restrictions: 1) each student was allowed to take the same test only once a day (this restrictive facility is provided by Siette and is configured during the test elicitation process). 2) Once the test was finished, the corrections were not shown and only the final score was supplied to the student. This restriction was included to force the students to try to complete the test, rather than simply copying the correct answers to the items, a strategy adopted by many students in other experiments. Instead of doing the actual test themselves, they wanted only to see the questions and the correct answers.

The test consisted of 20 multiple-choice questions, each one composed of a stem and a set of three choices. In each question, a blank response was allowed and only one choice was correct. All students always took the same set of questions but these were posed randomly each time.

6.2 Data Analysis and Results

A total of 233 sessions were collected from 103 individuals who participated in this experiment. Before any analysis, we computed Cronbach's alpha, which is commonly used as an estimator of the test result internal consistency. We obtained a value of 0.72, which is higher than the minimum validity threshold (0.70). Next we took the information from the students' performance and applied all the steps described in section 5: First, the performance data were expressed in vectors, identified by a session id, and formed by question-answer pairs. Then, we removed the pairs corresponding to correct answers and those which represented blank responses. After that, we applied the Apriori association rule inference algorithm with a minimum support of 25% and obtained the itemset collection and the association rules.

Regarding the itemsets (with more than one item), we only got itemsets of two items. More specifically, a total of 26 two-item sets were found. We sorted them in terms of their support, in reverse order. After that, we showed the results to the teachers graphically, i.e. the pair of questions of each itemset was displayed, and the corresponding answers included in the itemset were highlighted. Using this information, the teachers identified the seven misconceptions shown in Table 1. The first three columns contain the misconception description that the teachers produced after analyzing the itemsets, the itemset support and the concept which it is related to.

Table 1. Misconceptions identifies for the Programming Fundamentals domain

Misconception	Support	Concept	% fail
Misunderstanding of the use and management of the class method parameters	45	Data Abstraction Introduction	97%
Inability to seperate the implementation and the interface of a DAT	38	Abstract Data Types	89%
Misundestanding when differentiating between a Positional List (i.e. a DAT) and a linked list	36	Abstract Data Types	81%
Misundestanding of the mechanism of procedure calls in recursive algorithms	32	Recursion	68%
Misundestanding of the diference between binary and text files	31	Files	87%
Misunderstading of the way in which the elements of a Positional List can be modified	31	Abstract Data Types	81%
Misunderstanding of the assignment between objects of a class	28	Data Abstraction Introduction	89%

In addition, we computed the percentage of success and failure of those students who selected each itemset, i.e., for each itemset, we took the sessions of those students whose answers corresponded to the itemset. Then we calculated their test score and the percentage of sessions in which the score was less than 50%. As can be seen in the last column of Table 1, in all the cases but one, the percentage of students who failed was greater than 80%. This result suggests that, as well as considering the itemset support, the percentage of failure could help the teachers to determine the itemsets that actually correspond to misconceptions.

Concerning the association rules, we have not found any useful rule. Finally, we should mention that we conducted other executions of the association rule inference algorithm, reducing the threshold of the minimum support. As a consequence, we obtained more itemsets (a superset of the previous itemsets).No new misconceptions were discovered however.

7 Conclusions and Future Work

In this paper, we have presented a semi-automatic technique for discovering domain misconceptions. In general, this is a difficult task, usually carried out manually by human experts in the particular domain, who have observed over time the incorrect behavior of a large number of students. In this work, however, we have tried to automate the first part of the process normally done manually by experts. In the same way that association rule inference algorithms were initially developed to discover which products most consumers bought together, incorrect student answers can be correlated to student misconceptions. Using this premise, we have applied the association rule algorithms to discover potential misconceptions.

The experiment we have conducted, suggests that this technique can help teachers to identify the domain misconceptions. For this purpose, we use the itemsets discovered by the association rule algorithm. The experiment also suggests that data concerning the itemset support and the percentage of students who both selected this itemset and did not pass the assessment session, could be used to automatically discard those itemsets not corresponding to misconceptions.

In addition to the experiment described in this paper, we have carried out others with similar results. In all of them, a set of misconceptions is discovered from the itemsets but, regarding the association rules, either the algorithm did not identify any relevant misconceptions, or else it only expressed the association among answers.

Our technique helps teachers discover student misconceptions and, as a consequence, relate them with the tasks (questions or exercises) which provide evidence about this incorrect knowledge. In this sense, we are developing a well-founded and quantitative assessment model (as an extension of our previous work [20]) for updating student models which include misconceptions. As a result, after an assessment session, our model will supply the knowledge level estimation for each domain concept and also a "*misknowledge*" estimation for each domain misconception.

In addition, we are also working on an analogous technique for domain concept discovery, based on association rules. The goal is not only to discover the concepts, but the relationships among them.

Acknowledgments. This work has been co-financed by the Spanish Ministry of Science and Innovation (TIN2007-67515) and by the Andalusian Regional Ministry of Science, Innovation and Enterprise (P09-TIC-5105). Authors want to thank prof. Eva Millán for her comments on the final version of this paper.

References

1. Mayo, M., Mitrovic, A.: Optimising ITS Behaviour with Bayesian Networks and Decision Theory. International Journal of Artificial Intelligence in Education 12, 124–153 (2001)
2. Self, J.A.: Bypassing the intractable problem of student modeling. In: Frasson, C., Gauthier, G. (eds.) Intelligent Tutoring Systems: at the Crossroads of Artificial Intelligence and Education, pp. 107–123. Ablex, Norwood (1990)
3. Holt, P., Dubs, S., Jones, M., Greer, J.: The state of student modelling. In: Greer, E.J.E., McCalla, G. (eds.) Student modelling: The key to individualized knowledge-based instruction, vol. 125, pp. 3–35. Springer, New York (1994)
4. Brown, J.S., Burton, R.R.: Diagnostic models for procedural bugs in basic mathematical skills. Cognitive Science 2, 155–192 (1978)
5. Brown, J.S., VanLehn, K.: Repair theory: A generative theory of bugs in procedural skills. Cognitive Science 4, 379–426 (1980)

6. Baffes, P., Mooney, R.: Refinement-Based Student Modeling and Automated Bug Library Construction. International Journal of Artificial Intelligence in Education 7(1), 75–116 (1996)

7. Sleeman, D., Hirsh, H., Ellery, I., Kim, I.: Extending domain theories: two case studies in student modeling. Machine Learning 5, 11–37 (1990)

8. Sison, R., Numao, M., Shimura, M.: Multistrategy Discovery and Detection of Novice Programmer Errors. Machine Learning 38, 157–180 (2000)

9. Han, J., Kamber, M.: Data Mining: Concepts and Techniques. Academic Press, USA (2001)

10. Agrawal, R., Imielinski, T., Swami, A.: Mining Association Rules Between Sets of Items in Large Databases. In: SIGMOD Conference, pp. 207–216 (1993)

11. Witten, I.H., Frank, E.: Data Mining: Practical Machine Learning Tools and Techniques, 2nd edn. Morgan Kaufmann, San Francisco (2005)

12. Romero, C., Ventura, S.: Educational Data Mining: A Survey from 1995 to 2005. Expert Systems with Applications 33(1), 135–146 (2007)

13. Baker, R.S.J.D., Yacef, K.: The State of Educational Data Mining in 2009: A Review and Future Visions. Journal of Educational Data Mining 1(1), 3–17 (2009)

14. Mislevy, R.J., Almond, R.G., Lukas, J.F.: A Brief Introduction to Evidence-Centered Design., CSE Report 632, National Center for Research on Evaluation, Standards and Student Testing (CRESST) (May 2004)

15. Shute, V.J., Graf, E.A., Hansen, E.: Designing adaptive, diagnostic math assessments for individuals with and without visual disabilities. In: PytlikZillig, L., Bruning, R., Bodvarsson, M. (eds.) Technology-based education: Bringing researchers and practitioners together, pp. 169–202 (2005)

16. Greer, J.E., McCalla, G.: Granularity-based reasoning and belief revision in student models. In: Greer, J.E., McCalla, G. (eds.) Student Modelling: The Key to Individualized Knowledge-Based Instruction, vol. 125, pp. 39–62. Springer, New York (1994)

17. Schank, R.C., Cleary, C.: Engines for Education. Lawrence Erlbaum Associates, Hillscale (1994)

18. Reye, J.: A belief net backbone for student modelling. In: Cerri, S.A., Gouardres, G., Paraguacu, F. (eds.) ITS 2002. LNCS, vol. 2363, pp. 596–604. Springer, New York (2002)

19. Guzmán, E., Conejo, R., Pérez-de-la-Cruz, J.L.: Improving Student Performance Using Self-Assessment Tests. IEEE Intelligent Systems 22(4), 46–52 (2007)

20. Guzmán, E., Conejo, R., Pérez-de-la-Cruz, J.L.: Adaptive Testing for Hierarchical Student Models. User Modeling and User-Adapted Interaction 17, 119–157 (2007)

Modeling Individualization in a Bayesian Networks Implementation of Knowledge Tracing

Zachary A. Pardos[*] and Neil T. Heffernan

Worcester Polytechnic Institute
Department of Computer Science
{zpardos,nth}@wpi.edu

Abstract. The field of intelligent tutoring systems has been using the well known knowledge tracing model, popularized by Corbett and Anderson (1995), to track student knowledge for over a decade. Surprisingly, models currently in use do not allow for individual learning rates nor individualized estimates of student initial knowledge. Corbett and Anderson, in their original articles, were interested in trying to add individualization to their model which they accomplished but with mixed results. Since their original work, the field has not made significant progress towards individualization of knowledge tracing models in fitting data. In this work, we introduce an elegant way of formulating the individualization problem entirely within a Bayesian networks framework that fits individualized as well as skill specific parameters simultaneously, in a single step. With this new individualization technique we are able to show a reliable improvement in prediction of real world data by individualizing the initial knowledge parameter. We explore three difference strategies for setting the initial individualized knowledge parameters and report that the best strategy is one in which information from multiple skills is used to inform each student's prior. Using this strategy we achieved lower prediction error in 33 of the 42 problem sets evaluated. The implication of this work is the ability to enhance existing intelligent tutoring systems to more accurately estimate when a student has reached mastery of a skill. Adaptation of instruction based on individualized knowledge and learning speed is discussed as well as open research questions facing those that wish to exploit student and skill information in their user models.

Keywords: Knowledge Tracing, Individualization, Bayesian Networks, Data Mining, Prediction, Intelligent Tutoring Systems.

1 Introduction

Our initial goal was simple; to show that with more data about students' prior knowledge, we should be able to achieve a better fitting model and more accurate prediction of student data. The problem to solve was that there existed no Bayesian network model to exploit per user prior knowledge information. Knowledge tracing

---

[*] National Science Foundation funded GK-12 Fellow.

P. De Bra, A. Kobsa, and D. Chin (Eds.): UMAP 2010, LNCS 6075, pp. 255–266, 2010.

(KT) is the predominant method used to model student knowledge and learning over time. This model, however, assumes that all students share the same initial prior knowledge and does not allow for per student prior information to be incorporated. The model we have engineered is a modification to knowledge tracing that increases its generality by allowing for multiple prior knowledge parameters to be specified and lets the Bayesian network determine which prior parameter value a student belongs to if that information is not known before hand. The improvements we see in predicting real world data sets are palpable, with the new model predicting student responses better than standard knowledge tracing in 33 out of the 42 problem sets with the use of information from other skills to inform a prior per student that applied to all problem sets. Equally encouraging was that the individualized model predicted better than knowledge tracing in 30 out of 42 problem sets without the use of any external data. Correlation between actual and predicted responses also improved significantly with the individualized model.

1.1 Inception of Knowledge Tracing

Knowledge tracing has become the dominant method of modeling student knowledge. It is a variation on a model of learning first introduced by Atkinson in 1972 [1]. Knowledge tracing assumes that each skill has 4 parameters; two knowledge parameters and two performance parameters. The two knowledge parameters are: initial (or prior) knowledge and learn rate. The initial knowledge parameter is the probability that a particular skill was known by the student before interacting with the tutor. The learn rate is the probability that a student will transition between the unlearned and the learned state after each learning opportunity (or question). The two performance parameters are: guess rate and slip rate. The guess rate is the probability that a student will answer correctly even if she does not know the skill associated with the question. The slip rate is the probability that a student will answer incorrectly even if she knows the required skill. Corbett and Anderson introduced this method to the intelligent tutoring field in 1995 [2]. It is currently employed by the cognitive tutor, used by hundreds of thousands of students, and many other intelligent tutoring systems to predict performance and determine when a student has mastered a particular skill.

It might strike the uninitiated as a surprise that the dominant method of modeling student knowledge in intelligent tutoring systems, knowledge tracing, does not allow for students to have different learn rates even though it seems likely that students differ in this regard. Similarly, knowledge tracing assumes that all students have the same probability of knowing a particular skill at their first opportunity.

In this paper we hope to reinvigorate the field to further explore and adopt models that explicitly represent the assumption that students differ in their individual initial knowledge, learning rate and possibly their propensity to guess or slip.

1.2 Previous Approaches to Predicting Student Data Using Knowledge Tracing

Corbett and Anderson were interested in implementing the learning rate and prior knowledge individualization that was originally described as part of Atkinson's model of learning. They accomplished this but with limited success. They created a two step process for learning the parameters of their model where the four KT parameters were learned for each skill in the first step and the individual weights were applied to those parameters for each student in the second step. The second step used a form of regression to fit student specific weights to the parameters of each skill. Various factors were also identified for influencing the individual priors and learn rates [3]. The results [2] of their work showed that while the individualized model's predictions correlated better with the actual test results than the non-individualized model, their individualized model did not show an improvement in the overall accuracy of the predictions.

More recent work by Baker et al [4] has found utility in the contextualization of the guess and slip parameters using a multi-staged machine-learning processes that also uses regression to fine tune parameter values. Baker's work has shown an improvement in the internal fit of their model versus other knowledge tracing approaches when correlating inferred knowledge at a learning opportunity with the actual student response at that opportunity but has yet to validate the model with an external validity test.

One of the knowledge tracing approaches compared to the contextual guess and slip method was the Dirichlet approach introduced by Beck et al [5]. The goal of this method was not individualization or contextualization but rather to learn plausible knowledge tracing model parameters by biasing the values of the initial knowledge parameter. The investigators of this work engaged in predicting student data from a reading tutor but found only a 1% increase in performance over standard knowledge tracing (0.006 on the AUC scale). This improvement was achieved by setting model parameters manually based on the authors understanding of the domain and not by learning the parameters from data.

1.3 The ASSISTment System

Our dataset consisted of student responses from The ASSISTment System, a web based math tutoring system for 7th-12th grade students that provides preparation for the state standardized test by using released math problems from previous tests as questions on the system. Tutorial help is given if a student answers the question wrong or asks for help. The tutorial help assists the student learn the required knowledge by breaking the problem into sub questions called scaffolding or giving the student hints on how to solve the question.

2 The Model

Our model uses Bayesian networks to learn the parameters of the model and predict performance. Reye [6] showed that the formulas used by Corbett and Anderson in their knowledge tracing work could be derived from a Hidden Markov Model or Dynamic Bayesian Network (DBN). Corbett and colleagues later released a toolkit [7] using non-individualized Bayesian knowledge tracing to allow researchers to fit their own data and student models with DBNs.

2.1   The Prior Per Student Model vs. Standard Knowledge Tracing

The model we present in this paper focuses only on individualizing the prior knowledge parameter. We call it the Prior Per Student (PPS) model. The difference between PPS and Knowledge Tracing (KT) is the ability to represent a different prior knowledge parameter for each student. Knowledge Tracing is a special case of this prior per student model and can be derived by fixing all the priors of the PPS model to the same values or by specifying that there is only one shared student ID. This equivalence was confirmed empirically.

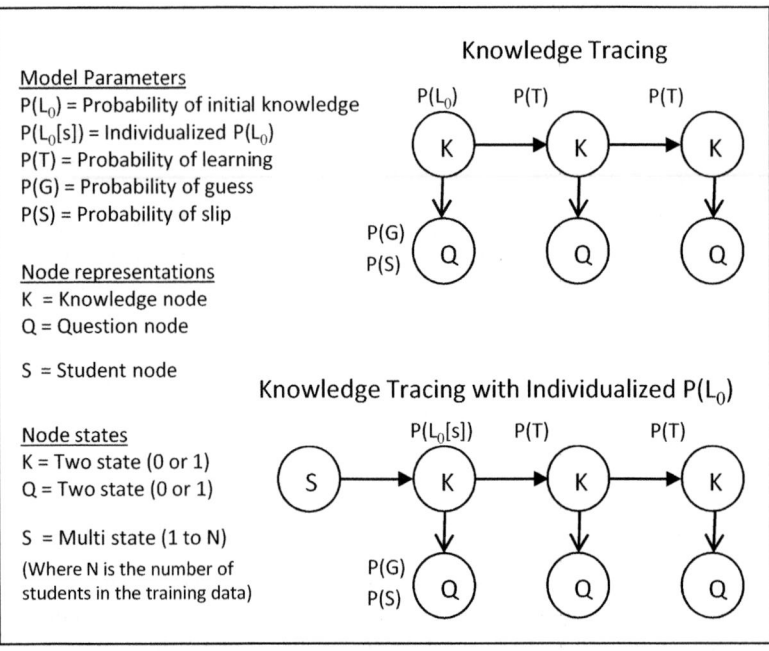

Fig. 1. The topology and parameter description of Knowledge Tracing and PPS

The two model designs are shown in Fig. 1. Initial knowledge and prior knowledge are synonymous. The individualization of the prior is achieved by adding a student node. The student node can take on values that range from one to the number of students being considered. The conditional probability table of the initial knowledge node is therefore conditioned upon the student node value. The student node itself also has a conditional probability table associated with it which determines the probability that a student will be of a particular ID. The parameters for this node are fixed to be 1/N where N is the number of students. The parameter values set for this node are not relevant since the student node is an observed node that corresponds to the student ID and need never be inferred.

This model can be easily changed to individualize learning rates instead of prior knowledge by connecting the student node to the subsequent knowledge nodes thus training an individualized P(T) conditioned upon student as shown in Fig. 2.

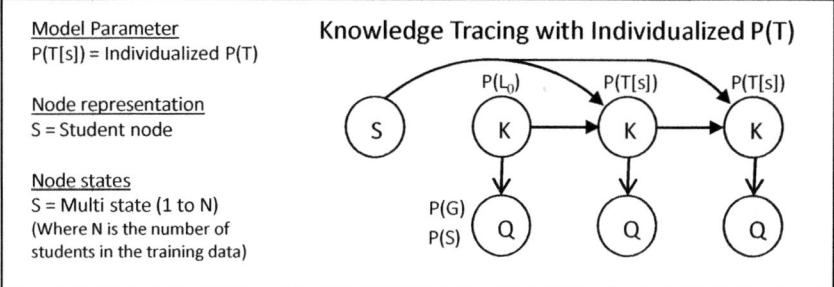

Fig. 2. Graphical depiction of our individualization modeling technique applied to the probability of learning parameter. This model is not evaluated in this paper but is presented to demonstrate the simplicity in adapting our model to other parameters.

2.2 Parameter Learning and Inference

There are two distinct steps in knowledge tracing models. The first step is learning the parameters of the model from all student data. The second step is tracing an individual student's knowledge given their respective data. All knowledge tracing models allow for initial knowledge to be inferred per student in the second step. The original KT work [2] that individualized parameters added an additional step in between 1 and 2 to fit individual weights to the general parameters learned in step one. The PPS model allows for the individualized parameters to be learned along with the non-individualized parameters of the model in a single step. Assuming there is variance worth modeling in the individualization parameter, we believe that a single step procedure allows for more accurate parameters to be learned since a global best fit to the data can now be searched for instead of a best fit of the individual parameters after the skill specific parameters are already learned.

In our model each student has a student ID represented in the student node. This number is presented during step one to associate a student with his or her prior parameter. In step two, the individual student knowledge tracing, this number is again presented along with the student's respective data in order to again associate that student with the individualized parameters learned for that student in the first step.

3 External Validity: Student Performance Prediction

In order to test the real world utility of the prior per student model, we used the last question of each of our problem sets as the test question. For each problem set we trained two separate models: the prior per student model and the standard knowledge tracing model. Both models then made predictions of each student's last question responses which could then be compared to the students' actual responses.

3.1 Dataset Description

Our dataset consisted of student responses to problem sets that satisfied the following constraints:

- Items in the problem set must have been given in a random order
- A student must have answered all items in the problem set in one day
- The problem set must have data from at least 100 students
- There are at least four items in the problem set of the exact same skill
- Data is from Fall of 2008 to Spring of 2010

Forty-two problem sets matched these constraints. Only the items within the problem set with the exact same skill tagging were used. 70% of the items in the 42 problem sets were multiple choice, 30% were fill in the blank (numeric). The size of our resulting problem sets ranged from 4 items to 13. There were 4,354 unique students in total with each problem set having an average of 312 students ($\sigma = 201$) and each student completing an average of three problem sets ($\sigma = 3.1$).

Table 1. Sample of the data from a five item problem set

Student ID	1st response	2nd response	3rd response	4th response	5th response
750	0	1	1	1	1
751	0	1	1	1	0
752	1	1	0	1	0

In Table 1, each response represents either a correct or incorrect answer to the original question of the item. Scaffold responses are ignored in our analysis and requests for help are marked as incorrect responses by the system.

3.2 Prediction Procedure

Each problem set was evaluated individually by first constructing the appropriate sized Bayesian network for that problem set. In the case of the individualized model, the size of the constructed student node corresponded to the number of students with data for that problem set. All the data for that problem set, except for responses to the last question, was organized into an array to be used to train the parameters of the network using the Expectation Maximization (EM) algorithm. The initial values for the learn rate, guess and slip parameters were set to different values between 0.05 and 0.90 chosen at random. After EM had learned parameters for the network, student performance was predicted. The prediction was done one student at a time by entering ,as evidence to the network, the responses of the particular student except for the response to the last question. A static unrolled dynamic Bayesian network was used. This enabled individual inferences of knowledge and performance to be made about the student at each question including the last question. The probability of the student answering the last question correctly was computed and saved to later be compared to the actual response.

3.3 Approaches to Setting the Individualized Initial Knowledge Values

In the prediction procedure, due to the number of parameters in the model, care had to be given to how the individualized priors would be set before the parameters of the network were learned with EM. There were two decisions we focused on: a) what initial values should the individualized priors be set to and b) whether or not those values should be fixed or adjustable during the EM parameter learning process. Since it was impossible to know the ground truth prior knowledge for each student for each problem set, we generated three heuristic strategies for setting these values, each of which will be evaluated in the results section.

3.3.1 Setting Initial Individualized Knowledge to Random Values

One strategy was to treat the individualized priors exactly like the learn, guess and slip parameters by setting them to random values to then be adjusted by EM during the parameter learning process. This strategy effectively learns a prior per student per skill. This is perhaps the most naïve strategy that assumes there is no means of estimating a prior from other sources of information and no better heuristic for setting prior values. To further clarify, if there are 600 students there will be 600 random values between 0 and 1 set for for each skill. EM will then have 600 parameters to learn in addition to the learn, guess and slip parameters of each skill. For the non-individualized model, the singular prior was set to a random value and was allowed to be adjusted by EM.

3.3.2 Setting Initial Individualized Knowledge Based on 1st Response Heuristic

This strategy was based on the idea that a student's prior is largely a reflection of their performance on the first question with guess and slip probabilities taken into account. If a student answered the first question correctly, their prior was set to one minus an *ad-hoc* guess value. If they answered the first question incorrectly, their prior was set to an *ad-hoc* slip value. *Ad-hoc* guess and slip values are used because ground truth guess and slip values cannot be known and because these values must be used before parameters are learned. The accuracy of these values could largely impact the effectiveness of this strategy. An *ad-hoc* guess value of 0.15 and slip value of 0.10 were used for this heuristic. Note that these guess and slip values are not learned by EM and are separate from the performance parameters. The non-individualized prior was set to the mean of the first responses and was allowed to be adjusted while the individualized priors were fixed. This strategy will be referred to as the "cold start heuristic" due to its bootstrapping approach.

3.3.3 Setting Initial Individualized Knowledge Based on Global Percent Correct

This last strategy was based on the assumption that there is a correlation between student performance on one problem set to the next, or from one skill to the next. This is also the closest strategy to a model that assumes there is a single prior per student that is the same across all skills. For each student, a percent correct was computed, averaged over each problem set they completed. This was calculated using data from all of the problem sets they completed except the problem set being predicted. If a student had only completed the problem set being predicted then her prior was set to the average of the other student priors. The single KT prior was also set to the average of the individualized priors for this strategy. The individualized priors were fixed while the non-individualized prior was adjustable.

3.4 Performance Prediction Results

The prediction performance of the models was calculated in terms of mean absolute error (MAE). The mean absolute error for a problem set was calculated by taking the mean of the absolute difference between the predicted probability of correct on the last question and the actual response for each student. This was calculated for each model's prediction of correct on the last question. The model with the lowest mean absolute error for a problem set was deemed to be the more accurate predictor of that problem set. Correlation was also calculated between actual and predicted responses.

Table 2. Prediction accuracy and correlation of each model and initial prior strategy

	Most accurate predictor (of 42)		Avg. Correlation	
P(L_0) Strategy	PPS	KT	PPS	KT
Percent correct heuristic	33	8	0.3515	0.1933
Cold start heuristic	30	12	0.3014	0.1726
Random parameter values	26	16	0.2518	0.1726

Table 2 shows the number of problem sets that PPS predicted more accurately than KT and vice versa in terms of MAE for each prior strategy. This metric was used instead of average MAE to avoid taking an average of averages. With the percent correct heuristic, the PPS model was able to better predict student data in 33 of the 42 problem sets. The binomial with p = 0.50 tells us that the probability of 33 success or more in 42 trials is << 0.05 (cutoff is 27 to achieve statistical significance), indicating a result that was not the product of random chance. In one problem set the MAE of PPS and KT were equal resulting in a total other than 42 (33 + 8 = 41). The cold start heuristic, which used the 1st response from the problem set and two *ad-hoc* parameter values, also performed well; better predicting 30 of the 42 problem sets which was also statistically significantly reliable. We recalculated MAE for PPS and KT for the percent correct heuristic this time taking the mean absolute difference between the rounded probability of correct on the last question and actual response for each student. The result was that PPS predicted better than KT in 28 out of the 42 problem sets and tied KT in MAE in 10 of the problem sets leaving KT with 4 problem sets predicted more accurately than PPS with the recalculated MAE. This demonstrates a meaningful difference between PPS and KT in predicting actual student responses.

The correlation between the predicted probability of last response and actual last response using the percent correct strategy was also evaluated for each problem set. The PPS model had a higher correlation coefficient than the KT model in 32 out of 39 problem sets. A correlation coefficient was not able to be calculated for the KT model in three of the problem sets due to a lack of variation in prediction across students. This occurred in one problem set for the PPS model. The average correlation coefficient across all problem sets was 0.1933 for KT and 0.3515 for PPS using the percent correct heuristic. The MAE and correlation of the random parameter strategy using PPS was better than KT. This was surprising since the PPS random parameter strategy represents a prior per student per skill which could be considered an over parameterization of the model. This is evidence to us that the PPS model may outperform KT in prediction under a wide variety of conditions.

3.4.1 Response Sequence Analysis of Results

We wanted to further inspect our models to see under what circumstances they correctly and incorrectly predicted the data. To do this we looked at response sequences and counted how many times their prediction of the last question was right or wrong (rounding predicted probability of correct). For example: student response sequence [0 1 1 1] means that the student answered incorrectly on the first question but then answered correctly on the following three. The PPS (using percent correct heuristic) and KT models were given the first three responses in addition to the parameters of the model to predict the fourth. If PPS predicted 0.68 and KT predicted 0.72 probability of correct for the last question, they would both be counted as predicting that instance correctly. We conducted this analysis on the 11 problem sets of length four. There were 4,448 total student response sequence instances among the 11 problem sets. Tables 3 and 4 show the top sequences in terms of number of instances where both models predicted the last question correctly (Table 3) and incorrectly (Table 4). Tables 5-6 show the top instances of sequences where one model predicted the last question correctly but the other did not.

Table 3. Predicted correctly by both

# of Instances	Response sequence
1167	1 1 1 1
340	0 1 1 1
253	1 0 1 1
252	1 1 0 1

Table 4. Predicted incorrectly by both

# of Instances	Response sequence
251	1 1 1 0
154	0 1 1 0
135	1 1 0 0
106	1 0 1 0

Table 5. Predicted correctly by PPS only

# of Instances	Response sequence
175	0 0 0 0
84	0 1 0 0
72	0 0 1 0
61	1 0 0 0

Table 6. Predicted correctly by KT only

# of Instances	Response sequence
75	0 0 0 1
54	1 0 0 1
51	0 0 1 1
47	0 1 0 1

Table 3 shows the sequences most frequently predicted correctly by both models. These happen to also be among the top 5 occurring sequences overall. The top occurring sequence [1 1 1 1] accounts for more than 1/3 of the instances. Table 4 shows that the sequence where students answer all questions correctly except the last question is most often predicted incorrectly by both models. Table 5 shows that PPS is able to predict the sequence where no problems are answered correctly. In no instances does KT predict sequences [0 1 1 0] or [1 1 1 0] correctly. This sequence analysis may not generalize to other datasets but it provides a means to identify areas the model can improve in and where it is most strong. Figure 3 shows a graphical representation of the distribution of sequences predicted by KT and PPS versus the actual distribution of sequences. This distribution combines the predicted sequences from all 11 of the four item problem sets. The response sequences are sorted by frequency of actual response sequences from left to right in descending order.

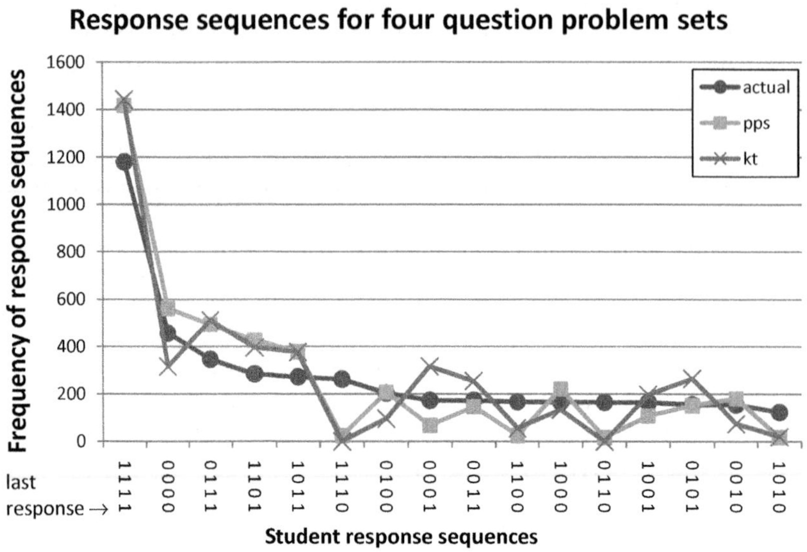

Response sequences for four question problem sets

Fig. 3. Actual and predicted sequence distributions of PPS (percent correct heuristic) and KT

The average residual of PPS is smaller than KT but as the chart shows, it is not by much. This suggests that while PPS has been shown to provide reliably better predictions, the increase in performance prediction accuracy may not be substantial.

4 Contribution

In this work we have shown how any Bayesian knowledge tracing model can easily be extended to support individualization of any or all of the four KT parameters using the simple technique of creating a student node and connecting it to the parameter node or nodes to be individualized. The model we have presented allows for individualized and skill specific parameters of the model to be learned simultaneously in a single step thus enabling global best fit parameters to potentially be learned, a potential that is prohibitive with multi step parameter learning methods [2,4].

We have also shown the utility of using this technique to individualize the prior parameter by demonstrating reliable improvement over standard knowledge tracing in predicting real world student responses. The superior performance of the model that uses PPS based on the student's percent correct across all skills makes a significant scientific suggestion that it may be more important to model a single prior per student across skills rather than a single prior per skill across students, as is the norm.

5 Discussion and Future Work

We hope this paper is the beginning of a resurgence in attempting to better individualize and thereby personalize students' learning experiences in intelligent tutoring systems.

We would like to know when using a prior per student is not beneficial. Certainly if in reality all students had the same prior per skill then there would be no utility in modeling an individualized prior. On the other hand, if student priors for a skill are highly varied, which appears to be the case, then individualized priors will lead to a better fitting model by allowing the variation in that parameter to be captured.

Is an individual parameter per student necessary or can the same or better performance be achieved by grouping individual parameters into clusters? The relatively high performance of our cold start heuristic model suggests that much can be gained by grouping students into one of two priors based on their first response to a given skill. While this heuristic worked, we suspect there are superior representations and ones that allow for the value of the cluster prior to be learned rather than set *ad-hoc* as we did. Ritter et al [8] recently showed that clustering of similar skills can drastically reduce the number of parameters that need to be learned when fitting hundreds of skills while still maintaining a high degree of fit to the data. Perhaps a similar approach can be employed to find clusters of students and learning their parameters instead of learning individualized parameters for every student.

Our work here has focused on just one of the four parameters in knowledge tracing. We are particularly excited to see if by explicitly modeling the fact that students have different rates of learning we can achieve higher levels of prediction accuracy. The questions and tutorial feedback a student receives could be adapted to his or learning rate. Student learning rates could also be reported to teachers allowing them to more precisely or more quickly understand their classes of students. Guess and slip individualization is also possible and a direct comparison to Baker's contextual guess and slip method would be an informative piece of future work.

We have shown that choosing a prior per student representation over the prior per skill representation of knowledge tracing is beneficial in fitting our dataset; however, a superior model is likely one that combines the attributes of the student with the attributes of a skill. How to design this model that properly treats the interaction of these two pieces of information is an open research question for the field. We believe that in order to extend the benefit of individualization to new users of a system, multiple problem sets must be linked in a single Bayesian network that uses evidence from the multiple problem sets to help trace individual student knowledge and more fully reap the benefits suggested by the percent correct heuristic.

This work has concentrated on knowledge tracing, however, we recognize there are alternatives. Draney, Wilson and Pirolli [9] have introduced a model they argue is more parsimonious than knowledge tracing due to having fewer parameters. Additionally, Pavlik et al [10] have reported using different algorithms, as well as brute force, for fitting the parameters of their models. We also point out that more standard models that do not track knowledge such as item response theory that have had large uses in and outside of the ITS field for estimating individual student and question parameters. We know there is value in these other approaches and strive as a field to learn how best to exploit information about students, questions and skills towards the goal of a truly effective, adaptive and intelligent tutoring system.

Acknowledgements. We would like to thank all of the people associated with creating the ASSISTment system listed at www.ASSISTment.org. We would also like to acknowledge funding from the US Department of Education, the National Science Foundation, the Office of Naval Research and the Spencer Foundation. All of the opinions expressed in this paper are those of the authors and do not necessarily reflect the views of our funders.

References

1. Atkinson, R.C., Paulson, J.A.: An approach to the psychology of instruction. Psychological Bulletin 78, 49–61 (1972)
2. Corbett, A.T., Anderson, J.R.: Knowledge tracing: modeling the acquisition of procedural knowledge. User Modeling and User-Adapted Interaction 4, 253–278 (1995)
3. Corbett, A., Bhatnagar, A.: Student Modeling in the ACT Programming Tutor: Adjusting a Procedural Learning Model with Declarative Knowledge. In: Jameson, A., Paris, C., Tasso, C. (eds.) Proceedings of the 6th International Conference on User Modeling, pp. 243–254 (1997)
4. Baker, R.S.J.d., Corbett, A.T., Aleven, V.: More accurate student modeling through contextual estimation of slip and guess probabilities in bayesian knowledge tracing. In: Woolf, B.P., Aïmeur, E., Nkambou, R., Lajoie, S. (eds.) ITS 2008. LNCS, vol. 5091, pp. 406–415. Springer, Heidelberg (2008)
5. Beck, J.E., Chang, K.M.: Identifiability: A Fundamental Problem of Student Modeling. In: Conati, C., McCoy, K., Paliouras, G. (eds.) UM 2007. LNCS (LNAI), vol. 4511, pp. 137–146. Springer, Heidelberg (2007)
6. Reye, J.: Student modelling based on belief networks. International Journal of Artificial Intelligence in Education 14, 63–96 (2004)
7. Chang, K.M., Beck, J.E., Mostow, J., Corbett, A.: A Bayes Net Toolkit for Student Modeling in Intelligent Tutoring Systems. In: Ikeda, M., Ashley, K.D., Chan, T.-W. (eds.) ITS 2006. LNCS, vol. 4053, pp. 104–113. Springer, Heidelberg (2006)
8. Ritter, S., Harris, T., Nixon, T., Dickison, D., Murray, C., Towle, B.: Reducing the knowledge tracing space. In: Proceedings of the 2nd International Conference on Educational Data Mining, Cordoba, Spain, pp. 151–160 (2009)
9. Draney, K.L., Pirolli, P., Wilson, M.: A measurement model for a complex cognitive skill. In: Nichols, P.D., Chipman, S.F., Brennan, R.L. (eds.) Cognitively diagnostic assessment, pp. 103–125. Erlbaum, Hillsdale (1995)
10. Pavlik, P.I., Cen, H., Koedinger, K.R.: Performance Factors Analysis - A New Alternative to Knowledge Tracing. In: Proceedings of the 14th International Conference on Artificial Intelligence in Education, Brighton, UK, pp. 531–538 (2009)

Detecting Gaming the System in Constraint-Based Tutors

Ryan S.J.d. Baker[1], Antonija Mitrović[2], and Moffat Mathews[2]

[1] Department of Social Science and Policy Studies, Worcester Polytechnic Institute
100 Institute Road, Worcester MA 01609, USA
rsbaker@wpi.edu
[2] Intelligent Computer Tutoring Group, Computer Science and Software Engineering
University of Canterbury, New Zealand
{tanja.mitrovic,moffat.mathews}@canterbury.ac.nz

Abstract. Recently, detectors of gaming the system have been developed for several intelligent tutoring systems where the problem-solving process is reified, and gaming consists of systematic guessing and help abuse. Constraint-based tutors differ from the tutors where gaming detectors have previously been developed on several dimensions: in particular, higher-level answers are assessed according to a larger number of finer-grained constraints, and feedback is split into levels rather than an entire help sequence being available at any time. Correspondingly, help abuse behaviors differ, including behaviors such as rapidly repeating the same answer or blank answers to elicit answers. We use text replay labeling in combination with educational data mining methods to create a gaming detector for SQL-Tutor, a popular constraint-based tutor. This detector assesses gaming at the level of multiple-submission sequences and is accurate both at identifying gaming within submission sequences and at identifying how much each student games the system. It achieves only limited success, however, at distinguishing different types of gaming behavior from each other.

Keywords: gaming the system, educational data mining, machine learning.

1 Introduction

In recent years, increasing attention has been paid to developing educational software which can recognize and adapt to when students game the system. A student games the system when they attempt to succeed in an educational task by systematically taking advantage of properties and regularities in the system used to complete that task, rather than by thinking through the material (cf. [4]). Detectors of gaming the system have been developed through both educational data mining/machine learning methods [5, 7, 9, 28] and through knowledge engineering approaches [1, 10, 16, 25]. These detectors have then been incorporated into interventions shown to reduce gaming and improve learning (cf. [2, 3]). However, despite the broad range of types of educational software where gaming the system has been observed (e.g. [11, 15, 23, 26, 29]), past detectors of gaming the system have been designed for fairly similar types of educational software, where students enter simple answers (a number, a word, or selecting from a set of options) with the

P. De Bra, A. Kobsa, and D. Chin (Eds.): UMAP 2010, LNCS 6075, pp. 267–278, 2010.

problem-solving process reified into individual steps (though sometimes the reification only occurs after an incorrect answer (cf. [28])). In addition, in these educational software systems, gaming behavior has consisted predominantly of systematic guessing and abuse of multi-level hints (with the exception of [10], where gaming behavior consisted of near-instantaneous responses to multiple-choice questions given a single time). Many other gaming behaviors have been reported in other educational software packages (cf. [11, 21]).

If gaming detection can be applied to a broader range of educational software, we can extend the benefits of gaming intervention to more topics and a greater number of students. In this paper, we study the feasibility of developing gaming detection for SQL-Tutor [19], a constraint-based tutor with design substantially different than environments for which gaming detection has previously been developed, and where students correspondingly engage in different gaming behaviors. Within SQL-Tutor, the "inner-loop" (e.g. [27]) of individual cognitive steps in the problem-solving process is not reified or responded to [27]; complete solutions are analyzed for correctness, making for a different gaming detection challenge than in previous work.

2   Learning Environment and Gaming Behaviors

SQL-Tutor is a constraint-based tutor that assists university-level students in acquiring the knowledge and skills necessary to create SQL queries [19]. It is a mature intelligent tutoring system, designed as a practice environment with the prerequisite that students be previously exposed to the SQL concepts in lectures. Students submit solutions which are sent to the student modeller for analysis. The student modeller identifies any errors and updates the student model accordingly to reflect the student's progress within the domain. To check the correctness of the student's solution, SQL-Tutor evaluates the student's solution against domain knowledge, represented as a set of 700 constraints. A preferred solution, entered by the instructor, aids in this evaluation process while still allowing for novel correct solutions from the student. A constraint consists of a relevance condition C_r, which checks whether the constraint is appropriate for a particular student's solution, and a satisfaction condition, C_s. A solution is correct if it satisfies the satisfaction conditions of all relevant constraints.

Once the student's solution is evaluated, the student model passes information to the pedagogical module which generates the appropriate feedback. If any constraints are violated, SQL-tutor will provide feedback on them. In the case where the solution is correct or the student requires a new problem to work on, the pedagogical module uses the information from the student model to select an appropriate problem.

SQL-Tutor provides feedback on demand only, when the student submits the solution. The system offers six levels of feedback, differing in the amount of detail provided to the student. On the first attempt, the system only informs the student whether the solution is correct or not. All other feedback levels provide feedback on errors. The second level (*Error Flag*) points to the part of the solution that is incorrect. The third level (*Hint*) provides a description of one error, pointing out where exactly the error is, what constitutes the error (performing blame allocation) and referring the student to the underlying domain principle that is violated (revising student's knowledge). The hint message comes directly from the violated constraint. The automatic progression of feedback levels ends at the hint level; to obtain higher levels of feedback, the student needs to explicitly request them. For example, the student can ask for the hint message for all violated constraints (*All Errors*), a partial solution (showing the correct version of one part of the solution that is wrong), or a complete solution for the problem.

Within SQL-Tutor, several types of gaming behavior can be observed. As in many other intelligent tutors (cf. [1, 5, 16, 28]), systematic guessing is one way for students to game the system and complete problems. Figure 1 presents a sequence of submissions extracted from a student's interaction with SQL-Tutor, in which the

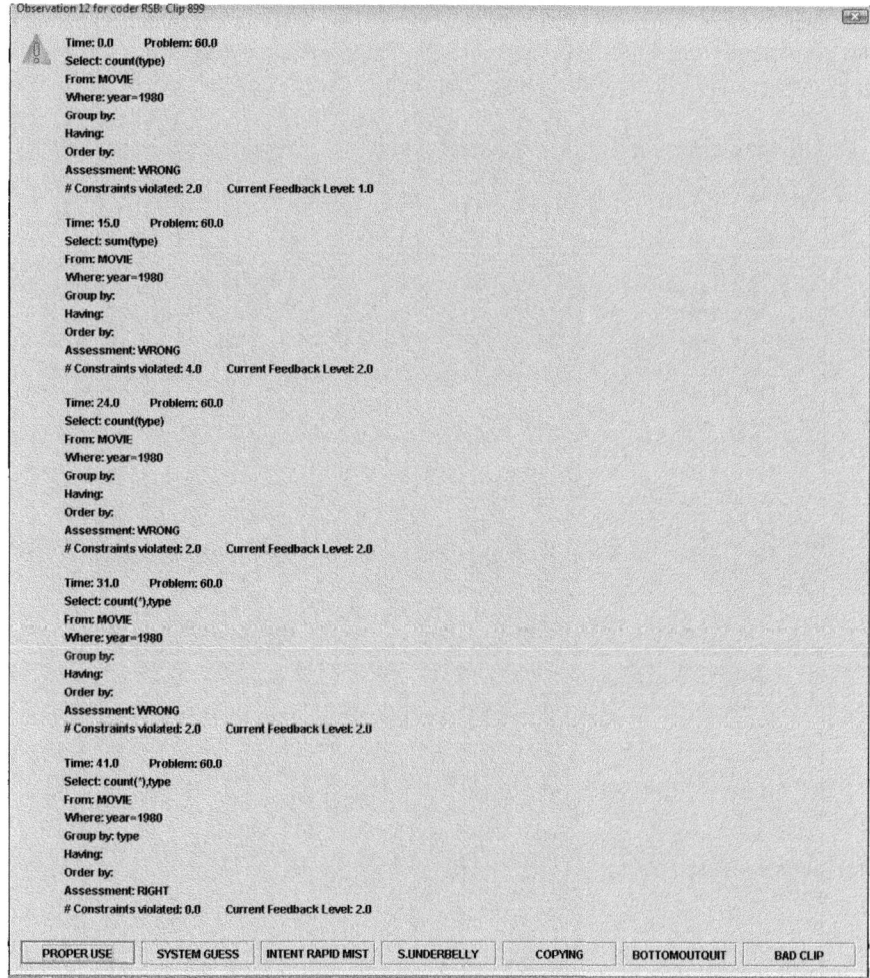

Fig. 1. A "text replay" of student interaction with SQL-Tutor showing systematic guessing

student appears to be repeatedly guessing in order to get the answer without having to think through why the answer should be what he/she is typing in. The student submitted an initial solution for problem 60, and after being informed that the solution is wrong, simply changed the function used in the SELECT clause. After being informed that the modified solution is still wrong, the student converted back to the previous solution.

However, as discussed, SQL-Tutor handles hints in a different way than many other intelligent tutoring systems, and hence, hint abuse occurs differently. Notably, hint abuse is replaced by requesting the full solution (feedback level 5), reading it and then quitting the problem (BOTTOM-OUT QUIT, illustrated in Fig. 2), a behavior also reported in [24].

Sometimes students ask for full solutions, and simply copy them (COPYING) into their own solution which they then submit. It has been previously found that students who consistently use high-level help in SQL-Tutor frequently employ a "guess then copy" strategy [17]. It is worth noting that both of these strategies could be followed

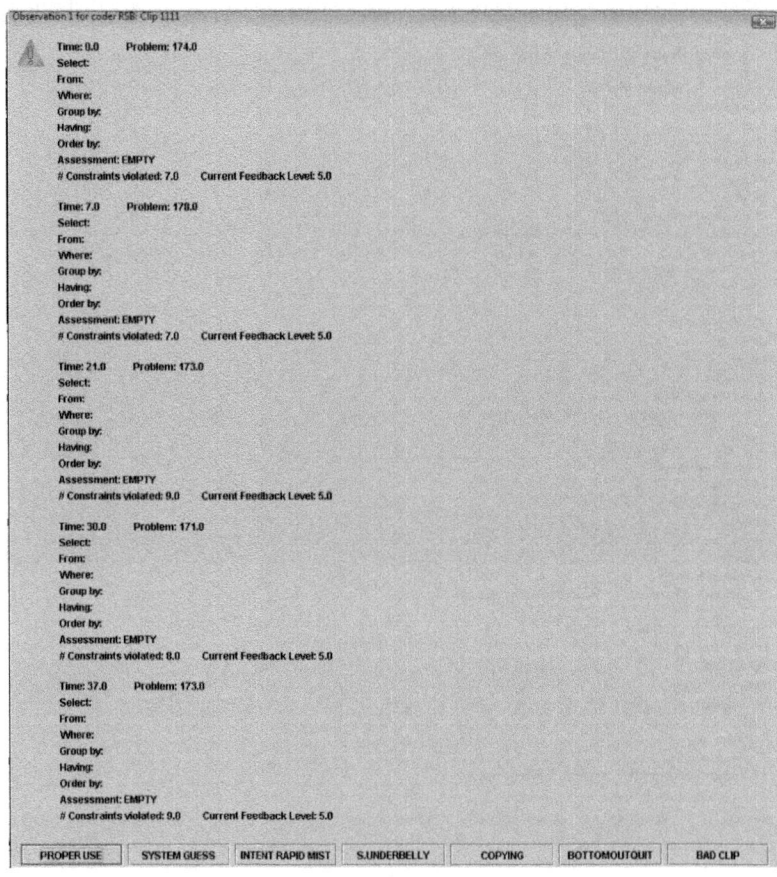

Fig. 2. An example of the "BOTTOM-OUT QUIT" behaviour

by self-explaining after reading the answer (cf. [25]). However, within SQL-Tutor's behavior logs, it is difficult to distinguish self-explanation from gaming, due to the lack of step-level reification. Other common gaming behaviors include intentional entering the same answer or a blank answer. In previous analyses within SQL-Tutor, these attempts, labeled as requests for help, have been manually distinguished from valid forms of help-seeking [17]. Intentional rapid mistakes have also been seen in the Andes learning environment [21]. Some students alternatively game the system through "soft underbelly" strategies, repeatedly switching problems to try to find an easier problem to work on.

3 Data and Data Labeling

Data was collected from 61 students using SQL-Tutor in 2003 as part of an introductory database course. The goal of the 2003 study was to investigate two different problem-selection strategies [20]. The students used SQL-Tutor over multiple days (M=2.4, SD=1.8) in sessions of varying length, up to three hours long. Students completed an average of 19.3 problems (SD=18.0), attempting to answer each problem an average of 3.7 times (SD=1.7). The total data set consisted of 4428 attempts to solve problems. It should be noted that the number of solution attempts per problem can be expected to be significantly lower in SQL-Tutor than in many other intelligent tutors, as SQL-Tutor requires complete answers (e.g. complete SQL queries) for each problem, rather than reifying each problem step individually. Hence, our data set consisted of roughly an order of magnitude less data per student-time than has been used in previous detectors of gaming behavior trained or developed in more reified intelligent tutors (cf. [5, 8, 16, 28]).

Log files were labeled via text replays with reference to whether the student was gaming [6, 7]. Text replays represent a segment of student behavior from the log files in a textual ("pretty-printed") form. A sequence of actions of a pre-selected duration (in this case, sets of 5 problem-solving attempts) is shown in a textual format that gives information about the actions and their context. In the portion of a text replay shown in Figs. 1 and 2, the coder sees each action's time (relative to the first action in the clip), the problem number, the input entered, how the system assessed the action (correct, incorrect), the number of constraints violated, and the current feedback level. The coder can then choose one of a set of behavior categories. Text replays provide limited information on student behavior, and require that the coder understand the user interface, in order to interpret contextual information. However, text replays offer several advantages: text replays can be classified extremely quickly (between 9 and 40 seconds per label (cf. [6, 7]), achieve acceptable inter-rater reliability [6], can be generated automatically from existing log files, agree well with assessments generated automatically by gaming detectors trained on field observation data [6], and have been previously used to train detectors of gaming the system [7].

Text replays were conducted by two labelers (the second and third authors) in two rounds. In the first round, a small set of text replays were conducted and inter-rater reliability was poor – Cohen's Kappa was 0.38 in differentiating gaming behavior from non-gaming behavior and 0.27 in distinguishing all behaviors from each other (e.g. treating different gaming behaviors as separate categories). After the first round, the two labelers recoded the observations where they had disagreed together and discussed their interpretation of the categories, and then conducted additional labeling. In the second set of text replays, Cohen's Kappa was 0.80 in differentiating gaming behavior from non-gaming behavior and 0.68 in distinguishing all behaviors from each other. In total, a randomly selected 637 sequences of student behavior were coded, accounting for 2297 of the problem-solving attempts.

According to the labels, 38% of the sequences of behavior within SQL-Tutor involved one or more of the categories of gaming the system. This is substantially more gaming than has been seen in previous observational or text-replay based research on the frequency of gaming the system in other intelligent tutors (cf. [4, 7, 13, 28]). This finding is likely to be at least partially due to differences in the logs – while the labeling method was identical to the one used in [7], the different grain-size of the logs likely altered the proportion of gaming observations. For instance, non-gaming behavior during successful intermediate steps of problem-solving is not reified in SQL-Tutor, and thus would not show up in the logs. By contrast, if gaming behavior occurred at the same point in the problem-solving process, it would appear in the logs.

However, it is not clear that the difference in logs accounts for all the difference in gaming frequency; in particular, most of the gaming behavior seen in SQL-Tutor consists of behaviors not seen in the previously studied learning environments. Within SQL-Tutor, requesting the answer and then copying it in, and requesting the answer and then quitting were the most common categories of gaming behavior (respectively accounting for 49.6% and 33.6% of gaming behavior noted). Intentional rapid mistakes, systematic guessing, and soft underbelly strategies were significantly rarer (accounting for 8.6%, 5.7%, and 2.4% of gaming behavior). This suggests that interface design differences may explain some proportion of the greater gaming observed in SQL-Tutor. In particular, the high frequency of answer-request-based gaming strategies may be due to their relative ease of execution, compared to other gaming strategies. It is also possible that SQL-Tutor's lack of reification increases difficulty, and therefore the motivation to game the system; the relatively brief problem scenarios in SQL-Tutor may also contribute to the incidence of gaming behavior (cf. [8]). Another possible explanation for greater gaming in SQL-Tutor may be the setting of use of SQL-Tutor. Whereas the previous environments studied were utilized in classrooms with teachers and other students present (potentially reducing opportunity to game the system), SQL-Tutor is used by students alone in their dorm rooms, or in computer labs where other students generally are not using SQL-Tutor.

4   Detecting Gaming

Once the text replays were obtained, we developed detectors of gaming the system, differentiating gaming behavior from all other categories of behavior (the following section describes our attempts to detect individual gaming behaviors).

The first step was to distil a set of features of student behavior from the log files, which could potentially be predictive of the choice to game the system. In doing so, we built off of prior efforts to detect gaming the system in Cognitive Tutors [5] and ASSISTments [28]. 40 features were distilled for each action in the log files for use by the classification algorithms, including features involving details about the action (Was it correct or incorrect? Was it the first attempt at the problem? Was the answer empty or a repeat of the previous answer? What was the current help level?), assessments of student knowledge (probability of acquisition of the constraints involved in the problem according to Bayesian Knowledge Tracing – (cf. [12]); how much progress is the student making towards the correct answer with each problem-solving attempt), information about the time the student took on this step (absolute, and relative to other students), and information about the student's previous interaction behaviors (How often has this student gotten this step wrong in the past?).

Having done this, the next task was to determine the grain size of the models of gaming. Past work with developing detectors of gaming the system from text replay data chose the approach of labeling all actions in a text replay according to the majority label [7]. However, since a set of attempts is labeled as gaming if any gaming at all occurs, this approach explicitly labels some non-gaming actions as gaming, and therefore necessarily reduces the assessed accuracy.

Instead, we decided to detect gaming at the grain-size of the sequence of actions labeled rather than at the grain-size of individual attempts. This approach has the disadvantage of delaying the software's response to gaming by 2-4 problem-solving attempts (until the entire 5 action window has completed), but has the advantage of increasing the precision of the detector and its validation (by better aligning exactly what was labeled with the detection process). To this end, for each labeled sequence we averaged each of the distilled variables together (in a running tutor this could be computed as a running window for each variable).

Detection of gaming was conducted in RapidMiner [18]. All reported validation is batch 6-fold cross-validation, at the student level (e.g. detectors are trained on five groups of students and tested on a sixth group of students). By cross-validating at this level, we increase confidence that detectors will be accurate for new groups of students.

Several algorithms were tried within RapidMiner, including J48 decision trees (as in [7, 28]), step regression (as in [5]), support vector machines, naïve bayes, and bagged decision stumps. We assessed the classifiers using five metrics. First, we used A' [14]. A' is the probability that if the detector is comparing two sequences, one involving gaming and one not involving gaming, it will correctly identify which sequence is which. A' is equivalent to both the area under the ROC curve in signal detection theory, and to W, the Wilcoxon statistic [14]. A model with an A' of 0.5 performs at chance, and a model with an A' of 1.0 performs perfectly. In these analyses, A' was used at the level of action sequences rather than students, a different grain-size than used in [5] or [7]. Second and third, we used precision/recall (at the level of action sequences), as assessments of whether the detector was appropriately balancing between identifying gaming and avoiding false positives. Fourth, we used Kappa (at the level of action sequences). Kappa assesses whether the detector identifies the correct action sequences as gaming, better than chance. A Kappa of 0 indicates that the detector performs at chance, and a Kappa of 1 indicates that the detector performs perfectly. Fifth, we used the correlation between a student's observed frequency of gaming the system and their predicted frequency of gaming the system, to assess whether the detector correctly determines how much each student gamed.

The step regression model produced the best performance overall (full details on other algorithms attempted are not given, due to space limitations). The model is:

```
G = 0.752 + 0.187(correct) - 0.187(incorrect)
- 0.752(blank answer) - 0.345 (repeat answer from previous)
+ 0.121 (number of last 8 attempts that were blank or repeat answers)
- 0.034(number of times database switched in last 3 problems)
- 0.012(number of errors on this problem so far)
- 0.222 (pct. of problems abandoned in last 3 sessions)
+ 0.1(total number of problems abandoned)
+ 0.519(first attempt at current problem)
- 0.133 (lowest probability known of any constraint relevant to problem)
+ 0.225 (average probability known of all relevant constraints)
- 0.083 (time taken on action, in SD from average time for all students
    attempting this problem-all attempts)
- 0.017 (time taken on action, in SD from average time for all students
    attempting this problem-first attempts only)
- 0.245 (current feedback level)
+ 0.081 (most frequent feedback level for this student)
- 0.178 ("Where" clause used in correct solution to current problem)
+ 0.044 (average number of attempts needed to reach correct answer on
    current problem, across all students)
```

Each feature is averaged across the previous five solution-attempts. The resultant value of *G* is thresholded, using the original cut-off used during training: values *below* 0.5 denote gaming, values equal to or above 0.5 denote non-gaming.

For this model, A' equaled 0.77, meaning that the detector could distinguish a sequence of behavior involving gaming 77% of the time. Precision = 76% and Recall = 87%, indicated fairly balanced performance, though with some bias in favor of indicating gaming. Kappa = 0.36, indicating that the detector was 36% better than chance at identifying gaming (Kappa for detecting gaming in Cognitive Tutors was previously found to be 0.40, a comparable value). The correlation between a student's observed frequency of gaming the system and their predicted frequency of gaming the system was 0.73, better than previous action-level detectors of gaming behavior in Cognitive Tutors (where r=0.44). While this detector was clearly not perfect, this level of accuracy is similar to that of previous detectors used in effective interventions given to gaming students (e.g. [2, 3]).

5   Differentiating Between Types of Gaming Behavior

Having developed a reasonably effective detector of gaming behavior overall, in this section we discuss our attempts to differentiate between types of gaming behavior. We attempted to differentiate gaming behaviors in three fashions: differentiating all categories at once, differentiating all of the gaming categories from each other in a sample not including any data originally labeled as non-gaming, and differentiating each gaming category from all other categories. In all cases, the same cross-validation method was used as above.

In differentiating all categories at once, the most effective algorithm was J48. The step regression approach used above was not possible, due to the multiple output classes. Kappa was a relatively poor 0.164 (correlation cannot be calculated for a multi-class problem, and calculating multi-class A' for J48 decision trees is highly non-trivial). The model was most accurate at capturing non-gaming behavior (Precision=69%, Recall = 78%). Performance was somewhat above zero at capturing quitting after requesting the answer (Precision = 27%, Recall = 19%) and requesting the answer and copying it in (Precision = 28%, Recall = 27%). Precision and recall were 0 for all other classes.

In differentiating each of the gaming categories from each other (in a data set omitting behavior originally labeled as non-gaming), the most effective algorithm was J48. Kappa was a relatively poor 0.142. The classes best captured were quitting after requesting an answer (Precision = 49%, Recall = 48%) and requesting the answer and copying it in (Precision = 58%, Recall = 63%). The other classes had Precision and Recall under 15%.

In capturing single gaming categories, the most effective algorithm in all cases was J48 (performing better than step regression). The classes best captured were once again quitting after requesting the answer (A'=0.76, Kappa = 0.22, Precision = 36%, Recall = 26%) and requesting the answer and copying it in (A' = 0.61, Kappa = 0.14, Precision = 32%, Recall = 25%). The model was not able to achieve Precision or Recall over zero for any of the other classes.

Considering that the two categories involving requesting the complete answer were predicted better than the other categories, it may be worth separating out the gaming involving answer requests and the gaming not involving answer requests. If we attempt to differentiate answer-request-based gaming from all other behavior, performance is relatively decent, though still not as good as the performance for identifying gaming in general (A'=0.72, Kappa = 0.30, Precision = 55%, Recall = 47%). If we compare gaming involving answer requests to two other categories, gaming not involving answer requests, and non-gaming behavior, Kappa drops to 0.235. The precision and recall for the unified answer-request gaming category is also worse in this case (Precision = 49%, Recall = 43%).

In general, it is worth considering why machine learning was so successful at identifying gaming the system at a general level, but unable (for the most part) to successfully distinguish between types of gaming beyond distinguishing answer requests from other types of gaming. One possibility is that different gaming behaviors, while looking different within the text replays, are actually quite similar in terms of how they appear in our feature set – occurring in similar contexts, and manifesting in similar ways (e.g. lots of rapid errors or help requests). It therefore may be possible to do better at distinguishing types of gaming by improving the feature set, potentially by including a greater breadth of low level information on the responses, to try to create a closer match between the information used by the labelers and the information used by the machine learning algorithm.

At the same time, it is useful information that the gaming behaviors that were in themselves most easily distinguishable were the two most common gaming behaviors. Hence it may simply be that the other gaming behaviors were not common enough in the data set to form the basis of an effective single-behavior gaming detector. In the whole data set, there were only 13 observations of systematic guessing, 5 observations of soft underbelly strategies, and 17 observations of intentional rapid mistakes, and these behaviors were clustered in small numbers of students. Hence, replicating with a larger data set may improve detection of some of the rarer gaming behaviors.

Another reason why distinguishing gaming may be difficult is that many clips showed evidence for more than one kind of gaming. Commonly, these clips showed strong evidence for one type of gaming, while having weaker evidence for another type of gaming – for instance, when a student tried two answers quickly and then repeated the second answer multiple times until the system gave the correct answer. In this case, the coders both identified this sequence of actions as intentional rapid mistakes, but it is understandable that the machine learning algorithm might incorrectly label what type of gaming was predominant in this clip.

6 Conclusions

In this paper, we have presented an accurate detector of gaming the system for SQL-Tutor, a constraint-based tutor. This marks the first detector of gaming for an intelligent tutoring system that does not reify student thinking at the "inner loop" of problem-solving performance (e.g. [27]) – where students are only evaluated at the level of complete answers, rather than being evaluated at multiple steps of the problem-solving process. Constraint-based tutors differ from other types of intelligent tutors for which gaming detectors have been developed in other fashions as well: student answers are assessed according to a larger number of finer-grained constraints, and feedback is split into levels rather than an entire help sequence being available at any time, though complete answers are available upon request. Correspondingly, gaming behaviors differ, with high frequency for behaviors such as requesting the answer, copying it in, and quitting.

The detector we present, developed using a combination of text replays and educational data mining/machine learning methods, assesses gaming behavior across sequences of five problem-solving attempts. It can accurately identify both when a student is gaming (action-level A' = 0.77, Precision = 76%, Recall = 87%, Kappa = 0.36), and how much each student games the system (correlation = 0.73), performance in line with previous detectors of gaming the system. As such, it should be usable to drive interventions when students game the system (cf. [2, 3]). The detector was also able to differentiate gaming behavior involving directly requesting the answer from other types of gaming behavior, albeit with lower success.

However, detector performance was considerably worse at differentiating between each type of gaming behavior, potentially because gaming behaviors occur in similar contexts and share similarities in their features (e.g. fast actions, requesting the answer). One possibility is that the feature set, in only looking at high-level differences between responses, may be missing more subtle patterns that human coders were able to utilize to identify different types of gaming. It is worth noting that there has been previous success at a different type of differentiation, differentiating the same gaming behaviors in different situations (such as gaming on poorly known skills versus well-known skills; or gaming and then self-explaining) [5, 25]. At the same time, this type of differentiation also might be more difficult to produce in SQL-Tutor, due to the higher-level logging (which occurs because intermediate problem-solving steps are not reified). For instance, Shih and colleagues [25] were able to differentiate gaming from self-explaining after a bottom-out hint because of the short time until the next reified problem-solving step observed in most cases; this same type of differentiation would not be possible within SQL-Tutor's logs.

However, an alternate possibility is that in this study the rarer categories were poorly identified due to a lack of sufficient training examples. Hence, one potential future step would be to obtain more labels of those categories. In doing this, it might be valuable to sample the data for human labeling in a fashion biased against action sequences which have a high confidence of being non-gaming or involving the two more common gaming behaviors (a process analogous to automated biased sampling procedures – (cf. [22])).

Acknowledgements. The development of SQL-Tutor was supported by the University of Canterbury research grant. We thank all members of ICTG for their support. This research was supported by the Pittsburgh Science of Learning Center (National Science Foundation) via grant "Toward a Decade of PSLC Research", award number SBE-0836012.

References

1. Aleven, V., McLaren, B., Roll, I., Koedinger, K.: Toward meta-cognitive tutoring: A model of help seeking with a Cognitive Tutor. Artificial Intelligence and Education 16, 101–128 (2006)
2. Arroyo, I., Ferguson, K., Johns, J., Dragon, T., Meheranian, H., Fisher, D., Barto, A., Mahadevan, S., Woolf, B.P.: Repairing Disengagement with Non-Invasive Interventions. In: Proc. 13th Int. Conf. Artificial Intelligence in Education, pp. 195–202 (2007)
3. Baker, R.S.J.d., Corbett, A.T., Koedinger, K.R., Evenson, S., Roll, I., Wagner, A.Z., Naim, M., Raspat, J., Baker, D.J., Beck, J.E.: Adapting to When Students Game an Intelligent Tutoring System. In: Ikeda, M., Ashley, K.D., Chan, T.-W. (eds.) ITS 2006. LNCS, vol. 4053, pp. 392–401. Springer, Heidelberg (2006)
4. Baker, R.S., Corbett, A.T., Koedinger, K.R., Wagner, A.Z.: Off-Task Behavior in the Cognitive Tutor Classroom: When Students Game The System. In: Proc. ACM CHI 2004: Computer-Human Interaction, pp. 383–390 (2004)
5. Baker, R.S.J.d., Corbett, A.T., Roll, I., Koedinger, K.R.: Developing a Generalizable Detector of When Students Game the System. User Modeling and User-Adapted Interaction 18(3), 287–314 (2008)
6. Baker, R.S.J.d., Corbett, A.T., Wagner, A.Z.: Human Classification of Low-Fidelity Replays of Student Actions. In: Ikeda, M., Ashley, K.D., Chan, T.-W. (eds.) ITS 2006. LNCS, vol. 4053, pp. 29–36. Springer, Heidelberg (2006)
7. Baker, R.S.J.d., de Carvalho, A.M.J.A.: Labeling Student Behavior Faster and More Precisely with Text Replays. In: Proc. 1st Int. Conf. Educational Data Mining, pp. 38–47 (2008)
8. Baker, R.S.J.d., de Carvalho, A.M.J.A., Raspat, J., Aleven, V., Corbett, A.T., Koedinger, K.R.: Educational Software Features that Encourage and Discourage Gaming the System. In: Proc. 14th Int. Conf. Artificial Intelligence in Education, pp. 475–482 (2009)
9. Beal, C.R., Qu, L., Lee, H.: Mathematics motivation and achievement as predictors of high school students' guessing and help-seeking with instructional software. Journal of Computer Assisted Learning 24, 507–514 (2008)
10. Beck, J.: Engagement tracing: using response times to model student disengagement. In: Proc.12th Int. Conf. on Artificial Intelligence in Education, pp. 88–95 (2005)

11. Cheng, R., Vassileva, J.: Design and evaluation of an adaptive incentive mechanism for sustained educational online communities. User Modeling and User-Adapted Interaction 16(3/4), 312–348 (2006)
12. Corbett, A.T., Anderson, J.R.: Knowledge tracing: Modeling the acquisition of procedural knowledge. User Modeling and User-Adapted Interaction 4, 253–278 (1995)
13. Gobel, P.: Student Off-task Behavior and Motivation in the CALL Classroom. International Journal of Pedagogies and Learning 4(4), 4–18 (2008)
14. Hanley, J.A., McNeil, B.J.: The Meaning and Use of the Area under a Receiver Operating Characteristic (ROC) Curve. Radiology 143, 29–36 (1982)
15. Heilman, M., Eskenazi, M.: Language Learning: Challenges for Intelligent Tutoring Systems. In: Proc. Workshop on Intelligent Tutoring Systems for Ill-Defined Domains, Proc. 8th Int. Conf. Intelligent Tutoring Systems (2006)
16. Johns, J., Woolf, B.: A Dynamic Mixture Model to Detect Student Motivation and Proficiency. In: Proc. 21st National Conference on Artificial Intelligence (AAAI-06), pp. 163–168 (2006)
17. Mathews, M., Mitrovic, A.: How does students' help-seeking behaviour affect learning? In: Woolf, B.P., Aïmeur, E., Nkambou, R., Lajoie, S. (eds.) ITS 2008. LNCS, vol. 5091, pp. 363–372. Springer, Heidelberg (2008)
18. Mierswa, I., Wurst, M., Klinkenberg, R., Scholz, M., Euler, T.: YALE: Rapid Prototyping for Complex Data Mining Tasks. In: Proc. 12th ACM SIGKDD International Conference on Knowledge Discovery and Data Mining (KDD 2006), pp. 935–940 (2006)
19. Mitrovic, A.: An Intelligent SQL Tutor on the Web. Artificial Intelligence in Education 13(2), 173–197 (2003)
20. Mitrovic, A., Martin, B.: Evaluating adaptive problem selection. In: De Bra, P.M.E., Nejdl, W. (eds.) AH 2004. LNCS, vol. 3137, pp. 185–194. Springer, Heidelberg (2004)
21. Murray, R.C., Van Lehn, K.: Effects of dissuading unnecessary help requests while providing proactive help. In: Proc. 12th International Conference on Artificial Intelligence in Education, pp. 887–889 (2005)
22. Palmer, C.R., Faloutsos, C.: Density biased sampling: an improved method for data mining and clustering. In: Proc. 2000 ACM SIGMOD Int. Conf. Management of Data, pp. 82–92 (2000)
23. Rodrigo, M.M.T., Baker, R.S.J.d., Lagud, M.C.V., Lim, S.A.L., Macapanpan, A.F., Pascua, S.A.M.S., Santillano, J.Q., Sevilla, L.R.S., Sugay, J.O., Tep, S., Viehland, N.J.B.: Affect and Usage Choices in Simulation Problem Solving Environments. In: Proc. 13th Int. Conf. Artificial Intelligence in Education, pp. 145–152 (2007)
24. Schofield, J.W.: Computers and Classroom Culture. Cambridge University Press, Cambridge (1995)
25. Shih, B., Koedinger, K., Scheines, R.: A Response Time Model for Bottom-Out Hints as Worked Examples. In: Proc. 1st Int. Conf. on Educational Data Mining, pp. 117–126 (2008)
26. Tait, K., Hartley, J., Anderson, R.C.: Feedback procedures in computer-assisted arithmetic instruction. British Journal of Educational Psychology 43, 161–171 (1973)
27. VanLehn, K.: The Behavior of Tutoring Systems. International Journal of Artificial Intelligence in Education 16(3), 227–265 (2006)
28. Walonoski, J.A., Heffernan, N.T.: Detection and Analysis of Off-Task Gaming Behavior in Intelligent Tutoring Systems. In: Ikeda, M., Ashley, K.D., Chan, T.-W. (eds.) ITS 2006. LNCS, vol. 4053, pp. 382–391. Springer, Heidelberg (2006)
29. Wood, H., Wood, D.: Help Seeking, Learning, and Contingent Tutoring. Computers and Education 33, 153–169 (1999)

Bayesian Credibility Modeling for Personalized Recommendation in Participatory Media

Aaditeshwar Seth[1], Jie Zhang[2], and Robin Cohen[3]

[1] Department of Computer Science and Engineering, IIT Delhi, India
[2] School of Computer Engineering, Nanyang Technological University, Singapore
[3] School of Computer Science, University of Waterloo, Canada
{aseth}@cse.iitd.ernet.in

Abstract. In this paper, we focus on the challenge that users face in processing messages on the web posted in participatory media settings, such as blogs. It is desirable to recommend to users a restricted set of messages that may be most valuable to them. Credibility of a message is an important criteria to judge its value. In our approach, theories developed in sociology, political science and information science are used to design a model for evaluating the credibility of messages that is user-specific and that is sensitive to the social network in which the user resides. To recommend new messages to users, we employ Bayesian learning, built on past user behaviour, integrating new concepts of context and completeness of messages inspired from the strength of weak ties hypothesis, from social network theory. We are able to demonstrate that our method is effective in providing the most credible messages to users and significantly enhances the performance of collaborative filtering recommendation, through a user study on the digg.com dataset.

1 Introduction

In the context of participatory media where web messaging is becoming increasingly prevalent, users are faced with a plethora of messages to view. Current techniques such as RSS feeds are not personalized and users often have to sift their way through hundreds of messages each day. In this paper, we aim to show how artificial intelligence techniques can be effectively introduced in order to assist users in their processing of messages. Our central theme is that fields such as sociology, political science and information science can be instrumental in developing a model for recommending credible messages to users. In particular, the modeling of a user's social network becomes a critical element and the approach of learning about each specific user's messaging preferences is essential in the successful recommendation of messages. We outline the motivating multi-disciplinary research, present our model for determining the credibility of messages to users and then introduce experimental results from a user study on the digg.com dataset (where users view and rate messages), to confirm the value of our proposed approach and its use in recommender systems.

Various researchers have proposed to model credibility as a multi-dimensional construct. Fogg and Tseng [1] reason about credibility criteria used by people

P. De Bra, A. Kobsa, and D. Chin (Eds.): UMAP 2010, LNCS 6075, pp. 279–290, 2010.

to judge the credibility of computerized devices and software, and propose to include the modeling of (a) first-hand experience, (b) bias of a user towards categories of products, and (c) third-party reports about products. A model with similar distinctions is developed in [2] to evaluate the trustworthiness of users in an e-commerce setting. Here, the authors distinguish *witness reputation* (i.e. general public opinion) from *direct reputation* (i.e. opinion from a user's own experience) and include as well *system reputation* (i.e. the reputation from the role of a user, as buyer, seller or broker). These interacting users are modeled as being embedded in a social network of relationships that may be pre-declared or inferred based on the past history of interactions.

From sociology, the *strength-of-weak-ties* hypothesis [3] states that social networks of people consist of clusters with *strong* ties among members of each cluster, and *weak* ties linking people across clusters. Whereas strong ties are typically constituted of close friends, weak ties are constituted of remote acquaintances. The hypothesis claims that weak ties are useful for the diffusion of information and economic mobility, because they connect diverse people with each other. People strongly tied to each other in the same cluster may not be as diverse.

One among many studies based on the *strength-of-weak-ties* hypothesis, [4] traces the changes in political opinion of people before and after the 1996 presidential elections in USA, observed with respect to the social networks of people. It is shown that weak ties (identified as geographically dispersed ties of acquaintances) are primarily responsible for the diffusion of divergent political opinion into localized clusters of people having strong ties between themselves. As indicated by the *strength-of-weak-ties* hypothesis, this reflects that local community clusters of people are often homogeneous in opinion, and these opinions may be different from those of people belonging to other clusters. Furthermore, people have different propensities to respect opinions different from those of their local community members. This reflects that the personal characteristics of people also influence the extent to which they would be comfortable in deviating from the beliefs of their immediate local cluster.

From these studies, we learn that (a) there is value to look at the special case of third-party reporting within a user's cluster or local community, and (b) it is important to allow users to have different weights on the importance of different types of credibilities. Note that this last insight is reinforced by studies in information science [5], which argue that users have different preferences for different types of credibilities discussed so far. Inspired by these studies, we develop and operationalize a multi-dimensional subjective credibility model for participatory media as described next.

2   Bayesian Credibility Model

Knowledge Assumptions: Suppose that we wish to predict whether a message m_k about a topic t and written by user u_j, will be considered credible by user u_i. We consider a scenario where all older messages about topic t written in the past are labeled with the author of each message. In addition, a message may have also been assigned ratings by various recipient users, whenever users would have read the message, based on the credibility of the message for the recipient. The set of credibility ratings of any message are also assumed to be available.

Users may declare a subset of other users as their "friends". We refer to an explicitly declared relationship between two users as a *link* between them, and assume to have knowledge of the social network graph formed by all users and the links between pairs of users. Users may also declare topics of interest to them. We use this information, and the social network graph, to derive the *topic specific social network graph* for topic t, as the induced subgraph of the overall social network graph consisting only of those users and edges between users who are interested in topic t.

For each topic specific social network graph, community identification algorithms such as [6] can identify dense clusters of users and links. We use the definition of *strong* and *weak* ties proposed by [3], and refer to *strong* ties as links between users in the same cluster, and *weak* ties as links between users in different clusters. We use V_{it} to denote the local cluster of users strongly tied to user u_i with respect to topic t.

These assumptions are reasonable in contexts such as the website digg.com, which allows users to construct social networks by declaring some users as their friends. Information about message authorship and ratings given by users to messages is also available. We will show that we can use this knowledge to quantify different types of credibilities for each message with respect to each user. Then, based on ratings given by a particular user to older messages, we can use a Bayesian model to learn preferences of the user towards these different kinds of credibilities of messages. Finally, we can use this learned model to predict whether or not the new message m_k will be considered credible by user u_i.

Bayesian Network: We use the notion of strong and weak ties to develop two characteristics of messages: *context* and *completeness*. We assume that strong ties of a user, ie. close friends in the same social network cluster, share the same context, and hence their opinions contribute to the context of a message. On the other hand, completeness is assumed to be influenced by public opinion and not just the immediate social network cluster Based on this premise, the different types of credibilities that we choose to model are as follows:

- s_{ikt} = *cluster credibility*: This is based on the ratings given by other users in cluster V_{it}, that is, the cluster of user u_i. It denotes the credibility associated by the cluster or local community of u_i to the message m_k written by u_j, based on the belief of the members of the cluster about m_k. We assume that opinions of users in the same cluster will contribute only to adding context to messages; their contribution to completeness is already accounted for through public credibility explained next.
- p_{kt} = *public credibility*: This is based on ratings by all the users, and reflects the public opinion about the credibility for the message m_k written by u_j. Public credibility contributes only to the completeness of messages across all

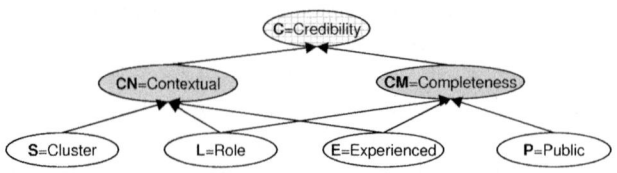

Fig. 1. Bayesian Credibility Model

users, including the users who's opinions have already been accounted in the cluster credibility construct.

- e_{ikt} = *experienced credibility*: This is based only on ratings given by user u_i in the past, and denotes the credibility that u_i associates with the message m_k written by u_j, based on u_i's self belief about u_j. We distinguish between the contributions experienced credibility would make to adding context to the message, or adding completeness.
- l_{ikt} = *role based credibility*: This denotes the credibility that u_i associates with the message m_k written by users having the same role as that of u_j; for example, based on whether the messages' authors are students, or professors, or journalists, etc.

Each of these credibilities can be expressed as a real number $\in [0,1]$, and we propose a Bayesian network to combine them into a single credibility score. The model is shown in Fig. 1. Our aim is to learn the distribution for $P_{it}(\mathbf{C}|\mathbf{E,L,S,P})$ for each user and topic based on ratings given by various users to older messages; here, $\{\mathbf{E,L,S,P}\}$ are evidence variables for the four types of credibilities for a message, and $\mathbf{C}$ is a variable denoting the credibility that u_i associates with the message. Thus, for each topic t, a set of messages M about t will be used during the training phase with samples of $(c_{ik}, e_{ik}, l_{ik}, s_{ik}, p_k)$ for different messages $m_k \in M$ to learn the topic specific credibility models for u_i. Assuming that a user's behavior with respect to preferences for different kinds of credibilities remains consistent over time, the learned model can now be used to predict c_{ix} for a new message m_x about topic t, that is, $P_{it}(c_{ix}|e_{ix}, l_{ix}, s_{ix}, p_x)$. We also introduce two hidden variables, to help make the model more tractable to learn, and to capture insights about messages that we developed in prior work [7] – context and completeness, defined as follows:

- *Context* relates to the ease of understanding of the message, based on how well the message content explains the relationship of the message to its recipient. *Simplification* of the meaning of the message [8], can be considered as an outcome of the amount of *context* in the message.
- *Completeness* denotes the depth and breadth of topics covered in the message. The *scope* of the message, or the *opinion diversity* expressed in the message [8], can be considered as outcomes of the degree of *completeness* of the message.

Note that our modeling method has some interesting design features: the model takes into account personal and contextual opinions of people that may influence their credibility judgements; the model is learned in a personalized manner, and allows accommodating varying degrees of propensities of users to respect opinions of other users; different model instances are learned for different topics, making credibility judgements topic-specific.

3   Credibility Computation

We begin with the following axioms:

- *A-1*: A message is credible if it is rated highly by credible users.
- *A-2*: A user is credible if messages written by her are rated highly by other credible users.
- *A-3*: A user is also credible if ratings given by her are credible, that is, she gives high ratings to messages that appear to be credible to credible users, and low ratings to messages that appear to be non-credible.
- *A-4*: A user is also credible if she is linked to by other credible users.

We henceforth assume that we are operating within some topic t, and drop the subscript for simplicity. We begin with the following information:

- **A[k,n]**: A matrix for k messages and n users, where $a_{ij} \in \{0,1\}$ indicates whether message m_i was written by u_j
- **R[k,n]**: A ratings matrix for k messages and n users, where $r_{ij} \in \{0,1\}^2$ indicates the rating given to message m_i by user u_j
- **N[n,n]**: A social network matrix where $n_{ij} \in \{0,1\}$ indicates the presence or absence of a link from user u_i to user u_j. We also assume that the clustering algorithm can identify clusters of strong ties among users, connected to other clusters through weak ties.

Our goal is to find a method to compute the evidence variables for the Bayesian model using the axioms given above. The evidence variables can be expressed as the matrices **E[n,k]**, **L[n,k]**, **S[n,k]**, and **P[k]**, containing the credibility values for messages. Here, p_k is the public credibility for message m_k authored by user u_j. e_{ij} and l_{ij} are the experienced and role based credibilities respectively for message m_k according to the self-beliefs of user u_i. Similarly, s_{ij} is the cluster credibility for message m_k according to the beliefs of the users in u_i's cluster V_i. Once these evidence variables are computed for older messages, they are used to learn the Bayesian model for each user. Subsequently, for a new message, the learned model for a user is used to predict the credibility of the new message for the user. We begin with computation of the evidence variable matrix for public credibility **P**; we will explain later how other credibilities can be computed in a similar fashion. Detailed algorithms can be found in [9].

---

² We assume in this paper that the ratings are binary. However, our method can be easily generalized to real-valued ratings as well.

1. Let $\mathbf{P'[n]}$ be a matrix containing the public credibilities of users, and consider the credibility of a message as the mean of the ratings for the message, weighted by the credibility of the raters (A-1): $p_k = \sum_i r_{ki} \cdot p'_i / |r_{ki} > 0|$. This is the same as a matrix multiplication $\mathbf{P} = \mathbf{R}_r \cdot \mathbf{P'}$, where $\mathbf{R}_r$ is the row-stochastic form of $\mathbf{R}$, ie. the sum of elements of each row = 1.

2. The credibility of users is calculated as follows:

2a. Consider the credibility of a user as the mean of the credibilities of her messages (A-2): $p'_i = \sum_k p_k / |p_k|$ (or written as $\mathbf{P'} = \mathbf{A}_c^T \cdot \mathbf{P}$), where $\mathbf{A}_c$ is the column-stochastic form of $\mathbf{A}$; and $\mathbf{A}_c^T$ is the transpose of $\mathbf{A}_c$.

2b. The above formulation indicates a fixed point computation:

$$\mathbf{P'} = \mathbf{A}_c^T \cdot \mathbf{R}_r \cdot \mathbf{P'} \tag{1}$$

Thus, $\mathbf{P'}$ can be computed as the dominant Eigenvector of $\mathbf{A}_c^T \cdot \mathbf{R}_r$. This formulation models the first two axioms, but not yet the ratings-based credibility (A-3) and social network structure of the users (A-4). This is done as explained next.

2c. Perform a fixed-point computation to infer the credibilities $\mathbf{G[n]}$ acquired by users from the social network (A-4):

$$\mathbf{G} = (\beta \cdot \mathbf{N}_r^T + (1-\beta) \cdot \mathbf{Z}_c \cdot \mathbf{1}^T) \cdot \mathbf{G} \tag{2}$$

Here, $\beta \in (0,1)$ denotes a weighting factor to combine the social network matrix $\mathbf{N}$ with the matrix $\mathbf{Z}$ that carries information about ratings given to messages by users. We generate $\mathbf{Z}$ by computing z_i as the mean similarity in credibility ratings of user u_i with all other users. The ratings similarity between a pair of users is computed as the Jacquard's coefficient of common ratings between the users. Thus, z_i will be high for users who give credible ratings, that is, their ratings agree with the ratings of other users (A-3). In this way, combining the social-network matrix with ratings-based credibility helps to model the two remaining axioms as well. Note that $\mathbf{Z}_c[n]$ is a column stochastic matrix and $\mathbf{1[n]}$ is a unit column matrix; augmenting $\mathbf{N}$ with $\mathbf{Z}_c \cdot \mathbf{1}^T$ provides an additional benefit of converting $\mathbf{N}$ into an irreducible matrix so that its Eigenvector can be computed[3]

2d. The ratings and social network based scores are then combined together as:

$$\mathbf{P'} = (\alpha \cdot \mathbf{A}_c^T \cdot \mathbf{R}_r + (1-\alpha) \cdot \mathbf{G}_c \cdot \mathbf{1}^T) \cdot \mathbf{P'} \tag{3}$$

Here again $\mathbf{1}$ is a unit column matrix, and $\alpha \in (0,1)$ is a weighting factor. The matrix $\mathbf{P'}$ can now be computed as the dominant Eigenvector using the power method.

3. Once $\mathbf{P'}$ is obtained, $\mathbf{P}$ is calculated in a straightforward manner as $\mathbf{P} = \mathbf{R}_r \cdot \mathbf{P'}$.

The processes to compute cluster $\mathbf{S[n,k]}$, experienced $\mathbf{E[n,k]}$, and role based $\mathbf{L[n,k]}$ credibilities are identical, except that different cluster credibilities are calculated with respect to each cluster in the social network, and different experienced and role based credibilities are calculated with respect to each user.

---

[3] This step is similar to the Pagerank computation for the importance of Internet web pages [10].

The cluster credibilities $\mathbf{S[n,k]}$ are computed in the same manner as the public credibilities, but after modifying the ratings matrix $\mathbf{R}$ to contain only the ratings of members of the same cluster. Thus, the above process is repeated for each cluster, modifying $\mathbf{R}$ in every case. For each users u_i belonging to cluster V_i, s_{ik} is then equal to the cluster credibility value for message m_k with respect to u_i. The matrix $\mathbf{Z}$ in the computation on the social network matrix is also modified. When computing the cluster credibilities for cluster V_i, element z_j of $\mathbf{Z}$ is calculated as the mean similarity of user u_j with users in cluster V_i. Thus, z_j will be high for users who are regarded credible by members of cluster V_i because their ratings agree with the ratings of the cluster members.

The experienced credibilities $\mathbf{E[n,k]}$ are computed in the same manner as well, but this time for each user by modifying the ratings matrix $\mathbf{R}$ to contain only the ratings given by the user. The matrix $\mathbf{Z}$ is also modified each time by considering z_j as the similarity between users u_i and u_j, when calculating the experienced credibilities for u_i.

Role based credibility is computed as the mean experienced credibilities of users having the same role. However, we do not use role based credibility in our evaluation because sufficient user profile information was not available in the digg dataset used by us. Henceforth, we ignore $\mathbf{L[n,k]}$ in our computations.

Model Learning: Once the various types of credibilities for messages are calculated with respect to different users, this training data is used to learn the Bayesian model for each user and topic of interest to the user using the Expectation-Maximization (EM) algorithm. The model parameters are learned to predict for user u_i interested in topic t, the probability $\mathrm{P}_{it}(c_{ix}|e_{ix}, s_{ix}, p_x)$ that u_i will find a new message m_x to be credible.

Inference: Now, for a new message m_x, the evidence variables are calculated with respect to a recipient user u_i in one of two ways as described next, and the learned model is used to produce a probabilistic prediction of whether u_i would find m_x to be credible.

- *Authorship*: The four types of credibilities of the message are considered to be the same as the credibilities of its author with respect to u_i.
- *Ratings*: The cluster and public credibilities are calculated as the weighted mean of ratings for the message given by other users and the credibilities of these users with respect to u_i. The experienced and role based credibilities are the same as the corresponding credibilities of the message author wrt u_i.

As we will show in the evaluation, the ratings method performs better than the authorship method. This allows new users to popularize useful messages written by them because their own credibility does not play a role in the computations. It also allows credible users to make mistakes because the credibility of the author is not taken into account. Given the evidence variables for the new message, and the learned Bayesian model, the probability of u_i finding the message to be credible is computed using standard belief propagation methods such as Markov-Chain-Monte-Carlo (MCMC).

4 Evaluation

We evaluate our method over a dataset of ratings by real users obtained from a popular knowledge sharing website, digg.com [11]. The website allows users to submit links to news articles or blogs, which are called *stories* by the website. Other users can vote for these stories; this is known as *digging* the stories. Stories that are *dugg* by a large number of users are promoted to the front-page of the website. In addition, users are allowed to link to other users in the social network. Thus, the dataset provides us with all the information we need:

- Social network of users: We use this information to construct the social network link matrix between users $\mathbf{N[n,n]}$. The social network is clustered using MCL, a flow-stochastic graph clustering algorithm [6], to produce classifications of ties as strong or weak. The cluster of users strongly connected to user u_i is referred to as V_i.
- Stories submitted by various users: We use this information to construct the authorship matrix $\mathbf{A[k,n]}$. Since all the stories in the dataset were related to technology, we consider all the stories as belonging to a single topic.
- Stories dugg by various users: We use this information to construct the ratings matrix $\mathbf{R[k,n]}$. We consider a vote of 1 as an evidence for credibility of the story, and a vote of 0 as an evidence of non-credibility.

Although the dataset is quite large with over 200 stories, we are able to use only 85 stories which have a sufficiently large number of ratings by a common set of users. This is because we require the same users to rate many stories so that we have enough data to construct training and test datasets for these users. Eventually, we assemble a dataset of 85 stories with ratings by 27 users. A few assumptions we make about the validity of the dataset for our experiments are as follows:

- The submission of a story to Digg may not necessarily be made by the author of the story. However, we regard the submitting user as the message author because it distinguishes this user from other users who only provide further ratings to the messages.
- The ratings provided on the Digg website may not reflect credibility, but rather usefulness ratings given to messages by users. We however consider them to be equivalent to credibility and do not include users who rate more than 65 stories as all credible or all non-credible. We argue that in this pruned dataset, all the users are likely to be interested in the topic and hence all the stories; therefore, the only reason for their not voting for a story would be its credibility.

We use an open-source package, OpenBayes, to program the Bayesian network. We simplify the model by discretizing the evidence variables $\mathbf{E,S,P}$ into 3 states, and a binary classification for the hidden variables $\mathbf{N}$, $\mathbf{M}$, and the credibility variable $\mathbf{C}$. The discretization of the evidence variables into 3 states is performed by observing the Cumulative Distribution Frequency (CDF) and Complementary CDF (CCDF) of each variable with respect to the credibility rating of users. The lower cutoff is chosen such that the product of the CDF for rating=0 and CCDF for rating=1 is maximum, and the upper cutoff is chosen such that the CCDF for rating=0 and CDF for rating=1 is maximum. This gives a high discrimination ability to the classifier because the cutoffs are selected to maximize the pair-wise correlation of each evidence variable with the credibility rating given by the user.

Metrics: We evaluate the performance of the model for each user by dividing the 85 stories into a training set of 67 stories and a test set of 17 stories (80% and 20% of the dataset respectively). We then repeat the process 20 times with different random selections of stories to get confidence bounds for the cross validation. For each evaluation, we use two kinds of performance metrics [12], *Matthew's correlation coefficient* (MCC) and *TPR-FPR*. The MCC gives a single metric for the quality of binary classifications. TPR-FPR plots on an XY-scale the true positive rate (TPR) with the false positive rate (FPR) of a binary classification. The random baseline is TPR=FPR. Points above the random baseline are considered to be good.

All experiments are performed with $\alpha = 0.5$ (eqn. 3) and $\beta = 0.85$ (eqn. 2) which were found to be robust values [9], and also convey our message that all of authorship, ratings, and social networks provide valuable credibility information.

Inference Methods: Figure 2 shows the TPR-FPR plot for ratings and authorship based evidence variable computation when $\alpha = 0.5$ and $\beta = 0.85$. As can be seen visually, the ratings-based method performs better than the authorship-based method. The former gives MCC = 0.156 (σ=0.073), while the latter gives MCC = 0.116 (σ=0.068). However, the authorship performance is still successful for a majority, which is encouraging. This indicates that authorship information may be used to solve the problem of cold-start for new messages that have not acquired a sufficient number of ratings. Similarly, ratings may be used to solve cold-start for new authors who have not acquired sufficient credibility.

Comparison: We next compare our method with other well known methods for trust and reputation computation meant for different applications.

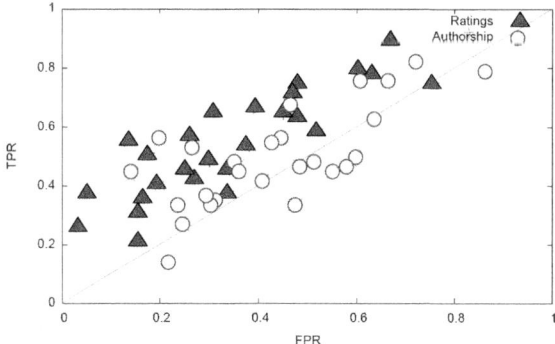

Fig. 2. Performance of Bayesian Credibility Model

An Eigenvector computation on $\mathbf{A}_c^T.\mathbf{R}_r$ by leaving out the social network part (eqn. 1), is identical to the Eigentrust algorithm [13]. The best choice of parameters could only give a performance of MCC = -0.015 ($\sigma = 0.062$). Eigentrust has primarily been shown to work in P2P file sharing scenarios to detect malicious users that inject viruses or corrupted data into the network. The P2P context requires an objective assessment of the trustworthiness of a user, and does not allow for subjective differences, as desired for participatory media.

An Eigenvector computation on the social network matrix (eqn. 2), personalized for each user, is identical to the Pagerank algorithm used to rank Internet web pages [10]. However, this too performs poorly with an MCC = 0.007 ($\sigma = 0.017$). This suggests that users are influenced not only by their own experiences, but also by the judgement of other users in their cluster, and by public opinion.

In conclusion, these and other methods we compared perform close to random, even with personalization. We believe this to be due to a fundamental drawback of these methods: they try to form an objective assessment of credibility for users and messages, which is not appropriate for participatory media. Our approach which subjectively model credibility, allowing users to be influenced in different ways by different sources, perform better than objective modeling approaches.

5   Use in Recommender Systems

Our method for credibility computation can be used in two ways to improve recommender systems: (i) Since our method serves to predict the probability of a user finding a message to be credible or non-credible, it can be used as a pre- or post-filtering stage with existing recommendation algorithms. (ii) It can also be adapted to integrate closely with recommendation algorithms; we show how to do this with collaborative filtering (CF) [14] in this section.

A basic CF algorithm works in two steps. First, similarity coefficients are computed between all pairs of users, based on the similarity of message ratings given by each pair. Second, to make a decision whether or not to recommend a new message to a user, the mean of the message ratings given by other similar users is computed, weighted on the coefficients of similarity to these users. If the mean is greater than a threshold, the message is recommended; else it is rejected.

The drawback of the CF method is that it only learns the average user behavior. However, as we have argued, user behavior can be different in different circumstances. We therefore develop an adaptation of our method. Rather than computing a single similarity coefficient between each pair of users, we compute four similarity coefficients based upon whether messages are believed to be highly contextual by both users, or highly complete by both users, or contextual by the first user and complete by the second user, or vice versa. Essentially, we break down the average user behavior into four components based upon the context and completeness of messages to users, as follows:

1. For each user, we run the EM algorithm on training set to learn the model.
2. We use the learned model to infer the probabilities of the hidden variables of context and completeness for each story in the training set: $P_i(\mathbf{CN}|\mathbf{E,S,P,C})$

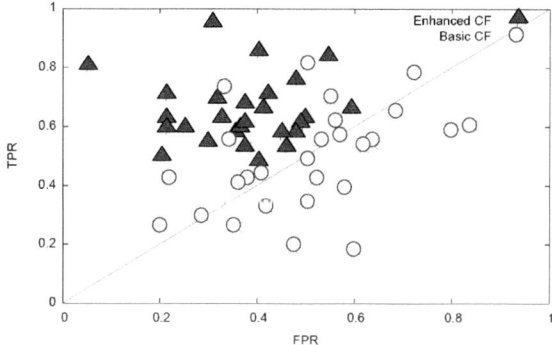

Fig. 3. Enhancement of Collaborative Filtering

and $P_i(\mathbf{CM}|\mathbf{E,S,P,C})$ shown in Fig. 1. That is, for each story m_j, we infer $P(cn_{ji}{=}0,1|e_{ji}, s_{ji}, p_{ji}, c_{ji})$ and $P(cm_{ji}{=}0,1|e_{ji}, s_{ji}, p_{ji}, c_{ji})$.

3. We then discretize the probabilities for $\mathbf{CN}$ and $\mathbf{CM}$ in same way as we did earlier, by finding cutoffs that maximized the product of the CDF for $c_{ji}{=}0$ and CCDF for $c_{ji}{=}1$. This gives us samples of $(c_{ji} \in \{0,1\}$, $cn_{ji} \in \{0,1\}$, $cm_{ji} \in \{0,1\})$, that is, which stories appear contextual or complete to a user, and the rating given by the user to these stories.
4. For every pair of users, their samples are then compared to produce four similarity coefficients on how similar the users are in their contextual opinion, completeness opinion, and cross opinions between messages that appear contextual to one user and complete to the other, or vice versa.
5. Finally, when evaluating the decision to recommend a test message to a user, the mean of the message ratings is computed over all the four coefficients of similarity, rather than over a single coefficient as in the basic CF algorithm.

Figure 3 shows the performance of the basic CF scheme and our enhanced version. The basic scheme performs worse than random for many users, but when enhanced with breaking up the average user behavior into contextual and completeness components, the performance improves considerably. The mean MCC for the basic scheme is 0.017 ($\sigma = 0.086$), and for the enhanced scheme is 0.278 ($\sigma = 0.077$), a sixteen-fold improvement. We consider this to be a huge improvement over the existing methodologies for recommendation algorithms, especially to build applications related to participatory media.

6   Conclusions

In this paper, we made use of insights from sociology, political and information science, and HCI, to propose a subjective credibility model for participatory media content. We formulated the model as a Bayesian network that can be learned in a personalized manner for each user, making use of information about the social network of users and ratings given by the users. We showed that our method works better than existing methods on trust and reputation computation. In addition, an adaptation of our method to recommendation algorithms such as collaborative filtering (CF) was able to improve CF on our dataset. This encourages the use of sociological insights in recommender system research.

References

1. Fogg, B., Tseng, H.: The elements of computer credibility. In: Proceedings of SIGCHI, pp. 80–87 (1999)
2. Sabater, J., Sierra, C.: Regret: A reputation model for gregarious societies. In: Proceedings of the Fifth International Conference on Autonomous Agents Workshop on Deception, Fraud and Trust in Agent Societies, pp. 61–69 (2001)
3. Granovetter, M.: The strength of weak ties. American Journal of Sociology 78(6), 1360–1380 (1973)
4. Baybeck, B., Huckfeldt, R.: Urban contexts, spatially dispersed networks, and the diffusion of political information. Political Geography 21(2), 195–220 (2002)
5. Rieh, S.: Judgement of information quality and cognitive authority on the web. Information Science and Technology 53(2), 145–161 (2002)
6. Dongen, S.: MCL: A Cluster Algorithm for Graphs. PhD Thesis, University of Utrecht (2000)
7. Seth, A., Zhang, J.: A social network based approach to personalized recommendation of participatory media content. In: Proceedings of ICWSM (2008)
8. Bryant, J., Zillman, D.: Media Effects: Advances in Theory and Research. Lawrence Erlbaum Associates, Mahwah (2002)
9. Seth, A.: An infrastructure for participatory media. PhD thesis, University of Waterloo (2009)
10. Brin, S., Page, L.: The pagerank citation ranking: Bringing order to the web. Technical Report Stanford University (2001),
    http://dbpubs.stanford.edu:8090/pub/1999-66
11. Lerman, K.: Social information processing in news aggregation. IEEE Internet Computing 11(6), 16–28 (2007)
12. Davis, J., Goadrich, M.: The relationship between precision-recall and roc curves. In: Proceedings of ICML, pp. 233–240 (2006)
13. Kamvar, S., Scholsser, M., Garcia-Molina, H.: The eigentrust algorithm for reputation management in p2p networks. In: Proceedings of WWW, pp. 640–651 (2003)
14. Adomavicius, G., Tuzhilin, A.: Toward the next generation of recommender systems: A survey of the state-of-the-art and possible extensions. IEEE Trans. Knowledge and Data Engineering 17(6), 734–749 (2005)

A Study on User Perception of Personality-Based Recommender Systems

Rong Hu and Pearl Pu

Human Computer Interaction Group
Swiss Federal Institute of Technology in Lausanne (EPFL)
CH-1015, Lausanne, Switzerland
{rong.hu,pearl.pu}@epfl.ch

Abstract. Our previous research indicates that using personality quizzes is a viable and promising way to build user profiles to recommend entertainment products. Based on these findings, our current research further investigates the feasibility of using personality quizzes to build user profiles not only for an active user but also his or her friends. We first propose a general method that infers users' music preferences in terms of their personalities. Our in-depth user studies show that while active users perceive the recommended items to be more accurate for their friends, they enjoy more using personality quiz based recommenders for finding items for themselves. Additionally, we explore if domain knowledge has an influence on users' perception of the system. We found that novice users, who are less knowledgeable about music, generally appreciated more personality-based recommenders. Finally, we propose some design issues for recommender systems using personality quizzes.

Keywords: Recommender System, Personality, Domain Knowledge, User Study, User Modeling.

1 Introduction

Recently, researchers suggested that personality characteristics can be used to build user profiles in recommender systems, inspired by the findings in psychological studies [8]. Studies have indicated that there is a significant connection between personality and people's tastes and interests. For example, Rentfrow and Gosling [14] revealed that musical preferences are associated with individual differences in personality, ability and self-perception. Moreover, due to their ability of lessening the cold start problem associated with commonly adopted collaborative filtering recommender systems, personality-based recommender systems are increasingly attracting the attention of researchers and industry practitioners [5].

Based on different recommendation recipients, recommender systems can be classified into two groups: one suggests items to the active users; the other predicts items for the people who don't directly interact with the system. The former has been studied widely in the literature [1]. The latter is known as gift finders that help users identify ideal gift for others, such as their friends [3].

P. De Bra, A. Kobsa, and D. Chin (Eds.): UMAP 2010, LNCS 6075, pp. 291–302, 2010.

The absence of recipients' interaction with systems makes recommendation more difficult, since the information which can be acquired to predict recipients' preferences is not as sufficient as that for active users. To resolve this issue, intelligent gift finders were developed based on recipients' personal characteristics, such as gender, age, occasion, life styles, and personalities [3]. Even though an increasing number of well-known online shopping websites, e.g., Yahoo shopping, Ebay.com, have tried to incorporate intelligent gift finders into their systems, the subject remains an open area of study in this field. Due to the important role of gift finders in commercial websites, it is believable that the related studies could greatly benefit the e-commerce society.

Our previous research showed that using personality quizzes is a viable and promising way to build users' profiles for the recommendation of entertainment products [8]. The tested personality quiz based system was preferred to a baseline rating-based system mainly due to its prominent merit on ease of use. In the present study, we are trying to gauge the values of using personality quizzes to build profiles not only for an active user but also his or her friends as a personality-based gift finder. On the other hand, prior research has revealed that prior domain knowledge is a crucial attribute which influences the way in which users interact with a system and further impacts their perceptions [11, 12]. We therefore are wondering whether the prior domain knowledge have an influence on users' perception to the personality-based recommender system in both cases.

In this paper, we first proposed a general personality-based recommendation method in music domain on the basis of the results reported in [14]. To our knowledge, no previous work has explicitly dealt with this issue in the literature. Then, an in-depth within-subject user study was conducted to compare two ways of building user profiles using personality quizzes in recommender systems: for active users and for their friends. We further explored the influence of domain knowledge on user perceptions of the tested personality-based recommender system in both cases. The evaluation criteria include objective ratings of the recommended songs and subjective measurements on perceived usefulness, perceived ease of use, attitudes and behavior intentions [4]. The results reveal several designing issues for recommender systems using personality quizzes.

The remainder of this paper is organized as follows. In Sect. 2 we present background and related work; then we describe the method of user modeling and musical preference inference framework in Sect. 3. And we describe our user study in detail on experiment setup, hypotheses, produce and design, experiment results and discussion in Sect. 4 followed by conclusion and future work.

2   Background and Related Work

2.1   Personality-Based Recommender Systems

According to Burger [2], personality is defined as a "consistent behavior pattern and intrapersonal processes originating within the individual". It is relatively stable and predictable. Studies also show that personalities influence human decision making process and interests [9, 14]. Drawing on the inherent inter-related patterns between users' personalities and their interests/behaviors, personality-based recommenders are designed to provide personalized services. Lin and Mcleod [10] proposed a temperament-based filtering model incorporating human factors, especially human temperaments (Keirsey's theory), into the processing of an information recommendation service. Their model categorizes the information space into 32 temperament segments. Combining with the content-based filtering technique, their method aims at recommending the information units which best matched both users' temperaments and interests. Even though the system utilizes personalities to model user profiles, they don't really take the psychological relation between human personalities and information items into account. In [13], authors applied the relation between musical preferences and personality traits found in [14] to recommend music. However, they didn't explain how it works in detail. In the current work, we propose a general algorithm for personality-based music recommender systems on the basis of the results reported in [14]. It can be easily generalized to other domains.

2.2 Personality and Musical Preferences

In the landmark work [14], according to the statistical results on a large-scale dataset, Rentfrow and Gosling found four musical preferences groups: *reflective and complex*(e.g., blues, jazz, classical and folk); *intense and rebellious* (e.g., rock, alternative, and heavy metal music); *upbeat and conventional* (e.g., country, religious, and pop music); and *energetic and rhythmic* (e.g., rap/hip-hop, soul/funk, and electronic/dance music). Most importantly, they empirically revealed that the four musical preferences are not only associated with the level of complexity, emotionality and energy of musical compositions, but also individual differences in personality, ability and self-perception.

More specifically, the fascinating pattern of links between musical preferences and personality (Big-Five model was used to measure personality in the work) revealed from the correlation analysis showed that the reflective and complex dimension is positively related to Openness to New Experience; the intense and rebellious dimension was positively related to Openness to New Experiences; the upbeat and conventional dimension has positive correlations with Extraversion, Agreeableness and Conscientiousness, and a negative correlation with Openness to New Experience; the energetic and rhythmic dimension is positively related to Extraversion and Agreeableness. More explanations can refer to [14].

2.3 Influence of Prior Domain Knowledge

Research has shown that prior domain knowledge is a critical attribute which influences the way in which users interact with a system and further impacts their perceptions [6, 11, 12]. It has been revealed that domain expertise enhances search performance. Expert users with a higher level of domain knowledge tend to find information in a more flexible and efficient way [11, 12]. However, domain novice users rely more on the simple searching functions of the information-seeking tools systems provide. On the other hand, research has indicated that domain novice users can get more help on reinforcing their knowledge when using information systems [12]. Contrastingly, domain expert users require more advanced information to satisfy their advanced needs. We are trying to investigate its influences in recommender systems.

3   User Modeling and Musical Preference Inference

Generally, the recommendation problem can be formalized as follows [1]. Let U be the set of all users and let P be the set of items or categories to be recommended (In our application, it is the set of musical preference dimensions). PF is defined as a prediction function that measures the possibility of one item p is liked by user u, i.e., $PF : U \times P \to R$, where R is a totally ordered set (e.g., nonnegative integers or real numbers within a certain range). Then for each user $u \in U$, we want to choose such item $p'_u \in P$ that maximizes the inferred preference value. More formally,

$$\forall u \in U, p'_u = \text{argmax}_{p \in P} PF(u, p). \tag{1}$$

In the following, we propose a general algorithm framework for inferring liked musical preference in terms of user personalities. The possibility of one musical dimension p is liked by user u is predicted by considering two factors: the personality of user u and the relations between the personality and four musical preference dimensions. More specifically, we present personality characteristics described in the Big-Five model as a vector $\mathbf{ps_u} = (ps_u^o, ps_u^c, ps_u^e, ps_u^a, ps_u^n)^T$ for user u. Here, $ps_u^o, ps_u^c, ps_u^e, ps_u^a$ and ps_u^n represent the values in the dimension of openness to new experience, conscientiousness, extraversion, agreeableness and neuroticism respectively. Their values are normalized to the range [-1, 1]. The user preference model is described as $\mathbf{mp_u} = (mp_u^{rc}, mp_u^{ir}, mp_u^{uc}, mp_u^{er})^T$, where, $mp_u^{rc}, mp_u^{ir}, mp_u^{uc}$ and mp_u^{er} represent the extent to which the user u like reflective and complex, intense and rebellious, upbeat and conventional, and energetic and rhythmic music respectively. Therefore, user preference model $\mathbf{mp_u}$ can be calculated as,

$$\mathbf{mp_u} = W \times \mathbf{ps_u}$$

where,

$$W = \begin{bmatrix} w_{11} & \cdots & w_{15} \\ \vdots & \ddots & \vdots \\ w_{41} & \cdots & w_{45} \end{bmatrix}.$$

W is a 4-by-5 weighting matrix. The value of w_{ij} means the strength of the relation between personality trait j and musical preference dimension i, and w_{ij} is also normalized in the range of [-1,1]. The positive value represents a positive relationship between a personality trait and a musical preference dimension. That is, a user have a higher value in the personality trait j, will like the music preference dimension i with a higher possibility, vice versa. On the other hand, the negative w_{ij} indicates a negative relationship between personality trait j and musical preference dimension i. The magnitude of w_{ij} represents the strength of such relations. The larger is this value, the stronger dominates the personality trait j on the musical preference dimension i. In our prototype system, we assign w_{ij} with the correlation value between music preference dimension i and personality trait j reported in Rentfrow.

Then, the musical preference dimension mp_u^k, $k \in \{rc, ir, uc, er\}$, with the maximal value is the one which user u might most like. As an alternative to increase recommendation diversity, the musical preference dimensions whose corresponding values in $\mathbf{mp_u}$ are more than a defined threshold r are picked out and the number of recommended songs in these dimensions are in proportion to the predicted preference values. In our study, we adopted the second strategy.

4 Experiment: User Study

We conducted a user study aiming to compare two ways of using personality quizzes to build user profiles in recommender systems, with the help of an music recommender implemented based on the algorithm described above. We also investigated the influence of domain knowledge on user perceptions of recommender systems in both cases.

4.1 Hypotheses

As mentioned above, it is greatly difficult to select an item as gift, since the recipient's preferences are often not well known. We assume that the personality-based recommendation technology using personality quizzes to build profiles is more appreciated by users when it is used to suggest songs as gifts, compared to the situation when it is used to recommend songs for users themselves. Additionally, as we can imagine, if one user is an expert on music, he or she could easily choose songs to listen to. Contrastively, it is reasonable to assume that domain novice users with a low level of knowledge on music can get more help from our system. Therefore, they would give higher evaluation scores. On the other hand, we assume expert users can also perceive the usefulness when they try to find songs for their friends, considering the lack of knowing well about their friends' preferences on music. We therefore formulate the following two hypotheses:

H1: *The personality quiz based music recommender system is perceived more positively when it is used to recommend music for friends compared to when it is used to recommend music for users themselves.*

H2: *Users with low level of music knowledge perceive more usefulness of the system when they use personality quizzes to find songs for both themselves and their friends, compared to those with high level of music knowledge.*

H3: *Users with high level of music knowledge can also perceive the usefulness of the system when they use personality quizzes to find songs for their friends.*

4.2   Experiment Setup

We prepared 1,581 songs (1956 - 2009) covering all 14 genres in the four musical preferences. We assigned genre labels to all songs by consulting several popular music websites (last.fm, new.music.yahoo.com, and itunes.com). After finding out user preferred musical preferences using the method described in the previous section, in each preferred musical preference dimension, recommended songs were selected evenly from all included genres and randomly in each genre.

TIPI (Ten Item Personality Inventory) [7] was used to assess users' personalities (Big Five Personality Traits) on a 7-point scale ranging from 1 (strongly disagree) to 7 (strongly agree). The acquisition process takes about a few minutes.

4.3   Experiment Design and Procedure

This user study was conducted in a within-subject design. All subjects were required to use personality quizzes to find songs for themselves and also for one of their friends as gifts. To minimize the carryover effects (both practice and fatigue effects), all subjects were randomly assigned to two experimental conditions. Each condition had a different order in usage scenarios. That is, half the subjects in one condition tried to use personality quizzes to find songs for themselves first and then for one of their friends. The other condition had the reverse sequence. To avoid any possible confusion during evaluation, subjects were told that they would evaluate two systems. One only recommends songs for users themselves and the other for friends.

The user study was launched online. An online procedure containing instructions, evaluated systems and post-study questionnaires was implemented so that participants could follow easily. In the first page of the online user study, participants were debriefed on the objective of the experiment and the upcoming tasks, and then they started the evaluation. The main user tasks include:

1. Answer a list of background question, such as gender, music knowledge etc.
2. Accomplish the TIPI for self/friend.
3. Listen to 20 recommended songs. If these songs are suggested to themselves, subjects are asked to rate them on a 5-point scale from 1 (dislike it very much) to 5 (like it very much). Otherwise, they are asked to send this list to the friend whom these songs are found for. It can be easily done with one function implemented in our online procedure.
4. Fill in a post-questionnaire on subjective perceptions of the evaluated system. Each question is to be answered on a 5-point Likert scale ranging from 1 (strongly disagree) to 5 (strongly agree).

Two versions of post-study questionnaire were designed for two usage scenarios, and they cover 10 measurements. All questions are listed in Table 1. If the questions on one measurement use the same sentence for two scenarios, only one is presented.

Table 1. Subjective evaluation questions (building profile for active users/ for friends)

Measurements	Questions
perceived effectiveness	This website was effective in recommending songs for me. / This website was effective in recommending songs for my friend.
perceived accuracy	The songs suggested to me corresponded to my taste. / I was confident that the songs suggested to my friend correspond to his(her) taste.
perceived helpfulness	The website helped me discover music for myself. / The website helped me to discover music for my friends.
enjoyment	I enjoyed using personality quizzes to get recommendations. / I enjoyed using personality quizzes to find songs for my friends.
ease to use	I found this site easy to use.
satisfaction	I am satisfied with the overall functions of this website.
use intention	If this were a real website, I would use it to get music recommendations. / If this were a real website, I would use it to find songs for my friends.
purchase intention	If necessary, I would buy the recommended songs. / If necessary, I would buy the recommended songs for my friends as gift.
return intention	I will use this type of recommender systems again.
reference intention	I will introduce this website to my friends

4.4 Participants

A total of 80 subjects (32 females) were recruited in our user study. Most of them (69 out of 80) are students at a university in Switzerland and others work in the related fields, including software engineers, designers, music promoters and graphic designers. All subjects listen to music frequently. To make the group as diverse as possible, the subjects were selected from a variety of nationalities (17 countries) and varying levels of music knowledge. The distribution of users' domain knowledge levels is described in the experiment results section. 55 subjects have the experience of using music recommenders. As the reward, all subjects were told to have a chance to win one novel generation iPod Shuffle (4G) valued at 99 CHF in a final lottery draw.

5   Results and Discussion

5.1   Objective Measure

The objective measure aims at evaluating the recommendation quality. All participants were asked to rate the 20 songs recommended for themselves on a 5-point scale from 1 (dislike it very much) to 5 (like it very much). On the other hand, participants were encouraged to send a "gift" with 20 songs to the friends whom they found these songs for, and to ask their friends to rate these "gift" songs on the same scale. Eventually, 21 "friends" accomplished the rating task.

The results show that there is no significant difference between the ratings to the songs recommended to active users themselves and those to the songs recommended to friends (Independent t-test: t = 1.369, p = 0.171). Regarding recommendations based on profiles built by users themselves, on average, 14.14 out of 20 songs are rated as acceptable (rated higher than "it's ok"), and wherein 8.29 songs are considered to be liked. For recommendations based on profiles built by friends, 14.57 out of 20 songs are acceptable, and 9.09 songs are thought to be liked.

5.2   Subjective Measure

The average scores of users' responses to the subjective measurements are shown in Fig. 1. Paired t-test was conducted to find significant differences. As we can see, subjects scored significantly higher on perceived accuracy when finding songs for friends than for themselves (mean = 2.8, SD = 0.925 vs. mean = 3.1, SD = 0.941, respectively; t = -2.287, p = 0.025). Subjects also gave significantly higher scores on enjoying using personality quizzes to find songs for themselves than for friends (mean = 3.6, SD = 0.951 vs. mean = 3.2, SD = 1.178, respectively; t = 3.001, p =0.004), and they are both higher than the median value (3.0). Subjects scored significantly higher on return the system to find songs for themselves than for friends (mean = 3.2, SD = 0.986 vs. mean = 3.0, SD = 1.012, respectively; p = 0.047).

The results show a peak in the perceived ease of use in both scenarios, which indicates that participants strongly felt that this personality quiz-based system is easy to use. However, there is no significant difference between two scenarios (mean = 4.3, SD = 0.89 for finding songs for self vs. mean = 4.25, SD = 0.89 for finding songs for friends; t = 0.49, p = 0.626).

As for the helpfulness of the personality-based music recommender, surprisingly, the difference between finding songs for users themselves and for friends is not significant by paired T-test (mean = 3.1, SD = 1.105 vs. mean = 2.9, SD = 0.917, respectively; t = 1.454, p = 0.15). Similarly, the comparisons on satisfaction (both are slightly higher than 3.5) and behavior intentions show no significant difference.

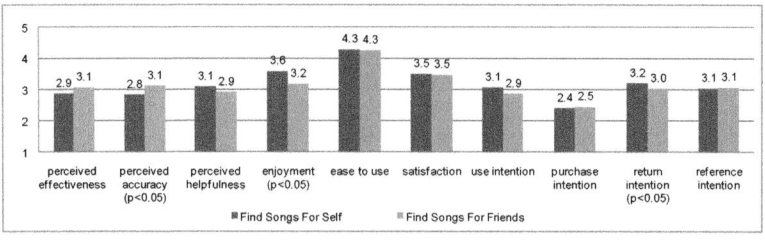

Fig. 1. Subjective evaluation comparison between finding songs for self and for friends

5.3   Influence of Domain Knowledge

Among all participants, 17 subjects strongly agreed that they have knowledge about music (expert users), 32 subjects agreed (medium users), 23 subjects had a neutral opinion (novice users) and 8 subjects didn't think they have any prior knowledge about music. According to [6], users do not have a well defined perception until they reach some degree of expertise. Additionally, only 8 out of 80 subjects thought they didn't have any knowledge of music. Their responses are not sufficient for statistical analysis. We therefore decided to eliminate their responses from our analysis. One-way ANOVA was conducted with prior knowledge levels as IVs and subjective scores as DVs. Then, we conducted post-hoc pairwise comparisons (Bonferroni) to identify how the three levels of prior knowledge varied from one another on the measurements with overall significant differences among three levels.

Figure 2 shows the mean scores of subjective measurements in terms of three different levels of music knowledge given the scenario of finding songs for themselves. The ANOVA results indicate significant differences in participants' perceived effectiveness ($F(2, 69) = 7.173$, $p = 0.001$), perceived accuracy ($F(2, 69) = 3.147$, $p = 0.049$), perceived helpfulness ($F(2, 69) = 6.333$, $p = 0.003$) and use intention ($F(2, 69) = 5.273$, $p = 0.007$). Pairwise comparison results show that medium users scored significantly higher than expert users on perceived effectiveness (mean: 3.34, SD: 0.827 vs. mean: 2.29, SD: 1.105, respectively; $p = 0.001$), perceived accuracy (mean: 3.03, SD: 0.933 vs. mean: 2.35, SD: 0.996, respectively; $p = 0.044$), perceived helpfulness (mean: 3.34, SD: 1.035 vs. mean: 2.29, SD: 0.985, respectively; $p = 0.004$) and use intention (mean: 3.31, SD: 1.061 vs. mean: 2.35, SD: 1.057, respectively; $p = 0.308$). In addition, we found that novice users scored significantly higher than expert users on perceived helpfulness (mean: 3.30, SD: 1.105 vs. mean: 2.29, SD: 0.985, respectively; $p = 0.011$) and use intention (mean: 3.21, SD: 0.951 vs. mean: 2.35, SD: 1.057, respectively; $p = 0.031$).

Figure 3 shows the mean scores of subjective measurements in terms of three different music knowledge levels given the scenario of finding songs for friends. The ANOVA result indicate significant differences in participants' perceived effectiveness scores ($F(2, 69) = 3.490$, $p = 0.036$) and return intention ($F(2, 69) = 3.617$, $p = 0.032$). Pairwise comparison results show that novice users scored significantly higher than expert users (mean: 3.43, SD: 0.843 vs. mean: 2.59,

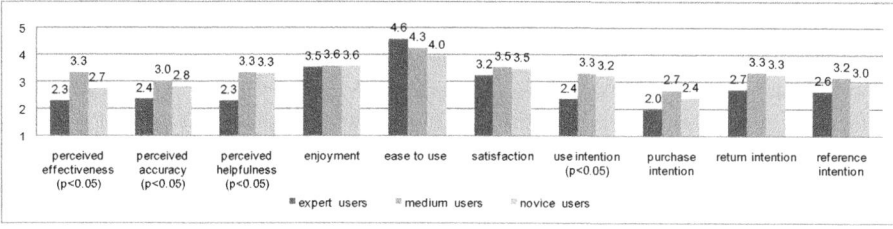

Fig. 2. Subjective evaluation comparison among different music knowledge levels in the scenario of finding songs for self

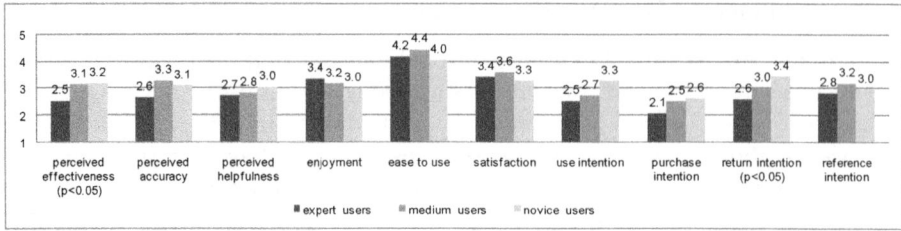

Fig. 3. Subjective evaluation comparison among different music knowledge levels in the scenarios of finding songs for friends

SD: 0.870, respectively; p = 0.028) on return intention. However, no significant differences are found between each two of knowledge levels on perceived effective, although the overall comparison is significantly different among all levels of domain knowledge. Furthermore, we found that domain expert users scored moderately higher on perceived helpfulness when using the system to find songs for friends than doing it for themselves (paired t-test: mean = 2.71, SD = 0.686 vs. mean: 2.29, SD: 0.985, respectively; p = 0.090).

5.4   Discussion

From the results of objective measure, we could see that in both cases, about 3/4 of the recommended songs are rated as acceptable and half of them are rated as to be liked. There is no significant difference between two compared scenarios. Regarding the subjective evaluation, participants, in general, expressed that users enjoyed using personality quizzes to get recommendations and satisfied with the overall functions in both cases. The finding that the personality quiz-based recommender systems were strongly perceived to be easy to use was revalidated in the present study. It is surprising to see that, while active users perceive the recommended items to be more accurate for their friends, they enjoy more using the system to find songs for themselves. It doesn't support our hypothesis *H1* that personality-based technology is more appreciated by users when they try to find songs for friends.

It can be seen that the scores on behavior intentions (intention to use the system, purchase the recommended songs, return to the system and introduce to friends) are low. It suggests some challenges of the acceptance of personality-based recommender systems, especially when they are used as gift finders. One important obstacle is privacy issues. As stated in [13], users were really worried about how their personal information can be used and whom this private information will be shown to. Similar phenomena happened in our experiment. Furthermore, when they were asked to use personality quizzes to find songs for friends, some participants expressed that they were worried about that their counterpart friends would see how they evaluated them, especially on some sensitive measurements, such as neuroticism. These might explain why the scores were pushed down on behavior intentions. This sheds light on design issues on how to reduce such negative effects inherent to personality-based recommender systems. For example, it might be a possible solution to design interfaces with explanations to tell who will see this personal information and what it will be used for, provide more benefits to users in order to trade off the fear of the risk from disclosing personal information [13].

The results also show that prior domain knowledge do influence users' perception of the system. Domain novice and medium users had significantly more positive perceptions than expert users, and further higher intention to use and return to the system. On the other hand, domain expert users scored moderately higher on perceived helpfulness when using the system to find songs for friends than for themselves. Therefore, our hypothesis *H2* is perfectly supported and *H3* is somehow sustained. It is indicated that the way of using personality quizzes to build user profiles cannot satisfy the advanced needs of domain expert users. There is a need to design adaptive personality-based recommender systems. For example, for expert users, we could integrate other modeling methods to increase user control, e.g., leveraging rating-based methods to update user preference models [1]. For novice users, systems with simple operating interfaces and low requirements on domain knowledge could be more helpful.

6   Conclusion and Future Work

With the help of an implemented personality-based music recommender system based on the general algorithm framework we proposed in this paper, an in-depth within subject user study was conducted to investigate user perceptions of two ways of using personality quizzes to build user profiles, for an active user and for his or her friends. We further explored the influences of domain knowledge to the user perception of the system. The results show that users enjoyed using personality quizzes to get recommendations and satisfied with the overall functions in both cases. However, while active users perceive the recommended items to be more accurate for their friends, they enjoy more using the system to find songs for themselves. On the other hand, users with low level of music domain knowledge scored more higher on perceived usefulness than domain expert users. However, expert users perceived slightly more helpful when trying to find songs for friends. These results shed light on several design issues for personality-based recommender systems. As the further work, it is worth comparing the personality-based modeling method with other existed modeling technologies, such as collaborative filtering and content-based methods.

Acknowledgments. We thank the EPFL and Chinese Government for sponsoring the reported research work. We are grateful to the participants for their patience and time.

References

1. Adomavicius, G., Tuzhilin, A.: Toward the Next Generation of Recommender Systems: A Survey of the State-of-the-Art and Possible Extensions. IEEE Trans. Knowledge and Data Eng. 17(6), 734–749 (2005)

2. Burger, J.M.: Personality, 7th edn. Thomson/Wadsworth, Belmont (2008)
3. Chan, F.Y., Cheung, W.K.: A Knowledge-based Recommender System for Customized Online Shopping. In: Proceedings of the World Congress on Mass Customization and Personalization 2001, MCPC'01. HKUST, TUM, Hongkong (2001)
4. Davis, F.D., Bagozzi, R.P., Warshaw, P.R.: User acceptance of computer technology: A comparison of two theoretical models. Management Science 35(8), 982–1003 (1989)
5. Dunn, G., Wiersema, J., Ham, J., Aroyo, L.: Evaluating Interface Variants on Personality Acquisition for Recommender Systems. In: Houben, G.-J., McCalla, G., Pianesi, F., Zancanaro, M. (eds.) UMAP 2009. LNCS, vol. 5535, pp. 259–270. Springer, Heidelberg (2009)
6. Frias-Martinez, E., Chen, S.Y., Macredie, R.D., Liu, X.: The role of human factors in stereotyping behavior and perception of digital library users: a robust clustering approach. Journal of User Modeling and User-Adapted Interaction 17(3), 305–337 (2007)
7. Gosling, S.D., Rentfrow, P.J., Swann Jr., W.B.: A very brief measure of the Big-Five personality domains. Journal of Research in Personality 37, 504–528 (2003)
8. Hu, R., Pu, P.: A comparative user study on rating vs. personality quiz based preference elicitation methods. In: Proceedings of the 13th international conference on Intelligent User Interfaces, pp. 367–372 (2009)
9. Jung, C.: Psychological Types. Princeton, New Jersey (1971)
10. Lin, C., McLeod, D.: Temperament-Based Information Filtering: A Human Factors Approach to Information Recommendation. In: Proc. of the IEEE International Conference on Multimedia and Exposition, vol. 2, pp. 941–944 (2000)
11. Lazonder, A.W., Biemans, H.J.A., Wopereis, I.G.J.H.: Differences between novice and experienced users in searching information on the World Wide Web. Journal of the American Society for Information Science 51(6), 576–581 (2000)
12. Mitchell, T.J.F., Chen, S.Y., Macredie, R.D.: Hypermedia learning and prior knowledge: Domain expertise vs. system expertise. Journal of Computer Assisted Learning 21(1), 53–64 (2005)
13. Perik, E.M., Ruyter, B., de Markopoulos, P., Eggen, J.H.: The Sensitivities of User Profile Information in Music Recommender Systems. In: Proceedings of Private, Security, Trust, pp. 137–141 (2004)
14. Rentfrow, P.J., Gosling, S.D.: The do re mi's of everyday life: The Structure and Personality Correlates of Music Preferences. Journal of Personality and Social Psychology 84, 1236–1256 (2003)

Compass to Locate the User Model I Need: Building the Bridge between Researchers and Practitioners in User Modeling

Armelle Brun[1], Anne Boyer[1], and Liana Razmerita[2]

[1] LORIA-Nancy Université
615, rue du jardin botanique - 54506 Vandœuvre les Nancy
{armelle.brun,anne.boyer}@loria.fr
[2] Copenhagen Business School, CBS, ISV,
Dalgas Have, 15, DK-2000 Frederiksberg, Denmark
lr.isv@cbs.dk

Abstract. User modeling is a complex task, and many user modeling techniques are proposed in the existing literature, but the way these models are presented is not homogeneous, the domain is fragmented and these models are not directly comparable. Thus there is a need for a unified view of the whole user modeling domain and of the applicability of the models to specific applications, contexts or according to specific requirements, type of data, availability of data, etc. A common question companies may ask when they want to build and exploit a user model in order to implement different kinds of personalization or adaptive systems is: "Given my specific requirements, which user modeling technique can be used?". No obvious answer can be given to this question. This article aims to propose a topic map of user modeling in connection with input data, data types, accessibility, approach, specific requirements and users' data acquisition methods. This schema/topic map is aimed to help practitioners and researchers as well to answer the above mentioned question. Furthermore the article provides two concrete scenarios in the area of recommender systems and shows how the topic map may be used for these scenarios and real world applications.

Keywords: user model, user modeling, recommender systems, personalization.

1 Introduction

Adaptive features have proven their value and personalization has become associated with a next generation of web services combining/interacting in a seemingly intelligent manner. These new intelligent services will enable a more personalized experience also known as Web 3.0. Personalized recommendation systems will feed us with news, new music, new products, targeted advertisements, according to preferences, moods, interests of the users etc. Given the increasing demands on personalization in real world applications and in particular on the web, there is a growing need for a classification of the different user modeling techniques, their characteristics and their applicability to different specific contexts and application

P. De Bra, A. Kobsa, and D. Chin (Eds.): UMAP 2010, LNCS 6075, pp. 303–314, 2010.
© Springer-Verlag Berlin Heidelberg 2010

scenarios. User modeling and personalization techniques in real world applications are still in a development phase even though the field has progressed considerably in the last few years and the various techniques have reached a considerable mature level. Personalized services may be vital for organizations in terms of branding and Customer-Relationship Management (CRM) but it can also be an optional feature if it is too expensive.

Personalization, recommender systems, and adaptive features are often dependent on user modeling techniques. Industrial applications have specific requirements and often do not fit in most of the models proposed in the existent research articles. Furthermore user modeling techniques are dependent on specific requirements of various application domains and specific requirements of different application scenarios.

Most of the research articles in the area of recommender systems focus only on one specific feature of user models (e.g. accuracy, privacy, etc.) but many real-cases applications require several features at the same time according to the specific usage scenario. Based on this statement, the question is which user model fits best the requirements of a given application?

This article proposes a topic map of user modeling and presents some concrete scenarios of usage in the area of recommender systems. This schema is necessary and can be used to guide both practitioners and researchers to position each other according to their specific focus of work or/and specific requirements. It will not only guide practitioners to find the best solution given their specific requirements but also guide researchers to place their models among the set of existing models in the user modeling literature.

Section 2 introduces related work in user modeling in the area of recommender systems and constitutes a basis for the construction of the proposed topic map. Section 3 details two real-cases scenarios in the frame of a banking group. Section 4 is dedicated to the presentation of the different concepts of the topic map. Section 5 demonstrates the use of the user modeling topic map for these two real-case scenarios and two models referred in the state of the art. The last section concludes this work and presents perspectives of future work developments.

2   Related Work

The user modeling domain is complex and can contribute to many different types of applications. User modeling can be either an objective in itself, or be exploited in many various application fields (e-commerce, e-learning, targeted advertising, etc.). In the marketing domain, for example, some studies have focused on different client segments – "usually people buying beer might buy other products" – generic profiles of users that can fit in a certain category which can be used to propose new services that meet customers' needs and expectations. User modeling has also been used in other domains, and specifically as a means to achieve better personalization in targeted services, recommender systems (e.g. e-commerce application) and achieve better customer-relationship management.

As it is not realistic to consider all the possible use cases, we decide to focus, in this paper, on user models in the framework of recommender systems. In this application domain, the user model highly influences the characteristics of the recommender; in the same way, the application domain, i.e. recommendation, influences the model.

Several user models have been studied and proposed in the literature in recommender systems. At the same time, several taxonomies of recommender systems have been proposed [13, 8, 12, 18, 11]. The user model component is included in these taxonomies.

However, the proposed taxonomies have different points of view and several precision degrees; the dimensions of the corresponding user models are not easily comparable. In the same way, the user models proposed in studies on recommender systems focus on specific dimensions; some of them being not kept in taxonomies. Some dimensions of user models are nevertheless recurrent in these studies and taxonomies, for example:

- The representation can be: history-based, vector-space, semantic network, user-rating matrix, demographic features, classifier, etc. [12, 13, 19].
- The persistence (also called the term) of the used data can be: short-term/long-term? [16, 17, 10].
- The input (knowledge sources) can be: purchase data, ratings, user factual data, transactional data: explicit (ratings) or implicit (behavior), item factual data, etc. [17, 8, 9].
- The granularity can be: individual or group modeling [16, 14].
- The distribution of the user model can be: centralized or distributed [3].

We can notice that most of the taxonomies are not very elaborated ([17] and [18] for example present a taxonomy with only two-features (dimensions)) and few of them specifically focus on the user model; user modeling is scattered in the whole area of recommender system without being enough elaborated. These taxonomies mainly focus on the following aspects related to the exploitation of the recommender system:

- The recommendation approach (also called recommendation technique or information filtering method) can be: demographical, content-based, knowledge-based, collaborative or hybrid [1, 11, 5, 20].
- The output of the system can be: absolute ratings, top N items, or top M users [1, 17, 13, 19].
- The delivery mode can be: active push, active pull or passive [19].
- The supported task can be: annotation in context, find good items, find all good items, receive sequence of items [7, 12].

In summary, on the one hand we can find some taxonomies of recommender systems that include the user modeling dimension, this user model part is too high level: its dimensions are not precise enough and are not easily comparable. On the other hand, studies on user models in recommender systems focus on a limited number of features. There is thus a lack of a general overview of user models for recommender systems.

In this paper we propose a user modeling topic map that can be used as a reference for both researchers and practitioners. Practitioners could describe their needs and researchers locate their contributions. We also present two concrete application scenarios in the domain of recommender systems, to show the way the topic map can be used.

3 Scenarios

We present in this section two concrete use cases we are currently working on for a multinational banking group. These use cases are presented in terms of industrial requirements, as they have been described by our industrial partner.

- **Scenario 1.** *Personalization of the information provided to a given user.* The banking group offers to all its employees and main customers a document management portal, where thousands of documents are available, many documents dealing with the same or similar topic. The documents are uploaded by bank employees on the portal and no indexing is performed. The question is how to deliver the adequate information to each bank employee. For example, about the financial crisis, even if a customer and a trader are interested in this topic, they are not looking for the same information. Thus, the objective is to determine which document is relevant for each user. The challenge is to help specific users to access the right document, given their specific profile. The requirements of the banking group emphasize that no explicit information should be asked to the users. Relevant documents have to be pushed to the users at the right moment (not through an explicit query from users). The bank wants to guarantee privacy of users whereas they are identified. Of course, as this data is not sensitive, it is stored only during a limited duration.
- **Scenario 2.** *Personalization of the public website for the large audience.* In this scenario, personalization is related to the navigational behavior and tastes of the users. When a user has just read information about stocks, the system should propose him to buy some. Personalization has to be performed for any user who surfs on the website. Typically users are not identified (anonymous), their usage data can be used only during the active session. Afterwards the usage data is stored anonymously. Thus there are no security requirements, data and models can be located anywhere. The website is relatively static, and all the pages have been indexed.

These two real use cases of a single banking group both require a user model, however each scenario leads to a different type of user model. This is the reason why it is very difficult for a practitioner who faces these scenarios to determine which solution(s) to choose: a single model, two different types of models? which one(s)?

So the problem could be summarized as: which kind of user model (in the literature) is adequate for a given scenario? Given the state of the art of user models, no direct answer is available. This paper proposes a topic map of user models that includes the main parameters and characteristics of user models. Then, given a specific scenario and its requirements, we propose to formalize these requirements according to the proposed topic map represented in Fig. 1.

Furthermore, this article proposes to also classify the different user models according to the user modeling topic map. Then a comparison of the requirements of the scenario and the characteristics of the models will be easily possible.

In the following section we present the user model topic map. This topic map is applied in the frame of recommender systems.

4   The User Modeling Topic Map

The user modeling topic map proposed in this paper is not only a compilation of the most important features (privacy, input, location, etc) from the state of the art, it is also a refinement of some of them and some new features are proposed.

This topic map, presented in Fig. 1, is divided into four main dimensions: the approach, the accessibility, the input and the type. All the identified user modeling features are classified along these four main dimensions.

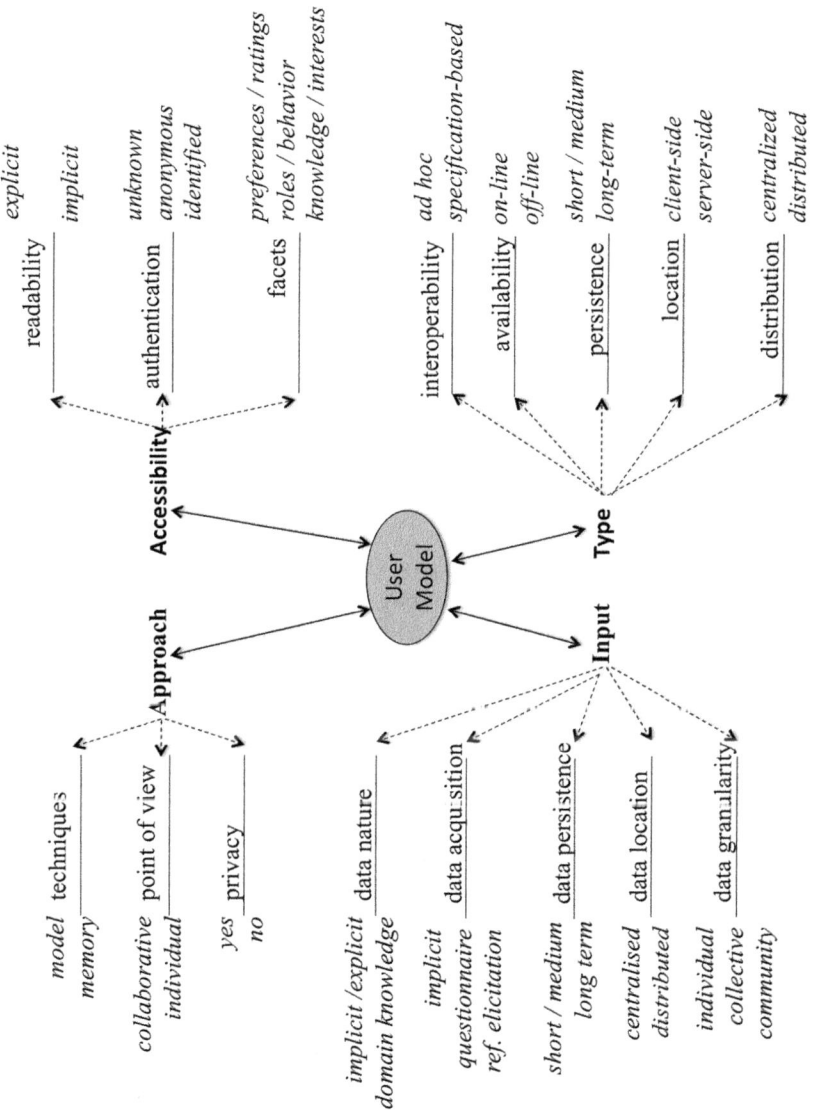

Fig. 1. Topic map of user modeling

4.1 Input

The input dimension represents the characteristics of the data available and used to build the model. This dimension is the basis for building a user model and contains the five following features:

Data acquisition. There exist several ways to acquire data about users to build a user model. 1) implicitly: the data is inferred from the user behavior (the user may not know that his actions are registered, the acquisition process may not interfere with the users' tasks, it is non-intrusive). This acquisition process is system-driven. 2) Explicitly: the data is explicitly given by the user about his interests (topics of interests, ratings, keywords in a request). The topic map we propose refines the explicit data acquisition by dividing it into form and preference elicitation. This acquisition process is user-driven. A proposition of an overview of the methods to acquire knowledge about users is proposed in [6].

Data nature. The nature of the data available is highly dependent of the data acquisition process. Explicit information (role, ratings, preferences) are more likely to be explicitly given by the user by using a form or preference elicitation technique. At the opposite, implicit data is usually usage data. Some data about the domain and users may be known *a priori*, independently of the user, they represent the knowledge *a priori*. Implicit data can be post-processed so as to be transformed into explicit data. Let us notice that there are differences between what a user declares to prefer and what he really likes, as deduced from his behavior. The nature of data has also been studied in [5] where a complex taxonomy of knowledge sources is presented.

Data persistence. The persistence of the data [17] may range from short term (no persistence) to long term (persistence). When short term data is used, information about the current user session is exploited by the model, whereas long term data represents data about the user in the long term. The middle term is an in between term user model where long term data is removed. One can notice here that when no data about users is stored, the resulting user model is independent of the user.

Data location. The data used to learn the model may be stored in several places, either on the server the model will be computed and stored (centralized) or on the client side, *i.e.* on the computer of each user (decentralized). Let us notice that storing data on the client-side is one way to be privacy compliant. For practice reasons, most studies store data on the server side, few of them store on the client side [3].

Data granularity. The data stored to compute the model can have several levels of granularity: it can either have an individual granularity: data is stored for each user, it can also have a collective granularity: data is stored for all the users (the data is less precise), or community granularity: data is stored at the community level (one data represents the users of the community, not a single user.) Let us notice that here again, in the case of community or collective granularity privacy is preserved: information is not stored at the user level. If data is stored at the collective granularity, no personalization can be performed as similar information is stored for each user.

4.2 Accessibility

The accessibility represents the way the model can be interpreted, what it represents.

Readability. A user model can be explicit or implicit. When a model is explicit, a human can read, understand and interpret the model, he can know what are the tastes and habits of a user. When a model is implicit, a human cannot interpret the model and cannot deduce the tastes of the user. When a model is implicit, it can be either autonomous (i.e. domain independent) or domain-dependent (it cannot be used in applications of other domains).

Authentication. According to the authentication, a user model can be: unknown, anonymous or identified. When a user model is unknown or anonymous, we cannot identify which user it corresponds to. When anonymous, we know to which user id it corresponds, but we cannot identify the user (we know only his id). When the model is identified, the identity of the user is known. From the privacy point a view an anonymous or unknown type of user model are more likely to be privacy-enhanced.

Facets. The facets of a user model are what the model represents. It can either model the behavior of a user [2], his roles, preferences, interests, knowledge or ratings. The facet of a model highly depends on the input of the model, and a model can have several facets according to the usage scenario and the personalization techniques.

4.3 Type

Taking into consideration the type, user models can be classified according to its availability, interoperability, persistence, location and distribution.

Availability. A model can be available either on-line or off-line. An on-line model is computed when the user is on-line and when the model is needed. The model is not available at any other moment. The advantage of such a model is that it is always up to date (new data is always included in the model). However the computation of the model may be time consuming and the user model may not be available when required. An off-line user model is computed off-line and is available at any moment. As there is no constraint about the computation time, the associated user model can be more sophisticated and more time-consuming algorithms can be used for this purpose. However, the model is not always up to date, the most recent input may not have been used to build the model. Some models can combine both on-line and off-line algorithms: creating an off-line user model and updating it on-line when new data is available.

Interoperability. A user model can be developed "ad hoc", so it can be used for one specific application and its specific requirements and may thus not comply with existing standards or existing specifications. Such a model is usually non-interoperable. Most of the user models are build "ad-hoc". Contrary to "ad hoc" type of model, user models can be developed to comply with specific standards or specifications so it can be shared or used in other applications [21]. Such standards include Information Management Systems Learner Information Package (IMS LIP) and IEEE PAPI.

Persistence. The persistence of a model represents the continuing existence of the model: it can be long-term (the model is available during a long time), middle-term or a short-term model (are available for short time, e.g. during active session).

Location. The user model can be stored either on the client side (generally for privacy purposes) or on the server side. When it is on the server side, the model can be stored on a single server, or distributed on several servers.

Distribution. A model can be either centralized or distributed. For example, [3] proposes a peer to peer model, in the frame of collaborative filtering, the model is stored on several computers.

4.4 Approach

The approach dimension is made up of the general characteristics that have to be chosen when building a model; the choice of these characteristics highly influences the resulting model.

Technique. The user modeling techniques can be either memory or model. When memory, the input of the system is not pre-processed, the model is made up of the data as it is. At the opposite, when the technique is model, the data is pre-processed to build a model. When a model is computed, it is not only lighter (in terms of space complexity) but also generally less complex in terms of computation time when the model is exploited. Obviously, when the model is memory, it is available off-line as it is made of raw data.

Point of view. When building the model, one can either exploit information about the user the model is dedicated to (individual), or exploiting also information about other users (collaborative), this additional information is used to build a more solid model.

Privacy. When building a user model, the resulting model may be privacy compliant or not. When a model is privacy compliant, it is not possible, given the model, to guess which user it represents. Thus, given a user, and a set of models, no link between them can be made. Contrary, when a model is not privacy compliant, information about a user can be obtained by using a model.

One can notice that many of the concepts of the topic map are linked and some of them are incompatible. For example, a user model cannot be located on the client side and be centralized. Second, some of the characteristics may have several values. For example, a model can be available off-line and on-line (built off-line and then updated online); a model can have several facets: it can represents a user's behavior and roles.

5 Exploitation of the User Modeling Topic Map

Given specific requirements of a usage scenario, a practitioner will ask which user model fits best his/her requirements. Given a user model, a researcher will ask how to compare it to other existing models? In this section we present the way we suggest to exploit the topic map to answer these questions.

We propose to represent the topic map under the form of a grid where the features of the topic map are the columns of the grid. Given a scenario, each column of the grid will be filled in with the adequate value. For example, about the distribution feature, it can be either centralized or distributed. If no requirement is specified about one feature in the scenario, no value will be assigned.

Table 1 presents the way the two scenarios described in Sect. 3 can be represented. As the topic map contains many features, the table presented below contains only a subset of them.

When considering scenario 1, few requirements are specified by the banking group. The data acquisition process will be based on usage data as no information has to be explicitly asked to users. As users are known, information *a priori* about users is known. The persistency of data may be medium or long term, which is possible as users are connected. Likewise, the persistence of the model may be medium or long term as in this scenario the profile of a user should not reflect only his short term behavior. The facets of the model will be behavior and role. No requirement has been made about the location of the model.

In scenario 2, the users do not need to be identified, thus data is short term, and the user modeling process has to be transparent for the user. User model data will be usage-based data, and the model will be a behavior model. No requirement is made about the location of the model.

To demonstrate the use of the topic map, two models referred in the state of the art are also represented in Table 1. We have chosen two models we designed, as we know all their features. The first model (called M1), presented in [2], is a model of anonymous users behavior and is built by exploiting only usage traces. This model has been developed to answer the research question of designing a tractable

Table 1. Grid representing the topic map, to be filled with scenario requirements and models characteristics

Scena rio/ model	Input		Accessibility		Type		Approach	
	Data acquis ition[1]	*Data Nature*[2]	*Authent ication*[3]	*Facets*[4]	*Persis tence*[5]	*Distri bution*[6]	*P. of View*[7]	*Priv acy*
Sc. 1	Impl.	Impl. I.U.	Anon.	Behav. Role	MT LT	?	?	Yes
Sc. 2	Impl.	Impl. I.I.	Unk.	Behav.	ST	?	Coll.	?
M1	Impl.	Impl.	Unk.	Behav	LT	Centr.	Coll.	Yes
M2	Impl.	Impl. I.U.	Anon.	Ratings Roles	LT	Centr.	Ind.	?

---

[1] "Impl." means Implicit, "Pref. El." means Preference Elicitation.

[2] "I. I." means Information *a priori* about items, "I.U." Information *a priori* about users and "E.I." Explicit information about users.

[3] "Anon." means anonymous, "Identif." identified and "Unk." Unknown.

[4] "Behav." means Behavior, "Pref." Preference.

[5] "ST" means Short Term, "MT" Medium Term and "LT" "Long Term".

[6] "Centr" means Centralized.

[7] "Ind." means Individual, "Coll." means Collective.

context-dependent model. The second model (called M2), presented in [4], is a model of both ratings and behaviors. The data exploited by this model is made up of traces of usage, implicit ratings are estimated based on these traces; demographical information is also exploited. To build and exploit this model, users have to be logged. It has been designed to model navigational behavior of users in a collaborative filtering framework.

The question is if one of the two candidate models can be used for one of the two scenarios. If a model has similar feature values than those of the scenario, then we can deduce that it can be used for this scenario.

In Table 1, we can first notice that the characteristics of each candidate model do not fit all the requirements of any described scenarios: at least one of the features has a value different between the requirements of the scenario and the characteristics of the model. For example, scenario 1 has behavior and role facets and the model M2 has ratings and roles facets. Moreover, the firm wants a medium term model whereas the model is long term.

In this table, given a scenario, there is thus no ideal model that fits all its requirements. The firm has to choose, among its requirements, which ones are the most important and which ones can be relaxed so as to fit a model. The question to ask to the firm is which requirements it accepts to adapt? Between the two models, M2 seems to better fit scenario1 as more features have identical values. Will the firm accept to have a model of ratings instead of a model of behavior and roles? It is a delicate question as the resulting model is really different, which is not the case for persistency feature as a mid term model is a refinement of a long term model. Depending on the answers, M1 will be used or not as the user model for scenario1.

We can notice here that when no requirement is specified by the firm about one feature, whatever is the value of this feature in a model, this feature will match.

6 Conclusion and Future Work

Personalization techniques, and specifically recommender systems, rely on user modeling and specific usage scenarios. Many user models have been proposed in the recommender systems literature, but the way these models are presented is not homogeneous, they are thus not easily comparable. This article proposes a user modeling topic map. This schema is necessary to guide both practitioners and researchers to position each other work according to their specific focus or/and requirements. It is aimed to guide practitioners to find the best user model given their specific requirements but also guide researchers to place their models among other existing models in the user modeling literature.

Furthermore this article presents two concrete usage scenarios in the area of recommender systems and presents the way the topic map is used to determine if a given model of the state of the art fits the requirements of the scenario.

This topic map is a first step towards a user modeling ontology and it will be extended for use in other application domains, although it may already be used, as it is, in other application areas of user modeling.

References

1. Adomavicius, G., Tuzhilin, A.: Toward the next generation of recommender systems: A survey of the state-of-the-art. IEEE Transactions on Knowledge and Data Engineering 17(6), 734–749 (2005)
2. Bonnin, G., Brun, A., Boyer, A.: A low-order markov model integrating long-distance histories for collaborative recommender systems. In: Proceedings of the ACM Int. Conf. on Intelligent User Interfaces (IUI'09), Sanibel Islands, USA, February 2009, pp. 57–66 (2009)
3. Castagnos, S., Boyer, A.: Modeling preferences in a distributed recommender system. In: Conati, C., McCoy, K., Paliouras, G. (eds.) UM 2007. LNCS (LNAI), vol. 4511, pp. 400–404. Springer, Heidelberg (2007)
4. Esslimani, I., Brun, A., Boyer, A.: Enhancing collaborative filtering by frequent usage patterns. In: 1st Int. Workshop on Recommender Systems and Personalized Retrieval, RSPR (2008)
5. Felfernig, A., Burke, R.: Constraint-based recommender systems: technologies and research issues. In: Fensel, D., Werthner, H. (eds.) 10th Int. Conf. on Electronic Commerce (EC'08), vol. 342 (2008)
6. Hanani, U., Shapira, B., Shoval, P.: Information filtering: Overview of issues, research and systems. User Modeling and User-Adapted Interaction 11, 203–259 (2001)
7. Herlocker, J.L., Konstan, J.A., Terveen, L.G., Riedl, J.T.: Evaluating collaborative filtering recommender systems. ACM Transactions on Information Systems 22(1), 5–53 (2004)
8. Huang, Z., Chung, W., Chen, H.: A graph model for e-commerce recommender systems. Journal of the American Society for Information Science and Technology 55(3), 259–274 (2004)
9. Lee, T., Park, Y., Park, Y.: A time-based approach to effective recommender systems using implicit feedback. Expert Systems with Applications 34(4), 3055–3062 (2008)
10. Liu, K., Chen, W., Bu, J., Chen, C.: User modeling for recommendation in blogspace. In: IEEE Int. Conf. on Web Intelligence and Intelligent Agent Technology (WI-IAT), pp. 79–82 (2007)
11. Lousame, F.P., Sanchez, E.: A taxonomy of collaborative-based recommender systems. In: Castellano, G., Jain, L., Fanelli, A. (eds.) Web Personalization in Intelligent Environments. SCI, vol. 229, pp. 81–117. Springer, Heidelberg (2009)
12. Manouselis, N., Costopoulou, C., Sideridis, A.: Introducing recommender systems for agricultural e-commerce applications. In: Int. Conf. on Inf. Systems in Sustainable Agriculture, Agroenvironment and Food Technology (2006)
13. Montaner, M., Lopez, B., De La Rossa, J.: A taxonomy of recommender agents on the internet. Artificial Intelligence Review 19, 285–330 (2003)
14. Park, Y., Chang, K.: Individual and group behavior-based customer profile model for personalized product recommendation. Expert Systems with Applications 36, 1932–1939 (2009)
15. Prassas, G., Pramataris, K., Papaemmanouil, O., Doukidis, G.: A recommender system for online shopping based on past customer behaviour. In: 14th Bled Electronic Commerce Conf., pp. 766–782 (2001)
16. Rich, E.: Users are individuals: individualizing user models. Int. Journal of Man-Machine Studies 18, 199–214 (1983)
17. Schafer, J., Konstan, J., Ridel, J.: Recommender systems in e-commerce. In: Proceedings of 1st ACM E-Commerce Conf., pp. 158–166 (1999)

18. Yu, L., Dong, M., Wang, R.: Taxonomy for personalized recommendation service. In: Int. Symp. on Electronic Commerce and Security, pp. 657–660 (2008)
19. Schafer, J., Konstan, J., Riedl, J.: E-commerce recommender applications. Data Mining and Knowledge Discovery 5(1/2), 115–152 (2001)
20. Burke, R.: Hybrid recommender systems: Survey and experiments. User Modeling and User-Adapted Interaction 12(4), 331–370 (2002)
21. Razmerita, L.: Modeling Behavior of Users in Semantic-enhanced Information Systems: The role of a User Ontology, in Adaptive Hypermedia. In: Proc. of Authoring of Adaptive and Adaptable Hypermedia Work, Hannover (2008)

myCOMAND Automotive User Interface: Personalized Interaction with Multimedia Content Based on Fuzzy Preference Modeling

Philipp Fischer[1] and Andreas Nürnberger[2]

[1] Mercedes-Benz Research & Development North America, Infotainment & Telematics, HMI Realization Group, 850 Hansen Way, Palo Alto, CA 94304
philipp.fischer@daimler.com
[2] Otto-von-Guericke-University Magdeburg, Faculty of Computer Science, Data & Knowledge Engineering Group, Universittsplatz 2, 39106 Magdeburg, Germany
andreas.nuernberger@ovgu.de

Abstract. myCOMAND case study explores the vision of an interactive user interface (UI) in the vehicle providing access to a large variety of information items aggregated from Web services. It was created for gaining insights into applicability of personalization and recommendation approaches for the visual ranking and grouping of items, composed as interactive UI layout components (e.g. carousels, lists). Quick access to preferred and important items can support less distracting interaction with a large web-based content collections and smaller screen size. Content gets aggregated on the server and then synchronized to an onboard module. Ranking for each data item is annotated based on a user profiles with a fuzzy preferences and a shared taxonomy on content categories. Preference values are implicitly learned from user interaction, but can be set explicitly by the user too. A circular UI component for browsing Internet radio stations is described, which dynamically groups items into categories during scrolling. Items are ranked according to the users preferences and item novelty. A visual overview mode helps to quickly review the structure of large content collections.

Keywords: Fuzzy Preference Modelling, Content-based Recommendation, Graphical User Interfaces, Haptic I/O, Prototyping, User Interface Framework Patterns, Automotive Human Machine Interaction.

1 Introduction

Most drivers are used to query trip-related content and services, e.g. driving directions, sights, weather information, restaurants, gas stations, online videos, or Internet radio streaming on their connected desktop computers or mobile devices. The myCOMAND system targets drivers that expect the same connected functionality and convenience, but integrated with their vehicle and specifically adapted to the requirements and constraints in the automotive domain. Some of the specific constraints for automotive user interfaces are the smaller screen size

P. De Bra, A. Kobsa, and D. Chin (Eds.): UMAP 2010, LNCS 6075, pp. 315–326, 2010.

(7-9 inch color display, aspect ratio 16:9) and requirements to be met for drive-save interaction, e.g. text labels have to retain a certain minimal size. Given those limitations, around 5-7 items can fit on the screen once at a time, but each additional interaction produces a higher cognitive load and distraction for the driver due to glances at the display for orientation.

On the contrary, Internet-based content collections usually contain of a huge number of information entities. Such variety of content is challenging to view with the constraints in screen real estate and input modalities in the automotive domain. An adequate and delightful user interaction is still highly desirable and access to web content collections and services essential for most use cases. A couple of concepts have proven to help dealing with distraction issues: voice-based and multimodal interaction, one-shot search queries, intelligent structuring of browsing taxonomies, grouping and structuring of list views, and recommendation and ranking of content based on novelty, personal preference, global popularity or context. In this paper, we will address the last two concepts.

Presentation of web-based information and connectivity features will increasingly drive the need for high system modularity, updatability and loose bindings between components in most vehicular infotainment systems. This flexibility is not yet common - most automotive systems and HMIs are deployed as monolithic blocks of embedded hardware and software that remain unchained for the entire lifetime of a vehicle platform. Flexible frameworks known from mobile platforms can be used for creation and prototyping of user interface and application in the automotive domain as well. In the myCOMAND system, user interface components, interaction flow and all content data (maps, internet radio streams, user media library etc.) are hosted off-board to explore the feasibility of such concepts. Only a small runtime resides in the client.

myCOMAND graphics design aims for creating a visual rich look-and-feel exposing the multimedia nature of web-based content and services. The 3D main menu is targeted to guide the user through the layers of the menu by 3D camera transitions, rather than having a clean and minimal design approach, as it would be the goal for a production design. Nevertheless, the user interface can be used for investigating recommendation approaches and impact on interaction in the same way because the number of items on screen is still limited.

1.1   Challenges and Key Issues

A goal of the study is to explore the applicability of an interaction module, that is able to rank and recommend content items to the user and thereby reduce the shortcomings of the limited number of items that can be shown on screen. Important items are ranked higher to gain better access. In addition we aim for close integration of the recommendation approach with the user interface framework to provide a very flexible visualization of content collections. The approach allows to forward the ranking of content collections into the UI layout components, even if the concrete graphical design and behavior would change. The prototype applies a fuzzy personalization and recommendation approaches for adaptive ranking and grouping of multimedia items. Applications functionality, user interface and menu flow is updated on every startup of the system and are not bound to a specific vehicle generation or hardware.

2   Related Work

In 2007, the SmartWeb HMI system was presented. It allows personalized access to web information and enables interaction with web content by natural language speech queries. An extension to classical command-based speech dialog systems [1] was proposed, where information semantics get extracted from web pages and pre-processed into speech-enabled packages. Each packages relates to specific topic areas (e.g. weather information). A set of new information packages gets frequently downloaded into the vehicle. A fuzzy recommendation approach ranks this list of incoming speech-enabled topics for the user according to a users explicit and implicitly preferences [2]. Fuzzy preference structures are based on the construction of preference, indifference and incomparability relations [3–6]. Multiple representations acquired from user interaction (preference ordering, ratings, fuzzy relations) can be integrated into an uniform information model of fuzzy preferences [7].

Recommender systems specifically address the problem of calculating a ranked set of content items to the user based on a user preference model and content similarities. An overview and characterization of content-based, collaborative and hybrid approaches is given in [8]. Mobasher and Anand describe web recommender system from a perspective, where content is hosted and preprocessed on a server and explores data-mining approaches for recommendation and implications [9].

Perny and Zucker described a synthesis of recommendation approaches and fuzzy modeling: a hybrid content-based and collaborative approach for content recommendation [10]. The system recommends other content items based both on fuzzy similarity measures for item-related features (content-based approach) as well as fuzzy similarity of user profiles with holding the rating scheme (collaborative approach). This framework is extended to a more general model of preferences, e.g. positive and negative preferences can be modeled and the recommendation approach is refined by Cornelis [11].

Ranked lists and flexible data models are eventually shown to the user and require a very flexible approach to modeling the user interface and processing visual state changes. The user interface framework derives several software engineering patterns from concepts, originally defined for data visualization frameworks by Heer and Agrawala [12]. Especially the reference model pattern and operator pattern [13] proved useful to evaluate expressive animations and transitions to support the user's decision making.

3   Prototype Design

The hardware components (automotive computer system, display, central haptical control element) were integrated and tested inside a concept car "Concept

Fascination" and two production vehicles. The user interacts with the system by turning and pressing a round controller knob integrated into the center console. Touch screen interaction is often pointed out as a potential alternative input method but is particular difficult in bigger vehicles due to the distance between driver and screen. The driver has to move forward towards the screen and also has to directly aim at the target on screen.

A software stack (operating system, graphics sub-system and framework) has been deployed to each of those cars. Only a slim module loader resides inside each vehicle, which downloads the actual application on every startup. In the context of this prototype, we assume that the client inside the vehicle is always connected to the Internet and gets guaranteed access to an offboard server. The server provides and maintains content in the form of data models, concept class hierarchies and new user interface components.

Most web services (Weather channel, YouTube, Google Maps) get queried directly from the client module, but some content items (e.g. vTuner) are aggregated and cached on a central content server that acts as a proxy to the vehicle. The proxy server indexes data collections and allows pre-fetching, synchronization and caching for clients with intermittent connection. An interaction profile for each driver is stored on the server if the driver approves to store his/her personal data. A aggregated collection of user interaction data is the basis to apply collaborative filtering approaches. Storage of user data always needs to be carefully considered and privacy issues need to be addressed in a production scenario by transparency.

3.1   Features

Especially those features were prototyped, which are based on user interaction with offboard data:

- **Streaming Internet Radio:** Set of radio stations streamed over Internet are available for browsing and filtering by genre and location and search. Personal favorites can be stored.

- **Online Media Library:** Personal Library of music content can be stored online and streamed to the vehicle.

- **Off-board Maps:** Internet maps and navigation solution. Panning and zooming, POI-related information and guidance (restaurant booking, etc.)

- **Map Information Overlays:** Display weather information, Wikipedia entries, ratings and reviews for particular POIs, free parking lots.

- **VoIP Telephone:** Leverages the existing IP data connection for Voice over IP telephony.

3.2   User Experience Storyboard

The small illustrated sequence of interaction with the system shows a couple of use cases and the user interface screen flow:

1. Florian is currently visiting from San Francisco and drives to Berlin over the weekend to meet his friend Martin. He uses the round input control element mounted to the center console of the car (CCE). The core application categories (Media, Web, Communication, Settings, Navi) are arranged in a 3-dimensional circular layout around a planet metaphor, and animate synchronously to the CCE movements.

2. By selecting a category, the main menu transitions to a sub-menu with 2-4 applications. The transition to the next level is animated by a 3D camera zooming and movement. After selection of an application, the camera movement continues to zoom further into the actual content screen. Florian is selecting the "navi" category and then the "map" application.

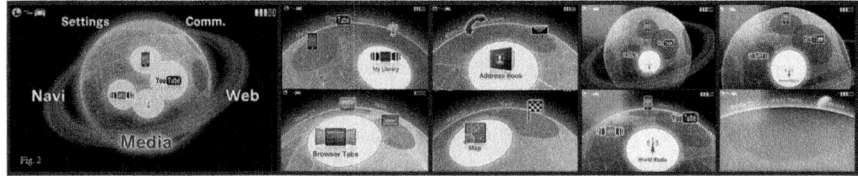

3. Florian looks at search results showing sights as selectable points on the map and one sight catches Florian's interest: Berlins Helium Balloon near Checkpoint Charlie. Switching through the map overlays shows sunny weather - Wikipedia entry, and user reviews look promising. Florian finds Martin's number in the online address book and calls using the Voice-over-IP for confirmation. He books 2 tickets using TripAssists online booking feature. In Google StreetView, street parking does not look promising, but the parking map overlay shows a nearby garage with currently 34 free spots.

4. Florian switches to the World Radio View. The radio stations are represented in a circular list. The content items can be scrolled and browsed by turning the central haptic controller too. He likes a particular Jazz radio station from San Francisco and would like to listen to it. He finds his favorite genres, locations and stations in prominent position and starts the stream. By pressing a special hardkey, an option menu transitions into the view. Each entry for the option menu has a preview list on the right side, which shows the next set of options or gives a preview on the filtered view of content items. In the internet radio screen, a genre and location hierarchy

can be browsed by using the option menu. The selection of genre or location filters main content list of all the radio stations.

3.3   User Interaction Framework

The framework supports the creation of composite user interface components which encapsulate a certain visual look and interaction behavior for a collection of data and item renderers. The internals of those components implement functionality to process user input events and translate them into visual changes of a component. Controllers strictly separate between user-driven input events (e.g. turning CCE, pressing hard-key buttons) and system-driven events originating from the underlying system function core more independently from the user (e.g. new data available, web data loading complete). Each of those events are mapped to certain functional actions (e.g. start station playback) or visual operators (e.g reconfigure menu layout) and get chosen based on the current state of the active user interface component. Each user interface component can forward focus to other components.

Flexible software design patterns are needed to bind data model, personalization model and user interaction to the visual views and transitions based on the changing nature of ranking-based layouts. Several patterns from data visualization frameworks naturally deal with such flexibility given the domain challenges [12]. While visual operators in data visualization frameworks usually highly depend on the data values alone, we extend those operators to be interactive and adaptive. Turning the CCE device results in subsequent events that get processed by the software controller component of the active user interface item that currently is set in focus.

Interactive Operators. The Operator pattern is used to compose visual data processing as a set of operators and enabling flexible and reconfigurable visual mappings. Each operator supports transition assignments. Transitioners update a visual state from the previous value to the next value over time based on interpolation (easing). 2 kinds of events mostly trigger operators: a) the user turns the central controller knob and b) the data model has changed or annotated preferences were updated. The operator updates the visual model, i.e. the visual properties of all children elements of a particular user interface component. Later on, all the visual items are rendered into the display list based on those visual properties. The interactive operators therefore base their functionality on input events, interaction state, recommender engine, and data values associated to the component's data collection.

4   Information Modelling

The interaction module, preference model and user interface framework are closely bound together to provide personalized recommendations to the user and also visualize them through animations and transitions. The personalization approach is based on 4 aspects: 1) a data model for content and content structuring , 2) preference model as user profile, 3) a recommendation engine, 4) a mechanism to bind specific interactions to updates of the preference model. Each of those aspects is described in the next sections.

4.1   Personalization Framework

We understand personalization as the adaptive process of filtering and prioritizing information based on the user's preferences and context, both changing over time. The user's preferences for unseen items are generalized from those that he/she has rated in the past.

The goal of the recommendation task is to order a number of items $I' \subseteq I$ from a set $I = \{i_j \mid 1 \leq j \leq n\}$ (Fig. 1, bottom layer) for the active user u_{active} under one or more criteria. In a multi-user case, u_{active} might be one user out of a set of users $U = \{u_k \mid 1 \leq k \leq m\}$.

Content Items and Concept Classes. The client module loads an index with references to a set of content items I into the client data model. This triggers events and registered operators of the user interface framework to process an update of the visual layouts and show indication to the user. Content items are songs, videos, Internet radio streams, points of interest for navigation, restaurants, new installable applications modules or anything else that we might be displayed to the user for selection.

For performing content-based filtering, a method for determining similarity of content items has to be defined. Each new item is matched to one or more categories $c_1, c_2, ..., c_m$ out of the set of categories $C = \{c_l \mid 1 \leq l \leq o\}$. Any

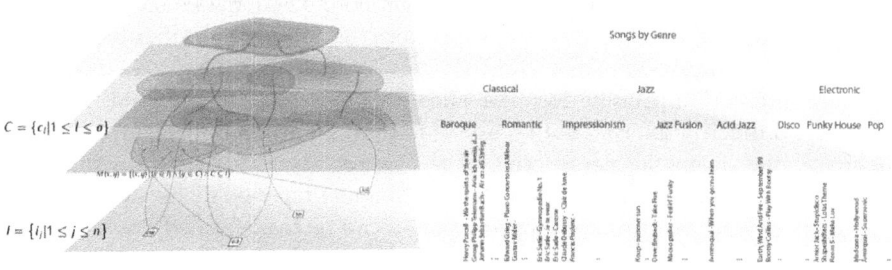

Fig. 1. Visualization of the content structure. Several content items (bottom layer) are gradually mapped to one or more concepts / clusters (upper layers)

concept class itself can be a child of another concept class. Thus, a graph-like structure resides in the model. For example, a musical genre taxonomy would be used as category hierarchy to classify Internet radio streams. The degree of membership expresses the degree to what extend a particular item matches the semantics of the concept class.

$$M(i, c) = \{(i, c) \mid (i \in I) \wedge (c \in C)\}$$

The client module relies on meta data assigned to each content item, which describe its relation to concept classes with a certain membership value. The concept hierarchy is shared among server and client.

The actual matching process for to concept classes (clustering, categorization) can be done by state-of-the-art methods for multimedia information retrieval on the server. For example, audio items might be processed on the server to extract acoustic features used by clustering algorithms. Term-frequency can be used to categorize text-based content (e.g. news articles). The myCOMAND client relies on meta data annotated by the supplying web service. For example, the Internet radio streams are pre-annotated by the vTuner web service API with one or more origin and genre class. Synchronization of concept hierarchies between server and vehicle client is managed independently from changes and updates to the data set of content items.

User Profile and Preference Modeling. A preference structure is used to describes the user's preference for any pair of alternatives $(a, b) \in A^2$ with 3 types of relations: $P(a, b) \Leftrightarrow$ the user prefers a over b, $I(a, b) \Leftrightarrow$ a and b are indifferent to the user, and $J(a, b) \Leftrightarrow$ if the user is unable to compare them. The triplet $< P, I, J >$ satisfies certain symmetry, reflexivity and completeness properties (see [3]). A fuzzy relation R in A is defined as an $A^2 \rightarrow [0, 1]$ mapping. Using fuzzy relations therefore allows to model a preference degree for two alternatives.

The user profile can store a number of fuzzy preference relations for a any content item or concept class. Preferences for one item add to the preferences of associated classes during recommendation process. Preferences values are acquired based on user interaction (implicit) or specifically expressed by the user (explicit). The strength of preference is stored along with the type (implicit, explicit), the specific method how it the preference has been inferred by the system and an uncertainty value attached to this method. The model also stores formal data such as the user identification number, the unique identifier of the item, and a time stamp.

User interaction usually does not allow to infer relations between two items directly. Implicitly inferred values are usually fuzzy utility values ore ratings for a single item from the perspective of the active user. Fuzzy rating values are stored using triangle distributions for the linguistic terms dislike, low, middle, highly preferred, favorite item. Rating values can be transformed into fuzzy relations using a transformation function and the ratio of the preference intensity for rating r_a and fuzzy rating r_b for two alternatives (a, b). We use the transformation function as described in [3] to get the preference relation for all alternatives $(a, b) \in A^2$, based on fuzzy ratings.

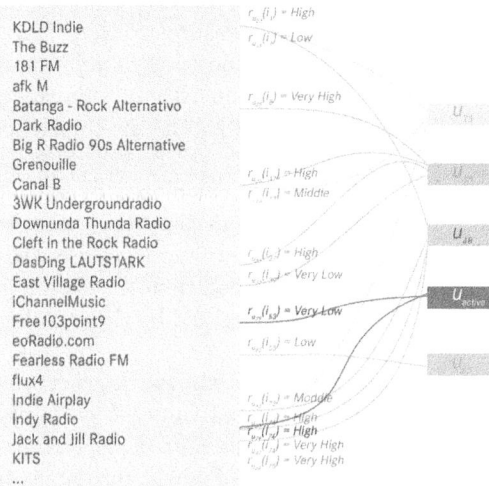

KDLD Indie
The Buzz
181 FM
afk M
Batanga - Rock Alternativo
Dark Radio
Big R Radio 90s Alternative
Grenouille
Canal B
3WK Undergroundradio
Downunda Thunda Radio
Cleft In the Rock Radio
DasDing LAUTSTARK
East Village Radio
iChannelMusic
Free103point9
eoRadio.com
Fearless Radio FM
flux4
Indie Airplay
Indy Radio
Jack and Jill Radio
KITS
...

Fig. 2. This visualization shows a couple of fuzzy ratings for a set of content items

$$t(\frac{r_a}{r_b}) = \frac{\frac{r_a}{r_b}}{\frac{r_a}{r_b} + \frac{r_b}{r_a}} = \frac{(r_a)^2}{(r_a)^2 + (r_b)^2}$$

Fuzzy ratings and fuzzy relations can be both aggregated into a single preference model by using the transformation function t. Fig. 2 illustrates a preference model for the active user with two preference values. Other users are shown with their ratings as well, this information is only available on the server and is not part of the preference model in the vehicle.

Recommendation. Content-based filtering uses features or properties of any item to calculate similarities between two content items. Thus it is based on the semantic structure of content items and related categories/clusters for recommendation of similar items. Collaborative filtering is based on similarity measures between two users or a clique of k-most similar users in the multi-user case. The approach is an hybrid approach which scales between both strategies based on the use case and availability of the server via connectivity. The membership values $M(i, c)$ can be used to calculate a similarity relation between any two items by fuzzy aggregation operators.

A T-Norm $T(M(i_x, c), M(i_y, c))$ can be used for "pessimistically" combining two membership relations M for two items i_x an i_y. The value describes to which extend each of those items belongs to a common category c. With a $max - min$ composition the following content-based similarity measure is used for each pair of content items i_x and i_y: $S(i_x, i_y) = \text{argmax}_{\forall l}(T(M(i_x, c_l), M(i_y, c_l)))$

The overall preference score for each item is calculated by aggregating individual and global preferences. Individual preferences are the preferences of a particular user and get calculated from explicit user ratings and implicit indicators by content-based filtering. Based on the similarity measure S, it gets possible to select the k best similarity measures $S(i_q, i_k)$. An algebraic product operator is used as implicator, to combine those similarity measures with the corresponding preference value of the item $R(u_{active}, i_k)$ for each content item i_k. Each output value is then aggregated by the max-operator for each content item.

The influence of global preferences is calculated by a collaborative filtering algorithm, based on Perny and Zucker's method [10]. It calculates the k most influential users by comparing the preference model of users with each other. The framework uses the Pearson correlation coefficient [8]. The individual preferences for the k most influential users is then weighted by the correlation coefficient and propagated to the active user.

After the calculation, three values are known for each content item: an explicit value, an implicit value and a global value. A certainty measure weights the values for aggregation. The certainty measure is dependent on rules about the information density and applicability of each recommendation strategy in the current situation. For example, if there are new explicit values given by the user, then explicit values are mainly used. If there were many user interactions recently, implicit preferences are used preferably. The three preferences are accumulated by a maximum-operator (T-conorm).

5   Interaction Concept

5.1   Adaptive Content Layouts

The default view of the Internet radio stations list (see Fig 1c) simply positions items in a circular arrangement one after another for example alphabetically. Most of the items are offscreen, since the screen should hold only up to 5 items. The list of items can be scrolled one by one. An improvement to interaction with those circular interactive layouts/lists is implemented by making their partitioning dependent on the users preferences. The user also has the ability to switch into an overview mode (Fig 3) where he/she can view the whole list, its groupings and similarities at once. Since the list layouts items 1-dimensional (one after another), a specific total order of items has to be calculated, taking into account item categories, preference values for items & categories and similarities of categories. The size of each segment relates to the preference rating for this category of the active user, i.e. important categories take more space. Important categories are the structured with more subcategories than others. For example, a person that prefers classical music will find more sub-genres when scrolling through this part of the list.

Interactive Grouping. While scrolling through the list, the user gets a sense of the amount of items that he/she is interacting with in a specific category since personal categories are bigger and more fine-grained. If the user scrolls faster, then the interface will dynamically merge items into groups and scroll groups rather than items. Group entities still take a more time to scroll through to the

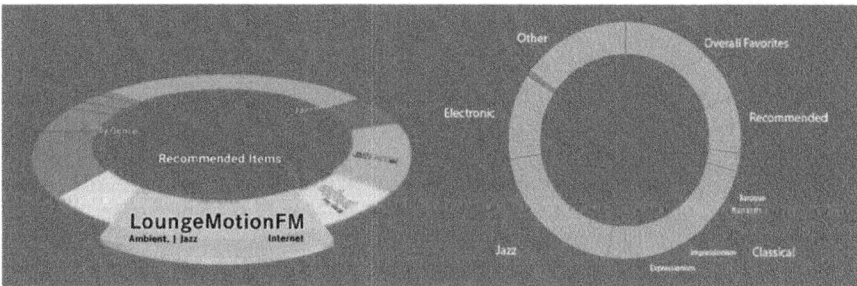

Fig. 3. (left) Shows the radio station view as a carussel list UI component. The stations are ordered by user preference. (right) overview mode for the station view.

next group it is just a visual helper. A very slow scrolling will let the user access each particular item at any time. A middle speed of scrolling will summarize songs/stations into sub-genres such as Romantic, Baroque, and Impressionism while a very fast speed will scroll through top categories only (Classical, Jazz, Electronic).

6   Conclusion

We discussed a prototype case study of an automotive user interface with personalized content ranking. A fuzzy preference model and recommendation approach proved to be applicable for the myCOMAND user interface. It seems to be abstract and general enough to be used for storing and ranking web-based content collections. We were able to integrate the ranking scheme into a user interface framework without scarifying the separation between model, view model and controller. This was done by using software engineering patterns described in [12], originally targeting data visualization frameworks. Especially the operator and decomposable architecture of composite user interface components proved useful and allows flexible bindings and updates.

For future work, the User Profile and Vehicle Setting Management could be implemented as a more central point for reviewing and adjusting privacy settings and profile information. Trust, organization and transparency plays an important role to let users feel comfortable with recommender systems and connected systems in general [14, 15]. The proprietary annotation with meta-data provided by the web service could be replaced by an automatic data mining approach. We can improve the recommendation quality by integrating collaborative recommendation approaches, though they are more difficult to test and evaluate without an existing user base.

Acknowledgments. We would like to thank the whole myCOMAND team for the tremendous efforts and dedication: Chris Lorenz, Hamza Lakhani, Michael Cheng, Michelle Cheung.

References

1. Berton, A., Regel-Brietzmann, P., Block, H.U., Schachtl, S., Gehrke, M.: How to integrate speech-operated internet information dialogs into a car. In: Proceedings 8th Annual Conference of the Int. Speech Communication Association, Interspeech (2007)
2. Fischer, P., Nüernberger, A.: Adaptive and multimodal interaction in the vehicle. In: IEEE International Conference on Systems, Man and Cybernetics, SMC 2008, Singapore, pp. 1512–1516 (2008)
3. Fodor, J., Roubens, M.: Fuzzy Preference Modelling and Multicriteria Decision Support. Kluwer Academic Publishers, Dordrecht (1994) ISBN 0-7923-3116-8
4. de Walle, B.V., Baets, B.D., Kerre, E.: Characterizable fuzzy preference structures. Annals of Operations Research 80, 105–136 (1998)
5. Georgescu, I.: Fuzzy preference relations. In: Fuzzy Choice Functions. Studies in Fuzziness and Soft Computing, vol. 214, pp. 49–74. Springer, Heidelberg (2007)
6. Fodor, J., de Baets, B.: Fuzzy preference modelling: Fundamentals and recent advances. In: Fuzzy Sets and Their Extensions: Representation, Aggregation and Models, pp. 207–217 (2008)
7. Chiclana, F., Herrera, F., Herrera-Viedma, E.: Integrating three representation models in fuzzy multipurpose decision making based on fuzzy preference relations. Technical report, ETS de Ingenieria Informatica, Universidad de Granada (1996)
8. Adomavicius, G., Tuzhilin, A.: Toward the next generation of recommender systems: A survey of the state-of-the-art and possible extensions. IEEE Transactions on Knowledge and Data Engineering 17(6), 734–749 (2005)
9. Mobasher, B.: Recommender systems. Kunstliche Intelligenz, Special Issue on Web Mining 3, 41–43 (2007)
10. Perny, P., Zucker, J.D.: Collaborative filtering methods based on fuzzy preference relations. In: Proc. of the EUROFUSE-SIC'99, pp. 279–285 (1999)
11. Cornelis, C., Guo, X., Lu, J., Zhang, G.: A fuzzy relational approach to event recommendation. In: Proc. of 2nd Indian Int. Conf. on Artificial Intelligence (IICAI 2005), pp. 2231–2242 (2005)
12. Heer, J., Agrawala, M.: Software design patterns for information visualization. IEEE Transactions on Visualization and Computer Graphics 12(5), 853–860 (2006)
13. Chi, E.H.h., Riedl, J.: An operator interaction framework for visualization systems. In: INFOVIS '98: Proceedings of the 1998 IEEE Symposium on Information Visualization, Washington, DC, USA, pp. 63–70. IEEE Computer Society, Los Alamitos (1998)
14. Pu, P., Chen, L.: Trust building with explanation interfaces. In: IUI '06: Proceedings of the 11th international conference on Intelligent user interfaces, pp. 93–100. ACM, New York (2006)
15. Sinha, R., Swearingen, K.: The role of transparency in recommender systems. In: Extended Abstracts of Conference on Human Factors in Computing Systems, CHI'02 (2002)

User Modeling for Telecommunication Applications: Experiences and Practical Implications

Heath Hohwald, Enrique Frías-Martínez, and Nuria Oliver

Data Mining and User Modeling Group
Telefonica Research, Madrid, Spain
{heath,efm,nuriao}@tid.es

Abstract. Telecommunication applications based on user modeling focus on extracting customer behavior and preferences from the information implicitly included in Call Detail Record (CDR) datasets. Even though there are many different application areas (fraud detection, viral and targeted marketing, churn prediction, etc.) they all share a common data source (CDRs) and a common set of features for modeling the user. In this paper we present our experience with different applications areas in generating user models from massive real datasets of both mobile phone and landline subscriber activity. We present the analysis of a dataset containing the traces of 50,000 mobile phone users and 50,000 landline users from the same geographical area for a period of six months and compare the different behaviors when using landlines and mobile phones and the implications that such differences have for each application. Our results indicate that user models for a variety of applications can be generated efficiently and in a homogeneous way using an architecture based on distributed computing and that there are numerous differences between mobile phone and landline users that have relevant practical implications.

1 Introduction

User Modeling is a key process in a wide variety of (telco) telecommunication applications in which knowledge of individual users is key for providing a better service and anticipating user needs. The most relevant applications include: (1) churn prediction, i.e. the ability to anticipate users that are at risk of leaving the company, (2) information spreading processes, such as viral and targeted marketing, which include a variety of techniques to spread information in the network and the ability to identify key users that can influence others in their decision making process, (3) fraud detection, which focuses on identifying users that will exhibit fraudulent behavior, and (4) network design and planning, which seeks to adapt and plan a network to meet the needs of the users and the design of pricing plans.

Although these applications are very different in nature they typically generate user models from a common data source. The features of the different user

P. De Bra, A. Kobsa, and D. Chin (Eds.): UMAP 2010, LNCS 6075, pp. 327–338, 2010.

models frequently overlap and the architecture used to generate the models can be shared. Regarding data sources, CDRs (Call Detail Records) are used as a primary source of information for constructing user models for telco applications since they implicitly contain the behavior of each customer, from calling patterns, to consumption, terminal changes or characteristics of the social network. In some cases other extra information of each customer, such as gender, can be used. As for the dimensions used for the user models, in general there is a set of features, such as total talk time or total degree, that are relevant for a wide variety of applications. These common factors imply that the same architecture can be used for generating user models for a variety of applications.

Telco applications can be divided into two main areas: mobile and landlines. While mobile phones are in widespread usage and are typically used by just one individual, the number of landlines is much smaller and their use is typically shared by more than one individual. The user models generated for both cases use the same set of features, although the relative importance and implications of each feature differs.

In general the architectures used for generating user models for telco applications have to be very data intensive in order to process the amount of data available (typically several months) for all the customers (typically several million). The main differences between applications are not so much in the way user models are generated or in the features of the user models, but in the training sets used to construct the classifiers, i.e. while the training set for churn prediction will include users that have churned, for fraud prediction they will include users that have committed fraud. This implies that the same architecture can be used to generate different user models for different applications, and that there is no need for ad-hoc solutions.

In this paper we present our experience in generating user models from real CDR traces for telco applications. Also we compare the differences between landline and mobile phones for each feature and the practical implications that those differences have. The rest of the paper is organized as follows: after presenting the related work, we detail the construction of user models and typical features used for telco applications. Section 4 presents the Methodology for User Modeling and Sect. 5 the lessons learned and the implications for different applications of the features studied. We conclude in Sect. 6.

2   Related Work

The literature reports a wide variety of studies related to telco applications. Most of the work has focused on studies using mobile phone data [1–4], while landline data has received less attention [5, 6]. Churn prediction algorithms have been implemented for landlines[7] and mobile phones [2, 8, 9]. Traditionally, churn prediction has been solved with classification techniques that predicted in which group (churner or non-churner) a given user was included. User models were constructed using implicit information provided by CDR data such as calling patterns [10] or social network patterns [2]. The techniques used for creating the classifier encompass typical machine learning techniques such as: neural networks [11], classification trees [12], SVM [9] and genetic algorithms [8]. Information spreading algorithms originally appeared in social sciences [13] and are based on the idea of using a social interaction network to model the flow of information and influence. The concept groups a variety of algorithms that model the pervasive word of mouth behavior and are typically based on the spreading activation method used in cognitive psychology. These family of algorithms have been successfully used in a variety of telco applications, including viral marketing [14], churn prediction [2], and modeling of trust [15]. Fraud detection in the telco context aims at detecting individuals that acquire a mobile phone and do not intend to pay their contract [16]. Typical approaches focus on classifying users according to their level of risk by calculating deviations from standard behaviors [17, 18]. Telco user models have also been effectively used for the improvement of the network infrastructure, including the design of pricing plans, an application where mobility data has proven extremely relevant. For example [19] modeled number of calls, number of cells visited and the entropy of user locations for voice, data and SMS in order to improve paging efficiency in cellular networks.

Our work, when compared to previous approaches, presents three main novel elements: (1) the techniques used for each solution are typically developed ad-hoc, but we consider that although the applications are very different, the fact that they share the data source and a lot of dimensions implies that the same architecture can be used to generate the user models needed, (2) in general previous approaches use a limited number of users, while we consider one of the key challenges of user modeling is going to be the ability to obtain conclusions from massive datasets, and (3) we present the first analysis of the differences between landlines and mobile phones and the implications that those differences have for telco applications.

3   Generating User Models for Telco Applications

3.1   Telco Data Acquisition

Mobile phone networks are constructed using base transceiver stations (BTS) that are in charge of communicating mobile phones with the network. The area covered by a BTS is called a cell. Call Detail Records (CDRs) are generated when a mobile phone connected to the network makes or receives a phone call or uses a service (SMS, MMS, etc.). In the process, the information regarding the connection is stored in the form of a Call Detail Record, which includes the originating phone number, the destination phone number, the time and date of the call, the total length of the call and the BTS used for the communication. The originating and destination numbers are encrypted to preserve privacy. The BTS gives an indication of the geographical position of the user, but no indication of the position of a user within the cell is known. CDR data for landline subscribers is acquired in a similar fashion but without the need for a BTS. Typically CDRs for a given period of time are stored in more than one file, for example one file per day, which facilitates the generation of user models when using data-driven architectures (See Sect. 3.3).

3.2 Features of Telco User Models

In this section we present a set of features that have been found to be generally useful for generating both landline and mobile user models across a range of applications.

Total Number and Total Duration of Calls. Two of the most basic metrics that can be computed for each user are the total number of calls and the total talk time over a specified time period. The number of calls and total talk time can each be further restricted according to direction of call, where each subscriber's incoming and outgoing calls are considered separately. From an application perspective, variations of these features are very relevant, for example the ratio between national and international calls or between calls made within the provider and outside the provider are very relevant for churn [8, 9] and fraud detection[17, 18]. Also these two variables are relevant to viral marketing as users that have a lot of connections are more capable of spreading information [13]. As for network design, these are key features used to balance the network [19].

Calling Behavior for Each Day. While features such as total number of calls and total duration capture a user's aggregate activity level, a vector of temporal features can be used to capture the variation in calling behavior during the course of the day or week. Considering first daily behavior, for each user two vectors of length seven record the total number of calls and total talk time for each day of the week. The same features can be computed for the reciprocal call CDR data sets. A day-by-day comparison between landline and mobile reciprocal call data is indicative of what day of the week each set of users tends to speak with members of their social circle and can be an important factor in targeted advertising campaigns. Also this information is very relevant for fraud as it is used to generate the user model that describes normal behavior [17] and churn[10].

Calling Behavior for Each Hour of the Day. Similar to the features that segment activity by day of the week, it is possible to calculate the number of calls and total talk time for each user based on the time that each call was initiated. Typically, the time intervals considered are 24 one-hour long bins beginning at the start of each hour. For each user, two vectors of length 24 can be constructed in order to capture the total number of calls and talk time for each hour of the day, aggregating over all days in the data. The vectors provide insight into understanding what time of the day each user tends to have most of their calls and speak the most. The percentage of calls and talk time coming from *reciprocal* talk partners indicates the time of day when each user is most likely to be speaking with members of their social circle, a key element for designing viral marketing campaigns. Considering the aggregate results for the entire population is useful for network planning, since the network operator must plan for the different peaks in usage for landline and mobile networks [19]. As in the previous case, this information is very relevant for fraud detection[17] and churn[10].

Social Network Features. The concept of degree is one of the fundamental metrics in social network analysis. A graph $G_D = (E, V)$ that represents the social network of the callers present in the data may be derived from CDR data D. Each node $v \in V$ corresponds to a different phone number and each directed edge $e = (v_1, v_2) \in E$ corresponds to a call from node v_1 to node v_2. In this context, the degree of a node v, denoted $Deg(v)$, corresponds to how many distinct talk partners subscriber v has and is given by the number of edges incident with node v. The in-degree of a node v corresponds to the number of distinct individuals that call v while the out-degree is given by the number of distinct individuals called by v. Reciprocal degree of a node v corresponds to the total number of edges incident with ode v in the reciprocal graph and is a measure of the total number of talk partners in a user's true social circle. The higher the degree, the larger the social circle. It is important to maintain customers with large social circles since they can exert influence on a large number of other subscribers, potentially causing them to churn [20]. Recently this information has also been included in churn prediction models [2]. The reciprocal degree is key element for viral marketing, an in general for diffusion information processed, because provides a way of identifying strong ties [13].

3.3   Construction of User Models

As illustrated above, there are a large number of features that can be calculated from a given set of CDR data for either landline or mobile subscribers and that are relevant for a variety of applications. The construction of telco user models is complicated by the fact that often CDR records usually contain several months of data with hundreds of millions of records for tens of millions of users. Rather than constructing each user model for each application area, we have developed ARBUD [21], an terabyte architecture for automating the user model construction process that is based on a distributed computing paradigm and typically run on a computer cluster. One of the components of ARBUD is a library containing reusable modules for constructing different features of a user model in an efficient way. All of the features mentioned in Sect. 3.2 have been added as modules in the ARBUD library, typically using the MapReduce programming paradigm [22]. A metamodel is used for specifying the desired features, location of the data and any other relevant parameters and ARBUD then interprets the metamodel and constructs the desired user models.

4   Experimental Setup

In order to compare residential mobile and landline users, two random samples of $50,000$ mobile and $50,000$ landline subscribers were drawn from the same metropolitan area. The sample of landline users is denoted as S_L and mobile users as S_M. Any overlap between S_L and S_M was arbitrary and not identifiable. A set of CDR data was obtained for the subscribers in S_L and S_M during the same six-month period. The CDR data associated with S_L and S_M are denoted by D_L and D_M and their size by $|D_L|$ and $|D_M|$, respectively. All calls for S_M were recorded, even when the subscriber left the geographic region from which the sample was drawn. The total number of calls (in millions) was $|D_L| = 50.3$ and $|D_M| = 41.8$. The information present in each CDR includes the encrypted originating phone number, the encrypted destination phone number, the duration of the call in seconds, and the time and date when the call originated. The sets of all reciprocal calls made and received by the subscribers in S_L and S_M are denoted by $D_{L,R}$ and $D_{M,R}$ (and their sizes by $|D_{L,R}|$ and $|D_{M,R}|$) respectively. The total number of reciprocal calls (in millions) was $|D_{L,R}| = 27.3$ and $|D_{M,R}| = 29.9$.

ARBUD was used to build user models from both the landline and mobile datasets for the users in S_L and S_M, including all the features mentioned in Sect. 3.2. The construction process was carried out on a cluster with 5 machines, each with 16 GB of RAM, 4 hard drives each with 1 Terabyte storage capacity, and 4 quad core processors. The nodes were all connected with a fast gigabit network switch. Both models were constructed in less than 24 minutes.

5   Results and Discussion

Using the user models generated in Sect. 4, this section studies the features presented, contrasting the differences between landline and mobile users and their implications from a practical perspective.

5.1   Total Number and Total Duration of Calls

Figure 1a depicts the empirical CDFs (Cumulative Distribution Function) for four different distributions: the number of incoming and outgoing calls for land-line and mobile subscribers. Looking at the median for each distribution, which corresponds to the horizontal line where $F(X) = 0.5$, we observe that of the four distributions, the lowest number of calls corresponds to outgoing calls from mobile subscribers with about 300 outgoing calls over the 6-month period, or a little less than 2 outgoing calls a day. For the region in question, mobile subscribers only pay for calls they place, so unsurprisingly mobile users make fewer calls than they receive. Landline users make and receive more calls than mobile users. In the case of the 10% of users from each population that have the most calls, three of the distributions have a similar number of calls, while the distribution for landline outgoing calls has a higher value, with the top 10% of the landline sample making at least 1500 calls during the 6-month period. This information is very relevant to offer and design new plans to subscribers. Also the top users of the distribution are key individuals as they can play a relevant role in the information spreading process needed for viral marketing and churn (other features such as degree are also related).

Figure 1b represents the empirical CDFs for the total talk time. Four different distributions are depicted: the total talk time of incoming and outgoing calls for

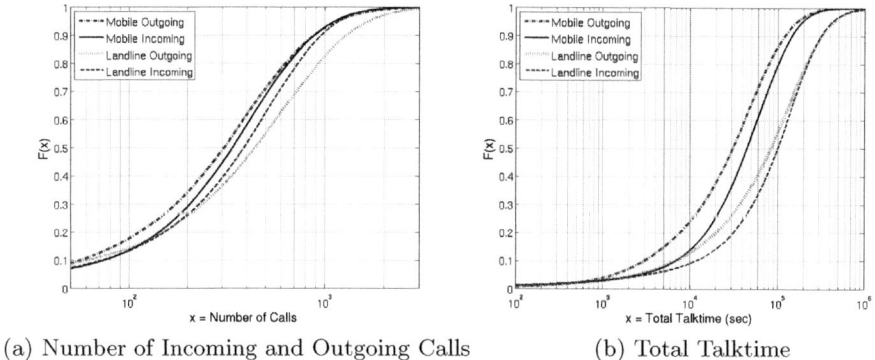

(a) Number of Incoming and Outgoing Calls   (b) Total Talktime

Fig. 1. CDFs for Number of Incoming and Outgoing Calls (left) and Total Talktime (right)

mobile and landline users. In this case, for the median subscriber the smallest amount of talk time corresponds to mobile subscribers making calls (outgoing), followed by mobile subscribers receiving calls (incoming) and landline subscribers making calls (outgoing). Landline subscribers receiving calls (incoming) have the highest talk time of all groups, even though the largest number of calls corresponded to outgoing calls from landline subscribers. Looking again at the median of each distribution, a total talk time of about 30, 000 seconds (8.3 hours) is seen for outgoing mobile calls, while a talk time of about 100, 000 seconds (27.7 hours) is seen for incoming landline calls: at the median of each distribution, landline subscribers spend more than 3 times more talking on the phone when receiving calls than mobile subscribers when making calls.

Figure 2 presents the CDFs of average call duration of S_L and S_M users for incoming and outgoing calls. The median value for the average call duration for incoming landline calls is more that twice the call duration of outgoing mobile calls, likely indicating sensitivity to different pricing structures. Mobile users in the top 10% of talk time are seen to talk for more than 30 times as much as the bottom 10%, so there is very heterogeneous behavior which helps explain the success of numerous pricing plans.

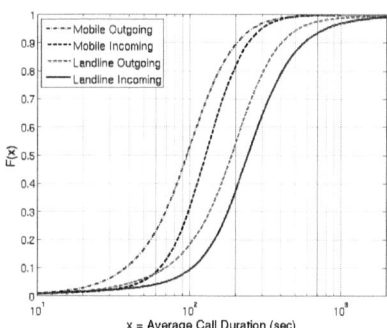

Fig. 2. CDFs for Average Call Duration for Incoming and Outgoing Calls

5.2   Calling Behavior for Each Day of the Week

The total number of calls (the sum of all the made/received calls) by all subscribers in the landline (S_L) and mobile (S_M) samples aggregated over all user models built for D_L and D_M for each day of the week are shown in Fig. 3a. Since both the mobile and landline samples are for 50,000 subscribers during the same time period, the aggregate number of calls and talk time for each day of the week can be directly compared on a day-by-day basis. Figure 3a also depicts the total number of calls with only reciprocal talk partners for each day of the week, obtained by summing for each day over $D_{L,R}$ and $D_{M,R}$. It can be seen that landline users make and receive more calls than mobile users for every day of the week. Interestingly, Monday is the day with most calls for landline users, while Friday is the day with most calls for mobile users. Both populations make fewer calls on the weekend. When considering only reciprocal calls, the differences between landline and mobile users decrease significantly and there are more mobile than landline phone calls on Fridays. These results are indicative of the culture of the sampled region, where Friday is the day when mobile users are most likely to make plans with their social circle and Monday is the day when landline subscribers make most non-social calls. These results can help inform a targeted advertising campaign, suggesting, for example, that socially oriented advertising directed to mobile subscribers may be best received on Fridays.

When looking at total talk time for each day of the week (Fig. 3b), the results are notably different. While Sunday is the day with the fewest calls for both landline and mobile users (see Fig. 3a, Saturday is the day with the least talk time in both cases. On Sunday, both populations have relatively few calls but calls tend to last longer, as evidenced by a rise in talk time despite a drop in the number of calls. Mobile users spend less time on the phone than landline users, considering both the full set of CDRs (D_M and D_L) and the reciprocal CDRs ($D_{M,R}$ and $D_{L,R}$). These differences indicate that for fraud detection the definition of a standard behavior is dependent on the type of communication (landline or cell phone) and of cultural elements.

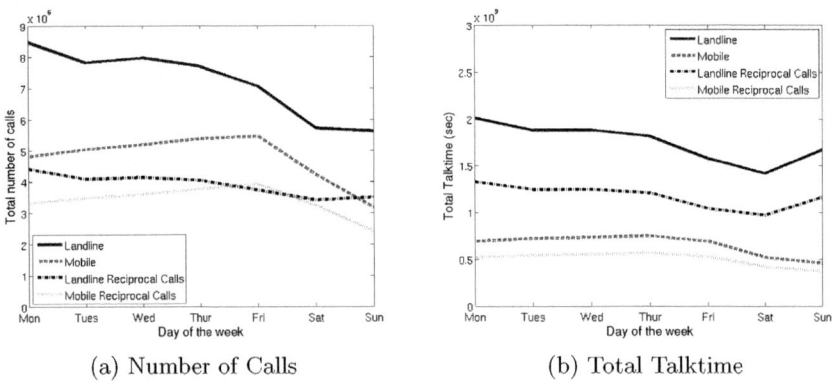

(a) Number of Calls   (b) Total Talktime

Fig. 3. Number of Calls and Total Talktime by Day of the Week

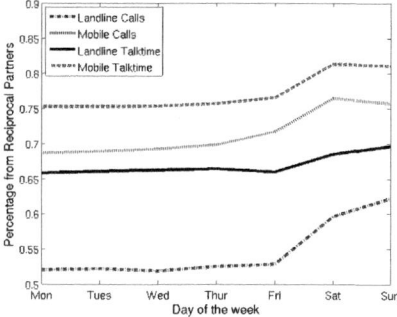

Fig. 4. Percentage of Calls and Talktime from Reciprocal Partners for Different Days of the Week

Looking at what percentage of all calls for each day of the week are reciprocal calls (Fig. 4), it can be seen that, both landline and mobile users, the weekend tends to be the time when users are most likely to have calls with their reciprocal partners. These results indicate that for the geographical region under consideration, the weekend is the time when the highest percentage of phone usage is dedicated to speaking with one's social circle.

5.3   Calling Behavior for Each Hour of the Day

The number of calls as well as the number of reciprocal calls for landline and mobile subscribers for each hour of the day are depicted in Fig. 5a. Both landline and mobile users show a similar trend, with very few calls in the early morning (from 0.00 to 9.00)and a significantly larger number of calls during the day (from 9.00 to 22.00). While the global maximum for mobile subscribers is at 19.00 to 20.00, the maximum for landline users takes place 2 hours later (from 21.00 to 22.00). The daily rhythm observed is indicative of the culture of the region sampled and would likely be different for other cultures. This information is also key for defining standard behavior in fraud detection and complements the information of the previous section. The fact that both groups have similar rhythms suggests that mobile and landline phones are not competing technologies, but rather complement each other. It can also be observed that landline users place and receive more calls than mobile subscribers at every hour of the day. This may be a result of the generally higher tariffs for mobile phones or the fact that landlines tend to be shared among several users while mobile phones tend to be exclusively used by a single person. Similar behavior is found when considering calls to reciprocal call partners.

The total talk time for each hour of the day (Fig. 5b) follows a similar two-peak pattern. However, the peaks in mobile talk time are significantly less pronounced than the peaks in landline talk time and the amount of (reciprocal and non-reciprocal) landline talk time is significantly larger than the mobile talk time: mobile users seem to use their devices uniformly during the day, particularly in terms of talk time, whereas landline users tend to talk longer in the evening.

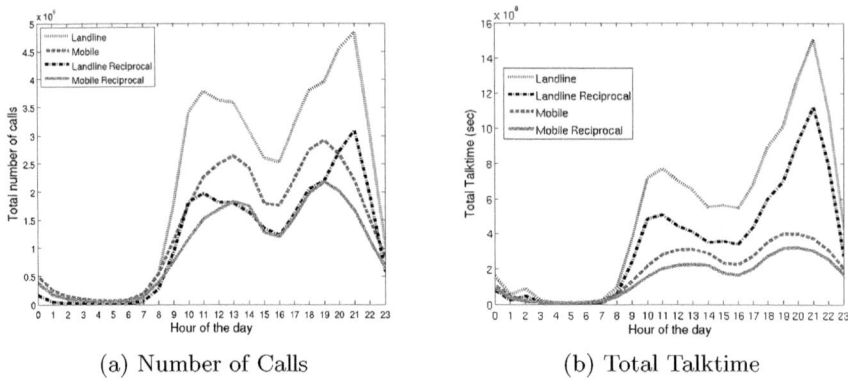

(a) Number of Calls   (b) Total Talktime

Fig. 5. Number of Calls and Total Talktime by Time of Day

Figures 5a and 5b indicate also that planning for peak usage in mobile and landline networks requires focusing on different time windows.

5.4   Size of Each User's Social Circle

The analysis of number of calls and talk time reveals patterns of behavior at the individual level. However, they do not capture much about the social networks of each individual. In order to see how landline and mobile users may differ in the size of their respective social circles, in and out degrees were calculated for all landline and mobile subscribers, resulting in four empirical distributions, which are plotted in Fig. 6a. Looking at the median of each distribution, the smallest degree is for outgoing calls from mobile subscribers, with the median mobile subscriber making calls to 40 different subscribers, while the median landline subscriber receives calls from 70 different subscribers.

The shortcoming of focusing on in or out degree computed over the full data set is that it does not take into account the *strength* of social connections. If degrees are calculated over the reciprocal data sets, however, a more accurate picture of the true size of each user's social circle is obtained. Figure 6b depicts empirical CDFs for reciprocal degrees calculated over the reciprocal CDR data sets $D_{L,R}$ and $D_{M,R}$. Note that when calculating degree over the reciprocal CDR data sets, it no longer makes sense to speak of in or out degree but of *reciprocal* degree. In addition, 3,956 landline and 934 mobile subscribers from S_L and S_M respectively did not have any reciprocal relationships and were not included in the empirical CDFs depicted in Fig. 6a. As seen in the figure, the distributions are significantly different than those seen in Fig. 6a. In the reciprocal case, the median of the landline and mobile phone degree distributions are almost identical and lower than the degrees shown in Fig. 6a, even though landline users tend to make/receive more calls and a landline is typically used by multiple individuals. In our dataset, the median size of the social circle –inferred from the reciprocal degree distributions– of landline and mobile users is 20. The top users of the reciprocal degree distribution are very important because with a social circle of size 50 or more each

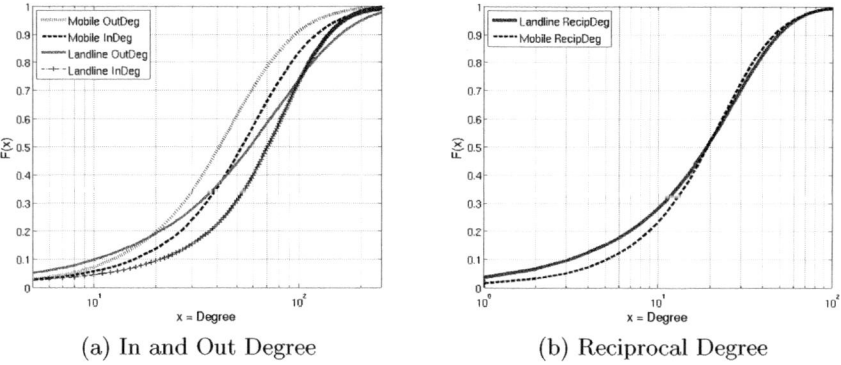

(a) In and Out Degree   (b) Reciprocal Degree

Fig. 6. CDFs for In, Out, and Reciprocal Degree

user can exert large influence on other subscribers, influencing their propensity to churn and/or for implementing viral marketing campaigns.

6   Conclusions

Many different application areas in the telco domain rely upon user models. In this paper, we presented a set of features common to many telco applications and indicated why certain features are particularly relevant in certain applications. Because the main telco applications use the same data source (CDR), and the same set of features to construct user models, we proposed a general data-driven architecture to build user models for any telco application. In order to illustrate typical behavior found when building user models for real datasets, we analyzed and compared the behavior of the mobile and landline traces of 50,000 anonymized individuals in a metropolitan area during 6 months. We considered three factors: (1) aggregate individual behavior, (2) temporal behavior, and (3) social network. The analysis identified, among others, that usage patterns depend on the time of day and the day of week, with landline users making/receiving more calls and talking longer than mobile users.

References

1. Seshadri, M., Machiraju, S., Sridharan, A., Bolot, J., Faloutsos, C., Leskovec, J.: Mobile call graphs: Beyond power-law and lognormal distributions. In: KDD '08, pp. 596–604 (2008)
2. Dasgupta, K., Singh, R., Viswanathan, B., Chakraborty, D., Mukherjea, S., Nanavati, A.A., Joshi, A.: Social ties and their relevance to churn in mobile telecom networks. In: EDBT '08, pp. 668–677. ACM, New York (2008)
3. Nanavati, A.A., Gurumurthy, S., Das, G., Chakraborty, D., Dasgupta, K., Mukherjea, S., Joshi, A.: On the structural properties of massive telecom call graphs: findings and implications. In: CIKM '06: Proceedings of the 15th ACM international conference on Information and knowledge management, pp. 435–444. ACM Press, New York (2006)

4. Onnela, J.P., Saramaki, J., Hyvonen, J., Szabo, G., Lazer, D., Kaski, K., Kertesz, J., Barabasi, A.L.: Structure and tie strengths in mobile communication networks. Proceedings of the National Academy of Sciences 104(18), 7332–7336 (2007)

5. Cortes, C., Pregibon, D., Volinsky, C.: Communities of interest. In: Hoffmann, F., Adams, N., Fisher, D., Guimarães, G., Hand, D.J. (eds.) IDA 2001. LNCS, vol. 2189, pp. 105–114. Springer, Heidelberg (2001)

6. Abello, J., Pardalos, P., Resende, M.G.: On maximum clique problems in very large graphs. In: Abello, J.M., Vitter, J.S. (eds.) Extrernal Memory Alogrithms. Dimacs Series In Discrete Mathematics and Theoretical Computer Science, vol. 50, pp. 119–130. American Mathematical society, Boston (1999)

7. Qi, J., Zhang, Y., Shu, H., Li, Y., Ge, L.: Churn prediction with limited information in fixed-line telecommunication. In: Proc. 5th Int. Symp. Communication Systems Networks and Digital Signal Processing, pp. 423–426 (2006)

8. Ferreira, J., Vellasco, M., Pacheco, M., Barbosa, C.: Data mining techniques on the evaluation of wireless churn. In: ESANN 2004 European Symposium on Artificial Neural Networks, Citeseer, pp. 483–488 (2004)

9. Archaux, C., Laanaya, H., Martin, A., Khenchaf, A.: An SVM based churn detector in prepaid mobile telephony. In: Int. Conf. Information & Communication Technologies (ICTTA), pp. 19–23 (2004)

10. Wei, C., Chiu, I.: Turning telecommunications call details to churn prediction. In: Expert Systems with Applications, vol. 23, pp. 4103–4112 (2002)

11. Au, W., Chan, K., Yao, X.: A novel evolutionary data mining algorithm with applications to churn prediction. IEEE Trans. Evol. Comp. 7(6), 532–545 (2003)

12. Bin, L., Peiji, S., Juan, L.: Customer Churn Prediction Based on the Decision Tree in Personal Handyphone System Service. In: 2007 Int. Conf. Service Systems and Service Management, pp. 1–5 (2007)

13. Goldenberg, J., Libai, B.: Talk of the network: A complex systems look at the underlying process of word-of-mouth. Marketing Letters 12(3), 211–223 (2001)

14. Richardson, M., Domingos, P.: Mining knowledge-sharing sites for viral marketing. In: Proc. 8th ACM SIGKDD, p. 70. ACM, New York (2002)

15. Ziegler, C., Lausen, G.: Spreading activation models for trust propagation. In: Proc. IEEE Int. Conf. on e-Technology, e-Commerce, and e-Service, Citeseer, pp. 83–97 (2004)

16. Estévez, P.A., Held, C.M., Perez, C.A.: Subscription fraud prevention in telecommunications using fuzzy rules and neural networks. Expert Systems with Applications 31(2), 337–344 (2006)

17. Xing, D., Girolami, M.: Employing Latent Dirichlet Allocation for fraud detection in telecommunications. Pattern Recognition Letters 28(13), 1727–1734 (2007)

18. Hilas, C., Sahalos, J.: User profiling for fraud detection in telecommunication networks. In: 5th Int. Conf. technology and automation, pp. 382–387 (2005)

19. Zang, H., Bolot, J.C.: Mining call and mobility data to improve paging efficiency in cellular networks. In: MobiCom '07, pp. 123–134. ACM, New York (2007)

20. How to generate customer loyalty in mobile markets. Technical report, Nokia Siemens Networks (March 2009)

21. Hohwald, H., Frias-Martinez, E., Oliver, N.: ARBUD: A Reusable Architecture for Building User Models from Massive Datasets. In: UMAP 2010 (submitted, 2010)

22. Dean, J., Ghemawat, S.: Mapreduce: Simplified data processing on large clusters, 137–150 (2004)

Mobile Web Profiling: A Study of Off-Portal Surfing Habits of Mobile Users

Daniel Olmedilla, Enrique Frías-Martínez, and Rubén Lara

Telefónica Research & Development
Emilio Vargas 6, 28043
Madrid, Spain
{danieloc,efm,rubenlh}@tid.es

Abstract. The World Wide Web has provided users with the opportunity to access from any computer the largest set of information ever existing. Researchers have analyzed how such users surf the Web, and such analysis has been used to improve existing services (e.g., by means of data mining and personalization techniques) as well as the generation of new ones (e.g., online targeted advertisement). In recent years, a new trend has developed by which users do not need a computer to access the Web. Instead, the low prices of mobile data connections allow them to access it anywhere anytime. Some studies analyze how users access the Web on their handsets, but these studies use only navigation logs from a specific portal. Therefore, very little attention (due to the complexity of obtaining the data) has been given to how users surf the Web (off-portal) from their mobiles and how that information could be used to build user profiles. This paper analyzes full navigation logs of a large set of mobile users in a developed country, providing useful information about the way those users access the Web. Additionally, it explores how navigation logs can be categorized, and thus users interest can be modeled, by using online sources of information such as Web directories and social tagging systems.

1 Introduction

Nowadays, millions of users in the world access daily information on the World Wide Web. We have transitioned from a Web available only to the academic world to a large source of data available to almost everyone. This transition has produced a new range of services combined with new business models (e.g., online advertising). A significant amount of effort has been dedicated to analyze and classify the activity users perform on given web sites (on-portal) or within their whole navigation sessions (off-portal) in order to profile and improve their experience (e.g., through personalization) and the web site service quality (e.g., sponsored links, recommendations, targeted vs. non-targeted advertisement). This work includes generic studies of surfing behavior and patterns [1–3], and more specific ones such as extracting profiles from Web navigation logs, link analysis for personalized Web Search or recommendations [4–6].

P. De Bra, A. Kobsa, and D. Chin (Eds.): UMAP 2010, LNCS 6075, pp. 339–350, 2010.

Lately, mobile phones have become a part of our daily life (in fact there are around 4 billion users subscriptions in the world[1] and around a billion new phones are bought each year[2], including new subscriptions and handset replacements). With new 3G technology and the reduction of data connection tariffs, not only users can access the Web from their handsets, but they are doing it constantly and in new set of situations not possible before (on-the-go). However, due to the lack of available data, only work with on-portal (logs generated within a portal) on user web navigation analysis and profiling has been performed. Very little work has been performed in order to analyze the usage users make of their mobile Web navigation capabilities in a broader spectrum, that is, analyzing their navigation from the moment they connect to Internet to the point where they disconnect, independently of how many portals they have visited.

This paper shows a first effort on trying to understand users off-portal mobile navigation behavior. We analyzed the logs of three months of web navigation (52 million visits to Web domains by 283,000 users) in a developed country and identified, among others, which type of web sites users accessed, which distribution these visits followed and which main categories users are interested in. This information is very valuable in order to personalize services (e.g., better knowledge of the customers in order to improve on-line recommendations and advertisements) as well as to improve existing ones (e.g., parental control services which filter out or warn users when accessing sites with adult or inappropriate content). However, extracting the main categories a domain belongs to is not an easy task.

In this paper, instead of text mining web pages, we tried to use collective intelligence available on the Web in order to automatically classify web sites. First, we relied on the Open Directory Project[3], which provides the largest manually annotated Web directory, and classifies web pages within a total of 17 top categories. Additionally, we also accessed information available on social tagging systems such as YahooMyWeb[4], as well as existing meta-tags available in the accessed web pages. Combining this two sources of information (categories and tags) allows us to identify a set of representative tags per category, which can be used to classify new domains for which tags exist, but no manual categorization has been performed.

The rest of the paper is organized as follows: Sect. 2 presents previous related work in the area. The dataset used and the first analysis made over it are presented in Sect. 3. Section 4 introduces a categorization scheme of pages found in the navigation logs based on the Open Directory Project, and analyzes its results. Web page Meta-Tags and social tagging systems are exploited in Sect. 5

---

[1] http://www.ngrguardiannews.com/compulife/article02/141009,
http://www.cellular-news.com/story/printer/32073.php

[2] http://communities-dominate.blogs.com/brands/2009/02/
bigger-than-tv-bigger-than-the-internet-understand-mobile-of-4-billion-
users.html

[3] http://www.dmoz.org/

[4] Information of YahooMyWeb was obtained before its shut down in March 18th, 2009.

in order to characterize pages, and Sect. 6 combines category and tag information in order to infer new categories from pages that otherwise would not be categorized. Finally, Sect. 7 concludes the paper and outlines future ideas we plan to explore in the future.

2   Related Work

The WWW has already been studied and characterized for online access in a variety of studies [1–3]. However, little is known about the behavior of users when navigating the Web through their mobile phones.

The literature reports a variety of studies that characterize user navigation, search strategies and content for mobile Internet. One of the main characteristics of such studies is that the data used for the studies typically comes from just one portal. In this context, [7] shows that the law of web surfing [3] (developed for traditional on-line access) holds true also for mobile web access, using an extensive data set coming from one web site. The studies presented in [8, 9] show the dynamics of mobile access to a commercial web portal, finding, as previous studies did, that the majority of client request are for a reduced number of documents (i.e. the navigation patterns are very similar for all mobile users). The work presented in [10] focuses on studying search patterns in Google for mobile users. Their conclusions indicate that the diversity of queries in mobile access is far less than in desktop, and that although users for the best part search similar content in both environments, the percentage of Adult queries is vastly larger in mobile access. Based on these results a variety of applications have been developed for predicting user navigation[11] and adapting content for mobile users [12].

A characteristic of mobile access characterization studies is that, while desktop access can be considered homogeneous, mobile access is done with phones with different capabilities (ranging from the size of the screen to the data input method) that deeply affects the analysis. For example, [13] found out that although searches in mobile phones are much shorter than in computers, searches done from iPhones were very similar to the ones performed through a computer, being this conclusion also true when evaluating the variety of queries.

There are some studies that use more than one portal to characterize the mobile WWW, nevertheless those studies focus on characterizing content. For instance, [14] studied over one million mobile pages, and found that from the three content types (WML, C-HTML and XHTML) WML was dominant.

To the best of our knowledge our study is the first one that characterizes and studies mobile user navigation using navigational data originating from the user not from a portal. This implies that for each given user our analysis reflects the different navigation sessions over the portals that the user has accessed for the period of time considered. In this context our study complements previous results in analyzing and characterizing user behavior that have focused (due to the complexity of obtaining the data) on an individual portal.

3   Mobile Web Navigation Analysis

As the basis for the analysis that is described in this paper, information from users off-portal access to mobile Internet via handsets has been used. This dataset includes a total of 52 million accesses (visits hereafter) to more than 45,000 different domains by more than 283,000 users for a time period of 3 months. This usage data belongs to a developed country and the information used from these logs include for each entry an anonymised user id, the domain accessed (not the whole URL in order to preserve privacy) and the time when the access took place.

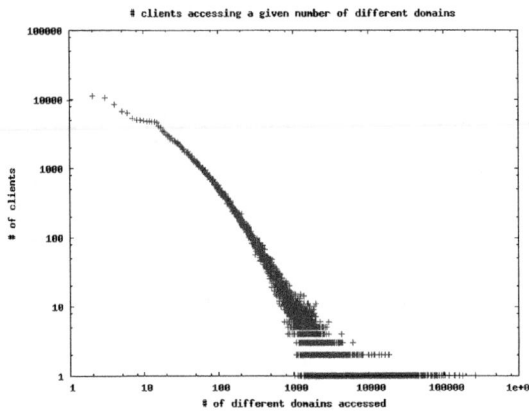

Fig. 1. Distribution of number of domain accesses by clients

Figure 1 and Table 1 respectively shows the distribution of domains visited by users and describes some basic statistics in order to better understand the nature of this dataset. In Fig. 1 we can observe a linear relation in a logarithmic scale between the number of users and the number of domain accessed, representing in a linear scale the typical long tale behavior, i.e. there is a core set of domains heavily accessed. This general behavior is also true when accessing the internet from laptops, and is in agreement with the results presented in [7].

The domains listed in the mobile web logs include a large number of domains that are only accessible from the handset, that is, via the operator network and not from a regular computer Internet connection. These domains typically represent portals belonging to the operator itself (or related advertising banners) or versions of websites adapted to mobile access. For the rest of the study, these domains are ignored, since there will not be any possibility to link them with the category and social tagging sources of information available on the Web. Therefore, we also eliminate the data associated to the users that only accessed those domains we are leaving out of the study, which amounts to 4.49% of our dataset. These customers are likely to have connected by mistake or in order to just test the connectivity, so they reached only the operator portal shown as

Table 1. Generic statistics associated to the analyzed mobile web navigation logs

Description	Amount	% over total
Total Users	283,198	
Total Accesses (visits)	52,297,157	
Total Domains Visited	45,103	
Web accessible domains	34,791	77.14%
Only-mobile domains	10,312	22.86%
Users visiting Web accessible domains	189,283	66.84%
Users visiting only-mobile domains	273,311	96.51%
Users visiting only-mobile domains	93,915	4.49%

starting page (also suggested by the fact that 96.51% of the users have visited at some point one of such operator domains and 90.14% have visited the operator starting portal).

4   Categorizing Web Domains

When analyzing user behavior within a portal, the categorization is typically performed by first categorizing the web pages (or areas) of the portal and then classifying each user visit according to the category assigned to that page. This can be done within a portal since the owner of the portal knows and has control over the content displayed on it. In these cases, user profiles can easily be constructed as a set of:

- (a) categories the user is interested in and
- (b) a weight for each category, typically based on the frequency pages on each category are accessed as a function of time (so recent visits are considered more important than older ones)

However, off-portal implies that users will access pages that are (most likely) not known, and therefore not classified, in advance. In these situations, user profiles typically consist of a set of top keywords extracted from the pages she has visited, and sometimes, there is an effort to classify those keywords in a set of categories. In this paper, we have decided to follow the opposite process. We will first try to analyze those webpages for which we know the category they belong to, and later on try to analyst which keywords are typically representing such categories.

In order to classify web pages on the World Wide Web, we relied on the Open Directory Project (ODP), "the most comprehensive human-reviewed directory of the web", which contains manual classification (by more than 85,000 editors) of more than 4 million sites over 590,000 categories (organized as a tree with 16 top categories). Therefore, we searched within the ODP database for the domains users were navigating. Table 2 describes the results of this analysis. As it is presented, only 15% of domains are annotated with at least one category within the ODP directory.

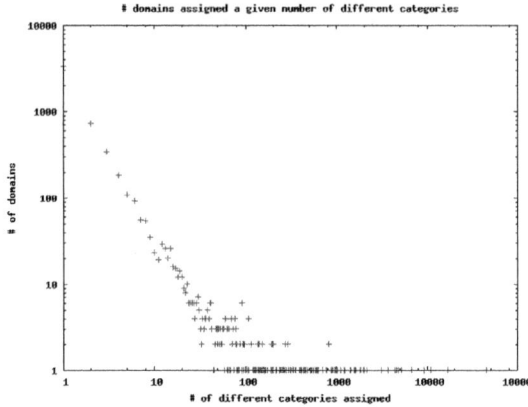

Fig. 2. Distribution of number of domains with N categories assigned

Table 2. Categories matches for domains and users accessing those domains

Description	Amount	% over total
Different Categories assigned to a domain	283,198	
Total Domains with a category assigned	5,483	15.76%
Total Users accessing a domain with category	37,826	19.98%

Additionally, since a domain might fall into more than one category, we analyzed how this distribution look like. Figure 2 shows the distribution of domains that are assigned a given number of categories, which as we can observe, follows a power-law distribution, i.e. a core of domains fall under a great number of categories (capturing probably news portals and such), while the best part of domains are assigned a reduced number of categories indicating also that the content of the portal is more focussed.

5   Page Meta-tags and Social Tagging Systems

In the past, the Web 1.0 provided a small number of people with the ability to share information with the whole world being able to access it. Website administrators (or some users via appropriate content management systems) could upload content to the Web and make it publicly available. Nowadays, the boom of the Web 2.0 allows users to act not only as consumers of information but also as providers. Wikis, blogs, community sites or photo sharing sites are just some examples where users create and publish content to be shared online. Among these there exist social tagging sites, where users are able to assign labels (tags hereafter) to resources other people publish or reference.

In this paper, we try to exploit the information web administrators provide on web pages (via web page meta-tags) as well as the wisdom of the crowds, that

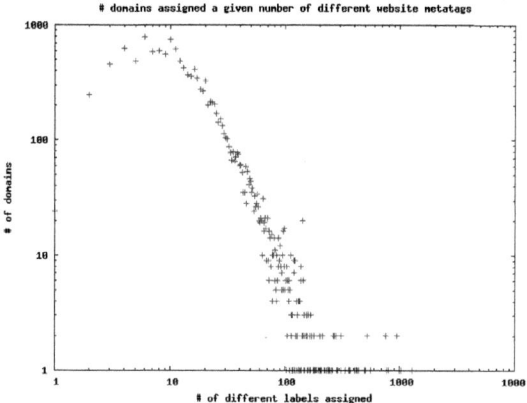

Fig. 3. Distribution of number of domains with N meta-tags assigned

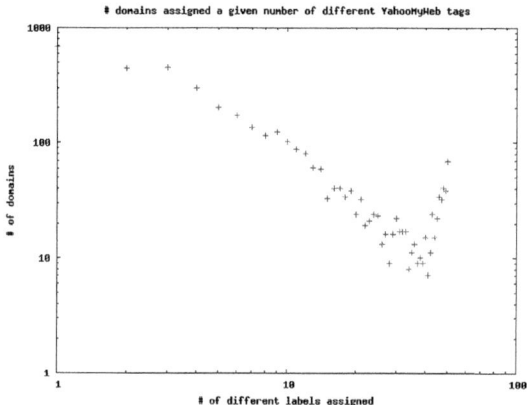

Fig. 4. Distribution of number of domains with N social tags assigned

is, the tags users assign to webpages on social tagging systems. In particular, we have built extractors that are able to receive meta-tags from any website and tags from systems such as YahooMyWeb. Table 3 describes the main characteristics of the data obtained through this process. It shows that the number of different social tags is still very reduced in comparison with the vocabulary used by web administrators. Additionally, only 37% of the domains visited by users were annotated by web administrators, and only 11% had any type of social annotation. It is interesting to see that social annotation already provides information from around a 5% of domains for which no meta-tags exist.

Figures 3 and 4 depict respectively the distribution of domains according to how many meta-tags they contain and the distribution of domains according to how many social tags they have been assigned in a logarithmic scale. Additionally, Fig. 5 shows the combined information also in a logarithmic scale, that is,

Table 3. Meta-tag and social-tag extraction statistics

Description	Amount	% over total
Different Tags assigned	108,851	
Different Meta-Tags assigned	102,188	
Different Social Tags assigned	12,497	
Total Tags assigned	321,533	
Total Meta-Tags assigned	281,090	
Total Social Tags assigned	40,353	
Total Domains with a tag assigned	14,436	41.49%
Total Domains with a Meta-Tag assigned	12,847	36.93%
Total Domains with a Social Tag assigned	3,839	11.03%
Total Users accessing a domain with tag	45,581	24.08%

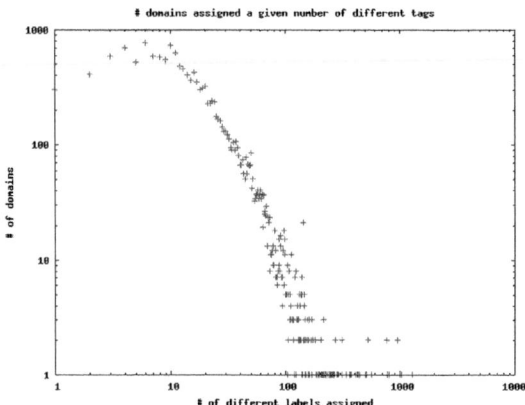

Fig. 5. Combined distribution of number of domains with N meta or social tags assigned

the distribution of domains for which we have any type of tag information. In Fig. 3 we can observe an inverse quadratic relation between the number of domains and the number of labels assigned, i.e. when the number of labels assigned to domains is small there is a quadratic increase, until 10 labels are reached, and after that the number of domains assigned a number of tags higher that 10 is quadratically reduced. Nevertheless, in Fig. 4 we can observe an almost linear reduction of social tags assigned, up to 60 tags, when there is a linear increase. That second part of the graph probably groups highly popular sites that are heavily tag by users. The combined information showed in Fig. 5 maintains the quadratic relation.

6   Combining Classification and Tagging Information

The combination of the two sources of information from sections above allows us to do two things: profile users both with categories and keywords for annotated pages and possibly predict categories of non-annotated ones based on

Table 4. Information available for domains and users visiting them

Description	Amount	% over total	Users	%
Domains with a category or social tag	15,949	45.84%	51,636	27.28%
Domains with category but no social tag	1,513	4.35%	14,350	7.58%
Domains with no category but social tag	10,466	30.08%	29,086	15.37%
Domains with neither category nor tag	3,970	11.41%	29,981	15.84%

the representative tags for each category. This section presents some results of the information we obtained from the combination of navigation logs, manual categorization of web pages and social tagging systems.

Table 4 shows some statistics about domains for which we can extract information based on the methods defined above and the number of users we are able to profile based on that (almost reaching a 30%, which already provides a very good amount taking into account the existence of many users accessing a very small number of domains or even seldom using mobile Internet). We also expect these numbers to increase when using information from other social tagging systems (e.g., delicious[5]) and more accurate navigation information (e.g., obtaining the full URL instead of only the domain). Additionally, we believe it would be good to perform the same analysis periodically in order to observe how the categories and social tagging systems adapt to the evolution of the navigation mobile users perform. Our intuition is that, since the trend for many users is to rely more and more on mobile Web navigation (especially after the emergence of flat rates in most developed countries), the coverage will increase, therefore providing more precise data.

physics science news space engineering nasa astronomy automotive car design

Fig. 6. Tag cloud for category Science

Based on the information gathered, there is a subset of domains for which we have both category and tags assigned. We have used this information to explore the most representative tags assigned to each category. For this, we applied the Inverse Category Frequency (same as Inverse Document Frequency but checking the number of categories where a tag appears):

$$ICF_i = log\frac{|C|}{|\{c : t_i \in c|} \tag{1}$$

where ICF_i represents the Inverse Category Frequency of tag i, C is the number of top categories, and $|\{c : t_i \in c|$ is the number of categories for which the tag t_i has been assigned (through a URL). Basically, what this formula provides us is a classification of which tags are representative in order to characterize a

---

[5] http://delicious.com/

web page, based on whether the same tag is used for many different categories and are therefore not at all representatives[6], or whether it is used consistently only for one category[7].

Based on the results of this process, we were able to extract the most representative tags for each one of the top categories based on the known classified webpages. The interesting aspect of these observations is that we might be able to classify web pages for which no categorization data exists, simply based on the existence of "representative" tags.

To this aim, we selected all keywords with the maximum ICF, that is, they were assigned consistently to only one category and use all these tags to classify domains for which no category exists. The number of representative tags for each category are shown in Fig. 7, and Figs. 6 and 8 show some examples.

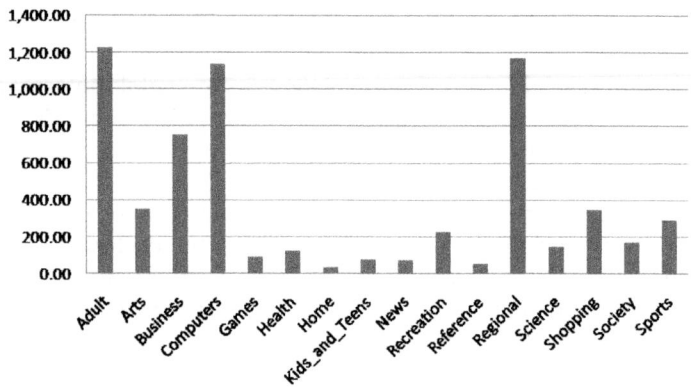

Fig. 7. Number of representative tags per category

Using these representative tags in our dataset allow us to classify a total of 5,617 new domains that were otherwise without category. This amount represents a 102,44% increase on domain classification with respect to the one made using only ODP categorization, therefore duplicating our coverage of domains, from 15,76% to 31,90%.

web aol online internet providers service blog blogging

Fig. 8. Tag cloud for category Computers

With all the information collected, it is possible to improve and personalize the services provided to users. For instance, Figure 7 shows how adult content might be identified even if no previous knowledge of the site exists. This should

---

[6] Examples in our dataset include i.e. "noticias", "online", "free", "imported delicious", "internet", "photos" or "software".

[7] Examples in our dataset included "liverpool" or "pga" for category Sports, "blog" or "aol" for category Computers, or "science" for the category with the same name.

be combined with other approaches analyzing the content of the page in order to increase coverage, but even in that case, the tags extracted and classified as representative for adult content could be used as an input vocabulary for state of art approaches using text mining.

Additionally, user profiling is a key area for companies in order to enhance customer experience, and improve targeted advertisement and marketing among others. For instance, knowing the main categories a user is interested in (as our approach provides) allows to better select online recommendations on web portals or application stores.

7   Conclusions and Future Work

While very large amount of research has been dedicated to web profile analysis on the World Wide Web, and in the mobile world research has focused on navigation logs belonging to a web portal, there is to our knowledge no paper analyzing the navigation of users in a broader sense, having all their session activity independently of the portal or website they access. The advantage of this work is also that a handset typically corresponds to a single person, as opposed to desktop computers where it might not be the case.

This paper provides a first insight by analyzing the logs of three months of mobile web off-portal navigation (over 52 million accesses and 283,000 users) in a developed country and identified, among others, which type of web sites users accessed, which distribution these accesses followed or which main categories they are interested in. This information has also been matched with that of social tagging systems in order to better characterize users and categories, and the combination of these two approaches has been used to infer new categories for domains that were not previously classified.

The work shown in this paper gives an overview of the information contained in mobile navigation logs, but allows for much more advanced analysis. For instance, we plan to explore the usage performed according to time, location and demographic information of the users and see whether categories accessed vary on any of these dimensions.

Additionally, we hope to get navigation logs with the full URL (instead of only the domain) in order to better access categories and tags of visited pages, and also to include information of the web page content, as an additional input of data in order to try to improve the inference process. Since more data will be available, this will solve some already identified problems such as that of keywords appearing consistently in one category because of the use they are given, and not because of the real semantics. For instance, "Britney Spears" or "motel" were classified as adult, and "indoor" as sports because they were consistently used in those categories, but adding information from web page content might help increase accuracy. Also, analyzing co-occurrences of tags will help solve this problem, since there are tags with well-identified semantics (e.g., "pharmacy" or "drugstore" for health).

References

1. Albert, R., Jeong, H., Barabási, A.: The diameter of the world wide web. Nature 401, 130–131 (1996)
2. Huberman, B., Adamic, L.: Growth dynamics of the world wide web. Nature 401, 131 (1999)
3. Huberman, B., Pirolli, P., Pitkow, J., Lukose, R.: Strong regularities in world wide web surfing. Science 280(5360), 95–97 (1998)
4. Jeh, G., Widom, J.: Simrank: a measure of structural-context similarity. In: KDD, pp. 538–543. ACM, New York (2002)
5. Jeh, G., Widom, J.: Scaling personalized web search. In: World Wide Web, pp. 271–279 (2003)
6. Shen, D., Chen, Z., Yang, Q., Zeng, H.J., Zhang, B., Lu, Y., Ma, W.Y.: Web-page classification through summarization. In: Sanderson, M., Järvelin, K., Allan, J., Bruza, P. (eds.) SIGIR, pp. 242–249. ACM, New York (2004)
7. Halvey, M., Keane, M., Smyth, B.: Mobile web surfing is the same as web surfing. Communications of the ACM 49(3) (2006)
8. Adya, A., Bahl, P., Qiu, L.: Characterizing alert and browse services for mobile clients. In: USENIX Tech. Conf., Citeseer, pp. 343–356 (2002)
9. Adya, A., Bahl, P., Qiu, L.: Analyzing the browse patterns of mobile clients. In: Proceedings of the 1st ACM SIGCOMM Workshop on Internet Measurement, pp. 189–194. ACM, New York (2001)
10. Kamvar, M., Baluja, S.: A large scale study of wireless search behavior: Google mobile search. In: Proceedings of the SIGCHI conference on Human Factors in computing systems, p. 709. ACM, New York (2006)
11. Halvey, M., Keane, M., Smyth, B.: Predicting navigation patterns on the mobile-internet using time of the week. In: Special interest tracks and posters of the 14th international conference on World Wide Web, p. 959. ACM, New York (2005)
12. Anderson, C., Domingos, P., Weld, D.: Adaptive web navigation for wireless devices. In: International Joint Conference on Artificial Intelligence, vol. 17, pp. 879–884. Citeseer (2001)
13. Kamvar, M., Kellar, M., Patel, R., Xu, Y.: Computers and iPhones and Mobile Phones, oh my!, 801–809 (2009)
14. Timmins, P., McCormick, S., Agu, E., Wills, C.: Characteristics of mobile web content. Hot Topics in Web Systems and Technologies, 1–10 (2006)

Personalized Implicit Learning in a Music Recommender System

Suzana Kordumova[1], Ivana Kostadinovska[1], Mauro Barbieri[2],
Verus Pronk[2], and Jan Korst[2]

[1] Ss. Cyril and Methodius University
Faculty of Natural Sciences and Mathematics
1000 Skopje, R. Macedonia
{suzana.kordumova,ivana.kostadinovska}@gmail.com
[2] Philips Research, High Tech Campus 34,
5656 AE Eindhoven, The Netherlands
{mauro.barbieri,verus.pronk,jan.korst}@philips.com

Abstract. Recommender systems typically require feedback from the user to learn the user's taste. This feedback can come in two forms: explicit and implicit. Explicit feedback consists of ratings provided by the user for a number of items, while implicit feedback comes from observing user actions on items. These actions have to be interpreted by the recommender system and translated into a rating. In this paper we propose a method to learn how to translate user actions on items to ratings on these items by correlating user actions with explicit feedback. We do this by associating user actions to rated items and subsequently applying naive Bayesian classification to rate new items with which the user has interacted. We apply and evaluate our method on data from a web-based music service and we show its potential as an addition to explicit rating.

Keywords: implicit learning; recommender systems; user behavior; relief algorithm; naive Bayesian classification.

1 Introduction

Recommenders are becoming a popular tool to retrieve, from a vast amount of items such as from A/V content repositories, product catalogues, and the like, only those items a user or a group of users likes. These recommenders are typically offered as a stand-alone service (e.g. MovieLens [1]), or as an add-on to an existing service (e.g. Amazon, iTunes). They increasingly appear in consumer devices, such as the TiVo DVR [2] or the Media Center PC running the Watchmi software plug-in [3].

Recommender systems typically require feedback from the user to learn the user's taste. This feedback generally comes in two forms: *explicit* and *implicit*. Explicit feedback consists of a user providing ratings for a number of items, e.g. on a five point scale or in a binary like/dislike form. Implicit feedback comes from observing user actions such as purchases, downloads, and selections of items for playback or deletion. These user actions have to be interpreted by the recommender system and

P. De Bra, A. Kobsa, and D. Chin (Eds.): UMAP 2010, LNCS 6075, pp. 351–362, 2010.

translated into a rating. For example, typically, recommender systems interpret a purchase action or, in case of music services, a listening action as a positive rating and the skipping of a song in an album as a negative rating.

These types of user actions are quite rudimentary, and it is not always clear how to interpret them. For example, a skipped song may, but need not, indicate a negative rating. Furthermore, the interpretation may be user-dependent.

Some of the advantages of implicit learning are that it frees users from having to explicitly rate items and it allows continuous updating of user preferences.

In this paper we propose a method to learn, from explicit ratings, how to translate user actions on items to ratings on these items. We do this by associating user actions to rated items and considering these actions as features of these items. Naive Bayesian classification is then used to rate new items with which the user has interacted. These implicitly rated items provide an additional source of information for recommending new items. We present a preliminary evaluation of our method on data from a web-based music player. In particular, we show that it enables a form of implicit learning that adapts to each user individually.

The rest of the paper is structured as follows. Section 2 gives an overview of related work in the domain of implicit learning for recommender systems. Section 3 introduces our approach to implicit learning and the architecture we adopted. In Sect. 4 we describe the music listening behavior features. Evaluation and results are presented in Sect. 5. Section 6 concludes the paper.

2 Related Work

Given the advantages that implicit learning may offer, more and more recommender systems now rely on implicit profiles. Recommender systems in all areas of everyday life such as e-purchasing, digital news, music or TV recommenders and search engines, collect implicit knowledge.

In [4] and [5] a user's buying behavior is recorded to achieve more personalized and effective recommendations. In [4], the proposed system first collects user behavioral data, then analyzes this data to extract user preferences and, using these preferences, it creates recommendations. This system deducts the most important attributes for the user among various product attributes based on the ID3 algorithm. In [5], besides tracking the user's behavior, the interaction history of a group of similar users is analyzed. They find a group of customers having similar characteristics, and then analyze their transaction histories. Customer profiles are constructed using the product features and individual and group interests.

Because studies have shown that the vast majority of users are reluctant to provide any explicit feedback on search results and their interest [6], search engines also try to learn the user's preferences automatically. In [6] a user model is proposed to formalize web-pages and to correlate them with clicks on search results. Then, based on this correlation, the paper describes an algorithm to actually learn interests.

Many TV recommender systems have adopted the concept of implicit learning. In [7], the author is exploring different user actions that can be tracked and possible features are extracted from the user's behavior to model the user's preferences in TV shows. The paper also analyzes the most relevant machine learning solutions for adapting recommendations to users.

3 Implicit Learning Model

We propose an implicit learning model that uses music listening behavioral data and an explicit rating history to build an implicit rating history. The architecture of the model is presented in Fig. 1.

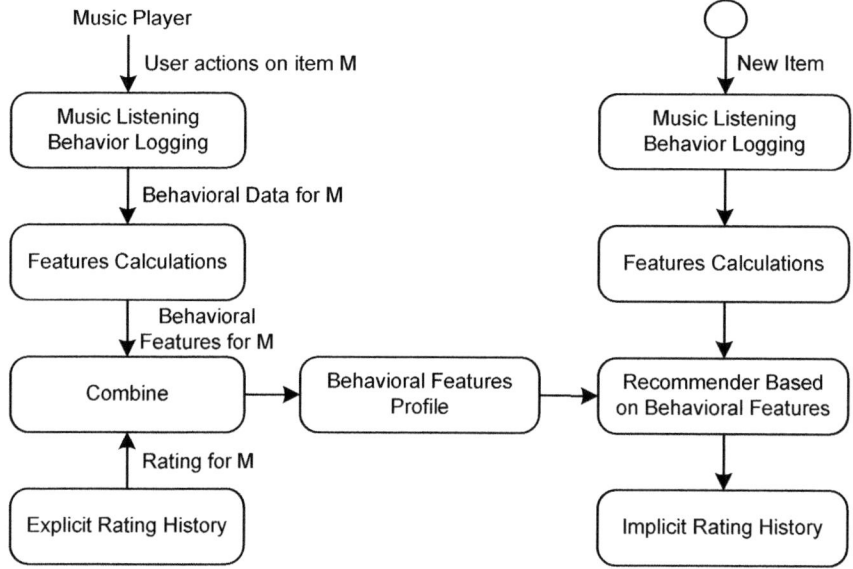

Fig. 1. Implicit learning model

The music player system logs the user's actions and generates behavioral data. The behavioral data corresponding to songs that have explicit ratings is used to calculate behavioral features. These features are combined with the explicit rating to build a behavioral features profile. This is a recommender specific representation of the rating history. Thus, in this architecture, explicit ratings are required, but they are here used to build a behavior-based recommender that can link user behavior to likes and dislikes.

When the explicit rating history and the features profile are rich enough, the behavior-based recommender can start working and build an implicit rating history. When the user interacts with a new song, this interaction is analyzed and behavioral features are extracted in the same way as the features calculation procedure for the explicitly rated items. Using these features and the features profile, the recommender classifies the new item and writes the item and its inferred class into the implicit rating history. In our model we work with two classes: positive and negative (likes and dislikes).

Note that the behavior-based recommender does not provide recommendations directly to the user. It is just used to classify items based on the user interaction. The idea is that the use of both rating histories results in faster learning when compared to only using an explicit rating history.

The implicit and explicit rating histories can be used in different ways to recommend new items with which the user has not yet had any interaction. One way is to construct separate user profiles from the implicit and explicit rating histories (see Fig. 2). Both profiles are then used by an *integrated recommender* to rate new items. This recommender is designed to combine two sources of information about the user and, based on them, make a distinction between future likes and dislikes. An example of integrated recommender is described in [8]. The behavior-based recommender and the integrated recommender are based on naive Bayesian classification.

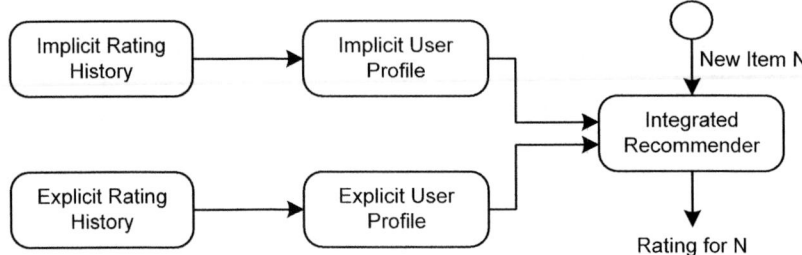

Fig. 2. Combination of implicit and explicit rating histories

4   Extracting Features from Music Listening Behavior

We have applied our model to a web-based music player called xStream [9] developed by Philips Research and used within the company for research purposes. xStream offers an interface to a music library with about 80000 songs from many artists and genres. Users can search or browse the library and select songs to play. When an artist is selected from the library, all his songs and albums are also displayed. It is important to note that the player plays all songs of an album sequentially unless the user decides otherwise by interacting with the player. After the user has selected a song, the player allows standard actions such as pause/resume, stop, seek, skip to next or previous song, repeat, shuffle, and volume adjustment. Additionally, the player allows rating songs, albums and artists on a five-point scale.

All interactions with the player are time stamped and saved in a database together with the song id and the user id. The usage data is then analyzed to extract, for each user and for each rated song, behavioral features. Table 1 gives an overview of the behavioral features we have defined.

The following paragraphs explain in more detail the behavioral features and the rationale for considering them.

Times played. Playing a song repeatedly is an indication that the user likes it; therefore the number of times a song has been played is potentially an important feature.

Table 1. Behavioral features

No.	Feature	Description
1	Times played	Number of times the song has been played
2	Percentage played	How much of the song has been actually played on average
3	Percentage played (multi-valued)	Same as feature 2, but multi-valued
4	Time when played	Time of the day the song was played
5	Date when played	Date the song was played
6	Day of the week when played	Day of the week the song was played
7	Times skipped	Number of times the song was skipped
8	Times next song	Number of times the song followed a previously played song from the same album

Percentage played. Sometimes users skip to another song before the current one is finished. The actual percentage of a song which has been played is an important feature to take into consideration. When a song is played more than once, this feature is averaged.

Percentage played (multi-valued). Most users play their favorite songs more than once. So the *percentage played* feature can also be treated as multi-valued feature [8] instead of averaging the percentage played.

Time, date and day of the week when played. Listening to songs in the morning while at work, or in the night, is different, and provides some behavioral information about the user, his habits and music listening mood during different times of the day, or different days of the week. These features allow taking into consideration part of the context in which the listening takes place. They are treated as multi-valued features.

Times skipped. The number of times a song is skipped is calculated indirectly by observing the user interacting with other songs of the same album. Because, without user intervention, the xStream system plays sequentially all songs of an album, we can identify different skipping behaviors schematically represented in Fig. 3:

1. **Times skipped in album more than two songs:** Corresponds to the user behavior shown in Fig. 3 (*i*). After playing song *a*, the user skips to song *b*, which is more than two songs further in the same album. The value of this feature for all songs between *a* and *c* is then increased.
2. **Times skipped in album at most two songs:** Fig. 3 (*ii*) and (*iii*) illustrate the listening behavior that corresponds to this feature. After playing song *a*, the user skips to song *b*, not more than two songs further in the same album. The value of this feature for all songs between *a* and *b* is then increased. This feature can be further separated into two features: *times one song skipped in album* and *times two songs skipped in album*. The first feature is extracted from the listening behavior illustrated in Fig. 3 (*ii*), and the second feature from the listening behavior shown

in Fig. 3 (*iii*). It is interesting to investigate if skipping just one song or skipping more songs can correspond to different degrees of dislike.

3. **Times skipped first play:** When the user starts the xStream music player, he has to choose one song manually. If this song is not the first track of an album, then we can consider all songs preceding that one as skipped. This situation is shown in Fig. 3 (*iv*), where the value of this feature is increased for all songs before *b*.

4. **Times skipped in other album:** This feature represents the behavior shown in Fig. 3 (*v*). The user skips from song *a* to a song *b* belonging to another album. The value of this feature is increased for all the songs of album X following *a* and for all the songs of album Y preceding *b*.

Mapping different skipping behaviors to different features allows building a more precise implicit user model. The disadvantage of this approach is a more complex model and a potentially larger number of features with missing values. As a trade off, we first calculate the features separately and then we combine them linearly into one feature using weights heuristically determined after interviewing 36 users on their skipping behavior.

Times next song. Considering that the xStream music player automatically plays the songs of an album following a song that a user has manually selected, we can maybe expect the users to take advantage of this feature or, at least, to have adapted their behavior to this characteristic of the player. Therefore, it is possible that a user could favor songs that are followed by other songs he likes. Given a song that is automatically played by the system, this feature counts how many times any of the preceding three songs were manually selected by the user.

To reduce the zero counts in calculating the implicit user profiles, some of the features described above are binned. For examples, the *percentage played* is mapped to 10 non-overlapping, equally distributed bins while *time when played* is non-uniformly mapped to intervals such as "morning", "afternoon", "late evening", etc.

4.1 Listening Sessions Characterization

The listening behavior of a user may vary from one listening session to another. People listen to music while working, when eating, at parties, when hanging out with friends and in many other situations. Depending on the situation, the interaction with the music player may differ considerably. For example, Fig. 4 shows two different listening sessions of about two hours occurring on different days and times of the day for a randomly chosen user.

The plusses represent events in which the user has manually selected a song, while the circles represent events in which a song is played automatically by the music player. The two listening sessions appear to be quite different. The first one has many plusses, indicating intensive interaction with the music player. Closer inspection of the data reveals that many songs are not played till the end, and most songs that are played are chosen manually by the user. The second listening session shows much less interaction and more songs automatically played. The two sessions correspond to two different times of the day and probably to two different listening situations.

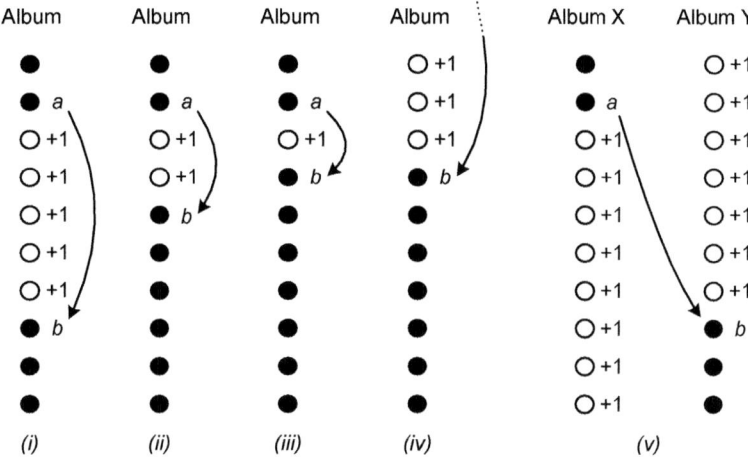

Fig. 3. Different behaviors in skipping songs. The feature values of the skipped tracks are increased by one.

Based on the observation of this and many other user session plots from different users, we decided to classify the listening sessions in two main categories: *active* and *passive* sessions. Active sessions are sessions in which the user wants to listen to some particular songs and choose them manually. There is a lot of interaction with the music player, songs are often skipped or not played until the end. Passive sessions are sessions in which songs are played mostly automatically and until the end. The user has little interaction with the music player and may occasionally skip some songs.

We use the definition of different session types to define new behavioral features. For example, skipping a song in an active or in a passive session indicates a different degree of dislike. Till now we did not made a distinction between listening behavior in different sessions. One approach for making such a distinction is weighing the values associated to the feature values based on the type of listening session. We define the following additional behavioral features based on listening sessions:

1. Times first played in a session.
2. Times last played in a session.
3. Percentage listened weighed based on the session type.
4. Times skipped weighed based on the session type.
5. Times next song weighed based on the session type.

Before calculating the session type, the behavioral data of a user is segmented into sessions depending on the timestamps of the interaction events. A new session is started every time a user interaction event occurs at least 20 minutes after the previous one. A session is labeled as active if the number of manually played songs exceeds the number of automatically played songs, passive otherwise.

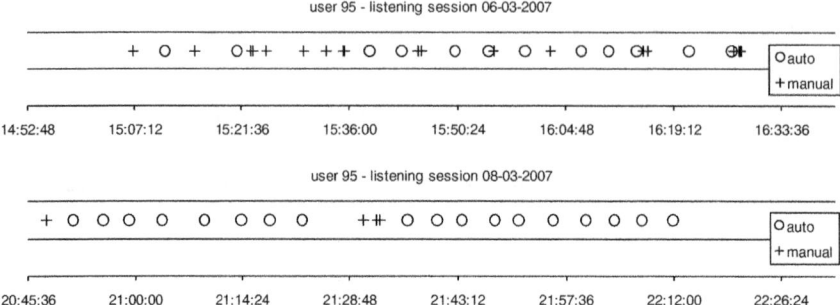

Fig. 4. Two different listening sessions for user 95

Times first and last played in a session. Users may start listening to music by choosing a song they definitely like. They may also stop their listening sessions after encountering a song they dislike.

Percentage listened weighed based on the session type. Not listening to an entire song may indicate a certain degree of dislike for that song. The degree of dislike may vary depending on the session type. This feature weighs the listened percentages of songs depending on the session type.

Times skipped weighed based on the session type. The skipping of a song in an active session is a typical behavior and it may not indicate the same degree of dislike as a skip in a passive session. This feature counts the number of times a song is skipped weighed depending on the session type: skipping a song in a passive session is weighed twice as heavy as skipping a song in an active session.

Times next song weighed based on the session type. Similar to skipping songs, the number of times a song has been played automatically may be interpreted differently for active and passive sessions. This feature counts how many times a song has been played automatically and weighs these counts depending on the session type.

5 Evaluation and Results

The goal of our evaluation was to evaluate the performance of the implicit learning model and its potential for a music recommender.

The experiments are conducted using data from the xStream database. From a total of 661 users who had 2626342 interactions in the period 08.12.2006 – 01.04.2009, we selected for our tests only the 58 users who had rated more than 50 songs. The number of rated songs for each user varies from 51 to 825.

To classify the instances into positive (like) and negative (dislike), we need to set a decision threshold for the Bayesian classifier. We have chosen to use symmetric classification (approximately same positive and negative classification accuracy) to balance the positive and negative error rates (see also [8]). For assessing the performance of the classifier, we have used leave-one-out cross-validation.

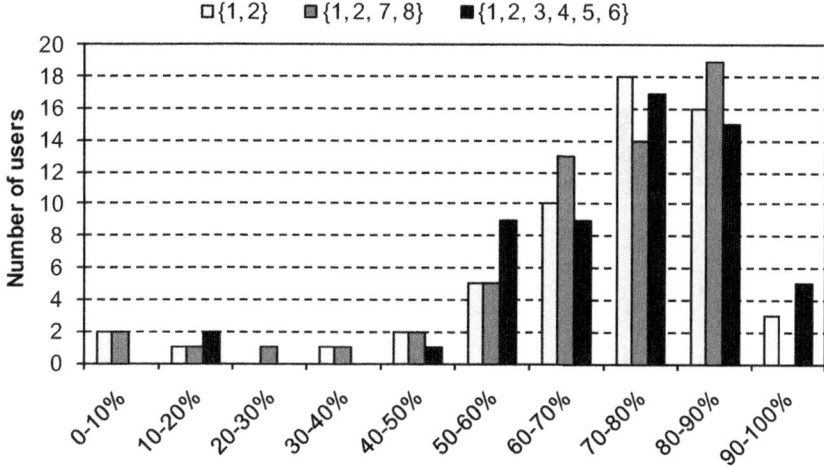

Fig. 5. Number of users with classification accuracy for various ranges

The first experiment was conducted using only two features: times played and percentage played. The classification accuracy for these two features was on average 0.70 with standard deviation 0.19, a promising result considering the use of only two features. For six users, the accuracy is particularly low, below 0.5. After observing the behavior of these users, it appears that they have interacted with the system only for one or two days and they have given most ratings after listening very briefly to the songs. Perhaps these users only wanted to test the system and the recommender or to have an idea of what songs were available. This could explain why the classification accuracy is so low.

If we remove these outliers, the average classification accuracy with merely two features increases to 0.76 with standard deviation 0.10 for the remaining 52 users. It is interesting to consider these six users as outliers because the system, based on a very low accuracy, could decide not to apply implicit learning to these users.

Fig. 5 shows a histogram of the number of users having symmetric classification accuracy in various ranges using three heuristically chosen subsets of behavioral features, namely subsets {1, 2}, {1, 2, 7, 8} and {1, 2, 3, 4, 5, 6}. These subsets were chosen manually as a first investigation into the dependency of the classification accuracy on the choice of feature subsets. See Table 1 for an overview of the behavioral features.

For all three feature subsets the average accuracy is around 0.76 with standard deviation 0.1. There are a few users for which the accuracy is very low and some user for which the accuracy is very high independently of the feature subset used. The third feature subset, however, has the least number of users (5%) with accuracy lower than 50%.

5.1 Optimization

The results discussed in the previous section indicate that different feature sets provide diverse classification accuracy results. For some users, one feature set

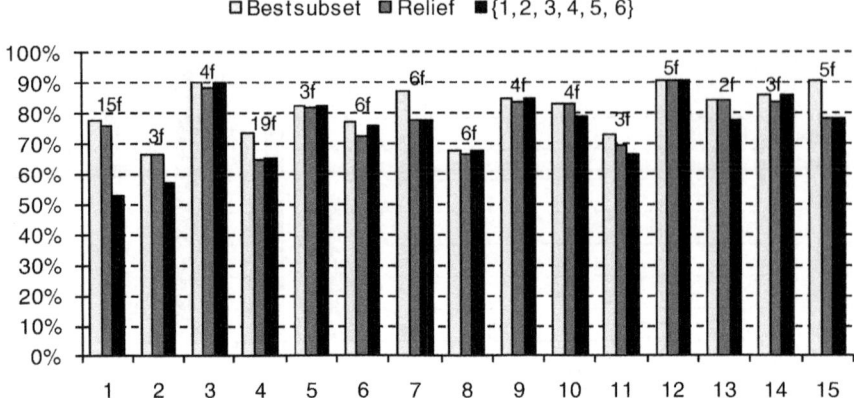

Fig. 6. Classification accuracy of the Relief feature subsets compared to the best feature subsets found

provides better results, for other users another feature set gives higher classification accuracy. In order to optimize the behavior-based recommender for each user, we have applied the Relief feature selection algorithm [10]. This algorithm estimates the importance of features depending on how much their values distinguish between instances that are near to each other, in the same and in different classes.

Figure 6 shows the classification accuracy of a set of 15 randomly chosen users when using the features with the highest weights calculated using the Relief algorithm compared to the best manually searched feature subset found for that user and to subset {1, 2, 3, 4, 5, 6}. Note that an exhaustive search for each user among all possible feature subsets would be impractical for 19 features. The figure also reports the number of features selected using the Relief algorithm.

For most users, the accuracy achieved with the Relief algorithm does not significantly decrease when compared to using the best feature subset found while, in most cases, the number of selected features is remarkably low. Using, for each user, the subset of features indicated by applying the Relief algorithm, provides an average symmetric classification accuracy of 0.75 with standard deviation 0.1. This is a remarkable result indicating that the music preferences of most users can be modeled using a few automatically chosen behavioral features with substantially the same level of accuracy achieved using all features.

In comparison to the heuristically chosen feature set {1, 2, 3, 4, 5, 6}, we can conclude that for most users Relief succeeds in reducing the number of features required without a significant decrease in accuracy. In a few cases, it achieves a better result than the heuristic choice, which was fixed for all users.

6   Conclusions and Future Work

In this paper we have investigated the use of implicit learning for personalization in a music listening context, as an extension to explicit learning. To this end, we have proposed an architecture in which the interaction of the user with user-rated items is used to create an implicit, user-tailored profile to rate items automatically. We have proposed a set of behavioral features based on the interaction of the user with a web-based music player. We have investigated the performance of a two-class naive Bayesian classifier with various feature subset selection methods.

Although our work is still ongoing, from our results we can conclude that the architecture proposed is feasible in that the behavior-based recommender provides encouraging results in terms of modeling explicit ratings through the use of behavioral features. The feature subset selection methods indicate that the subsets found are typically user-dependent, which justifies the use of a more complicated implicit learning architecture than using fixed behavioral features, independent of the user. However, for some users, implicit learning seems to work significantly worse than for others.

For future work, improved, efficient features subset selection methods may be devised to optimize the recommender. Also for further research is the precise embedding of the behavior-based recommender into the integrated recommender and the comparison of its recommendation performance with and without the implicit rating history. An important issue here is how and when to decide to add an implicitly rated song to the implicit rating history. An alternative to adding an implicitly rated song, for example if the behavior-based recommender cannot decide, is to solicit an explicit rating from the user for this song. The possibility of using active learning [11] could also be investigated.

Furthermore, initial results, not presented in this paper, indicate that the use of behavioral data can compete with music metadata, which warrants further research.

From an architectural point of view, a comparison should be made between a recommender based solely on explicit ratings and an integrated recommender based on both explicit ratings and implicit ratings based on behavioral data.

Another topic requiring future work is how to distinguish different types of listening sessions and to apply context-aware learning.

A further challenge is the gathering of a large body of detailed user interaction data, preferably using a real-life product or service over a long period of time. We are currently in the process of collecting such data set in the broadcast video domain.

Acknowledgments. We thank Marco Tiemann for giving us access to the xStream database, Tsvetomira Tsoneva and Igor Berezhnoy for reviewing this manuscript and all the xStream users for providing their behavioral data.

References

1. MovieLens Home Page, http://www.movielens.org/
2. TiVo Home Page, http://www.tivo.com/
3. Watchmi Home Page, http://www.watchmi.tv/
4. Oh, J., Lee, S., Lee, E.: A user modeling using implicit feedback for effective recommender system. In: ICHIT '08: Proceedings of the 2008 International Conference on Convergence and Hybrid Information Technology, pp. 155–158. IEEE Computer Society, Washington (2008)

5. Park, Y.-J., Chang, K.-N.: Individual and group behavior-based customer profile model for personalized product recommendation. Expert Syst. Appl. 36(2), 1932–1939 (2009)
6. Qiu, F., Cho, J.: Automatic identification of user interest for personalized search. In: WWW '06: Proceedings of the 15th international conference on World Wide Web, pp. 727–736. ACM, New York (2006)
7. Gadanho, S.: TV Recommendations based on implicit information. In: ECAI'06 workshop on Recommender Systems (2006)
8. Pronk, V., Verhaegh, W., Proidl, A., Tiemann, M.: Incorporating User Control into Recommender Systems Based on Naive Bayesian Classification. In: RecSys 2007: Proceedings of the 2007 ACM Conference on Recommender Systems, pp. 73–80. ACM, Minneapolis (2007)
9. Tiemann, M., Pauws, S., Vignoli, F.: A Modular Hybrid Recommender System. In: BNAIC 2009: Proceedings of the 21st Conference on Artificial Intelligence, Eindhoven, The Netherlands (2009)
10. Kira, K., Rendell, L.A.: The feature selection problem: Traditional methods and a new algorithm. In: AAAI, pp. 129–134. AAAI Press and MIT Press, Cambridge (1992)
11. Hanneke, S.: Theoretical Foundations of Active Learning, Ph.D. thesis, Carnegie Mellon University (2009)

Personalised Pathway Prediction

Fabian Bohnert and Ingrid Zukerman

Faculty of Information Technology, Monash University
Clayton, VIC 3800, Australia
{fabian.bohnert,ingrid.zukerman}@infotech.monash.edu.au

Abstract. This paper proposes a personalised frequency-based model for predicting a user's pathway through a physical space, based on non-intrusive observations of users' previous movements. Specifically, our approach estimates a user's transition probabilities between discrete locations utilising personalised transition frequency counts, which in turn are estimated from the movements of other similar users. Our evaluation with a real-world dataset from the museum domain shows that our approach performs at least as well as a non-personalised frequency-based baseline, while attaining a higher predictive accuracy than a model based on the spatial layout of the physical museum space.

1 Introduction

This paper proposes a personalised frequency-based model for predicting a user's pathway through a physical space, called *Personalised Transition Model (PTM)*. Our model utilises non-intrusive observations of users' previous movements to generate a pathway prediction, making the assumption that the movements of other like-minded users are more indicative of a target user's behaviour than those of dissimilar users. Specifically, *PTM* utilises personalised transition frequency counts to estimate a target user's transition probabilities between locations. We apply our model to a real-world scenario from the museum domain, by utilising it to predict a visitor's next few exhibits. Our results show that in our domain, *PTM* performs at least as well as a non-personalised frequency-based baseline. Additionally, our model attains a higher predictive accuracy than a transition model based on the spatial layout of the physical museum space.

Our application scenario is motivated by the need to automatically recommend exhibits to museum visitors, based on non-intrusive observations of visitors' movements in the physical space. Employing recommender systems in the museum domain is challenging, as predictions differ from recommendations (we do not want to recommend exhibits that visitors are going to see anyway). We will address this challenge by combining the prediction of a visitor's pathway through the museum with the exhibits predicted to be of interest to the visitor, e. g., [1]. This supports the recommendation of personally interesting exhibits that may be overlooked if the predicted pathway is followed.

2 Related Research

PTM extends our previous research on predicting a visitor's pathway through a physical museum space [2] by using *personalised* transition frequency counts to estimate a

P. De Bra, A. Kobsa, and D. Chin (Eds.): UMAP 2010, LNCS 6075, pp. 363–368, 2010.

visitor's transition probabilities from non-intrusive observations of visitors' previous movements. Other research projects that investigate techniques for personalising the museum experience include *PEACH* [3] for content presentation, *CHIP* [4] for exhibit recommendations based on explicit user input, and the system of Cantino *et al.* [5], which uses Markov decision processes to generate personalised tour proposals for museum visitors. Additional systems for predicting people's future pathways include [6–8]. Specifically, Han and Cho model and predict users' movements by combining Markov models with recurrent self-organising maps [6], Krumm's system uses a Markov model to make short-term route predictions for vehicle drivers [7], and Krumm and Horvitz's system utilises Bayesian inference to predict where a driver is going as a trip progresses, based on a history of the driver's destinations and data about driving behaviours [8].

3 Personalised Pathway Prediction from Non-intrusive Observations

Our *Personalised Transition Model (PTM)* predicts a user's next few locations (e. g., museum exhibits) from non-intrusive observations of users' previous movements. The model utilises a transition probability vector p^k which approximates the probabilities of moving between locations (k denotes the number of previously visited locations). Specifically, p^k's element $p^k(i)$ represents the probability of a user going from a current location i_k to location i (for all $i = 1, \ldots, n$, where n is the cardinality of the set I of all locations). More formally, the transition probability $p^k(i)$ approximates $\Pr(X_{k+1} = i \mid I_a^k)$, i. e., the probability that the current user a's $(k+1)$-th location is location i (where I_a^k is the sequence of locations visited by user a). The transition probabilities are updated whenever the current user moves to a new location. This section describes our approach of estimating p^k in a personalised frequency-based fashion.

For estimating the transition probabilities, we start by calculating the following similarity-weighted personalised frequency counts (for all $i = 1, \ldots, n$):

$$x_{i_k, i} = |N(a)| \frac{\sum_{u \in N(a)} sim_k(a, u) \, \mathbb{1}_u(i_k \to i)}{\sum_{u \in N(a)} sim_k(a, u)}, \tag{1}$$

where $N(a)$ is the set of nearest neighbours, $sim_k(a, u)$ is the similarity between users a and u, and $\mathbb{1}_u(i_k \to i)$ indicates whether user u went from location i_k to location i.

We calculate $sim_k(a, u)$ by comparing the sequences of visited locations of users a and u (called I_a^k and I_u respectively). To this effect, we first determine whether user u has visited user a's current location i_k. If this is the case, we identify the transitions between pairs of locations that occur in both I_a^k and I_u^l, where I_u^l denotes the beginning of I_u up to and including location i_k (otherwise, if $i_k \notin I_u$, we set $sim_k(a, u) = 0$). We then calculate a discounted count of the identical transitions, where a transition is discounted according to how long before location i_k the transition occurred in I_a^k and I_u^l (we use the inverse of the product of the number of visited locations from the transition until the current location i_k in I_a^k and I_u^l as the discounting factor). The discounting is motivated by the fact that identical transitions immediately preceding the current location i_k in I_a^k and I_u^l are more indicative of the users' pathway similarity $sim_k(a, u)$ around location i_k than identical transitions that occurred earlier. Finally, we normalise the resultant sum to the interval $[0, 1]$.[1]

The current user a's set of nearest neighbours $N(a)$ is constructed by selecting up to K_{NN} users that are most similar to the current user a, from those users whose similarity $sim_k(a, u)$ is above a certain non-negative threshold S.

To smooth out outliers, we apply *additive smoothing* to the personalised frequency counts $x_{i_k,i}$ (Equation 1) by adding a smoothing constant $\alpha > 0$ (except for x_{i_k,i_k}, which is 0). Further, we set to 0 the smoothed personalised frequency counts that correspond to the visited locations, and normalise the values so that their sum is 1. By doing this, we focus on unseen locations (e. g., museum visitors rarely return to previously viewed exhibits). The resultant normalised values correspond to the personalised transition probabilities $\tilde{p}^k(i)$, where $i = 1, \ldots, n$.

We employ *shrinkage to the mean* to regularise the personalised transition probabilities $\tilde{p}^k(i)$ by combining them with the transition probabilities $p_{TM}^k(i)$ delivered by a non-personalised frequency-based *Transition Model (TM)* [2]. This yields *PTM*'s shrunken personalised transition probabilities $p^k(i)$ as follows:

$$p^k(i) = p_{TM}^k(i) + \omega \left(\tilde{p}^k(i) - p_{TM}^k(i) \right),$$

where $\omega \in [0, 1]$ is the shrinkage weight. If the set of nearest neighbours is empty (i. e., a similarity-weighted prediction $\tilde{p}^k$ is not possible) or the current user a has visited less than M locations, we estimate the probabilities $p^k(i)$ using simply $p_{TM}^k(i)$.

In summary, *PTM* uses the following adjustable parameters when estimating p^k: (1) smoothing constant α, (2) the minimum number of visited locations M (personalised prediction), (3) the minimum similarity S (nearest neighbour), (4) the maximum number of nearest neighbours K_{NN}, and (5) shrinkage weight ω.

Having (re-)calculated the transition probabilities $p^k(i)$ after every move, we predict a user's next K locations using the *Sequence K* approach, which finds the sequence of the K unvisited locations $i_{k+1}, \ldots, i_{k+K} \in I \setminus I_a^k$ that maximises the probability

$$\Pr\left(X_{k+1} = i_{k+1}, \ldots, X_{k+K} = i_{k+K} \mid I_a^k \right)$$
$$= \prod_{m=1}^{K} \Pr\left(X_{k+m} = i_{k+m} \mid I_a^{k+m-1} \right) = \prod_{m=1}^{K} p^{k+m-1}(i_{k+m}).$$

Factorising the joint probability is possible due to X_{k+m} depending only on the past. This enables maximisation by recursively spanning a search tree of depth $K - 1$, and performing a search for a maximising path from its root to one of the leaves (we pruned the search tree by removing unlikely paths).

4   Evaluation

This section evaluates *PTM* with a real-world dataset of visitor pathways, which was obtained by manually tracking visitors at Melbourne Museum (Melbourne, Australia)

---

[1] We also experimented with similarity measures based on Hamming distance and Levenshtein distance, but their performance was inferior.

from April to June 2008. Specifically, we recorded 158 visitor pathways in the form of time-annotated sequences of visited exhibit areas, providing information of the type that may be automatically inferred from sensors. In total, the dataset (described in detail in [1]) contains 8327 stops at the 126 exhibit areas of Melbourne Museum.[2]

4.1   Experimental Setup

We implemented two baseline models to evaluate *PTM*'s performance: (1) a *Physical Distance Model (PDM)*, using the spatial layout of the museum space to estimate the transition probabilities (making the assumption that transitions to spatially close exhibits are exponentially more likely than those to exhibits that are farther away); and (2) a non-personalised frequency-based approach for estimating a visitor's location probabilities, called *Transition Model (TM)* [2].

Employing leave-one-out cross validation, we tested thousands of configurations of *PDM*, *TM* and *PTM* to assess the influence of the different parameters on the predictive performance of the models, and to determine the best-performing variants. Specifically, model assessment was done by comparing the *negative log probability (NLP)* scores of the various configurations (the NLP score represents the average of the negative logs of the probabilities with which the exhibits actually viewed next were predicted).[3]

We conducted two types of experiments for assessing the predictive accuracy of the best-performing model configurations in the *Sequence K* prediction mode (as above, we used leave-one-out cross validation):

- **Overall Visit (OV).** *OV* evaluates *overall* performance for a museum visit. For each visitor, we started with an empty visit, and iteratively added each viewed exhibit to the visit history. For each iteration, we predicted the next K exhibits, and added these predicted exhibits to a global set of predicted exhibits (ignoring duplicate predictions). At the end of a visit, we calculated *precision (Pre)*, *recall (Rec)* and *F-score* for the entire visit by comparing the accumulated set of all predicted exhibits to the set of actually viewed exhibits. The resultant values were averaged over all visitors.

- **Progressive Visit (PV).** *PV* evaluates *immediate* predictive model performance with the progression of a visit, i. e., as the number of viewed exhibit areas increases. For each *fraction of a visit*, we first predicted the next K exhibits (we used visit fractions rather than the actual number of viewed exhibits, because different visits have different lengths). We then measured immediate classification accuracy by calculating $CA(K) = |\mathcal{K} \cap \mathcal{M}|/K$ (i. e., the proportion of the predicted sequence $\mathcal{K}$ of the next K exhibits that appears in the sequence $\mathcal{M}$ of the next K actually viewed exhibits; this measure equals immediate recall and precision), and averaged the resultant values over all visitors for each visit fraction.

---

[2] For our experiments, we ignore travel time between exhibit areas, and collapse multiple viewing events of one area into one event.

[3] The *PTM* configuration that achieves the minimum NLP score is $\{\alpha = 0.04, M = 3, S = 0.00, K_{NN} = 58, \omega = 0.40\}$, where the symbols are explained at the end of Sect. 3. We omit further results of a sensitivity analysis of *PTM*'s parameters due to space limitations.

Table 1. Model performance for the *OV* and *PV* experiments

	Sequence 1			*CA(1)*	*Sequence 3*			*CA(3)*
	Pre	Rec	F-score		Pre	Rec	F-score	
PDM	59.03%	46.13%	51.73%	45.06%	54.27%	65.28%	59.08%	45.93%
TM	65.58%	56.87%	60.87%	54.29%	58.59%	71.52%	64.19%	52.59%
PTM	**67.09%**	**58.33%**	**62.35%**	**55.97%**	**59.12%**	**72.90%**	**65.06%**	**54.11%**

4.2 Results

This section presents the results of our evaluation for $K = 1$ and $K = 3$.

Evaluation of Pathway Predictions for K = 1. The results for $K = 1$ are summarised in the left-hand side of Table 1. For the *OV* experiment, the frequency-based models *TM* and *PTM* statistically significantly outperform the distance-based baseline *PDM*.[4] More importantly, *PTM* attains statistically significantly better results than *TM* with respect to all measures, which means that using personalised transition frequency counts is beneficial. For the *PV* experiment, all models perform at a relatively constant level with the progression of a visit, with *PTM* achieving the highest average *CA(1)* of 56% (averaged over 1000 equally-spaced visit fractions). Further, *TM* and *PTM* perform statistically significantly better than *PDM* for 66% and 82% of a visit respectively, and *PTM* performs consistently at least as well as *TM* (statistically significantly better than *TM* for 21% of a visit, while *TM* never outperforms *PTM*). These results indicate that other visitors' movements are better predictors of a visitor's next exhibit than the spatial layout of the museum. Additionally, personalisation aids prediction, as personalised *PTM* outperforms non-personalised *TM* for the *OV* experiment, and for some portion of a visit for the *PV* experiment.

Evaluation of Pathway Predictions for K = 3. The right-hand side of Table 1 summarises the results for $K = 3$. For the *OV* experiment (as for $K = 1$), the frequency-based models *TM* and *PTM* statistically significantly outperform the distance-based baseline *PDM*. Further, personalised *PTM* performs statistically significantly better than non-personalised *TM* with respect to all measures. For the *PV* experiment (as for $K = 1$), all models perform at a relatively constant level with the progression of a visit (*PTM* attains an average *CA(3)* of 54%). In addition, *TM* and *PTM* perform statistically significantly better than *PDM* for 63% and 77% of a visit respectively, and *PTM* attains a statistically significantly higher classification accuracy *CA(3)* than *TM* for 31% of a visit (*TM* never outperforms *PTM*). These results are consistent with those for $K = 1$. Comparing the *Sequence 3* variants of our models with the *Sequence 1* variants for the *OV* experiment, precision for *Sequence 3* is lower while recall is higher. This is because at each stage of a visit, predicting the next three exhibits leads to a larger accumulated set than predicting only the next exhibit (higher recall). However, the predictions for three exhibits are less likely to be correct than those for one exhibit (lower precision). For the *PV* experiment, *PTM* outperforms *TM* for a slightly longer portion of a visit for $K = 3$ compared to $K = 1$ (31% vs. 21%).

---

[4] The statistical tests performed are one-tailed paired t-tests (significance level $\alpha = 0.05$).

5   Conclusions and Future Work

This paper proposed a frequency-based *Personalised Transition Model (PTM)* for predicting a user's pathway through a physical space. Specifically, our model estimates a target user's transition probabilities between discrete locations utilising personalised transition frequency counts, which in turn are estimated from the movements of other like-minded users by means of a nearest-neighbour collaborative approach (using a pathway-based similarity measure). We evaluated *PTM* by predicting a museum visitor's next $K = 1$ and $K = 3$ exhibits, and showed that in our scenario (1) *PTM* and *TM* (a non-personalised frequency-based baseline) outperform a distance-based baseline, which means that other people's movements are better predictors of a visitor's pathway than the spatial layout of the museum; and (2) personalisation aids prediction, as *PTM* outperforms *TM* for the overall measures (recall, precision and F-score) and for at least some portion of a visit for the realistic *Progressive Visit* experiment.

Overall, *PTM* yields only a modest (yet statistically significant) improvement over *TM*. The small size of this improvement may be due to the sparsity problem, i. e., the small size of our dataset, which contains only a few visitors with very similar pathways. In the future, we intend to apply *PTM* to larger datasets to investigate this further. We also plan to investigate models for predicting longer location sequences.

Acknowledgements. This research was supported in part by grant DP0770931 from the Australian Research Council. The authors thank Museum Victoria for its assistance; and David Abramson and his team for their help with the computer cluster.

References

1. Bohnert, F., Zukerman, I.: Non-intrusive personalisation of the museum experience. In: Houben, G.-J., McCalla, G., Pianesi, F., Zancanaro, M. (eds.) UMAP 2009. LNCS, vol. 5535, pp. 197–209. Springer, Heidelberg (2009)
2. Bohnert, F., Zukerman, I., Berkovsky, S., Baldwin, T., Sonenberg, L.: Using interest and transition models to predict visitor locations in museums. AI Communications 21(2-3), 195–202 (2008)
3. Stock, O., Zancanaro, M., Busetta, P., Callaway, C., Krüger, A., Kruppa, M., Kuflik, T., Not, E., Rocchi, C.: Adaptive, intelligent presentation of information for the museum visitor in PEACH. User Modeling and User-Adapted Interaction 18(3), 257–304 (2007)
4. Wang, Y., Aroyo, L., Stash, N., Sambeek, R., Schuurmans, Y., Schreiber, G., Gorgels, P.: Cultivating personalized museum tours online and on-site. Interdisciplinary Science Reviews 34(2), 141–156 (2009)
5. Cantino, A.S., Roberts, D.L., Isbell, C.L.: Autonomous nondeterministic tour guides: Improving quality of experience with TTD-MDPs. In: Proc. of the 6th Intl. Joint Conf. on Autonomous Agents and Multi-Agent Systems (AAMAS-07), pp. 91–93 (2007)
6. Han, S.J., Cho, S.B.: Predicting user's movement with a combination of self-organizing map and Markov model. In: Kollias, S.D., Stafylopatis, A., Duch, W., Oja, E. (eds.) ICANN 2006. LNCS, vol. 4132, pp. 884–893. Springer, Heidelberg (2006)
7. Krumm, J.: A Markov model for driver turn prediction. In: Proc. of the Society of Automotive Engineers (SAE) 2008 World Congress (2008) Paper 2008-01-0195
8. Krumm, J., Horvitz, E.: Predestination: Inferring destinations from partial trajectories. In: Dourish, P., Friday, A. (eds.) UbiComp 2006. LNCS, vol. 4206, pp. 243–260. Springer, Heidelberg (2006)

Towards a Customization of Rating Scales in Adaptive Systems

Federica Cena, Fabiana Vernero, and Cristina Gena

Dipartimento di Informatica, Università di Torino Corso Svizzera 185, 10149 Torino, Italy
{cena,vernerof,gena}@di.unito.it

Abstract. In web-based adaptive systems, the same rating scales are usually provided to all users for expressing their preferences with respect to various items. It emerged from a user experiment that we recently carried out that different users show different preferences with respect to the rating scales to use in the interface of adaptive systems, given the particular topic they are evaluating. Starting from this finding, we propose to allow users to choose the kind of rating scale they prefer. This approach raises various issues; the most important is that of how an adaptation algorithm can properly deal with values coming from heterogeneous rating scales. We conducted an experiment to investigate how users rate the same object on different rating scales. On the basis of our interpretation of these results, as an example of one possible solution approach, we propose a three-phase normalization process for mapping preferences expressed with different rating scales onto a unique system representation.

1 Introduction

In modern adaptive web-based systems, users are active: they can constantly interact with the system (e.g. expressing preferences regarding things like privacy configuration settings, friends, and interests). This active user participation applies also in the personalization setting: in an increasing number of systems, users are allowed to inspect and modify their user model or to express some kind of preference towards the presented items. For expressing their preferences, users are provided with some kind of *rating scale*, which is usually the same for all users. However, we found in an experiment on preference evolution, conducted in collaboration with the Prevolution research unit[1], that users hold - and maintain over time - different opinions about the best rating scale for performing either a certain task or a certain kind of tasks[2]. No single rating scale can therefore be thought to satisfy the specific needs and preferences of all users. Given such heterogeneity in user preferences, offering all users the same rating scale could be a risky option: some users might find their experience with the system unsatisfactory and, in the extreme case, decide not to interact with it at all. We therefore

---

[1] The research described was conducted in collaboration with activities carried out by the targeted research unit Prevolution at FBK-irst, funded by the Autonomous Province of Trento, Italy. We thank Anthony Jameson and Silvia Gabrielli, who worked with us on the design of the research project to which this study belongs.

[2] For the specific results of this study, see
http://www.di.unito.it/~vernerof/experiment09.html

P. De Bra, A. Kobsa, and D. Chin (Eds.): UMAP 2010, LNCS 6075, pp. 369–374, 2010.

argue that adaptive systems should either offer customizable rating scales (thus allowing users to choose their favourite one for performing the various tasks in the system) or adapt them automatically, in order to improve user satisfaction. This flexibility, however, might come at a high price, since ratings expressed by means of different rating scales should be somehow normalized before the system makes use of them in the adaptation algorithm. Mapping may be complicated by the fact that rating scales may influence the kind of ratings given by users, because of their special features - in particular, we consider their *granularity* (i.e, their level of precision) and their *emotional connotation* (i.e. the emotions that they evoke). As for granularity, notice, for example, that a rating "3" in a scale from 1 to 3 does not necessarily correspond to the highest rating in another scale with more than 3 positions. As for emotional connotation, for example, users who perceive some ratings on a given rating scale as being rude may avoid using them. Moreover, we point out that users may attribute different meanings to the same rating made with the same rating scale, and they may exhibit idiosyncratic patterns of rating behaviour (e.g., some users may tend to give only positive ratings). Starting from these insights, we decided to carry out a further experiment with the goal of investigating users' rating behaviour and in particular the relationships among their ratings on different rating scales.

2 Experiment

The experiment was performed in the context of iCITY[3], a social recommender system in the cultural events domain which integrates adaptivity principles with Web 2.0 social features. Participants were asked to express their preferences for five different topics, evaluating each one with each of the three rating scales with which they were provided. Taking inspiration on interfaces often used in social websites, we selected thumbs, stars and sliders as rating scales since they differ with respect to the features that we wanted to consider (i.e., emotional connotation and granularity). Each rating scale conveys a different *metaphor*, which influences the emotional connotation: for the thumbs, the metaphor is related to human behaviour; for the stars, it relies heavily on cultural conventions (as with hotel ratings); for the sliders, it is technological (e.g., evoking measuring tools). Specific connotations were identified from the oral comments provided by users in our previous experiment: thumbs are "friendly" and young, but also "impolite" and too simple; stars are classical, familiar, cool; sliders are precise, cold, "detached" and boring. Regarding the *granularity*, the thumbs provide a coarse granularity, where only three different ratings are possible: negative, neutral/intermediate and positive; the stars provide a finer granularity with five positions and no explicitly negative ratings (the minimum rating being zero stars), while the sliders provide the finest granularity (the minimum rating being zero and the maximum being ten).

Hypothesis. We hypothesized that users do not always follow strict mathematical proportion when they map their assessments onto various rating scales (i.e. they sometimes map their assessments in an unexpected manner); and that this fact may be due to the differing granularity and emotional connotations of each rating scale, as well as to the users' partly idiosyncratic rating behaviours.

---

[3] http://icity.di.unito.it/dsa-en

Experimental Design. We employed a within-subject design in which each participant used each of 3 rating scales.

Subjects. We selected 16 participants, 19-55 years old, from colleagues and friends, according to an availability sampling strategy.

Measures and Material. A series of fifteen web pages was prepared, each one presenting the topic that was to be evaluated and containing one of the rating scales. For each topic, three web pages were devised, one for each rating scale. The participants' ratings were recorded and stored by the system.

Experimental Task. Participants had to read written instructions autonomously. They were asked to express their preferences for five different topics (corresponding to the categories of events in the taxonomy of iCITY, i.e. Art, Cinema, Theater, Literature, Music) using the three rating scales. Each participant used every rating scale for every topic. The order of presentation of the topics was randomized for each participant, and, given a certain topic, the order of presentation of the rating scales was also randomized.

Results. We found that 20% of the ratings expressed with the thumb rating scale were "thumbs down", 51% were "thumbs up", and the remaining 29% were in the intermediate position. As for the stars, 6% of the ratings were "0", 11% were "1", 18% were "2", 20% were "3", 20% were "4" and 24% were "5". As for the sliders, 1,5% of the ratings were "0", 6% were "1", 8% were "2", 8% were "3", 3% were "4", 14% were "5", 8% were "6", 12% were "7", 12% were "8", 9% were "9" and 15% were '10". Regarding users' mappings of ratings among the different scales, we found the following. "Thumbs down" rating was mapped by participants to a 0 (29%), to a 1 (50%) or to a 2 (21%) on the star rating scale; to a 0 (7%), to a 1 (29,5%), to a 2 (21,4%), to a 3 (21,4%) or to a 5 (21,4%) on the slider rating scale. The intermediate rating on the thumb rating scale was mapped either to 2 (45%), 3 (45%), or to 4 (10%) on a star rating scale. It was mapped to 2 (10%), 3 (10%), 4 (10%), 5 (30%), 6 (15%), 7 (10%), 8 (10%) or 9 (5%) on a slider rating scale. Then, we compared users' mappings among different scales with the corresponding mathematical mapping. Considering the ratings expressed on all three rating scales, 60% of ratings were examples of mathematical proportion: in particular, 24% of the ratings are mapped to a perfect mathematical proportion (e.g. a rating of 10 on the slider rating scale is mapped to a rating of 5 on the star rating scale and to a "thumbs up" on the thumb rating scale), while 36% of ratings can be considered a good approximation of mathematical proportion (e.g., a rating of 9 on the slider rating scale is mapped to a rating of 5 on the star rating scale). It is worth noting that 40% of ratings depart considerably from mathematical proportion, showing that mathematical proportion is not enough to make a mapping which is able to capture the actual meaning of user ratings. To give a better idea of what we mean, we report that two different users made quite opposite rating choices: the first assigned a certain topic a rating of "thumbs down" on the thumb rating scale, of 2 on the star rating scale and of 5 on the slider rating scale. The second user, by contrast, mapped a rating of "thumbs up" on the thumb rating scale to 2 on the star rating scale and 5 on the slider rating scale. Thus, the same two ratings on the two finest scales have opposite meanings for these two users. This example confirms our hypotheses that different users may attribute different meanings to the same rating made with a certain rating scale. With mathematical proportion alone, we would not be able to map these ratings according to the meaning that they really have for users. We also note that only one user showed perfect mathematical proportionality with respect to the lowest ratings, assigning a rating of 0 on all three rating scales when she assigned 0 on the thumb scale. None of the other users ever used the lowest rating on the sliders. This rating is apparently perceived as being more negative than the mathematically equivalent rating of 0 on the other scales.

Although conducted with a small number of participants, this experiment confirmed our idea that mathematical proportion alone is not sufficient to translate ratings from one scale to another - and consequently, to an internal representation.

3   Discussion of a Possible Normalization Process

If ratings could be exactly mapped from one scale to another by means of a simple mathematical proportion, they could also be mapped to a uniform representation - corresponding to the user model representation of a given system - with no effort. Unfortunately, it can be seen from our experiment that other factors have to be taken into account, such as emotional connotations and user features. We discuss here an algorithm for normalizing user ratings with respect to a uniform representation which considers emotional connotations and user features. It comprises the following steps: i) mathematical normalization; ii) connotation-based normalization; iii) user model-based adjustment. As explained in the following, the first two steps are mutually exclusive. We again refer to iCITY as a use case, giving example rules for the mapping of user ratings to the internal representation of such ratings, where user interests are coded in a scale in the range [0, 1].

Mathematical normalization. Mathematical proportion can still be used as a basis for converting ratings from the input rating scale to the uniform internal representation if such ratings tend not to deviate much from the mere mathematical proportion in the observed mapping from one scale to another. According to our data, proportionality could be used for i) extreme positive ratings and ii) intermediate ratings. As for the first case, a rating of 5 on the star rating scale is usually mapped to a 10 on the sliders (63%). Thus, such ratings can be mapped to a *normalized rating* of 1 in the internal representation [0,1], according to strict proportion (*rule 1*):

if ((user_rating = 5 and rating scale = 'stars') or (user_rating = 10 and rating scale = 'sliders')) *then* {*normalized_rating = 1;*}

As for the second case, intermediate ratings on the three rating scales (thumbs up for the thumbs, "2" or "3" for the stars, "5" for the sliders) are mapped according to strict proportion in 59% of the cases. Thus, they can be mapped to a *normalized rating* of 0.5 in the internal representation (*rule 2*).

Connotation-based normalization. The emotional connotation of the input rating scale should be considered for ratings where the mappings deviate from mathematical proportionality and show recognizable relationships among the scales. In our experiment, we observed some tendencies relating to extremely low ratings and based on the idea that these ratings given with a thumb or star scale are commonly perceived as less negative than if given with a slider rating scale. First, the lowest rating on the thumb rating scale is often mapped (71%) to ratings higher than strict mathematical proportion on the star and slider rating scale. Second, the lowest rating on the star rating scale tends to be mapped to ratings higher than the strict mathematical proportion on the slider rating scale (75%). Third, comparing how the lowest rating on the thumb and star rating scale tends to be mapped to ratings on the slider scale, the lowest rating on the thumb rating scale tends to be mapped to higher ratings than the lowest rating on the star rating scale. Thus, the lowest ratings on the thumb and star rating scale will be mapped to ratings slightly higher than 0 in the internal representation (*rule 3*). An example rule is:

if (user_rating = 0 and rating scale = 'thumbs') then {*normalized_rating = 0.2;*}
else if (user_rating = 0 and rating scale = 'star') then {*normalized_rating = 0.1;*}
else if (user_rating = 0 and rating scale = 'slider') then {*normalized_rating = 0;*}

User model-based adjustment. Normalized ratings can be further adjusted considering some features of the "user as rater", which can be inferred from her behavior in the recommender system. In our experiment, we observed that users differ as for their level of accuracy in rating and for their general attitude toward the evaluated topics (critical vs. enthusiastic users). Moreover, we assumed that users may also differ in terms of their reliability, that is, for their tendency to give ratings that actually correspond to their opinions. Accordingly, we could formulate the following heuristics. First, if the user is always very precise in rating (*accuracy*) we can consider her rating as it is. In case she is not very precise, her rating should be adjusted, for example by merging it with user interest as inferred by the system from user behavior (if available) (*rule 4*). Second, if the user is *reliable*, her rating can be considered as it is; otherwise, her rating should be adjusted, for example by merging it with inferred user interest, as suggested before (*rule 5*). Third, if the user is critical, her rating should be increased a little, while if she is enthusiastic, her rating should be slightly decreased (*rule 6*). Here is an example rule for the adjustment of normalized ratings from the previous steps. We consider the case of a user who is both accurate and reliable:

if (user_accurate = true and user_reliable = trues) then
{*if (user_attitude = 'critical') then* {*normalized_rating = normalized_rating + 0.05;*}
else if (user_attitude = 'enthusiastic') then {*normalized_rating = normalized_rating - 0.05;*}}

Let's look at an example to clarify these concepts. We can consider the case of a user (very precise, reliable, and enthusiastic) who rates an iCITY item with "2" on a star rating scale, and another one (very precise, reliable but critical) who rates the same item as "0" on a thumb rating scale. For the first user, we perform a mathematical normalization, which transforms the rating 2 on the star rating scale into the rating 0.5 (*rule 2*). For the second user, we instead perform the connotation-based normalization and the rating 0 is normalized to 0.2 (*rule 3*). Finally, it is necessary to consider specific user features. Since the first user is considered enthusiastic, the system lowers her rating to 0.45 (*rule 6*). By contrast, the rating of the other user is slightly increased to 0.25 since this user is considered very critical (rule 6).

4   Conclusion and Related Work

With this paper, we have made the first step towards a customization of rating scales in adaptive systems. The main contributions of this paper are i) raising the question of when it is desirable to use different scales for different people, ii) describing the problem of normalization that needs to be solved if this approach is taken, iii) discussing a possible solution for a not strictly mathematical normalization, and iv) providing some examples of rules of thumb based on the experiment we performed. The benefit is that designers of adaptive systems will now be aware of this problem and can take it into account in subsequent research.

Our results should be intended as initial heuristics that need to be further investigated. We are working on an experimental evaluation of the proposed approach. Furthermore, we will have to consider that inferring if a user is critic, enthusiastic or reliable is not simple. As a consequence, we may have to face the cold-start problem, requiring users to interact with the system for quite a long time before the user model-based part of the discussed normalization can be applied. Also, other approaches to the mapping of ratings should be considered, such as probabilistic approaches. Machine learning may also prove useful, since it could allow a system to automatically learn a model for normalizing the ratings, based on some training data. Notice that, in this case, no explicit heuristics about emotional connotations and user attitudes in rating might need to be applied - though it would still be important to have some understanding of the learned models, so as to be able to recognize the conditions under which they can be applied.

Considering similar or related research, another paper focusing on rating scales is [1], which defined the main elements that determine the design of rating scales aimed at collecting explicit user feedback. They also found that user preferences for scales were in poor agreement, in accordance with our findings. The study of rating scales can be framed into the larger domain of *option setting interfaces*. Since option setting is often considered a boring and time-consuming task ([2, 3]), it is particularly desirable that rating scales actually match user preferences. Notice that a first translation is needed whenever users have to express their preferences by means of some rating scale: [4] pointed out that the granularity of true user preferences, that is, the number of levels among which users wish to distinguish, may be different than the range and granularity provided by the available rating scales. [5] explicitly investigated how different rating scales affect user ratings. They compared a binary scale providing only thumbs up or down, a no-zero scale ranging from -3 to +3, and a 0.5 to 5 star scale, allowing half-star increments, with the original MovieLens five-position rating scale. They found that ratings on all three scales correlate strongly with original ratings on the five-position scale; however, they observed that users tended to give higher mean ratings on the binary and on the no-zero scales.

References

1. van Barneveld, J., van Setten, M.: Designing Usable Interfaces for TV Recommender Systems. In: Ardissono, L., Kobsa, A., Maybury, M. (eds.) Personalized Digital Television. Targeting programs to individual users. Kluwer Academic Publishers, Dordrecht (2004)
2. Page, S., Johnsgard, T., Albert, U., Allen, C.: User customization of a word processor. In: Proceedings of CHI 1996, pp. 340–346 (1996)
3. Trewin, S.: Configuration agents, control and privacy. In: Proceedings of the ACM Conference on Universal Usability, pp. 9–16 (2000)
4. Cosley, D., Lam, S., Albert, I., Konstan, J., Riedl, J.: Is seeing believing? how recommender system interfaces affect users' opinions. In: Cockton, G., Korhonen, P. (eds.) CHI, pp. 585–592. ACM, New York (2003)
5. Herlocker, J., Konstan, J., Terveen, L., Riedl, J.: Evaluating collaborative filtering recommender systems. ACM Trans. Inf. Syst. 22(1), 5–53 (2004)

Eye-Tracking Study of User Behavior in Recommender Interfaces

Li Chen[1] and Pearl Pu[2]

[1] Department of Computer Science, Hong Kong Baptist University
Hong Kong, China
lichen@comp.hkbu.edu.hk
[2] Human Computer Interaction Group, Swiss Federal Institute of Technology in Lausanne
(EPFL), Lausanne, Switzerland
pearl.pu@epfl.ch

Abstract. Recommender systems, as a type of Web personalized service to support users' online product searching, have been widely developed in recent years but with primary emphasis on algorithm accuracy. In this paper, we particularly investigate the efficacy of recommender interface designs in affecting users' decision making strategies through the observation of their eye movements and product selection behavior. One interface design is the standard list interface where all recommended items are listed one by one. Another two are layout variations of organization-based interface where recommendations are grouped into categories. The eye-tracking user evaluation shows that the organization interfaces, especially the one with a quadrant layout, can significantly attract users' attentions to more items, with the resulting benefit to enhance their objective decision quality.

Keywords: recommender systems, list interface, organization design, eye-tracking study, users' adaptive behavior.

1 Introduction

Although recommender systems have been widely developed in recent years to enable personalized decision supports (e.g., when users are searching for a movie, book, laptop, etc.), the focus has been mainly on the improvement of algorithm accuracy [1], less on studying the efficacy of interface designs from users' perspective. In fact, most of current recommender systems basically follow a ranked list structure, where all recommendations are listed one by one in the interface, according to the rank order of their predictive scores in matching users' interests as computed by the system.

Little is indeed known on the performance of such list interface and other possible display structures in influencing users' decision quality, although it has been claimed that users will likely adapt their decision strategies to the information presence on e-commerce sites, especially when they are confronted with a high-value product (e.g., computer, digital camera) for which the target product is not clear up front [7, 8]. For instance, Jedetski and Adelman have found that the amount of displayed alternatives will likely induce a significant effect [5]. When it is a small number (i.e., less than 30),

P. De Bra, A. Kobsa, and D. Chin (Eds.): UMAP 2010, LNCS 6075, pp. 375–380, 2010.

more compensatory strategies such as the weighted additive rule (WADD) [7] will be applied by users to produce more accurate decisions.

In this paper, we are particularly interested in exploring the effect of different recommendation displays on user behavior in the complex decision environment. Indeed, according to [3], users tend to focus on the top of a list, probably due to their cognitive limitations. Therefore, if such phenomenon also works when users face the list-based recommender interface, items that locate farther down in the list would attract little attention even though they may better match to the user's true interests. With this concern, we have attempted to understand: 1) how is users' actual visual searching pattern in the ranked list? And 2) are there more effective layout designs that can prompt users to consider more recommendations so as to potentially result in a more rigorous decision outcome?

We have accordingly designed a user study with the eye-tracker to answer the two questions. The experiment involved a comparison of the standard list view with category interfaces where recommendations are grouped into different categories and displayed in either vertical or quadrant layouts (see Fig. 1). The algorithm to generate the categories is called *preference-based organization method* that we have developed aiming to discover similar tradeoff properties among items (e.g., "these products are cheaper and lighter, but have slower processor speed") based on the association rule mining technique [2]. Prior simulation proved that this algorithm can obtain higher recommendation accuracy than related classification approaches due to its user-preferences focused clustering and selection strategies [2]. Thus, in this current work, we mainly aim at understanding whether and how the organization-based interface would in practice impact on end-users' cognitive searching process.

2   Experiment Setup

In our experiment, each user was asked to solve a decision problem (e.g., looking for a laptop to "buy"), for which the recommender was to assist them in locating interesting items and identifying the optimal choice. In the following, we will discuss in detail how the experiment was set up including materials used, participants recruited and experiment procedure, followed by results analysis.

Three recommender interfaces were prepared for this eye-tracking study. One is the standard list view where all recommendations are listed one by one, ranked by their satisfaction degrees according to user preferences (see Fig. 1.a "LIST"). More concretely, a set of 25 products (e.g., laptops), that have higher weighted utilities matched to the user's stated feature criteria as computed by the multi-attribute utility function [6], is returned as recommendations in this interface, and the highest ranked one is placed on the top, followed by others each with a "why" tool tip explaining the computational rational. The second one is the organization interface (see Fig. 1.b "ORG1"), where except for the ranked first item positioned as the top candidate, the remaining 24 products are organized into four categories. Each category is annotated with a title explaining how the attributes of products in that category provide benefits and compromises (i.e., tradeoff properties) in comparison with the top candidate (due to the space limit, please refer to [2] for detailed algorithm steps). In the third interface (see Fig. 1.c "ORG2"), instead of a vertical structure, the four categories are displayed in a quadrant arrangement with two categories laid out in parallel. The motivation for this new design actually came from [4], that indicates eye movements are likely to go to nearby objects. We were hence interested in knowing whether putting two categories at the same horizontal level would absorb attentions to more recommended products.

Each subject was randomly assigned one type of interface to evaluate. The main user task was to "*find a product that you would purchase if given the opportunity*" and the user was allowed to quit if s/he did not find any satisfactory product. A product catalog comprising 100 laptops extracted from a real e-commerce website was used for the generation of recommendations based on users' initially stated feature criteria (such as on the laptop's brand, price, processor speed, weight, etc.).

A Tobii 1750 eye-tracking monitor was used with a resolution setting of 1290x1024 pixels. It samples the position of the user's eyes by every 20ms. The monitor frame has a high resolution camera with near infra-read light-emitting diodes. This setting allows for more natural tracking of user behavior by not placing many restrictions on the participant. The ClearView software produces the log of eye-movements and user events, providing us with the screen coordinates for each area of interest (AOI).

Twenty-one participants (three females) volunteered to join in the study. They are students or employees in the university (ages ranging from 20 to 40) and from various countries (e.g., USA, China, Switzerland, Italy, Canada, India, etc.). More than half of them were interested in purchasing a laptop at the time of experiment, but no one was clearly certain of her/his targeted object before performing the experiment task.

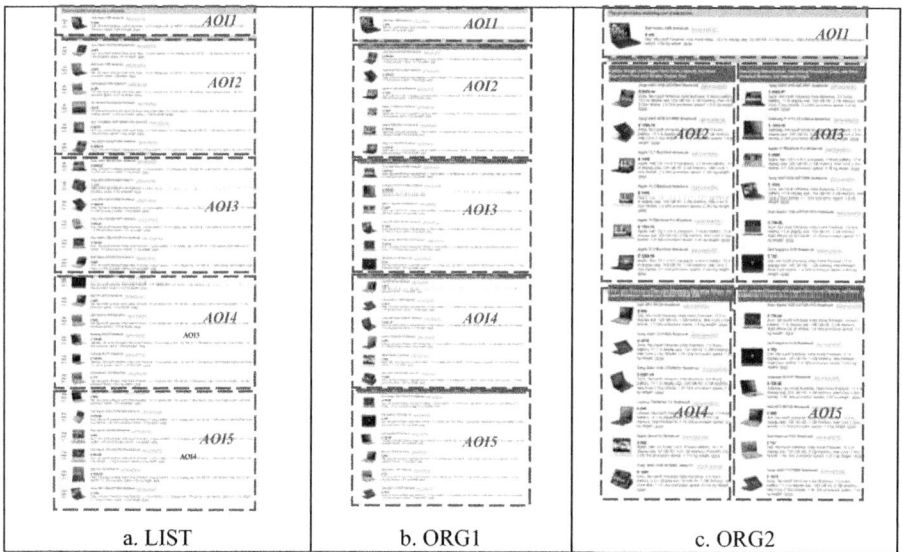

| a. LIST | b. ORG1 | c. ORG2 |

Fig. 1. Three recommender interface designs. Dashed boxes indicate Areas of Interest (AOIs).

3   Results

We mainly measured fixation frequency and duration. The fixation is gaze point with a minimum threshold of 100ms. Three participants were screened out from our analysis due to calibration difficulties or incomplete data with the eye-tracker, leaving us with eighteen participants for the analysis.

3.1   Areas of Interest

In total, 8389 gaze data points were recorded. Five major AOIs were defined on each recommender interface. In ORG1 and ORG2, each category (containing 6 products) represents an area of interest, in addition to the "top candidate" region (see Fig. 1). Accordingly, the list interface was also divided into five AOIs: the top candidate, 2nd to 7th recommended products, 8th to 13th ones, 14th to 19th ones, and 20th to 25th recommendations. The aim was hence to achieve the maximum comparability between the list interface and the organization interfaces.

The chance of looking at each AOI was first calculated. It shows that, although almost all studied users scanned over all AOIs on each interface, the focus was quite different as revealed by their fixation frequency and duration time (see Fig. 2.a). Specifically, in LIST, most of its average user's attentions were placed on AOI1 (the top candidate) and AOI2 (respectively 24.9s and 21.3s accumulated, covering 80.4% of the user's total duration time). It therefore indicates that users are likely to fixate on the top results in the rank list, though they were with the task goal of making a product choice among all displayed alternatives.

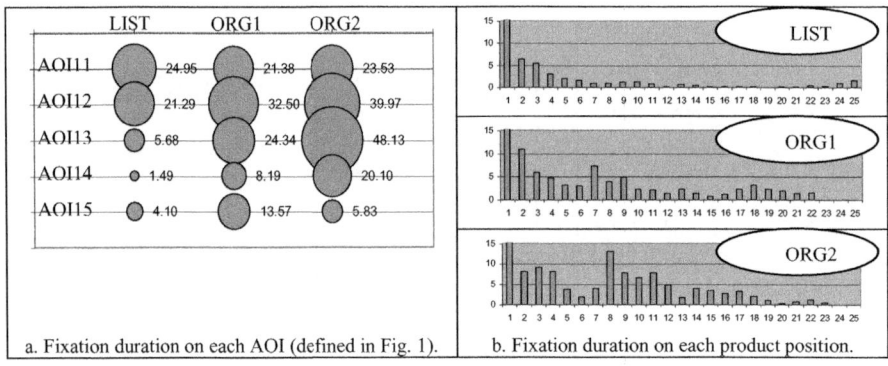

a. Fixation duration on each AOI (defined in Fig. 1).   b. Fixation duration on each product position.

Fig. 2. Mean fixation duration (in *seconds*) on each AOI (left figure) and on individual product position (right figure) in the three interfaces

However, the fixation was dramatically changed in ORG1 and ORG2. In fact, more fixations were observed in these two interfaces in terms of both frequency and duration time (average 484.6 fixations with 108.4s in ORG1, and 595.6 with 149.4s in ORG2, vs. 336.5 with 71.7s in LIST). Comparing respective gazes on AOIs reveals that all of the four areas (from AOI2 to AOI5) received more attentions, relative to those in the LIST interface. It is especially of significant differences w.r.t. AOI3 and AOI4 (respectively $F = 5.14$, $p = 0.02$; $F = 3.84$, $p = 0.045$, by ANOVA test regarding duration). Post-hoc multiple comparisons further tell that the durations through AO2 to AOI4 were all averagely higher and more equally distributed in ORG2. Another interesting phenomenon is that users paid more attentions to AOI5 than AOI4 in ORG1 and LIST (both with the vertical layout), whereas AOI4 got more gazes than AOI5 in ORG2 (with the quadrant arrangement).

3.2 Viewed and Selected Products

We further analyzed the average fixation duration on individual product position in the three interfaces. It indicates that users in fact carefully looked at more products (that exceed 1s duration) in both organization interfaces (average 12.3 and 15.2 products in ORG1 and ORG2, against 7.7 in LIST). The difference between ORG2 and LIST is even significant ($p = 0.046$). More specifically, in ORG1, except for the top candidate (at the 1st position), the first two categories were carefully examined in terms of both titles and contained products (product positions from 2 to 9, see Fig. 2.b "ORG1"). The fixations of products in the other two categories were relatively less, but still exhibited a certain amount of interests. In ORG2, products in the first four AOIs were given more in-depth examinations in respect of their details (i.e., product positions from 1 to 17 through AOI1 to AOI4, see Fig. 2.b "ORG2"). On the contrary, in LIST, only the top five products were fixated but in a linear reduction manner, and users rarely placed attention to the other products below (see Fig. 2.b "LIST").

To further measure the objective decision quality achieved in each interface, we counted number of users who have finally made the product choice. It shows that respective 71.43% and 100% users have chosen products in the two organization interfaces, relative to 50% users in LIST (meaning that the other 50% quitted without selecting any item in LIST). Because participants were allowed to save several products into their shopping cart, we found that more products were on average selected for the basket in ORG1 and ORG2 (1.86 and 3.2 respectively, vs. 1.33 in LIST). Moreover, in ORG1 and ORG2, the selected products came from almost all AOIs, whereas it was either the top candidate or from AOI2 in LIST (see Table 1). It thus infers that when users are motivated to review more options (i.e., by the category interfaces), they will likely carefully consider what they see and choose more items if satisfactory, which unfortunately will be ignored in the list view by chance.

Table 1. Actual product selections and their sources (i.e. AOIs) in the three interfaces

	% of users who made product choices	Average selections	AOI1 (% of selected products from this AOI)	AOI2	AOI3	AOI4	AOI5
LIST	50%	1.33	25%	75%			
ORG1	71%	1.86	23%	31%	15%	8%	23%
ORG2	100%	3.2	12.5%	37.5%	37.5%	12.5%	

4 Conclusion and Future Work

In conclusion, the eye-tracking results interestingly show that users did practically adapt their searching behavior to different recommendation displays. In the ranked list, most of attentions were paid to the top, whereas in the organization-based interfaces users were attracted to view more recommended items. As a result, above 70% users have made product choices in ORG1 and ORG2, against 50% in LIST. It hence suggests that the category structure can more likely lead to a rigorous consideration process, enabling users to make informed decisions at the end. More notably, the quadrant category layout was demonstrated more competent in prompting users' fixations and augmenting their decision quality in the experiment.

The findings therefore point to a promising direction, motivating us to conduct more studies in the future. One objective will be to recruiting more users from diverse origins (e.g., females, professions) to consolidate the results. Another area is to in-depth investigating users' perceptual processes and discovering the reason that causes their behavior difference between the quadrant category layout and the vertical one. We will also target to build predictive models of users' cognitive architecture through continuous collection of their eye gaze patterns.

References

1. Adomavicius, G., Tuzhilin, A.: Toward the Next Generation of Recommender Systems: a Survey of the State-of-the-Art and Possible Extensions. IEEE Transactions on Knowledge and Data Engineering 17(6), 734–749 (2005)
2. Chen, L., Pu, P.: Preference-based Organization Interfaces: Aiding User Critiques in Recommender Systems. In: Conati, C., McCoy, K., Paliouras, G. (eds.) UM 2007. LNCS (LNAI), vol. 4511, pp. 77–86. Springer, Heidelberg (2007)
3. Guan, Z., Cutrell, E.: An Eye Tracking Study of the Effect of Target Rank on Web Search. In: SIGCHI Conference on Human Factors in Computing Systems, pp. 417–420. ACM Press, New York (2007)
4. Halverson, T., Hornof, A.J.: A Minimal Model for Predicting Visual Search in Human-Computer Interaction. In: SIGCHI Conference on Human Factors in Computing Systems, pp. 431–434. ACM Press, New York (2007)
5. Jedetski, J., Adelman, L., Yeo, C.: How Web Site Decision Technology Affects Consumers. IEEE Internet Computing, 72–79 (2002)
6. Keeney, R.L., Raiffa, H.: Decisions with Multiple Objectives: Preferences and Value Trade-offs. Wiley, New York (1976)
7. Payne, J.W., Bettman, J.R., Johnson, E.J.: The Adaptive Decision Maker. Cambridge University Press, Cambridge (1993)
8. Pu, P., Chen, L., Kumar, P.: Evaluating Product Search and Recommender Systems for E-Commerce Environments. Electronic Commerce Research 8(1-2), 1–27 (2008)

Recommending Food: Reasoning on Recipes and Ingredients*

Jill Freyne and Shlomo Berkovsky

CSIRO, Tasmanian ICT Center
GPO Box 1538, Hobart, 7001, Australia
firstname.lastname@csiro.au

Abstract. With the number of people considered to be obese rising across the globe, the role of IT solutions in health management has been receiving increased attention by medical professionals in recent years. This paper focuses on an initial step toward understanding the applicability of recommender techniques in the food and diet domain. By understanding the food preferences and assisting users to plan a healthy and appealing meal, we aim to reduce the effort required of users to change their diet. As an initial feasibility study, we evaluate the performance of collaborative filtering, content-based and hybrid recommender algorithms on a dataset of 43,000 ratings from 512 users. We report on the accuracy and coverage of the algorithms and show that a content-based approach with a simple mechanism that breaks down recipe ratings into ingredient ratings performs best overall.

Keywords: Collaborative filtering, content-based, ingredient, recipes.

1 Introduction

The World Health Organisation [1] is predicting that the number of obese adults worldwide will reach 2.3 billion by 2015 and the issue is attracting increased attention. Much of this attention is being paid to online diet management systems, which have been replacing traditional pen-and-paper programs. These systems include informative content and services, which persuade users to alter their behaviour. Due to the popularity of diet monitoring facilities, these systems hold a vast amount of user preference information, which could be harnessed to personalize interactive features and to increase engagement with the system and, in turn, the diet program. One such personalized service, ideally suited to informing diet and lifestyle, is a personalized recipe recommender. This recommender could exploit explicit food ratings, food diary entries, and browsing behaviour to inform its recommendations.

---

* This research is jointly funded by the Australian Government through the Intelligent Island Program and CSIRO Preventative Health Flagship. The Intelligent Island Program is administered by the Tasmanian Department of Economic Development, Tourism, and the Arts. The authors acknowledge Mealopedia.com and Penguin Group (Australia) for permission to use their data.

P. De Bra, A. Kobsa, and D. Chin (Eds.): UMAP 2010, LNCS 6075, pp. 381–386, 2010.

The domain of food is varied and complex and presents many challenges to the recommender community. The content or *ingredients* of a meal is only one component, which impacts a user's opinion. Others include *cooking methods, ingredient costs* and *availability, complexity of cooking, preparation time, nutritional breakdown, ingredient combination effects,* as well as *cultural* and *social factors.* Add to this the sheer number of ingredients, the fact that eating often occurs in groups, and that sequencing is crucial, and the complexity of challenge becomes clear.

Initial efforts in addressing these challenges have resulted in systems, such as Chef [2] and Julia [3], which rely hugely on domain knowledge in their recommendation processes. Conversely, fuzzy logic [4] and active learning and knowledge sources techniques have been applied in [5] to generate recipes from ingredient sets without the need for expensive domain knowledge.

In this work we turn to the traditional recommender technologies to understand their applicability and accuracy in the food domain. We present a preliminary study into the suitability of recommender algorithms for recipe recommendation. The study is based on preferences provided by 512 users on a corpus of recipes. We examine the accuracy of collaborative, and content-based filtering algorithms, and compare them to hybrid recommender strategies, which break down recipes into their ingredients in order to generate more accurate recommendations. We show that solicitation of recipe ratings, which are transferred to ingredient ratings, is an accurate and effective method of capturing ingredient preferences, and that the introduction of simple intelligence can improve the accuracy of recommendations.

2   Recommender Strategies

The aim of this work is to develop recommender algorithms for personalized recipe recommendations. Figure 1 shows the simple recipe to ingredient relationship strategy adopted in this work. We ignore all cooking processes and combination effects and consider all ingredients to be equally weighted within a recipe. Also, we transfer ratings gathered on recipes equally to all its ingredients, and vice versa, from ingredients to their associated recipes. In contrast to previous work [6], here we solely investigate strategies, in which ratings are available on *recipes* and evaluate them on a much larger dataset.

In order to compare our recommender strategies, we implement a baseline algorithm *random*, which assigns a randomly generated prediction score to a recipe. We implemented five personalized recommender strategies. The first is a standard *collaborative filtering* algorithm assigning predictions to recipes based on the weighted ratings of a set of N *neighbours*. Briefly, N neighbours are identified using Pearson's correlation algorithm shown in Equation 1 and predictions for recipes not rated by user u_a are generated using Equation 2.

$$sim(u_a, u_b) = \frac{\sum_{i=1}^{k}(u_{a_i} - \overline{u_a})(u_{b_i} - \overline{u_b})}{\sum_{i=1}^{k}(u_{a_i} - \overline{u_a})^2 \sum_{i=1}^{k}(u_{b_i} - \overline{u_b})^2} \tag{1}$$

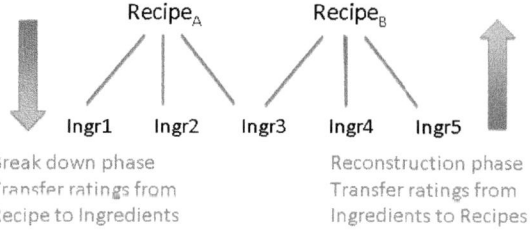

Fig. 1. Recipe - ingredient breakdown and reconstruction

$$pred(u_a, r_t) = \frac{\sum_{n \in N} sim(u_a, u_n)rat(u_n, r_t)}{\sum_{n \in N} sim(u_a, u_n)} \qquad (2)$$

The second is a *content-based* algorithm, which breaks down each recipe r_i rated by u_a into ingredients $ingr_1, ..., ingr_x$ (see Fig. 1) and assigns the ratings provided by u_a to each ingredient according to Equation 3. The strategy then applies a content-based algorithm shown in Equation 4 to predict a score for the target recipe r_t based on the average of all the scores provided by user u_a on ingredients $ingr_1, ..., ingr_j$ making up r_t.

$$score(u_a, ingredient_i) = \frac{\sum_{l \ s.t. \ ingr_i \in r_l} rat(u_a, r_l)}{l} \qquad (3)$$

$$pred(u_a, r_t) = \frac{\sum_{j \in r_t} score(u_a, ingr_j)}{j} \qquad (4)$$

We also implemented two *hybrid* strategies. Both of these break down each recipe rated by u_a into ingredients and exploit collaborative filtering techniques to reduce the data sparsity of the ingredient matrix by generating predictions for ingredients on which we have no information. The first strategy, $hybrid_{recipe}$, identifies a set of neighbours based on ratings provided on recipes as in Equation 1 and predicts scores for unrated ingredients using Equation 2 (applied to ingredient scores rather than recipe ratings). With the denser ingredient data, the content-based prediction shown in Equation 4 is used to generate a prediction for r_t. The second strategy, $hybrid_{ingr}$, differs from $hybrid_{recipe}$ only in its neighbour selection step. In $hybrid_{ingr}$, user similarity is based on the ingredients scores obtained after the recipe break down rather than on the recipe ratings as in $hybrid_{recipe}$.

3   Evaluation

We gathered a set of 43,893 recipe ratings from 512 users through the Amazon owned online HCI task facilitator Mechanical Turk (www.mturk.com). Online surveys, each containing 36 randomly selected recipes, were posted to the system and users were allowed to answer as many surveys as they choose.

3.1 Set-Up

The corpus of recipes used was sourced from the CSIRO Total Wellbeing Diet Books [7, 8] and the online meal planner Mealopedia (www.mealopedia.com). We extracted 404 recipes, which corresponded to 479 unique ingredients. On average, each recipe was made up of 9.52 ingredients (stdev 2.63) and the average number of recipes that each ingredient was found in was 8.03 (stdev 19.86). We gathered opinions of 512 users regarding the available recipes. Users were asked to provide their preferences on how much each recipe appealed to them. Each user provided at minimum 36 recipe ratings and also their demographical information. All ratings were captured on a 5-Likert scale, spanning from "not at all" to "a lot" (6274 recipes rated *not at all*, 6272 – *not really*, 8445 – *neutral*, 10873 – *a little*, and 12029 – *a lot*). In total 43,893 preferences were gathered with an average of 85.73 per user, such that the ratings matrix was 21.22% complete.

We conducted a traditional leave one out off-line analysis, which took each $\{u_i, r_t, rat(u_i, r_t)\}$ tuple from a user profile and used the algorithms presented in Sect. 2 to predict the rating $rat(u_i, r_t)$. A set of 20 neighbours were selected only once for each user, based on the entire set of ratings provided. The performance of the recommenders was evaluated using the normalized MAE measure [9] and coverage, i.e. their ability to generate recommendations.

3.2 Results

The *content-based* and both *hybrid* strategies obtained over 99% coverage. For *collaborative filtering*, the coverage was 95.9%, and for *random* recommendations it was obviously 100%. Hence, there is no significant coverage benefit gained by any approach on this dataset.

The lighter bars in Fig. 2 show the MAE score obtained for each strategy. As expected, the *random* algorithm performed worst with an MAE of 0.399. The poorest personalized strategy was the *collaborative filtering* algorithm with an MAE of 0.328. Producing MAE scores of 0.309 and 0.330 are the two hybrid strategies, $hybrid_{recipe}$ and $hybrid_{ingr}$, respectively. In comparison to the best performing, *content-based* algorithm, which obtained an MAE of 0.262, both *hybrid* strategies introduce noise in the ingredient scores during the collaborative filtering step. This finding is consistent with that of Melville *et al.*, who also used content-boosted collaborative filtering [10]. The differences in MAE are significant at $p < 0.05$ across all pairings.

A comparison between the *collaborative filtering* algorithm, which treats each recipe as one entity and ignores its ingredients, and the *content-based* algorithm, which considers the ingredients, shows that even the naive break down and re-construction rules applied here offer significant performance benefits in accuracy.

The decomposition of recipes into ingredients implemented in this experiment was simplistic: an ingredient score was computed by averaging the ratings of recipes in which it occurs. This lead to a large number of *mixed* ingredient scores. For example, consider two recipes containing an ingredient i: one is liked and one disliked, but i is not the cause of the dislike. Despite not being the cause

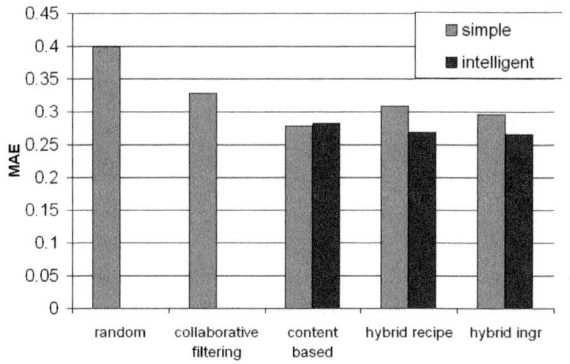

Fig. 2. Normalized MAE score

of the dislike, i will receive only a neutral score. To address this shortcoming, we implemented a more intelligent strategy, in which only the positive ratings for ingredients that receive mixed ratings are considered.

We see the impact of this assumption in the darker bars in Fig. 2. In the *content-based* algorithm, the impact is negligible. However, we see a positive impact of the intelligent break down on the MAE of the two *hybrid* strategies. Now, the intelligent hybrid strategies significantly outperform the *content-based* algorithm with an MAE of 0.269 and 0.265 for the intelligent versions of $hybrid_{recipe}$ and $hybrid_{ingr}$ strategies, respectively. The differences between the previously best performing *content-based* algorithm and both the intelligent versions of $hybrid_{recipe}$ and $hybrid_{ingr}$ strategies are significant at $p < 0.05$, as is the difference between both intelligent strategies. The collaborative filtering step used to generate predictions on the unknown food items has benefited the most from the introduction of this intelligent break down process.

Hence, the best performing algorithm is intelligent $hybrid_{ingr}$, which exploits both content-based and collaborative filtering techniques. Neighbours are determined based on the implied ratings of recipes, which have been transferred down to ingredient scores while reasoning on the presence of mixed ratings. Then, collaborative filtering is used to predict scores for unrated ingredients, and, finally, a recipe prediction is computed by averaging the scores of its ingredients.

4   Conclusions and Future Work

In this work we have investigated the applicability of recommender techniques to generate recipe recommendations. We found that high coverage and reasonable accuracy can be achieved through content-based strategies with a simple break down and construction used to relate recipes and ingredients. We found significant accuracy improvement through use of content-based techniques over a collaborative filtering algorithm. However, the optimal solution was obtained when we bootstrap the recommender process by breaking a recipe down into ingredients, computing ingredient scores, applying the collaborative filtering step to decrease the sparsity of the ingredient scores, and, finally, applying the content-based recipe rating prediction process by examining the scores of individual ingredients.

As noted earlier, there are many factors that influence a user's rating beyond a recipe content. Thus, our future work will focus on extraction of recipe features, such as complexity, time and cooking methods, to examine their impact on user ratings. Furthermore, here we implemented a simplistic idea of what a recipe recommender needs to achieve. We are, however, aware that generating recipe recommendations is a far more complicated task in reality, and we will investigate the issues of group recommendations, where varying social relationships can be at play. In particular, we aim to investigate how family roles and relationships affect compromise and satisfaction with menu plans. Complimentary to this, we need to examine applicability of sequential recommendations. Menu recommendations would not generally be provided in a single shot interaction, but rather users will plan meals over a period of time, such that diversity and satisfaction levels are complex, in particular when groups of users are involved.

References

1. World Health Organization: Chronic disease information sheet, http://www.who.int/mediacentre/factsheets/fs311/en/index.html (accessed January 2010)
2. Hammond, K.: CHEF: A Model of Case-Based Planning. In: Proceedings of the National Conference on Artificial Intelligence (1986)
3. Hinrichs, T.: Strategies for adaptation and recovery in a design problem solver. In: Proceedings of the Workshop on Case-Based Reasoning (1989)
4. Sobecki, J., Babiak, E., Slanina, M.: Application of hybrid recommendation in web-based cooking assistant. In: Gabrys, B., Howlett, R.J., Jain, L.C. (eds.) KES 2006. LNCS (LNAI), vol. 4253, pp. 797–804. Springer, Heidelberg (2006)
5. Zhang, Q., Hu, R., Namee, B., Delany, S.: Back to the future: Knowledge light case base cookery. Technical report, Dublin Institute of Technology (2008)
6. Freyne, J., Berkovsky, S.: Intelligent Food Planning: Personalized Recipe Recommendation. In: Proceedings of the International Conference on Intelligent User Interfaces, pp. 321–324 (2010)
7. Noakes, M., Clifton, P.: The CSIRO Total Wellbeing Diet. Penguin (2005)
8. Noakes, M., Clifton, P.: The CSIRO Total Wellbeing Diet Book, vol. 2. Penguin (2006)
9. Herlocker, J.L., Konstan, J.A., Terveen, L.G., Riedl, J.T.: Evaluating collaborative filtering recommender systems. ACM Trans. Inf. Syst. 22(1), 5–53 (2004)
10. Melville, P., Mooney, R., Nagarajan, R.: Content-boosted collaborative filtering for improved recommendations. In: Proceedings of the National Conference on Artificial Intelligence (2002)

Disambiguating Search by Leveraging a Social Context Based on the Stream of User's Activity

Tomáš Kramár, Michal Barla, and Mária Bieliková

Faculty of Informatics and Information Technology
Slovak University of Technology
Bratislava, Slovakia
kramar.tomas@gmail.com, {barla,bielik}@fiit.stuba.sk

Abstract. Older studies have proved that when searching information on the Web, users tend to write short queries, unconsciously trying to minimize the cognitive load. However, as these short queries are very ambiguous, search engines tend to find the most popular meaning – someone who does not know anything about cascading stylesheets might search for a music band called `css` and be very surprised about the results. In this paper we propose a method which can infer additional keywords for a search query by leveraging a social network context and a method to build this network from the stream of user's activity on the Web.

1 Introduction

Finding a relevant document based on few keywords is often difficult. Many keywords are ambiguous, their meaning varies from context to context and from person to person. Some words are ambiguous by nature, e.g., a *coach* might be a bus or a person, other words became ambiguous only after being adopted for a particular purpose, not to mention English nouns, which, apart from their natural meaning, also name a software, music band or any other entity. There are also words whose meaning depends on the person who is using them; clearly, architecture means different things to a processor designer than to an architect. Based on the previous observations, we might conclude that using short queries is not a good idea. Unfortunately, this is how we search [1].

The search engines work like databases: they crawl and index documents and respond to queries with a list of results. The order of documents depends on the adopted relevance function; the most widely used search engine today – Google – uses a PageRank relevance function: the more links to a document, the more likely it is to appear at the top positions. This ordering is however not always compatible with user's information needs: a programmer searching for *cucumber* probably does not want to make a salad.

We tackle the problem by implicitly inferring the context and modifying the user's query to include it. The original query is enriched with additional keywords which capture the user's focus. In case of the said programmer, the resulting query might be *cucumber testing* which provides much more valuable and relevant documents than the original query. We select additional keywords

P. De Bra, A. Kobsa, and D. Chin (Eds.): UMAP 2010, LNCS 6075, pp. 387–392, 2010.

following the social network or rather the virtual community the user belongs to in this network. The search thus becomes personalized – the same query for another user from another community might be *cucumber salad*.

The paper is structured as follows. In next section we talk about related work, Sect. 3 gives an overview of how the social network is built, how the communities are extracted and how they are used in the process of keyword inference. In Sect. 4 we talk about preliminary experiments and give conclusions.

2   Related Works

The concept of search disambiguation is certainly not novel. Haveliwala proposed an alternative method of document ranking – a topic-sensitive PageRank [2]. For each document, multiple rank values are calculated, each biased in the context of one root topic from Open Directory Project, ODP[1]. The search results are then biased towards the current topic determined from the words of the document which the user started the search from.

The disambiguation and personalization is often achieved by leveraging some kind of social connections. In [3], the authors proposed a method for personalising the search results by leveraging communities. First a network of users' sessions is constructed (offline) from available access logs. Then, this network is used as a basis for detection of user communities. Subsequently, for each available document an interest of each community is calculated. When a user starts searching, her session is matched to the communities using a cosine similarity and the matching documents are ranked using a Bayesian network computed from the degree of interest of the matched communities to the document. Bender et al. [4] exploit existing social networks for ranking documents with a model called UserRank. The documents tagged, bookmarked or rated by user's friends get higher ratings. This approach, however, does not solve the problem of actually getting the document into the list of results. It can only reorder this list once it was retrieved by other means.

Personalization of the document retrieval itself can be done by automatic query refinement (also called query expansion), which has been recognized long ago as an effective technique to improve search results. Many approaches exist, ranging from an analysis of the lexical affinities [5] to thesaurus based techniques [6]. These methods are however based on a static information which does not always accurately capture user's interest. We believe that query refinement could achieve deeper level of personalization and disambiguation by also analyzing the documents and behavior of similar users as was already shown in [7].

Our method extends and combines the social networks and query refinement methods. We link the users in a social network not only by analyzing URLs of the visited pages, but also by analyzing the content features of these pages. We later use these features to capture user's current interest when she is searching, and also to provide the basis for our query refinement methods.

---

[1] ODP, http://dmoz.org

3   Social-Context Driven Query Expansion

In order to be able to expand the user's query with additional keywords, we need to capture its context and find the most appropriate keywords representing this context. The process is driven by an underlying automatically constructed social network and communities found within this network. The network is constructed from the simple user model based on the stream of user's activity. It is created from content features extracted from pages the user visited and an implicitly acquired user rating of the page.

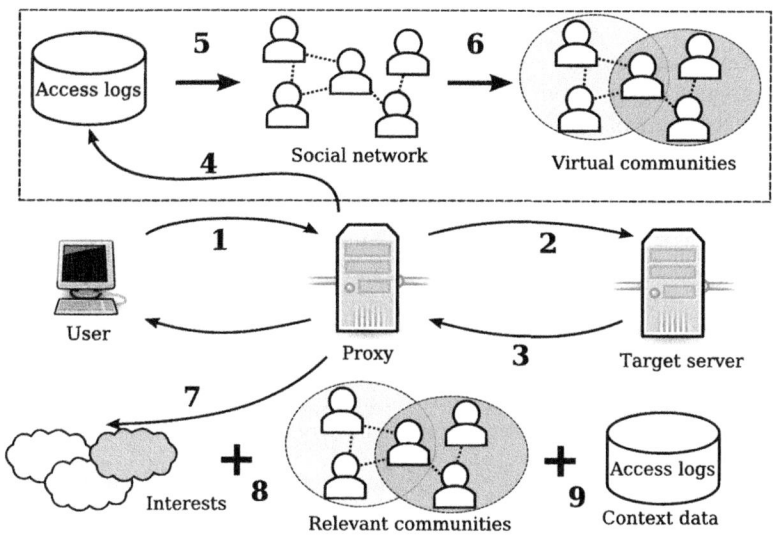

Fig. 1. Overview of the query expansion process

The overview of the process is depicted in Fig. 1. The user requests a page via proxy (step 1) configured in her browser. Proxy requests the page from the target server (step 2) and extracts the *characteristic document features* (step 4) – a vector of

- document keywords (using various keyword extraction algorithms and services such as tagthe.net or OpenCalais),
- tags from *delicious.com* (if available) and
- ODP category.

To capture user's *implicit rating* of the served webpage, a JavaScript code is inserted into every page, which detects user's scrolling and mouse movements and periodically updates the server's record about time the user has spent on the page. The implicit rating is subsequently calculated as a ratio between the time spent on page and the page size: $rating = 1 - \frac{1}{1+X}$, where $X = \frac{time\_on\_page}{document\_size}$.

To improve the accuracy of rating calculation and keyword extraction we extract and use the *cleartext*[2] version of the page, which holds only the core content, stripped of the markup and the navigational components.

Based on user's activity and the extracted features a *social network* is built (step 5), where a weight of an edge denotes a similarity of two users connected by this edge. The network is built in a sequence of steps:

1. New network is created, all users are connected with an edge of weight 0.
2. All documents which have implicit rating lower than the predefined minimal value are discarded.
3. For all visited domains (as in the Internet DNS system), a weight between the users who visited the same domain is incremented by a parameter **d**.
4. For all visited documents, a weight between the users who visited the same document is incremented by a parameter **p** (where **d** < **p**).
5. For each pair of users, a weight between them is incremented by the size of the overlap of the features extracted from the documents they visited.

The resulting graph represents the users and relationships among them. The stronger is their relationship, the more similar interests they have and the higher is the weight of the edge connecting them.

Next, a *community detection algorithm* is run (step 6), to partition the network into clusters of similar users (based on the stream of their activity). The algorithm is designed to take advantage of the weighted relations in the graph and produces overlapping communities, i.e., a user may belong to multiple communities at one time. This is an important property as a user might have multiple interests, each represented by one community. The community is created in the following steps:

1. Select a random vertex, not yet assigned to any community.
2. Spread the activation energy from the selected vertex to the rest of the network considering weights of the edges.
3. Create new community by collecting all vertices activated via the spreading.
4. If there is an unassigned vertex continue with step 1, otherwise end.

These two stages, the social network creation and community detection are performed periodically and offline. Found virtual communities then provide context for the search – the content features extracted from documents visited by the community or the actions carried on by its members.

In order to identify the *search context*, we capture user's current interest (step 7), which in our case is a set of documents features the user is currently interested in. We construct it dynamically – for every requested document, we search for an overlap of document's content features with the content features from the user's current interest. If an overlap is found, current interest is enriched by document's features. Otherwise, we consider that a new session (and thus a switch of interests) just got started. When we detect that a search has been

---

[2] We used custom implementation of the publicly available *readability* service, http://lab.arc90.com/experiments/readability/

initiated, the current interest helps us to determine all relevant (i.e., sharing at least one feature) communities (step 8). The top n communities are then considered as the *search context* and passed to the final stage of query expansion.

We use two approaches to infer new keywords, each using the data provided by the members of the communities (step 9) – *query stream analysis* and *keyword co-occurrence analysis*.

Query stream analysis follows a simple observation of how we do our searches. When a search query does not return relevant documents, it is redefined. The redefinition continues unless the user finds the information or gives up. A query stream represents one searching session, an uninterrupted succession of queries issued by the same user which have some common parts. A sample query stream is: jaguar, jaguar speed, jaguar car speed. We take all queries issued by users from the *search context* and search for query streams where at least one query matches the user's query. Query streams which did not lead to a successful retrieval (the documents visited from a search results page have low implicit rating) are discarded. The last query is extracted from each successful query stream and used to enrich the original query.

A *keyword co-occurrence analysis* is based on analyzing which additional keywords frequently occur with the words from the query in the documents viewed by the users from the current *search context*. The original query is enriched with the top n co-occurring keywords.

4   Preliminary Experiments and Conclusion

We evaluate the method of query refinement within a platform of an enhanced proxy server [8]. The proxy plays a crucial role in the experiment setup as it allows us to log each user request and further process and modify the response from a webserver. This way we modify the page before it is displayed and include the search results provided by the expanded query. We also use the modification features to insert scripts into pages to monitor user's activity. We do not need to cope with the user identification as this is all handled by the proxy itself.

The goal of the preliminary experiments was to verify that, given the context and the access logs, the proposed query expansion methods would give satisfactory results, generating queries which achieve subjectively better search results.

To evaluate the query stream analysis, we used AOL search engine data[3]. This dataset contains roughly 20M of queries from 650k users. Table 1 summarizes some queries and how they would be reformulated using this approach.

For the evaluation of the keyword co-occurrence, we used the data collected by the proxy server during its development, that is four users and 310 visited documents. Results provided by this method are also summarized in table 1.

The results of preliminary experiments are promising, as both approaches are capable of redefining the queries to rule out the ambiguity and to provide (subjectively) more relevant search results.

---

[3] AOL search engine logs, http://www.gregsadetsky.com/aol-data/

Table 1. Some examples of query reformulation using different approaches

Method used			
Query stream analysis		**Keyword co-occurrence**	
Original query	**Expanded query**	**Original query**	**Expanded query**
java history	history of java indonesia	passenger	passenger apache
jaguar	jaguar animal	branch	branch git
sphinx	sphinx cats	apache	apache server

The key parts of the method are based on the keywords and on the keyword overlap. We work on improvement of the keyword extraction process by extracting the parent keywords (hypernyms) as proposed in [9]. That should improve the chance for a match between two related documents. For example, documents with keywords ruby and python do not match, but when extended with their parent category programming they generate a match on programming.

The proposed method combines social networks and query expansion approaches based on the characteristic content features of the documents. User's current interest is mapped to the interests of the communities and the semantics of the query is inferred from the behaviour of these communities.

Acknowledgments. This work was partially supported by the Scientic Grant Agency of Slovak Republic, grant No. VG1/0508/09 and it is the partial result of the Research & Development Operational Programme for the project Support of Center of Excellence for Smart Technologies, Systems and Services II, ITMS 26240120029, co-funded by the ERDF.

References

1. Jansen, J., Spink, A., Saracevic, T.: Real Life, Real Users, and Real Needs: a Study and Analysis of User Queries on the Web. Information Processing & Management 36(2), 207–227 (2000)
2. Haveliwala, T.H.: Topic-sensitive Pagerank. In: WWW 2002, pp. 517–526. ACM, New York (2002)
3. Almeida, R.B., Almeida, V.A.F.: A Community-aware Search Engine. In: WWW 2004 (2004)
4. Bender, M., et al.: Exploiting Social Relations for Query Expansion and Result Ranking. In: ICDEW 2008, pp. 501–506. IEEE, Los Alamitos (2008)
5. Carmel, D., et al.: Automatic Query Refinement using Lexical Affinities with Maximal Information Gain. In: SIGIR 2002, pp. 283–290. ACM, New York (2002)
6. Liu, S., et al.: An Effective Approach to Document Retrieval via Utilizing WordNet and Recognizing Phrases. In: SIGIR 2004, pp. 266–272. ACM, New York (2004)
7. Kajaba, M., Návrat, P., Chudá, D.: A simple personalization layer improving relevancy of web search. Computing and Information Systems Journal 13, 29–35 (2009)
8. Barla, M., Bieliková, M.: "Wild" Web Personalization: Adaptive Proxy Server. In: Workshop on Intelligent and Knowledge Oriented Tech., WIKT 2009, pp. 48–51 (2009)
9. Barla, M., Bieliková, M.: On Deriving Tagsonomies: Keyword Relations Coming from Crowd. In: Nguyen, N.T., Kowalczyk, R., Chen, S.-M. (eds.) ICCCI 2009. LNCS (LNAI), vol. 5796, pp. 309–320. Springer, Heidelberg (2009)

Features of an Independent Open Learner Model Influencing Uptake by University Students

Susan Bull

Electronic, Electrical and Computer Engineering, University of Birmingham, U.K.
s.bull@bham.ac.uk

Abstract. Building on previous research with an independent open learner model in a range of university courses, this paper investigates features that may influence student choice about whether to use the environment in a particular course. It was found that some features are considered particularly important by students, but other features are less influential in students' decisions to use an independent open learner model. Recommendations for features to consider promoting uptake of this type of environment are given.

Keywords: Open learner model, learner preferences, adaptive e-learning.

1 Introduction

Adaptive learning environments typically model a user's knowledge or level of knowledge, and often also their difficulties. This allows the environment to personalise the interaction to the current needs of the learner, and is usually the main reason for modeling a user. However, systems are increasingly recognizing benefits of opening the learner model contents to the user, e.g. to: promote reflection and planning; support independent learning; foster collaboration or competition; help the learner take responsibility for their learning (see [1]).

Opening the learner model is not as straightforward as showing the user the system representations, as these are not designed for human interpretation. The model must be opened to the user in an understandable form. In principle, a learner model inferred using any technique where learner model data may also be useful *to the learner*, could be opened to them. For example, open learner models (OLM) have been used in systems with simple weighted numerical models [2]; models including conceptual or hierarchical relationships [3–6]; constraint-based models [7]; Bayesian models [8]. The method by which the model is shown to the user does not have to match the underlying complexity of the model. For example, skill meters have shown level of knowledge in a simple numerical model [2] and a constraint-based model [7]. Furthermore, simple and structured learner model views can be combined in a single system, presenting information from the same learner model data [5]. Previous work has shown improved learning with OLMs for adults [7, 9]. In this paper we investigate OLM features that may make it more likely for students to adopt it in their learning, as uptake is necessary for any resulting educational benefits to be recognised by users.

P. De Bra, A. Kobsa, and D. Chin (Eds.): UMAP 2010, LNCS 6075, pp. 393–398, 2010.

2   An Independent Open Learner Model in University Courses

Independent OLMs (IOLM) are used independently of a larger system: rather than the system providing personalised coaching/tutoring, the learner model is the focus of interactions, with the purpose of providing information about their knowledge and progress to the user *for them to use* to identify their understanding, and monitor and plan their learning [2]. This is in line with recommendations for formative feedback by the UK Higher Education Academy [10], and suggestions for students to take responsibility and control of their learning with OLMs [4].

2.1   Core OLMlets Features

The OLMlets IOLM [2] has been in use in Electronic, Electrical and Computer Engineering, University of Birmingham, since 2005. It is now available in 20 courses (ranging from courses such as 1st year "Introduction to Circuits, Devices and Fields" to 4th year "User Models and Models of Human Performance". Instructors input multiple choice questions separated into topics/concepts, and define corresponding misconceptions if applicable. A primary aim is to provide a consistent method of supporting formative assessment across the various courses a student is taking in their degree. However, it is not expected that all students should use OLMlets. Unless their learner model is assessed as part of a course, it is available optionally for those who find it a helpful complement to their existing successful approaches to study.

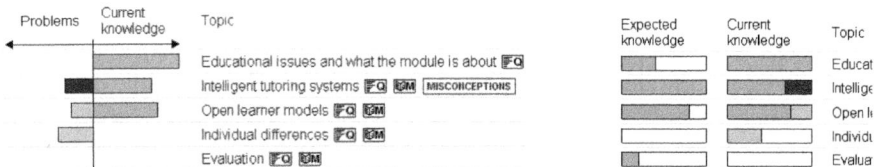

Fig. 1. Excerpt from graph and skill meter learner model views

OLMlets has 5 learner model views for the user to choose between. Figure 1 shows excerpts from the graph view, where the extent of current difficulties (misconceptions or general difficulties not related to specific misconceptions) are indicated on the left of an axis, and learner knowledge on the right; and skill meters showing the same information. The level of understanding of a topic is in green (medium shading in Fig. 1); misconceptions in red (dark shading); and general difficulties in grey (light shading). Brief misconception descriptions are obtained by clicking 'misconceptions' links. For example, in a course called "Adaptive Learning Environments": *you may believe that misconceptions are part of the domain model*. This misconception sometimes occurs if a student recognises misconceptions can be predefined (e.g. in a misconceptions library), and sees a predefined domain as a similar entity without considering that the domain/expert model will by definition not include misconceptions. The aim is that on being confronted with their misconceptions, a user will recognise that they have a problem and work to overcome it.

The excerpt from the skill meters view in Fig. 1 also shows knowledge expected for the present stage of the course, to allow the user to compare their progress against instructor expectations (available in all learner model views). This information is optionally input by the instructor. It can be likened to adaptive navigation support [11], but requires students to *themselves* make comparisons between their progress and expected progress, rather than the system indicating their readiness to access parts of a course. 'Q' icons lead to questions; 'M' icons to lecture notes, slides or other online materials. Another key feature of OLMlets is that users can choose to release their learner model to any or all other users, anonymously or with their name, and can view peer models that have been made available to them (if the instructor has permitted this in their course) [12]. Peer models are shown below the user's own model, accessed by scrolling, and can be individually included/excluded from display. A recent development allows users to edit their model (extended from t-OLM [13]).

2.2 Previous Findings

OLMlets has shown that in most cases students will consult misconceptions descriptions [2]. It has also been found that many will view peer models if available, and release their own model to peers to support their collaborative or competitive approaches to learning [12]. General use is quite high, ranging from 1/6 uptake, to uptake by all registered on a course; with use across all courses by 2/3 of students [2]. Use is optional in most courses in which OLMlets is deployed, suggesting that students perceive some benefit. Nevertheless, usage does vary across courses. This paper investigates the reasons for higher uptake of an IOLM in some courses than in others, to identify features that may be more likely to engender use of such an environment.

3 Uptake of an Independent Open Learner Model

This section gives questionnaire results on features of OLMlets across courses considered important by students; and examines model edits in the logs from one course.

3.1 Participants, Materials and Methods

Participants were 18 2nd and 3rd year students in Electronic, Electrical and Computer Engineering, University of Birmingham, taking a 'Computer Interactive Systems' degree. They completed a questionnaire with responses on a 5 point scale (strongly agree, agree, neutral, disagree, strongly disagree). Results are presented with 'strongly agree' and 'agree', 'disagree' and 'strongly disagree', combined. Questionnaire items relate to users' independent use (or non-use) of OLMlets in previous courses. Logs were examined for students' editing of their model in one course over 5 weeks, which had 18 topics (18 users). The final model state contributed 5% to the course mark.

3.2 Results

Figure 2 shows OLMlets features students considered important in their choice to use it in courses. Dark shading (agree) indicates a feature is important/useful; light shading (neutral) indicates the feature is neither important nor unimportant for students' decisions to use OLMlets; lack of shading (disagree) shows the feature was not important/useful. (Disagreement with a statement does not necessarily indicate that OLMlets was not used, or that a feature was not found useful.)

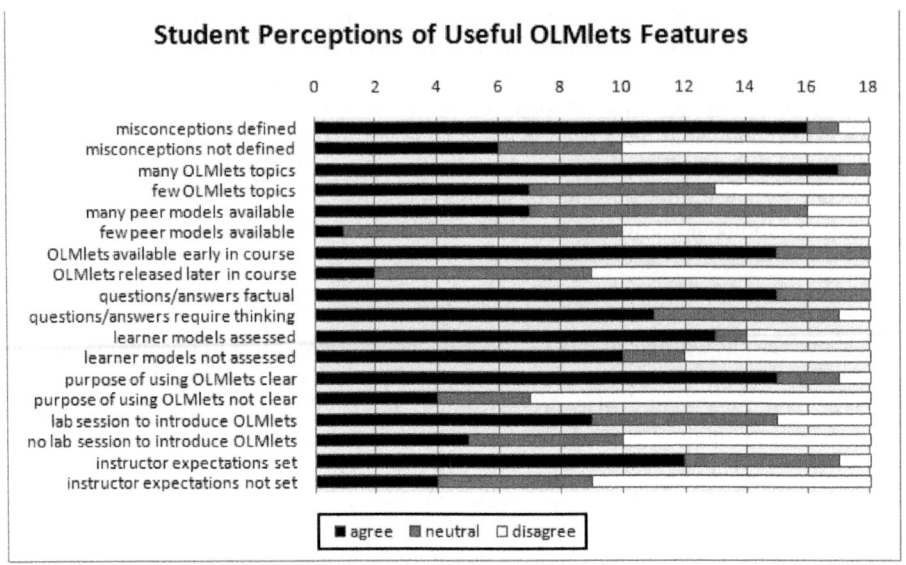

Fig. 2. Student perceptions of important OLMlets features in courses

Nearly all students considered misconception descriptions useful in OLMlets; however, 56% also considered OLMlets useful when misconceptions were not defined (or did not consider this important). Students found OLMlets more helpful in courses where there were a larger set of topics (range: 3-18 topics); though for 72%, the lack of such a breakdown was not a problem. Not all students use peer models, so the extent of their availability is only relevant for students who use them. Nevertheless, 39% considered the release of many peer models to be an important aspect of their use of OLMlets. OLMlets was much more likely to be used when it was available early in a course. There was relatively little difference in perceptions of the importance of whether questions require straightforward factual knowledge, or provoke thinking to arrive at a correct response. Assessment is a strong factor in use of OLMlets; nevertheless, summative assessment of the learner model did not strongly affect students' perceptions of its utility or importance in their learning. It was important that students understood the purpose of using OLMlets in a particular course, and it helped some students if there was part of a lab session to introduce it. The option to compare their own knowledge to instructor expectations was considered important by 3/4 of students.

Table 1 gives results for editing (changing) the learner model in a course where the model was assessed. 43% of edits were in the final week, which included 18% in the final 3 days. However, few students were editing their model by this stage. The range of questions attempted across the period was 211-1340 (mean 452, median 327).

Table 1. Editing the learner model in a course in which the model was assessed (18 topics)

OLM Edits	Total	Mean	Median	Range
Whole period	112	5.9	4	1-26
Final week	48	2.5	1	0-16
Final 3 days	20	1.1	0	0-9

3.3 Discussion and Recommendations

As stated above, OLMs can facilitate learning and OLMlets has enjoyed widespread use in many courses (see Sect. 2.2), indicating that students perceive benefit. To further demonstrate the likely advantages, the extent of editing the learner model was investigated. As most users did not extensively edit their model shortly before it was assessed, coupled with the fact that they answered many questions, it seems that they will for the most part *not* misuse an IOLM even when this opportunity is available (for course credit). Given previous positive results, we here focus on factors that may influence uptake of an IOLM. (The range of benefits available from an IOLM–e.g. raising learner awareness of their knowledge, progress, difficulties, position amongst the group; promoting metacognitive activities such as planning, self-monitoring; facilitating collaborative interactions and peer help–can only be gained if students actually use it. Many of these features are in the hands of the instructor or designer.

Based on questionnaire responses, it is suggested that the following may positively influence uptake of an IOLM (strength of recommendation is indicated by asterisks: *=recommended, **=strongly recommended, ***=very strongly recommended).

- o   Purpose of using the environment in a particular course is understood ***
- o   Availability of the environment from early in a course ***
- o   Presentation of the individual's misconceptions ***
- o   Allow users to compare their learner model to the instructor's expectations for the current stage of the course **
- o   Allow peers to release their learner models to each other **
- o   Use (part of) a lab session to introduce the environment *
- o   Breakdown of course into many topics/concepts *

In contrast, the type of question (factual, requires thinking) may be less influential in students' decisions to use an IOLM. This will maintain flexibility for instructors or designers to design questions in the style that suits their material and learning goals. While more students used OLMlets when the learner model was assessed, for 3/4 assessment was not critical to ensure use – a high figure considering that in most courses OLMlets is optional.

While OLMlets learner modelling is simple and the OLM is displayed in a simple manner, the results may be applicable to OLMs based on other modelling techniques and with different learner model displays (see Sect. 1), as suitable for the contexts in which they are used. Further investigation is recommended.

4   Summary

This paper has presented an IOLM used in a range of university courses, and identified some of the features that are important in students' decisions about whether to use it. While here investigated in the context of a specific environment, the findings may be applicable to other IOLMs for university students, and so could be considered as recommendations of features that may promote uptake of such an approach to promoting formative assessment and independent learning.

References

1. Bull, S., Kay, J.: Student Models that Invite the Learner. The SMILI Open Learner Modelling Framework. Int. J. of Artificial Intelligence in Education 17(2), 89–120 (2007)
2. Bull, S., Mabbott, A., Gardner, P., Jackson, T., Lancaster, M., Quigley, S., Childs, P.A.: Supporting Interaction Preferences and Recognition of Misconceptions with Independent Open Learner Models. In: Nejdl, W., Kay, J., Pu, P., Herder, E. (eds.) AH 2008. LNCS, vol. 5149, pp. 62–72. Springer, Heidelberg (2008)
3. Dimitrova, V.: StyLE-OLM: Interactive Open Learner Modelling. Int. J. of Artificial Intelligence in Education 13(1), 35–78 (2003)
4. Kay, J.: Learner Know Thyself: Student Models to Give Learner Control and Responsibility. In: Halim, Z., Ottomann, T., Razak, Z. (eds.) ICCE, AACE, pp. 17–24 (1997)
5. Mabbott, A., Bull, S.: Student Preferences for Editing, Persuading and Negotiating the Open Learner Model. In: Ikeda, M., Ashley, K.D., Chan, T.-W. (eds.) ITS 2006. LNCS, vol. 4053, pp. 481–490. Springer, Heidelberg (2006)
6. Perez-Marin, D., Pascual-Neito, I., Alfonseca, E., Rodriguez, P.: Automatically Generated Inspectable Learning Models for Students. In: Luckin, R., Koedinger, K.R., Greer, J. (eds.) Artificial Intelligence in Education, pp. 632–634. IOS Press, Amsterdam (2007)
7. Mitrovic, A., Martin, B.: Evaluating the Effect of Open Student Models on Self-Assessment. Int. J. of Artificial Intelligence in Education 17(2), 121–144 (2007)
8. Zapata-Rivera, J.D., Greer, J.E.: Interacting with Inspectable Bayesian Models. Int. J. of Artificial Intelligence in Education 14, 127–163 (2004)
9. Shahrour, G., Bull, S.: Interaction Preferences and Learning in an Inspectable Learner Model for Language. In: Dimitrova, V., Mizoguchi, R., du Boulay, B., Graesser, A. (eds.) Artificial Intelligence in Education, pp. 659–661. IOS Press, Amsterdam (2009)
10. Higher Education Academy/Juwah et al. Enhancing Student Learning Through Effective Formative Feedback (2004), http://www.heacademy.ac.uk
11. Yudelson, M., Brusilovsky, P.: NavEx: Providing Navigation Support for Adaptive Browsing of Annotated Code Examples. In: Looi, C.-K., McCalla, G., Bredeweg, B., Breuker, J. (eds.) Artificial Intelligence in Education, pp. 710–717. IOS Press, Amsterdam (2005)
12. Bull, S., Britland, M.: Group Interaction Prompted by a Simple Assessed Open Learner Model that can be Optionally Released to Peers. In: Brusilovsky, P., Papanikolaou, K., Grigoriadou, M. (eds.) PING Workshop, User Modeling (2007)
13. Ahmad, N., Bull, S.: Learner Trust in Learner Model Externalisations. In: Dimitrova, V., Mizoguchi, R., du Boulay, B., Graesser, A. (eds.) Artificial Intelligence in Education, pp. 617–619. IOS Press, Amsterdam (2009)

Recognizing and Predicting the Impact on Human Emotion (Affect) Using Computing Systems

David G. Cooper

University of Massachusetts, Department of Computer Science,
140 Governors Drive, Amherst MA 01003, USA
dcooper@cs.umass.edu
http://www.cs.umass.edu/~dcooper

Abstract. Emotional intelligence is a clear factor in education [1–3], health care [4], and day to day interaction. With the increasing use of computer technology, computers are interacting with more and more individuals. This interaction provides an opportunity to increase knowledge about human emotion for human consumption, well-being, and improved computer adaptation.

This research makes five main contributions. 1) Construct a method for determining a set of sensor features that can be automatically processed to predict human emotional changes in observed people. 2) Identify principles, algorithms, and classifiers that enable computational recognition of human emotion. 3) Apply this method to an intelligent tutoring system instrumented with sensors. 4) Apply and adapt the method to audio and video sensors for a number of applications such as a) detection of psychological disorders, b) detection of emotional changes in health care providers, c) detection of emotional impact of one person on another during video chat, and/or d) detection of emotional impact of one fictional character on another in a motion picture. 5) Integrate emotional detection technologies so that they can be used in more realistic settings.

Keywords: emotional interaction, multi-sensor affective processing, smart environments, actionable affect, social signal processing.

Approach

I intend to research affective processing in three domains: Intelligent Tutoring Systems (ITS), clinical voice analysis, and personal interaction. The method will consist of data exploration over a number of data sets. For the ITS domain, we have already collected data from five different schools with more than ten classrooms and over 600 students using between 0 and 4 sensors for one ITS. For the clinical voice analysis we have data from five studies. Three of the studies have a very consistent protocol as far as data collection, however the population and the reason for collection differ. The other two studies are not as consistently

P. De Bra, A. Kobsa, and D. Chin (Eds.): UMAP 2010, LNCS 6075, pp. 399–402, 2010.

controlled, and may have more artifacts. The studies are a multicultural study with 4 different cultures, 10 subjects from each culture balanced for gender and age [5]; a study with a Greek population where the individuals were shown pictures meant to elicit an emotion and spontaneous speech was collected; one additional study is with an examiner and a child with typical development, apraxia of speech, and autism conditions. For the personal interaction domain a study will be developed in either a nursing lab, an architecture critique, or an office interaction and will use audio and video processing as the source of affective features.

The first domain is using affect for the Intelligent Tutoring System (ITS) Wayang Outpost. For Wayang Outpost, data has been collected for more than 600 students using between 0 and 4 sensors in a classroom environment. Along with this data are a sparse set of emotional labels pertaining to 4 different emotional states (Frustration, Confidence, Interest, and Excitement). So far I have shown that linear classifiers can be created to get good Specificity for Confidence, and good Sensitivity for Interest and Excitement using basic statistics on a per problem basis. Results of a feature selection and ranking are summarized in Table 1. This is a follow-on study to [6].

Table 1. Classifier Ranking Using Validation data from the Spring of 2009. Parametric results and Non-parametric results are shown side by side.

Confident	Tukey HSD	NPMC
Specificity	$(confCameraA \sim confTutorA \sim confTutorM) > (confSeat \sim confTutorW) > confBasline$ $confCameraB > confTutorW > confBaseline$	$(confCameraA \sim confTutorA \sim confTutorM) > (confSeat \sim confTutorW) > confBasline$ $confCameraB > confTutorW > confBaseline$
Interested	Tukey HSD	NPMC
Sensitivity	$intCamera > intBaseline$	$intCamera > intBaseline$
Excited	Tukey HSD	NPMC
Sensitivity	$((excCamera > excTutor) \sim excCameraSeat) > excBaseline$	$excCamera > excCameraSeat > excTutor > excBaseline$

The next steps include feature improvement, such as finding event related sensor features, finding 'time series motifs' [7] based on the time-series data, and using other sensor specific methods; in addition, applying more advanced classifiers such as support vector clustering [8], the group method of data handling [9, 10], decision trees, and random forests will likely improve the current results.

In addition, in order to move to sensors that don't have to be on or near the body, I plan to integrate video and audio based emotional detection systems in order to run meaningful experiments for the detection of emotional impact. To that end, I intend to both extract new features from the video in the tutor data (e.g. head position, head motion, looking away) and utilize new video features from distal video, such as body and face position relative to another body, body and hand gestures and articulation, as well as audio features, such as prosody (rate of speech), inflection, and other acoustical changes in speech. The audio features will be explored in the clinical voice domain before they are used in the personal interaction domain.

Using the Viola-Jones face detector [11] implemented in OpenCV, the faces of the tracked people can be detected, extracted and sent to a facial feature tracker such as [12] used with the Wayang Outpost Tutor. The difficulties in this are getting a connected sequence of faces at a fast enough frame rate, and the resolution. Thus, it is likely that when the face is too far from any camera, that other features will need to be relied upon such as audio features and body gesture, head position and motion.

There are a number of ways that researchers have categorized the observation of emotion. The two most prevalent are 1) a two dimensional feature space, and 2) a discrete set of classes for emotion. The two dimensional feature space consists of valence (or the pleasantness of an experience) ranging from negative to positive and arousal, ranging from low to high [13]. This two dimensional space tends to be adequate for generating agreement when placing an affective label on it and has been used in connection with observing facial expressions and physiological features since 1954 [14, 15]. A similar two dimensional scale developed by Ralph Bierman was created for the purpose of personal interaction (PICI) [16]. The two dimensions are rejecting-accepting and passive-active. They relate to how one individual is interacting with another. Though we do not use the valence and arousal dimensions directly, they may become useful factors for audio. In addition the PICI may be a good first step for looking at personal interaction. In the case of a student interacting with an intelligent tutoring system, this scale may be useful at the extremes of accepting and rejecting, however the personal and impersonal parts of the scale may be skipped altogether. In the personal interaction domain, location of the two persons relating to each other to determine proximity could imply acceptance vs. rejection, and an ability to detect the amount of motion of each body may indicate activity. Looking at full body motion in video has been done for identification [17] and for estimating interaction cues such as head pose, fidgeting, body pose, etc. [18].

The time-line for this research is to perform a nursing student study over the next four to five months using a lab that is already instrumented and using pre-test and post-test emotional reports as labels. I will spend a few months developing emotional classification methods on the data from the ITS, and then I will spend another few months adapting and applying those methods to the clinical voice data. I will then apply both the voice analysis and the ITS based video interaction analysis to the Nursing Study. The goal is to finish this research by May of 2011.

I would appreciate advice on the details of the study to be performed. If there is a group with an instrumented room that might be interested in this research, then that would be a great help.

References

1. Lepper, M.R., Chabay, R.W.: Socializing the intelligent tutor: bringing empathy to computer tutors, pp. 242–257 (1988)
2. Lepper, M.R., Woolverton, M., Mumme, D.L., Gurtner, J.L.: Technology in education. In: Motivational techniques of expert human tutors: Lessons for the design of computer-based tutors, pp. 75–105. Lawrence Erlbaum Associates, Inc., Mahwah (1993)
3. Derry, S.J., Potts, M.K.: How tutors characterize students: a study of personal constructs in tutoring. In: ICLS '96: Proceedings of the 1996 international conference on Learning sciences, International Society of the Learning Sciences, pp. 368–373 (1996)
4. Ostir, G., Markides, K., Black, S., Goodwin, J.: Emotional well-being predicts subsequent functional independence and survival. Journal of the American Geriatrics Society 48(5), 473 (2000)
5. Andrianopoulos, M., Darrow, K., Chen, J.: Multimodal standardization of voice among four multicultural populations formant structures. Journal of Voice 15(1), 61–77 (2001)
6. Cooper, D.G., Arroyo, I., Woolf, B.P., Muldner, K., Burleson, W., Christopherson, R.: Sensors model student self concept in the classroom. In: Houben, G.-J., McCalla, G., Pianesi, F., Zancanaro, M. (eds.) UMAP 2009. LNCS, vol. 5535, pp. 30–41. Springer, Heidelberg (2009)
7. Lin, J., Keogh, E., Lonardi, S., Patel, P.: Finding motifs in time series. In: Proc. of the 2nd Workshop on Temporal Data Mining, pp. 53–68 (2002)
8. Ben-Hur, A., Horn, D., Siegelmann, H., Vapnik, V.: Support vector clustering. The Journal of Machine Learning Research 2, 125–137 (2002)
9. Ivakhnenko, A.: Heuristic self-organization in problems of engineering cybernetics (group method of data handling based on heuristic self organization for solving complex system problems, with applications to random processes prediction). Automatica 6, 207–219 (1970)
10. Ivakhnenko, A.: Polynomial theory of complex systems. IEEE Transactions on Systems, Man, and Cybernetics 1(4) (1971)
11. Viola, P., Jones, M.: Robust real-time object detection. International Journal of Computer Vision (2001)
12. el Kaliouby, R.: Mind-reading Machines: the automated inference of complex mental states from video. PhD thesis, University of Cambridge (2005)
13. Feldman, L.: Valence focus and arousal focus: Individual differences in the structure of affective experience. Journal of Personality and Social Psychology 69, 153 (1995)
14. Schlosberg, H.: Three dimensions of emotion. Psychological review 61(2), 81 (1954)
15. Russell, J.: A circumplex model of affect. Journal of personality and social psychology 39(6), 1161–1178 (1980)
16. Bierman, R.: The personal interaction coding inventory. J. Counc. Assn. Univ. Stud. Personnel Serv. V (Spring 1970)
17. Pratheepan, Y., Torr, P., Condell, J., Prasad, G.: Body language based individual identification in video using gait and actions. Image and Signal Processing, 368–377 (2009)
18. Chippendale, P., Lanz, O.: Optimised meeting recording and annotation using real-time video analysis. Machine Learning for Multimodal Interaction, 50–61 (2008)

Utilising User Texts to Improve Recommendations

Yanir Seroussi

Faculty of Information Technology, Monash University
Clayton, Victoria 3800, Australia
yanir.seroussi@infotech.monash.edu.au

Abstract. Recommender systems traditionally rely on numeric ratings to represent user opinions, and thus are limited by the single-dimensional nature of such
ratings. Recent years have seen an abundance of user-generated texts available
online, and advances in natural language processing allow us to better understand
users by analysing the texts they write. Specifically, sentiment analysis enables
inference of people's sentiments and opinions from texts, while authorship attribution investigates authors' characteristics. We propose to use these techniques to
build text-based user models, and incorporate these models into state-of-the-art
recommender systems to generate recommendations that are based on a more
profound understanding of the users than rating-based recommendations. Our
preliminary results suggest that this is a promising direction.

1 Introduction

Recommender systems deal with predicting people's opinions of items, usually in the
form of numeric ratings [1]. The two main approaches to predicting ratings are collaborative filtering (*CF*) and content-based recommendation (*CB*). CF employs a *target
user*'s previous ratings and ratings submitted by similar *training users* to predict the ratings that the target user will give to unrated items, while CB is based on the target user's
past ratings and an analysis of the commonalities between the previously rated items.
Recommender systems traditionally rely on numeric ratings to calculate user similarity (for CF) and to determine the users' opinions on items (for CB and CF). Ratings
may be explicit (e.g., MovieLens at www.movielens.org asks users to rate 15 movies
before it starts generating recommendations) or implicit (e.g., using viewing times of
museum exhibits [2]). In either case, the system's understanding of the users is limited
by the single-dimensional nature of ratings. For example, a user may assign the same
rating to two different movies, but for completely different reasons. This illustrates that
analysing user opinion based only on ratings has inherent limitations.

Sentiment analysis and authorship attribution are two research areas that can be
utilised to enhance a system's understanding of its users. Sentiment analysis deals with
inferring people's sentiments and opinions from texts [3]. Tasks in this field include:
inferring the overall positivity of texts; detecting sentiments towards aspects of a text's
topic (e.g., the acting in a movie); and determining the usefulness of a product review.
For example, the system can discover that a movie review is very positive and that the
text implies that the acting in the movie is superb, but also indicates that other users are
unlikely to find the review useful. Authorship attribution deals with inferring people's
characteristics from documents they have written [4]. As the name indicates, the main

P. De Bra, A. Kobsa, and D. Chin (Eds.): UMAP 2010, LNCS 6075, pp. 403–406, 2010.

task in this field is attributing texts to their original authors (e.g., when plagiarism is suspected). Other tasks include inferring whether texts were written by authors who belong to a certain age group, gender, etc. Therefore, authorship attribution techniques can be utilised to infer user demographics and discover latent similarities between users.

Online communication is often textual: users publish reviews, write emails, use instant messaging, and discuss issues on message boards. This research will utilise such texts to gain an in-depth knowledge of users and recommend items accordingly. We propose to utilise sentiment analysis and authorship attribution techniques to create user models that are more comprehensive than those based only on ratings, thereby addressing the limitations that are caused by using only ratings in recommender systems.

2   Research Questions and Related Work

The main research question is: *How can we utilise texts written by users to generate better recommendations?* We divide this question into three sub-questions:

Q1: *Can we improve the performance of sentiment analysis methods by considering the users who wrote the texts?* The first step is to investigate the relation between users and sentiments expressed in their texts. Traditional methods for sentiment analysis consider texts as standalone entities, yet several researchers found that authorship affects performance in sentiment analysis [5, 6]. Pang and Lee [5] found that a classifier trained on film reviews by one user and tested on reviews by a different user is likely to perform poorly. Lin *et al.* [6] obtained similar results with respect to a dataset that contains pro-Palestinian and pro-Israeli articles, half of them written by two editors and the other half by various guest writers. We therefore propose to harness cross-user similarity to improve performance in sentiment analysis, and gain insights on textual user modeling.

Q2: *Can we improve the predictive accuracy of CF systems by using texts to measure user similarity?* Predictive accuracy in CF is largely dependent on the way user similarity is measured. In most systems, user similarity is based on ratings, which are not always available and do not carry as much information as texts. It has been shown that the performance of CF can be improved by using other information sources than ratings, such as users' demographics and TV program preferences [7]. However, obtaining such information requires explicitly questioning the users – a process that many users may not be willing to go through. We therefore propose to obtain personal information from user texts in a non-intrusive way, using sentiment analysis and authorship attribution methods, and use it to model user similarity.

Q3: *Can we improve the predictive accuracy of CB systems by detecting sentiments towards item aspects in user reviews?* CB systems use numeric ratings to determine what items the users like, and from that they try to infer what the users like about the items. Textual reviews of items are a richer information source than ratings, since they contain sentiments towards specific aspects of the items. For example, Snyder and Barzilay [8] analysed restaurant reviews to infer the authors' sentiments regarding several aspects of the restaurant-going experience (e.g., food and service). We propose to use such analysis techniques to better understand the sentiments of users towards item aspects, utilise these sentiments to build richer user profiles, and thereby generate more accurate recommendations.

Recently, Jakob *et al.* [9] showed that is is possible to improve the accuracy of CF by automatically inferring the ratings of movie aspects from review texts and using these ratings as additional features. This is different from our proposed approach to CF, as they ultimately added finer-grained ratings, rather than enhancing the actual similarity model. An earlier attempt at incorporating texts into recommender systems was made by Aciar *et al.* [10], who aggregated opinions from product reviews and displayed a product score based on queries by the target users (nothing is known about the target users apart from their queries, which explicitly specify their preferences). This is also different from our research, as we propose to utilise various types of user-generated texts in well-known recommendation frameworks that consider the target user's history.

3   Approach and Preliminary Results

To address the research questions, we need a dataset of ratings and texts to empirically evaluate the methods we develop. Therefore, we created the *Prolific IMDb Users* dataset by collecting data from the *Internet Movie Database (IMDb)* at www.imdb.com in May 2009. Our dataset contains all the movie reviews and message board posts by 184 users who wrote at least 500 reviews. The large number of user reviews makes it possible to study the effect of different numbers of reviews on the user models.

So far we have worked on Q1: exploring the effect of taking users into account on inferring sentiment from texts. We introduced a nearest-neighbour collaborative framework: we trained user-specific classifiers, and considered user similarity to combine the outputs of the classifiers [11]. This approach decreases the error in cross-user sentiment analysis, while requiring less computational resources than user-blind methods.

To evaluate the performance of our methods we used a subset of the Prolific IMDb Users dataset, which includes 62 users with 1000 reviews each. The item/rating matrix of this subset is very sparse, and thus we observed that basing user similarity on pairwise comparison of reviews or ratings for co-reviewed items – the common approach in CF – did not perform as well as using the entire review sets for calculating user similarity. In addition, our experiments showed that basing similarity on review texts or on message board posts (when reviews are unavailable) yields better results than using explicit ratings. Moreover, this occurs for target users with a small number of reviews and even when a small number of prolific training users is available. This leads us to conjecture that the similarity measures we introduced can be used to address two problems that occur in CF: the item/rating matrix sparsity problem and the new user problem [1].

4   Planned Work and Conclusion

In the immediate future, we plan on finishing the evaluation of our collaborative model for sentiment inference. We already experimented with different numbers of prolific training users and available reviews by the target user. The missing element is experimenting with non-prolific training users – users who have not submitted many reviews.

The next step is applying to CF the user similarity models that yielded the best performance in sentiment analysis (Q2). We then plan to introduce similarity models that will utilise authorship attribution techniques to detect user characteristics from texts.

In addition, we will investigate using texts and ratings as joint sources for similarity calculation. We will also consider the analysis of item aspects in reviews for calculating user similarity in CF, as a preparatory step to applying aspect analysis to CB (Q3).

Another dimension that we may consider is automatic inference of quality of reviews and expertise of the review authors, which will enable assigning different weights to reviews and users based on their usefulness to the target user. If time allows, we will then apply our methods to different domains (e.g., books and electronic appliances) and use our text-based approach in hybrid recommender systems that combine CF and CB.

In conclusion, our work so far shows that modeling users based on their texts has the potential to overcome some of the known problems in recommender systems. We would like to receive advice on additional ways of employing user texts in recommender systems, as well as thoughts on possible problems with our proposed research directions.

References

1. Adomavicius, G., Tuzhilin, A.: Toward the next generation of recommender systems: A survey of the state-of-the-art and possible extensions. IEEE Transactions on Knowledge and Data Engineering 17(6), 734–749 (2005)
2. Bohnert, F., Zukerman, I.: Non-intrusive personalisation of the museum experience. In: Houben, G.J., McCalla, G.I., Pianesi, F., Zancanaro, M. (eds.) UMAP 2009. LNCS, vol. 5535, pp. 197–209. Springer, Heidelberg (2009)
3. Pang, B., Lee, L.: Opinion mining and sentiment analysis. Foundations and Trends in Information Retrieval 2(1-2), 1–135 (2008)
4. Juola, P.: Authorship attribution. Foundations and Trends in Information Retrieval 1(3), 233–334 (2006)
5. Pang, B., Lee, L.: Seeing stars: Exploiting class relationships for sentiment categorization with respect to rating scales. In: Proceedings of the 43rd Annual Meeting of the Association for Computational Linguistics (ACL), Ann Arbor, Michigan, pp. 115–124 (2005)
6. Lin, W., Wilson, T., Wiebe, J., Hauptmann, A.: Which side are you on? Identifying perspectives at the document and sentence levels. In: Proceedings of the 10th Conference on Natural Language Learning (CoNLL), New York, pp. 109–116 (2006)
7. Lekakos, G., Giaglis, G.M.: A hybrid approach for improving predictive accuracy of collaborative filtering algorithms. User Modeling and User-Adapted Interaction 17(1), 5–40 (2007)
8. Snyder, B., Barzilay, R.: Multiple aspect ranking using the Good Grief algorithm. In: Proceedings of the Joint Human Language Technology/North American Chapter of the ACL Conference (HLT/NAACL), Rochester, New York, pp. 300–307 (2007)
9. Jakob, N., Weber, S.H., Müller, M.C., Gurevych, I.: Beyond the stars: Exploiting free-text user reviews for improving the accuracy of movie recommendations. In: Proceedings of the 1st International CIKM Workshop on Topic-Sentiment Analysis for Mass Opinion Measurement, Hong Kong, China, pp. 57–64 (2009)
10. Aciar, S., Zhang, D., Simoff, S., Debenham, J.: Informed recommender: Basing recommendations on consumer product reviews. IEEE Intelligent Systems 22(3), 39–47 (2007)
11. Seroussi, Y., Zukerman, I., Bohnert, F.: Collaborative inference of sentiments from texts. In: Proceedings of the Conference on User Modeling, Adaptation, and Personalization (UMAP), Waikoloa, HI. LNCS, vol. 6075, pp. 195–206. Springer, Heidelberg (2010)

Semantically-Enhanced Ubiquitous User Modeling

Till Plumbaum

Technische Universität Berlin, DAI-Labor, Germany
till.plumbaum@dai-labor.de

Abstract. Semantically-enhanced Ubiquitous User Modeling aims at
the management of distributed user models and the integration into ontologies to share user information amongst adaptive applications for personalization purposes. To reach this goal, different problems have to be
solved. The collection of implicit user information by observing the user
behavior on dynamic web applications is important to better understand
the user interests and needs. The aggregation of different user models is
essential to combine all available user information to one big knowledge
repository. Additionally, the Semantic Web offers new possibilities to
enhance the knowledge about the user for better personalization.

1 Introduction

With the advent of the Web 2.0 and the growing impact of the Internet on our
every day life, people use more and more different web applications. Thereby,
they generate and distribute personal information like interests, preferences and
goals. This distributed and heterogeneous collection of user information, stored
in the user model (UM) of each application, is a valuable source of knowledge
for adaptive systems. Current adaptive systems take into account user features
like interest, plans and context such as the context of interaction, the device,
etc. The modeling of the user is usually done in the design phase of the system,
and therefore changes to the model, to adapt to changing requirements or user
characteristics, can not be implemented without major changes to the system.
Also the representation of the user model is in most cases strongly application
dependent and therefore not understandable and usable by other applications.
That implies that the knowledge about the users, which is buried deeply in
the databases of one adaptive system, cannot be shared with other systems to
provide better personalization and adaptation results.

In my dissertation, I focus on the combination of Semantic Web technologies
with adaptive systems and the use of shared ontologies to describe and model
knowledge about users. A UM that is based on shared and open ontologies can be
used to share the knowledge with other systems and moreover it can be extended
easily using additional ontologies.

P. De Bra, A. Kobsa, and D. Chin (Eds.): UMAP 2010, LNCS 6075, pp. 407–410, 2010.
© Springer-Verlag Berlin Heidelberg 2010

2   Identified Problems and Related Work

I focus on two major aspects of the user modeling process. Firstly, obtaining implicit information about user needs and interests by collecting information about the user behavior using semantic technologies. Secondly, managing the UMs, which includes the aggregation of models from UMs from different applications, taking care of the heterogeneity of the information, and representing the information about a user based on an ontology. An extension which I consider, is the enrichment if this ontology-based UM with data from the Semantic Web.

User Behavior Collection and Management. Web applications become more and more dynamic, and the way users can interact with them change. Therefore, the techniques to track the user behavior have to cope with these new challenges and have to be extended to collect fine-grained data from user interactions to provide better information for adaptive systems. Additionally, the collected data must be managed in ontologies to share user behavior information with other adaptive systems. Zhou et al. [1] focus on mining client-side access logs of a single user or client and then incorporate fuzzy logic to generate a usage ontology. Schmidt et al. [2] embed concepts into a portal which provide the context for JavaScript events, which are collected and used to adjust the portal. All relevant UI elements are linked to a concept ontology containing semantic information about the element. None of these approaches make full use of semantic technologies. First steps in the direction of semantic technologies are done but they still cannot be applied across applications and lack the necessary extensibility and dynamism.

User Model Management and Aggregation. Applications typically store their user information in a proprietary format. This leads to a distributed web model of a user with several partial UMs in different applications potentially duplicating information. Therefore, the challenge is to solve the heterogeneity of the user models. Current research on user model management and aggregation emphasizes two different strategies [3]. The first strategy introduced in [4] uses a generic user model mediation framework with the goal of improving the quality of recommendations. The actual UM mediation in the framework is done by specialized mediator components which translate the data between different models using inference and reasoning mechanisms. The second strategy focuses on the standardization of user models to allow data sharing between applications. Heckmann [5] proposes an ontological approach, the General User Model Ontology (GUMO), as a top level ontology for user models and suggest the ontology to be the standard model for user modeling tasks. Another standardization approach is to define a centralized user modeling system that is used and updated by all connected applications [6]. The shortcomings of the mediation layer approach is the effort needed to aggregate such heterogeneous user models, while standardized user models suffer from the lack of a common standard. As long as different application providers pursue different goals with strong commercial interest, a global standard for user model does not seem likely in the near future.

3   Open Research Questions and Proposed Solutions

The identified problems lead to open research questions, that I want to solve in my dissertation:

- How can I merge and manage user profiles from different applications?
- How can I use the collected user information to enrich the user profile?
- How can user data acquisition benefit from Semantic Web standards?
- How can I model and share the collected user behavior data between applications and different domains?

To solve these problems it is necessary to develop a user tracking mechanism which collects implicit user feedback from dynamic websites and to manage the data in a user behavior ontology which allows the collected information to be shared. Furthermore, the development of an ontology based user profile management framework is required. This framework should combine the presented ideas of UM management and aggregation presented here. The aggregation has to be done by specialized mediators focusing on automatic ontology matching approaches and fuzzy logic techniques to address the uncertainty of this process. The framework also has to support the goals of generality and extensibility to enable adaptive systems to manage and share user knowledge.

4   Work Done and Future Work Schedule

User Behavoir Collection and Management. A user tracking component which used Microformats was already implemented and presented in [7]. The usage of Microformats allows to add semantic information to web pages and, because it is an open standard, the semantic information can be used by other applications, too. The next steps are to extend the already existing user tracking component to support RDFa, and to evaluate the level of information that can be collected and its impact on adaptive systems. I'm currently working on an a user behavior ontology that manages the collected data and can be used to share information about the common behavior of a user between adaptive systems. Thus, applications are able to adapt to the user right from the start. A first version of the ontology exists already, and I expect to publish information in the near future.

User Model Management and Aggregation. A lot of work is done, most of it yet unpublished. We have implemented a framework (see Fig. 1), presented in [6], that manages user models from different applications. This framework was extended to support ontologies, using the JENA framework and open ontologies like FOAF. I developed a meta-ontology that allows connecting information from different applications. The framework also consists of a component which collects information from the Semantic Web and uses it to enrich the user models. My current work concentrates on merging the models of different applications. The main focus here is the usage of ontology matching approaches to automatically aggregate the models. A main open challenge is the evaluation of these methods due to fact that no corpus exists to measure the quality of the developed methods, as stated in [4]. Therefore, I plan to build such a corpus that can be used to measure the quality of these approaches.

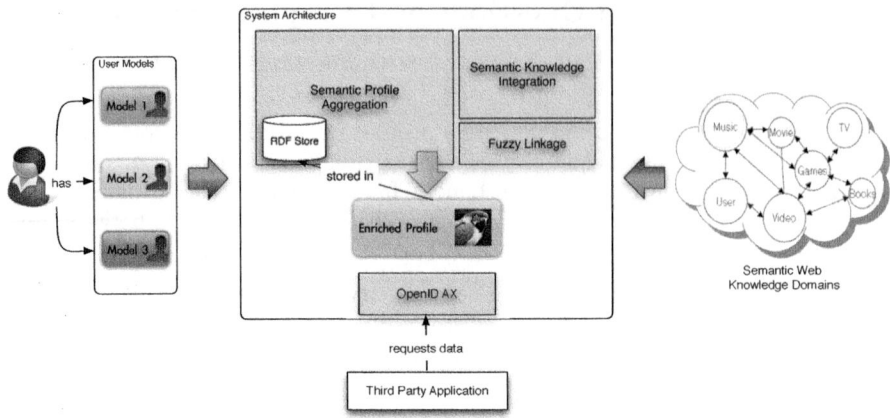

Fig. 1. System overview with components and the OpenID interface for data privacy

References

1. Zhou, B., Hui, S.C., Fong, A.C.M.: Web usage mining for semantic web personalization. In: Proc. of the Workshop on Personalization on the Semantic Web (PerSWeb'05), Edinburgh, Scotland, pp. 66–72 (2005)
2. Schmidt, K.U., Stojanovic, L., Stojanovic, N., Thomas, S.: On enriching ajax with semantics: The web personalization use case. In: Franconi, E., Kifer, M., May, W. (eds.) ESWC 2007. LNCS, vol. 4519, pp. 686–700. Springer, Heidelberg (2007)
3. Kuflik, T.: Semantically-enhanced user models mediation: Research agenda. In: Proc. of the 5th international workshop on ubiquitous user modeling (UbiqUM 2008), Gran Canaria, Spain (2008)
4. Berkovsky, S., Kuflik, T., Ricci, F.: Mediation of user models for enhanced personalization in recommender systems. User Modeling and User-Adapted Interaction 18(3), 245–286 (2008)
5. Heckmann, D., Schwartz, T., Brandherm, B., Schmitz, M., von Wilamowitz-Moellendorff, M.: Gumo - the general user model ontology. In: Ardissono, L., Brna, P., Mitrović, A. (eds.) UM 2005. LNCS (LNAI), vol. 3538, pp. 428–432. Springer, Heidelberg (2005)
6. Korth, A., Plumbaum, T.: A framework for ubiquitous user modeling. In: Proc. of the IEEE International Conference on Information Reuse and Integration, Las Vegas, USA, pp. 291–297. IEEE Systems, Man, and Cybernetics Society (2007)
7. Plumbaum, T., Stelter, T., Korth, A.: Semantic web usage mining: Using semantics to understand user intentions. In: Houben, G.J., McCalla, G.I., Pianesi, F., Zancanaro, M. (eds.) UMAP 2009. LNCS, vol. 5535, pp. 391–396. Springer, Heidelberg (2009)

User Modeling Based on Emergent Domain Semantics

Marián Šimko and Mária Bieliková

Institute of Informatics and Software Engineering, Faculty of Informatics and
Information Technology, Slovak University of Technology,
Ilkovičova 3, 842 16 Bratislava, Slovakia
{simko,bielik}@fiit.stuba.sk

Abstract. In this paper we present an approach to user modeling based
on the domain model that we generate *automatically* by resource (text)
content processing and analysis of associated tags from a social annotation service. User's interests are modeled by overlaying the domain
model – via keywords extracted from resource's (text) content, and tags
assigned by the user or other (similar) users. The user model is derived
automatically. We combine content- and tag-based approaches, shifting
our approach beyond flat "folksonomical" representation of user interests
to involve relationships between both keywords and tags.

Keywords: user modeling, emergent domain semantics, automatic domain model composition, folksonomy, text mining.

1 Introduction and Related Work

Recommendation in social systems, also referred to as collaborative filtering,
consists of *(i)* user similarity computation and *(ii)* relevant resource prediction.
The purpose of the first step is to find the most similar users with the "active"
user, often assuming their similar behavior during the process of information
search or navigation (visiting a page, buying a product). Similar behavior is interpreted as similar interest, which is a base for the second step, where resources
(pages, products) are predicted based on their relation to the most similar users.

Traditional approaches to the user similarity computation utilize methods of
usage mining [9]. Visiting the same web page or similar movie rating indicates
similar interest. The other group of approaches is based on social tagging. Tags
are promising source of information for recommendation as the number of tagging users all along increases. From the user modeling perspective, tags represent
user interest. Strictly speaking, the action of assigning a tag to a resource is what
is interpreted as user interest in tagged resource [2, 3]. Furthermore, different
tags from different users are analyzed to consider the context of tagging [7, 8].
In order to derive more accurate recommendation, contextualized score for resources is being computed for each user to reveal different purposes of tags.

In our work we primarily focus on recommendation of text-based resources
such as web pages or learning objects in collaborative learning environment. In

P. De Bra, A. Kobsa, and D. Chin (Eds.): UMAP 2010, LNCS 6075, pp. 411–414, 2010.

the thesis proposal we combine social-based collaborative filtering with content-based approach. We utilize tags from tag-based systems, but we shift the whole approach beyond flat folksonomical representation of resources, leveraging lightweight emergent semantics generated automatically based on resource analysis.

2   Emergent Semantics

When selecting users whose associated resources will be recommended, their models are compared in order to obtain user similarity level. User model composition is one of the most delicate parts of any method for personalized search or recommendation. We build on overlay user model that is based on the domain model [1]. The crucial part of our work and our contribution is automated user model generation based on the resource content and tag analysis. The acquired representation we refer to as resource metadata (see Fig. 1).

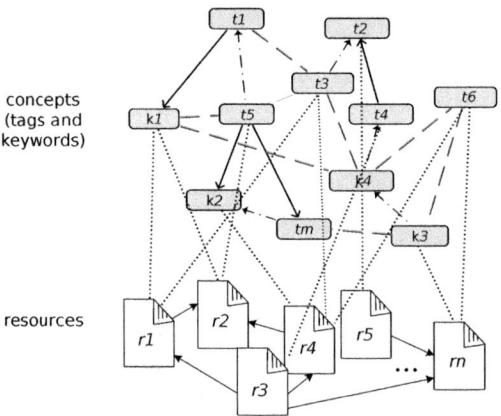

Fig. 1. User model representation (upper part) for a set of resources (pages) he visited. Entity kx represents keyword, ty represents tag and rz represents resource. Different relationship markup between entities reflects different semantic power.

Metadata consist of *concepts*[1] and *relationships* between concepts. We differentiate two types of concepts:

- keywords (content),
- tags (folksonomy).

Concepts represented by keywords extracted from resource content have different semantics than tags assigned by a user himself. While the first type of concept is

---

[1] Using term *concept* can be slightly misleading as someone can think of the concept only for representing conceptual knowledge, abstract or general ideas inferred or derived from specific instances. We use term *concept* because resource metadata serve exactly by the same way as conceptual knowledge.

added to the user model when visiting a page (similarly to [5]), the second type is added when the user tags a page. The latter action is more explicit and reflects into the higher weight of relation between a resource and a tag when computing concept and user similarities. We believe that considering tags together with the resource content is feasible, as research of social tagging showed that tags are to a certain, significant, extent dependent on resource's content (e.g. title) [4].

Relationships between concepts represent concepts' relatedness. We create them automatically by an underlying graph analysis utilizing the notion of node centrality. As they are derived from underlying domain model, they can be viewed as emergent domain semantics representing additional value to pure concept-based user model. The method for relationship generation we already evaluated in e-learning domain when discovering relationships between concepts extracted from learning objects [11]. Similarly to relationships, keywords are extracted automatically. We apply here the methods and techniques for automated term recognition [6].

Based on described representation, user similarity we compute considering following principles: users are more similar if

- the more similar concepts are assigned to resources they visit;
- the more similar tags are assigned (by other users) to resources they visit;
- the more similar tags they assign to same resources;
- the more similar relationships between concepts (keywords and tags) exist.

The user model is generated automatically and it is different for every user (it contains different concepts and different relationships between them, which are derived from users' actions). The user's context is considered as we track the way he accesses the resource: by visiting and/or by tagging.

The next step, resource prediction, is based on the user model similarity computation. For the most similar users (those exceeding a certain similarity threshold), a prediction score is computed. We consider two computational variants: statistical and topological, each representing different view on two user models. Variants can be mutually combined in order to achieve better recommendation.

3   Conclusions

In our work we focus on automatic composition of a user model. We proposed the method that builds the user model *combining* the content-based and tag-based approach. After a resource's content and assigned tags are analyzed, concepts represented by keywords and tags are added to user model. The relationships between them representing relatedness of entities are composed considering the user's context. This approach building on the domain model created automatically we consider the main contribution that the thesis aims to achieve. The created user model is used for user similarity computation and resource prediction computation for recommendation.

In the current stage of our research we have analyzed methods for automatic term extraction, we have analyzed methods for relationship discovery from the text (as a part of *ontology* learning field), and we have analyzed methods for concept relationship induction from folksonomies. We proposed a method for automatic relationship discovery based on underlying graph representation (the graph on Fig. 1 with no relationships between concepts yet) that we evaluated in the e-learning domain [10].

Acknowledgments. This work was partially supported by the grants VEGA 1/0508/09, KEGA 028-025STU-4/2010 and it is the partial result of the Research & Development Operational Programme for the project Support of Center of Excellence for Smart Technologies, Systems and Services, ITMS 26240120029, co-funded by the ERDF.

References

1. Brusilovsky, P.: Methods and Techniques of Adaptive Hypermedia. User Modeling and User-Adapted Interaction 6(2-3), 87–129 (1996)
2. Carmagnola, F., Cena, F., Cortassa, O., Gena, C., Torre, I.: Towards a Tag-Based User Model: How Can User Model Benefit from Tags? In: Conati, C., McCoy, K., Paliouras, G. (eds.) UM 2007. LNCS (LNAI), vol. 4511, pp. 445–449. Springer, Heidelberg (2007)
3. Durao, F., Dolog, P.: A personalized tag-based recommendation in social web systems. In: Tasso, D.A., Farzan, R., Kleanthous, S., Bueno Vallejo, D., Vassileva, J. (eds.) Proc. of Int. Workshop on Adaptation and Personalization for Web 2.0 (AP-WEB 2.0 2009) at UMAP2009, CEUR,, Trento, Italy, vol. 485 (2009)
4. Heymann, P., Koutrika, G., Garcia-Molina, H.: Can Social Bookmarking Improve Web Search? In: Proc. of the WSDM 2008, pp. 195–206 (2008)
5. Kajaba, M., Návrat, P., Chudá, D.: A Simple Personalization Layer Improving Relevancy of Web Search. Computing and Information Systems Journal 13(3), 29–35 (2009)
6. Knoth, P., Schmidt, M., Smrž, P., Zdráhal, Z.: Towards a Framework for Comparing Automatic Term Recognition Methods. In: Znalosti 2009, pp. 83–94 (2009)
7. Nakamoto, R., Nakajima, S., Miyazaki, J., Uemura, S.: Tag-based contextual collaborative filtering. Int. Journal of Computer Science, IAENG 34(2) (2007)
8. Niwa, S., Doi, T., Honiden, S.: Web page recommender system based on folksonomy mining for ITNG'06 submissions. In: Proc.of the 3rd Int. Conf. on Information Tech.: New Generations, Washington, DC, USA, pp. 388–393. IEEE, Los Alamitos (2006)
9. Pazzani, M.J., Billsus, D.: Content-Based Recommendation Systems. In: Brusilovsky, P., Kobsa, A., Nejdl, W. (eds.) Adaptive Web 2007. LNCS, vol. 4321, pp. 325–341. Springer, Heidelberg (2007)
10. Šimko, M., Bieliková, M.: Automated Educational Course Metadata Generation Based on Semantics Discovery. In: Cress, U., Dimitrova, V., Specht, M. (eds.) EC-TEL 2009. LNCS, vol. 5794, pp. 99–105. Springer, Heidelberg (2009)
11. Šimko, M., Bieliková, M.: Automatic Concept Relationships Discovery for an Adaptive E-course. In: Barnes, T., et al. (eds.) Proc. of Educational Data Mining 2009: 2nd Int. Conf. on Educational Data Mining, Cordoba, Spain, pp. 171–179 (2009)

"Biographic spaces": A Personalized Smoking Cessation Intervention in Second Life

Ana Boa-Ventura[1] and Luís Saboga-Nunes[2]

[1] University of Texas at Austin, Austin, USA
anaventura@mail.utexas.edu
[2] CIESP, Escola Nacional de Saúde Pública, UNL, Lisboa, Portugal

Abstract. In this paper we are proposing a proof-of-concept leveraging the use of 3D virtual worlds in addictive behavior interventions. We propose a model that we call biographic space, which embeds the successive stages that a smoker may go through while attempting to quit smoking including emotionally loaded aspects such as deciding to quit and post cessation withdrawal. The design of this space is informed by storytelling and explores the rich media affordance of virtual environments.

Keywords: smoking cessation; storytelling; virtual worlds; Second Life.

1 Introduction

Habitually smoking, drinking alcohol and consuming other drugs are often 'addictive behaviors'. An addiction can be defined as any activity or behavior that has become the most important focus of a person's life leading to the exclusion of other activities, or that has become harmful to the self or others either physically, mentally, or socially. Smoking cessation interventions have been implemented through a variety of media. Quitlines are telephone counseling interventions for smoking cessation and have been extensively studied. Recently, the replacement of landlines with cell phones has complicated the continuity of treatment. Many multi-session interventions (3 to 8 sessions) average only 1 to 2 completed sessions [1].

Interventions based on the web are called Web Assisted Tobacco Interventions (WATIs). In our work we are offering a WATI in Second Life (SL) for the treatment of addiction to nicotine. The site in SL that we have developed for the implementation of this proof-of-concept already provides assistance to smokers willing to quit, in the form of one-to-many and one-on-one interventions. We now want to incorporate in this virtual site a new, more flexible, type of intervention: biographic spaces. These spaces can be explored by smokers anytime, anywhere internet access is available, and provide information to assist smokers wanting to quit. This information is media-rich and addresses primarily the emotional component of the quitting process.

P. De Bra, A. Kobsa, and D. Chin (Eds.): UMAP 2010, LNCS 6075, pp. 415–418, 2010.
© Springer-Verlag Berlin Heidelberg 2010

2   The Model

We are proposing a 'biographic space' model in a virtual world, representing in a storytelling form the successive stages that a smoker goes through in her attempt to quit. The site in SL that we developed for the implementation of this model already provided assistance to smokers willing to quit smoking in the form of one-to-many and one-on-one interventions. We now want to incorporate in this virtual site a new, more flexible, type of intervention: the biographic space. These spaces can be explored by smokers anytime / anywhere internet access is available, and provide information to assist smokers wanting to quit. This information is media-rich and addresses primarily the emotional component of the quitting process.

The biographic space will illustrate the successive stages of the quitting process by telling the life of a character through the architectural space and the objects in that space. We are modeling these stages based on Prochaska and DiClementes Stages of Change Model [2]. This model was initially developed for stages of smoking cessation but has since then been adapted to the treatment of other addictive behaviors. The stages include pre-contemplation, contemplation, preparation, action and maintenance. Relapses may incur at any point of the loop following action, sending the individual back to the contemplation stage (See Fig. 1).

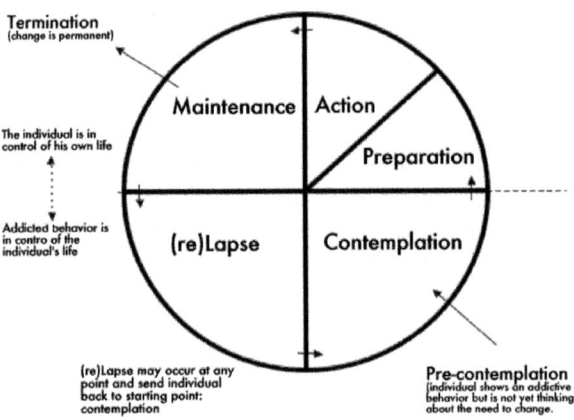

Fig. 1. The States of Change model

There are two ways in which we address user modeling in our proof-of-concept. The first regards the way studies addictive behaviors in the Public Health and specifically, smoking cessation, address user-centric design and, hence, user modeling. In Web Assisted Tobacco Interventions (WATIs), studies suggest that tailored interventions attain higher quit rates than targeted interventions [3–5]. In WATIs, the term tailored designates personalized interventions, designed for e,g, a 45 year-old African American male who just had his 1st relapse during his third attempt to quit smoking. Still in WATIs, a targeted intervention designates an intervention designed for a group of users (e.g. all individuals in the pre-contemplation phase of the Stages of Change Model). The second way in which this paper addresses user modeling is through the strategy adopted to tailor the virtual environment to the state of change of the user. Before entering the biographic space, the user is prompted by an avatar to answer a staging algorithm. The biographic space will then assume the stage adequate to the users stage of change as identified by algorithm (See Fig. 2). When interacted with, objects in this space will respond according to the script triggered for the stage of change identified: e.g., the photo of the house inhabitant will change according to state of change, as well as the series of voice mails left by friends and colleagues.

The user can access the biographic space at any time and engage the life of the fictional character at the users own stage of behavioral change (See Fig. 2). The scripted environment conveys the fictional characters life during her attempt to quit smoking through voice, written text, still images and movie clips all

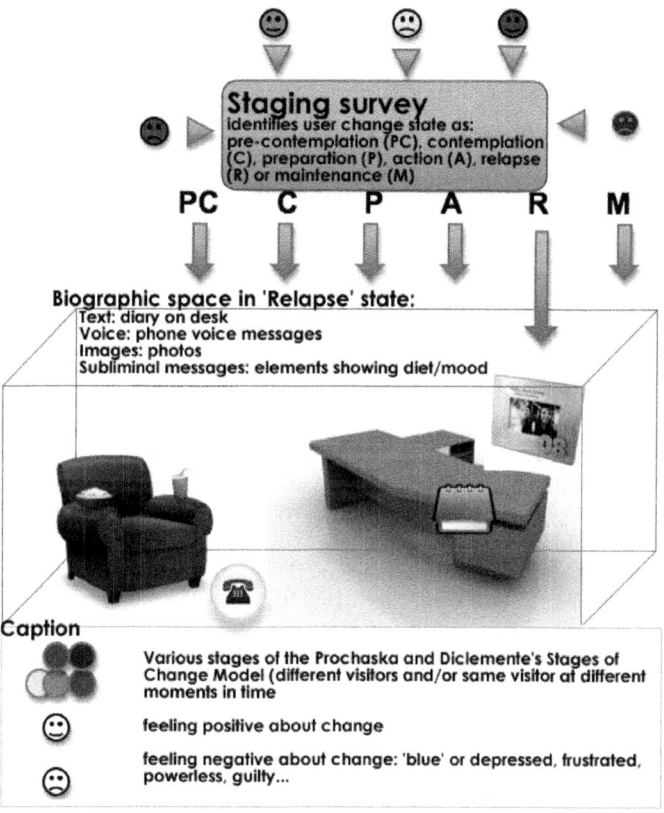

Fig. 2. The "biographic space" model

of these media channels document the emotional stages experienced during a typical attempt to quit smoking.

Users need to register to enter the space and unique IDs will be assigned. Users interactions with objects as well as number and duration of visits will be recorded using standard tools available in Second Life for these metrics. The change state as determined by the staging algorithm will also be recorded for each user. The data relative to change state and visit/interactions with objects will be recorded each time a user with a unique ID re-enters the system.

Our storytelling model was inspired by Myst, the first blockbuster in computer games, where due to technological restrictions at the time, the characters were represented through objects evocative of their personalities and actions [6].

Final Considerations

The proof-of-concept explores two features of Second Life that the authors deem key to the support of addictive behavior change: 24 hour support and media rich channels of information [7]. Until 2010, the American Cancer Society was the only organization offering smoking cessation assistance where a smoker seeking help can reach a live person 24 hours a day, 7 days a week [8]. Due to the economic crisis, during 2010 this aspect of the ACS service will be phased-out.

References

1. Rabius, V., Pike, K.J., Hunter, J., Wiatrek, D., McAlister, A.L.: Effectiveness of Frequency and Duration in Telephone Counseling for Smoking Cessation. Tobacco Control 16(suppl. 1), 71–74 (2007)
2. Prochaska, J.O., DiClemente, C.C.: Stages and processes of self-change of smoking: Toward an integrative model of change. Journal of Consulting and Clinical Psychology 51, 390–395 (1983)
3. Bock, B., Graham, A., Sciamanna, C., Krishnamoorthy, J., Whiteley, J., Carmona-Barros, R., Niaura, R., Abrams, D.: Smoking cessation treatment on the Internet: content, quality, and usability. Nicotine Tob. Res. 6(2), 207–219 (2004)
4. Lustria, M., Cortese, J., Noar, S., Glueckauf, R.: Computer tailored health interventions delivered over the web: review and analysis of key components. Patient Education and Counseling 74, 156–173 (2009)
5. Myung, S., McDonnell, D., Kazinets, G., Seo, H., Moskowitz, J.: Effects of Web- and Computer-Based Smoking Cessation Programs: Meta-analysis of Randomized Controlled Trials? Arch. Intern. Med. 169(10), 929–937 (2009)
6. Myst. Spokane. Cyan Worlds, WA (1993)
7. Calongne, C.: Educational Frontiers: Learning in a Virtual World. EDUCAUSE Review 43(5), 36 (2008)
8. All Quitline Facts (2009) North American Quitline Consortium, Web (January 31, 2010), http://www.naquitline.org/resource/resmgr/ql_about_facts/2008_quitline_facts_qa_final.pdf

Task-Based User Modelling for Knowledge Work Support

Charlie Abela[1], Chris Staff[1], and Siegfried Handschuh[2]

[1] Department of Intelligent Computer Systems,
University of Malta, Malta
{charlie.abela,chris.staff}@um.edu.mt
[2] Digitial Enterprise Research Institute,
National University of Ireland Galway, IDA Business Park, Galway, Ireland
{siegfried.handschuh}@deri.org

Abstract. A Knowledge Worker (KW) uses her computer to perform different tasks for which she gathers and uses information from disparate sources such as the Web and e-mail, and creates new information such as calendar events, e-mails, and documents (*resources*). This forms a Task Space (TS): an information space composed of all computer-based resources the KW uses in relation to a task. Furthermore, KWs may switch between multiple tasks, some of which may be suspended and resumed after some time. These effects compound the KW's ability to organise and visualise an accurate mental model of the individual TSs. We propose a Task-Based User Model (TBUM) that acts as the KW's mental model for each task by automatically tracking, relating and organising resources associated with that task. The generated TBUM can be used to support complex activities such as task-resumption, searching within a task-context, task sharing and collaboration.

Keywords: Task-Based Computing, Personal Knowledge Management, Task-Based User Model.

1 Introduction and Motivating Scenario

Multi-tasking operating systems manage tasks (or rather *processes*) by tracking resources that each running process is using or wants to use. Processes can be interrupted, suspended or resumed. Interrupting or suspending a process requires the operating system to take a snapshot of critical resources so that when the process is resumed these can be restored to the required state through a *context-switch*. On the other hand, multi-tasking performed by a KW normally involves the use of several applications and a KW may have multiple e-mail messages opened, representing several distinct tasks, together with several word processing documents opened for editing, some of which may be related to the e-mail messages. She may also have several web browser windows, where each window represents a different task. So, different documents "owned" by a single application may belong to different KW tasks, and a single KW task may involve the use of several documents "owned" by many different applications.

P. De Bra, A. Kobsa, and D. Chin (Eds.): UMAP 2010, LNCS 6075, pp. 419–422, 2010.

We are motivated by the operating system's analogy to process management to find a solution to task management at the conceptual level of a task, by representing each KW's task by a Task-Based User Model (TBUM). When a task is interrupted or suspended the TBUM can be used to restore the TS upon task resumption. The TBUM can be considered as a dynamic and persistent mental model of the resources that the KW uses to perform a task. Furthermore the TBUM can be used to support activities such as search, which can be restricted to a task-context, and sharing of task knowledge between collaborating KWs.

1.1 Motivating Scenario

Dirk is a typical KW and is switching between different tasks including a paper that he wants to submit with a colleague to UMAP 2010 and a set of slides for his adaptive systems' course. At times Dirk's work is interrupted by notifications from either his email or chat clients which require him to switch to these applications and to view and/or reply. Task tabs, rather than file tabs, are opened in his task-bar, each associated with one of his current tasks. Dirk clicks on a task and the resources associated with this task are displayed on his desktop while those related to the previous task are hidden in the background. Dirk is using a latex editor to edit his UMAP paper and through an adaptive-visual representation scheme, this takes a central role on his display, being the Most Significant Document (MSD). Supporting resources, such as digitized notes and other reference documents, are displayed in such a way that reflect their computed relatedness weighting to the MSD and to each other. This weighting is based on the semantic analyses between window titles and between the various document content and Dirk's window-switching behaviour.

Dirk can share tasks with his colleagues by dragging the task's MSD to the shared-tasks' space on his desktop. This automatically makes all resources associated with that task available via the shared space. The access to this space, as well as that of single resources pertaining to a task, are configurable.

Dirk receives an e-mail reply to a previously sent e-mail. The incoming e-mail is automatically associated with the task in which the original e-mail was sent. He is then interrupted by a notification from his calendar application that indicates that he has a project meeting in 15 minutes. Dirk attends to the alert by shelving his current task-context, searches for the meeting-task context and opens up the MSD for printing. When Dirk returns from the meeting, he switches back to the umap_2010 task-context to see whether his colleague Claudia has contributed to the paper.

2 Related Research

[1] uses *Activity-Based Computing* (ABC) to abstract the human-computer interaction in terms of tasks or activities. User-support is through various UI enhancements embedded in the Windows OS and provides for activities to be distributed across a network. Our proposed approach is similar. We will make use

of Task-Based Computing (TBC) and share models across users. We will explore a combination of unsupervised learning complimented by UI enhancements.

The work in [2] and [5], computes the relatedness between open application windows, which is based on the user's window switching behaviour. However in [5] the relatedness also depends on a semantic analysis of the application-windows' titles. We will take a similar approach to measure the relatedness between resources, but we intend to also consider semantic analysis of the document content. A supervised approach is used in [8] to tackle the task-assignment problem by automatically assigning observed actions to a predefined task classification. Although similar, the approach presented in [7] allows for more flexibility in the task-naming process and relies on an ontology-based task detection approach. Task-naming in our case will be related to the identification of a task's MSD. This is similar to *key resources* in [3] and represent the goals of a task.

From the modelling perspective, our approach is similar to [7] and [6], however neither support unsupervised learning. We agree with [9], which defines a task model as the "kernel" of a user model, suggesting that the *goals* defined by a user model can be represented through tasks. NEPOMUK [4] allows the linking of different information resources to be reflected within the KW's personal information and the task models. In our work we will investigate automating this linking process as much as possible.

3   Research Questions

The main research question we are attempting to address is the following:
Can user window-switching behaviour and semantic analysis of viewed documents be used to automatically generate Task-Based User Models to support the KW in managing information individually and in collaborations? This question motivates a number of challenges:

 i. *How to automatically discern between the KW's activities that pertain to different tasks, given that the KW might switch between one task and another at unspecified intervals?* We will identify which significant and supporting documents accurately represent a task by exploring unsupervised learning coupled by enhanced UI and minimally intrusive techniques.
 ii. *How to identify the Most Significant Document within a task?* The identification of the MSD is central to automatically labelling a task. Also, an MSD is likely to be an edited document, but this is not necessarily always the case. Some tasks may have multiple MSDs.
 iii. *How to share task-related information during collaborative tasks?* We will investigate and evaluate how TBUMs can be exploited by existing collaborative frameworks.

4   Conclusion and Expected Feedback

Our research is at an early stage and we will split our work into two phases. The first phase involves the definition of a high level architecture for task-based user modelling to identify the most suitable approach to use. The second phase considers issues related to the sharing of TBUMs across KWs through existing collaborative frameworks. Feedback is requested on the following:

i. Is the proposed approach suitable for the automatic generation of TBUMS?
ii. TBUMS are dynamic and will represent incomplete knowledge. How should maintenance of this model be affected?
iii. There are no standard domain and conceptual models to compare TBUMS. What evaluation methodology is appropriate?

References

1. Bardram, J.E., Bunde-Pedersen, J., Soegaard, M.: Support for activity-based computing in a personal computing operating system. In: Nichols, J.A., Schneider, M.L. (eds.) CHI '06: Proceedings of the SIGCHI Conference on Human Factors in Computing Systems, pp. 211–220. ACM Press, New York (2006)
2. Bernstein, M., Shrager, J., Winograd, T.: Taskpose: Exploring Fluid Boundaries in an Associative Window Visualization. In: Cousins, S.B., Beaudouin-Lafon, M. (eds.) UIST '08: Proceedings of the 21st annual ACM symposium on User Interface Software and Technology, pp. 231–234. ACM Press, New York (2008)
3. Chen, J., Guo, H., Wu, W., Wang, W.: An Associative Memory Based Desktop Search System. In: Cheung, D., Song, I.-L., Chu, W., Hu, X., Lin, J., Li, J., Peng, Z. (eds.) CIKM '09: Proceeding of the 18th ACM Conference on Information and Knowledge Management, pp. 731–740. ACM Press, New York (2008)
4. Groza, T., Handschuh, S., Moeller, K., Grimnes, G., Sauermann, L., Minack, E., Mesnage, C., Jazayeri, M., Reif, G., Gudjnsdttir, R.: The NEPOMUK Project-On the Way to the Social Semantic Desktop, in I-Semantics' 07. Journal of Universal Computer Science, 201–211 (2007)
5. Oliver, N., Smith, G., Surendran, A.C.: SWISH: Semantic Analysis of Window Titles and Switching History. In: Edmonds, E., Riecken, D., Paris, C.L., Sidner, C.L. (eds.) IUI '06: Proceedings of the 11th international conference on Intelligent User interfaces, pp. 194–201. ACM Press, New York (2006)
6. Ong, E., Riss, U.V., Grebner, O., Du, Y.: Semantic Task Management Framework: Bridging Information and Work. In: Pellegrini, T., Auer, S., Tochtermann, K., Schaffert, S. (eds.) Networked Knowledge-Networked Media, vol. 221, pp. 25–43. Springer, Berlin (2009)
7. Rath, A.S., Devaurs, D., Lindstaedt, S.N.: An Ontology-Based User Interaction Context Model for Automatic Task Detection on the Computer Desktop. In: Gomez-Parez, J.M., Haase, P., Tilly, M., Warren, P. (eds.) CIAO '09: Proceedings of the 1st Workshop on Context, Information and Ontologies, pp. 1–10. ACM Press, New York (2009)
8. Shen, J., Li, L., Dietterich, T.G., Herlocker, J.L.: A Hybrid learning System for Recognising User Tasks from Desktop Activities and Email messages. In: Edmonds, E., Riecken, D., Paris, C.L., Sidner, C.L. (eds.) IUI '06: Proceedings of the 11th International Conference on Intelligent User Interfaces, pp. 86–92. ACM Press, New York (2006)
9. Vassileva, J.: A Task-Centered Approach for User Modeling in a Hypermedia Office Documentation System. In: Brusilovsky, P., Vassileva, J. (eds.) Adaptive Hypertext and Hypermedia, Special Issue of User Modeling and User Adapted Interaction, vol. 6(2-3), pp. 185–223. Springer, Berlin (1996)

Enhancing User Interaction in Virtual Environments through Adaptive Personalized 3D Interaction Techniques

Johanna Renny Octavia, Karin Coninx, and Chris Raymaekers

Hasselt University - tUL - IBBT
Expertise Centre for Digital Media
Wetenschapspark 2, 3590 Diepenbeek, Belgium

Abstract. Leveraging interactive systems by integrating adaptivity is considered as an important key to accommodate user diversity and enhance user interaction. A virtual environment is a highly interactive system which involves users performing complex tasks using diverse 3D interaction techniques. Adaptivity has not been investigated thoroughly in the context of virtual environments. This PhD research is concerned with embedding intelligence to enhance user interaction in virtual environments (i.e. providing adaptive personalized 3D interaction techniques).

Keywords: virtual environments, adaptation, 3D interaction techniques.

1 Introduction and Research Problem

Research on virtual environments has been growing vastly both in terms of quantities and areas of real applications, ranging from education to entertainment [1]. However, the issue of engaging users naturally and intuitively while they interact in a virtual environment still motivates researchers. In a virtual environment, users typically perform complex tasks with highly interactive 3D user interfaces and use a variety of 3D interaction techniques ranging from a simple technique to a very complex one [1]. This situation likely increases the complexity for users and eventually impedes them to interact naturally in the environment.

My PhD research is aimed to integrate adaptivity into virtual environments and investigate to what extent adaptivity can enhance natural and intuitive user interaction. This contribution details the research on integrating adaptive and personalized 3D interaction techniques for enhancing user interaction in virtual environments by means of intelligent algorithms.

2 Research Approach

We recognize the significance of enhancing user interaction in virtual environments by means of integrating intelligence and establishing adaptive personalized

P. De Bra, A. Kobsa, and D. Chin (Eds.): UMAP 2010, LNCS 6075, pp. 423–426, 2010.

3D interaction techniques. At the start of this PhD research, we formulated a research approach to answer the challenge by proposing a conceptual framework for adaptive and personalized 3D interaction techniques as shown in Fig. 1(a). Throughout the research, I am conducting a series of user experiments to construct and validate the various building blocks of the framework.

Using the proposed framework, we intend to gather information and build up knowledge about users' working methods, performances, preferences and abilities when performing tasks and interacting in a virtual environment. This knowledge will later on be used to assess the adaptation and personalization of interaction techniques with the help of intelligent algorithms. This PhD research focuses on investigating intelligent algorithms to provide three types of adaptation in virtual environments: (1) *switching between interaction techniques*, which offers the most suited interaction technique for a user in a certain situation, (2) *adapting the interaction technique itself* that is adjusting parameters of an interaction technique to control how the user should perform it, and (3) *enhancing the interaction technique with modalities*, which adds multimodal feedback, such as visual, audio, and force feedback into the technique in order to provide more control for the user.

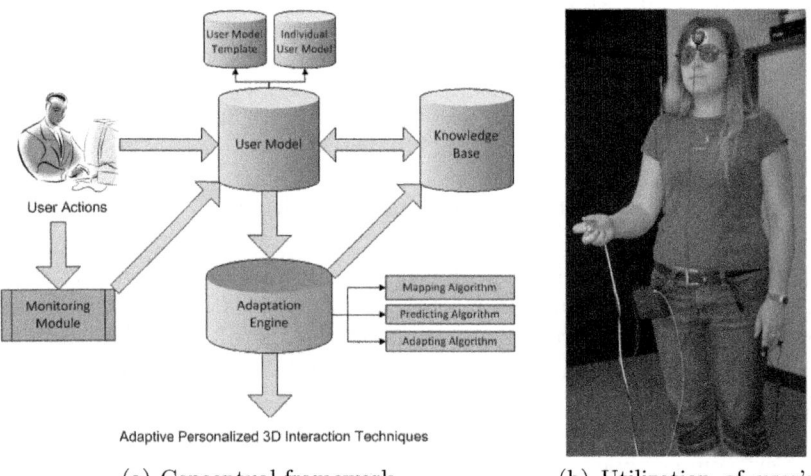

(a) Conceptual framework

(b) Utilization of user's physiological measures

Fig. 1. Towards adaptive 3D interaction techniques in virtual environments

3   Research Progress

After conducting an exhaustive literature study on user modeling and adaptive and intelligent (2D/3D) user interfaces, I started the realization of the conceptual framework by constructing the user model as the first building block. The user model, comprised of a user model template (general and group user models) and individual user models, is acquired by conducting user modeling activity through experiments. Two subsequent experiments were carried out with the objectives of constructing the user model template and establishing the individual user model. In both experiments, I focused on investigating user interaction when performing a 3D target acquisition task using two selection techniques, the bubble cursor and the depth ray [2].

The first experiment was conducted as an initial study to investigate the possibility of adaptation and personalization in virtual environments [3]. The experiment resulted in a *general user model* for 3D target acquisition tasks in virtual environments, which can be beneficial for novice users interacting in a virtual environment for the first time. In the second experiment, this user model template was verified to be favorable for enhancing first-time users' interaction in a virtual environment. Moreover, *individual user models* were successfully constructed which led to the establishment of the complete user model. After establishing the user models, I continued to the implementation of the adaptation engine as the next building block. The first type of adaptation, switching between interaction techniques, was implemented based on the user models and an algorithm built upon user's performance and preference of interaction technique.

In the second experiment, I also investigated the user's reaction to adaptation by gathering physiological data to assess user frustration with regard to adaptation. For this purpose, I utilized the ProComp Infiniti (see Fig. 1(b)) to collect two kinds of physiological data: galvanic skin response (GSR) and electromyography (EMG). Concerning the adaptation of interaction technique implemented in the second experiment, I found that users perform significantly better and experience less frustration when adaptation is incorporated during their interaction in virtual environments [4].

4   Future Work Plan

The work completed to date has concentrated on investigating two building blocks of the framework, namely the user model and the adaptation engine. There are two more components of the framework left, the monitoring module and the knowledge base, that still need to be researched in the remaining time of the PhD research. The research will also continue to investigate further on the establishment of several intelligent algorithms that support the role of the adaptation engine.

Currently, I am preparing the conduct of an experiment to construct the monitoring module, where users' physiological measures are recognized and monitored as triggers for adaptation. I intend to investigate the utilization of user frustration as an indicator to provide adaptation in a gaming virtual environment. User frustration will be induced and measured throughout the game sessions and whenever frustration is detected, adaptive feedback will be provided with the expectation that it will decrease user frustration and increase user performance. The third type of adaptation, enhancing the interaction technique with modalities, will be implemented in the adaptation engine. The interaction technique will be adapted with the use of multimodal feedback such as visual and force feedback. For instance, particular users may perform a 3D target acquisition task better when the selection technique is complemented by haptics (force feedback) as such.

Following this experiment, longer-term goals include the investigation of the second type of adaptation (adapting the interaction technique itself) to be incorporated in the adaptation engine and also the investigation of intelligent interaction techniques. I would like to work on these ideas further on with a specific group of users (e.g. users with limited motor abilities).

5 Conclusion and Expected Feedback

This doctoral consortium paper mainly describes the research problem and work progress to date of my PhD research, which is currently situated at the intermediate stage. The motivation behind this PhD research is that I envision that by integrating adaptation and personalization into 3D interaction techniques, natural, intuitive and enhanced user interaction in virtual environments can be achieved. With this paper, I seek constructive feedbacks and suggestions for improvement particularly on these following issues:

1. *The conceptual framework.* How can we justify the comprehensiveness, accuracy, generality and applicability of the proposed framework, including each element that serves as building blocks of the framework?
2. *The adaptation engine and intelligent algorithms.* How to determine the effectiveness and efficiency of the adaptation engine? What aspects define the intelligence of such adaptation algorithms?
3. *Adaptation determinants.* What other aspects of user interaction in virtual environments that are considered significant in the adaptation process, besides user's performance, preference and physiological measures?

References

1. Bowman, D.A., Chen, J., Wingrave, C.A., Lucas, J., Ray, A., Polys, N.F., Li, Q., Haciahmetoglu, Y., Kim, J.S., Kim, S., Boehringer, R., Ni, T.: New directions in 3d user interfaces. International Journal of Virtual Reality 5(2), 3–14 (2006)
2. Vanacken, L., Grossman, T., Coninx, K.: Multimodal selection techniques for dense and occluded 3d virtual environments. Int. J. Hum.-Comput. Stud. 67(3), 237–255 (2009)
3. Octavia, J.R., Raymaekers, C., Coninx, K.: Investigating the possibility of adaptation and personalization in virtual environments. In: Houben, G., McCalla, G., Pianesi, F., Zancanaro, M. (eds.) UMAP 2009. LNCS, vol. 5535, pp. 361–366. Springer, Heidelberg (2009)
4. Octavia, J.R., Raymaekers, C., Coninx, K.: Adaptation in virtual environments: Conceptual framework and user models. Multimedia Tools and Applications Journal, 1–20 (2010) (accepted)

Author Index

Abel, Fabian   16
Abela, Charlie   419
Aghasaryan, Armen   40
Ahn, Jae-wook   4
Arroyo, Ivon   135
Ashman, Helen   207

Bakalov, Fedor   219
Baker, Ryan S.J.d.   52, 267
Barbieri, Mauro   351
Barla, Michal   387
Berkovsky, Shlomo   381
Bernier, Cédric   40
Bieliková, Mária   387, 411
Boa-Ventura, Ana   415
Bohnert, Fabian   99, 195, 363
Boughanem, Mohand   171
Bouzid, Makram   40
Boyer, Anne   303
Brun, Armelle   303
Brusilovsky, Peter   4
Bull, Susan   393
Burleson, Winslow   135, 159

Cena, Federica   369
Chang, Edward Y.   3
Chen, Li   375
Chi, Min   147
Cohon, Robin   279
Conejo, Ricardo   243
Coninx, Karin   423
Cooper, David G.   135, 399
Corbett, Albert T.   52

Daoud, Mariam   171
Dimitrova, Vania   231
Dumais, Susan T.   28

Fischer, Philipp   315
Freyne, Jill   381
Frías-Martínez, Enrique   327, 339
Fry, Michael   111

Gálvez, Jaime   243
Gena, Cristina   369

Gerber, Simon   111
Gerdenitsch, Cornelia   75
Germanakos, Panagiotis   64
Giguere, Stephen   52
Gowda, Sujith M.   52
Guzmán, Eduardo   243

Handschuh, Siegfried   419
Heffernan, Neil T.   255
Henze, Nicola   16
Herder, Eelco   16
Hohwald, Heath   327
Hu, Rong   291

Ishii, Yutaka   87

Jordan, Pamela   147

Kaieda, Yohei   87
Kamahara, Junzo   87
Kapoor, Ashish   28
Kauffman, Linda R.   52
Kay, Judy   111
Kleanthous, Styliani   231
König-Ries, Birgitta   219
Kordumova, Suzana   351
Korst, Jan   351
Kostadinov, Dimitre   40
Kostadinovska, Ivana   351
Kramár, Tomáš   387
Krause, Daniel   16
Kummerfeld, Bob   111
Kump, Barbara   75

Lara, Rubén   339
Lekkas, Zacharias   64
Ley, Tobias   75
Litman, Diane   147

MacLaren, Benjamin A.   52
Marsella, Stacy   1
Mathews, Moffat   267
Mitchell, Aaron P.   52
Mitrović, Antonija   267
Mourlas, Costas   64
Muldner, Kasia   135, 159

Nagamatsu, Takashi   87
Nauerz, Andreas   219
Nürnberger, Andreas   315

Octavia, Johanna Renny   423
Oliver, Nuria   327
Olmedilla, Daniel   339

Pannell, Grant   207
Pardos, Zachary A.   255
Patterson, Donald J.   123
Picault, Jérôme   40
Pink, Glen   111
Plumbaum, Till   407
Pronk, Verus   351
Pu, Pearl   291, 375

Raymaekers, Chris   423
Razmerita, Liana   303

Saboga-Nunes, Luís   415
Samaras, George   64
Senot, Christophe   40

Seroussi, Yanir   195, 403
Seth, Aaditeshwar   279
Šimko, Marián   411
Staff, Chris   419

Tada, Masashi   87
Tamine, Lynda   171
Tang, Justin   123
Tsianos, Nikos   64

VanLehn, Kurt   147, 159
Vernero, Fabiana   369

Wagner, Angela Z.   52
Wasinger, Rainer   111
Welsch, Martin   219
White, Ryen W.   28
Wolfe, Shawn R.   183
Woolf, Beverly Park   135

Zhang, Jie   279
Zhang, Yi   183
Zukerman, Ingrid   99, 195, 363